P9-AGH-068

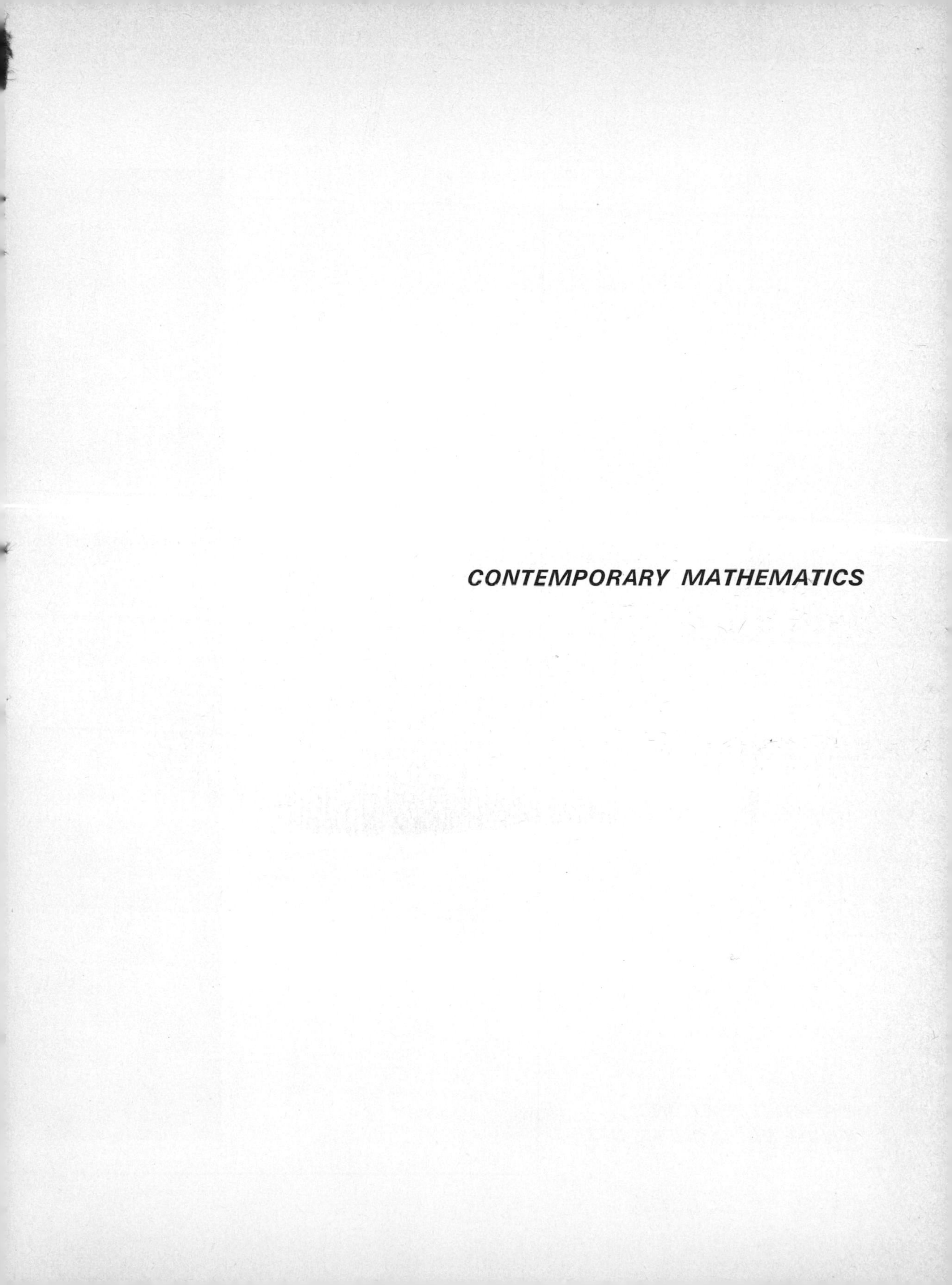

CONTEMPORARY MATHEMATICS

BRUCE E. MESERVE

Professor of Mathematics
University of Vermont

MAX A. SOBEL

Professor of Mathematics
Montclair State College

PRENTICE-HALL, INC.
ENGLEWOOD CLIFFS, N. J.

CONTEMPORARY MATHEMATICS

2nd Edition

Library of Congress Cataloging in Publication Data

MESERVE, BRUCE ELWYN, (1917)
Contemporary mathematics.

Includes bibliographies and index.
1. Mathematics—1961– I. Sobel, Max A., joint author. II. Title.
QA39.2.M47 1977 510 75-45170
ISBN 0-13-170092-8

CONTEMPORARY MATHEMATICS

2nd Edition

Bruce E. Meserve / Max A. Sobel

10 9 8 7 6 5 4 3

Printed in the United States of America

PRENTICE-HALL INTERNATIONAL, INC., London
PRENTICE-HALL OF AUSTRALIA PTY. LIMITED, Sydney
PRENTICE-HALL OF CANADA, LTD., Toronto
PRENTICE-HALL OF INDIA PRIVATE LIMITED, New Delhi
PRENTICE-HALL OF JAPAN, INC., Tokyo
PRENTICE-HALL OF SOUTHEAST ASIA PTE. LTD., Singapore

Contents

Contemporary Mathematics, second edition, has been written to serve the needs of both liberal arts students and prospective teachers of elementary mathematics in grades K through 8. It is designed primarily as a one-year text at the undergraduate level, although it may be used effectively for in-service courses as well.

The authors have prepared this material on the basis of their many years of experience at both the pre-service and in-service levels of instruction. They have also taken into account the numerous worthwhile suggestions made by users of the first edition of *Contemporary Mathematics*. The authors are grateful to everyone who has submitted suggestions and to the many students who have both directly and indirectly contributed to the conception and formulation of this text. Special recognition is given to Mrs. Dorothy T. Meserve for her careful analysis of the manuscript and her detailed work in checking answers for use in the text and in the *Instructor's Manual*.

A liberal arts approach suitable for the twentieth-century citizen is stressed in this text. Also included is an emphasis upon the basic mathematical concepts and skills that a prospective teacher needs to know in order to be able to teach the contemporary courses found in recent elementary school textbooks, and to enjoy the pleasure that comes from confidence and competence. However, this text does not neglect those aspects of mathematics that are as important today as they were prior to the "revolution" in school mathematics. Thus, careful attention is given to the development of the number system and to the various algorithms representing fundamental operations of arithmetic.

The key changes incorporated into the second edition include:

—A new Chapter 1 ("The Language of Mathematics") to provide a basic vocabulary for the text.
—Increased attention to the properties of, and operations with, whole numbers, integers, rational numbers, and real numbers.
—A new chapter on the metric (SI) system and measurement.
—A new chapter on computers and calculators.
—Substantial expansion of the "Pedagogical Explorations," since these were so favorably received in the first edition by both teachers and students.
—The addition of "Readings and Projects" at the end of each chapter.

Many of the last two items mentioned are specifically addressed to prospective teachers. Included are discussions related to the teaching of elementary school mathematics, suggestions for additional explorations and laboratory experiments, introductions to the publications of various professional organizations such as the National Council of Teachers of Mathematics (NCTM), and other items that help provide insight into the full role of a successful teacher of elementary school mathematics. A list of several NCTM Yearbooks is included here to facilitate the numerous references to them throughout this text. Both the number and the title of each yearbook are given, since they are referred to in the text by number only.

YEARBOOK NUMBER	YEARBOOK TITLE
24th	*The Growth of Mathematical Ideas, Grades K–12*
25th	*Instruction in Arithmetic*
26th	*Evaluation in Mathematics*
27th	*Enrichment Mathematics for the Grades*
28th	*Enrichment Mathematics for High School*
29th	*Topics in Mathematics for Elementary School Teachers*
30th	*More Topics in Mathematics for Elementary School Teachers*
31th	*Historical Topics for the Mathematics Classroom*
34th	*Instructional Aids in Mathematics*
35th	*The Slow Learner in Mathematics*
36th	*Geometry in the Mathematics Classroom*
37th	*Mathematics Learning in Early Childhood*

Further information about these and other publications of the NCTM may be obtained directly from:

The National Council of Teachers of Mathematics
1906 Association Drive
Reston, Virginia 22091

The structure of the text has been changed slightly in this revision to allow for greater flexibility. Chapters 1 and 3 through 7 can be used effectively for a one-quarter or a one-semester course on number systems. The inclusion of at least the first part of Chapter 8 (an introduction to the metric system) is recommended for any number systems course. The inclusion or omission of Chapter 2 (logic) provides one of several means for adapting the text to the mathematical maturity of the students. Chapters 8 through 13 are suitable for a one-quarter or one-semester course on elementary geometry, algebra, probability, and statistics. Chapter 14 (computers and calculators) may be considered at any time after Chapter 10 and provides flexibility. The text may also be used in courses for three quarters with a variety of structures according to the interests and abilities of the students.

This book has been prepared with the expectation that most of the students will have had only a minimal amount of previous mathematical experience. (However, additional challenges for the above-average student have also been included in the exercises and in the pedagogical explorations.) Furthermore, it has been anticipated that a significant number of students will have some fear of mathematics. Thus, every effort has been made to provide an interesting approach that will motivate the learning of the subject. To this end, the introduction is entitled "Explorations with Mathematics" and is offered as one means of starting a course with topics that can be studied independently of the remainder of the text and just for fun. It has been the experience of the authors that such an approach can do much at the outset of a semester to relieve anxieties and to begin the course in a light and somewhat informal manner.

It is the fond hope of the authors that both the instructors and students who use this book will enjoy it as much as the authors have enjoyed its preparation. To their many students—past, present, and future—the authors sincerely dedicate this book and hope that it may serve to stimulate interest in the study of mathematics.

BRUCE E. MESERVE
MAX A. SOBEL

Introduction: Explorations with Mathematics

Mathematics has numerous practical applications, ranging from everyday household uses to the charting of the astronauts' travels through outer space. Many people study the subject because such applications relate to their particular fields of interest. Mathematics may also be considered as part of mankind's great cultural heritage. The history of mathematics dates back many thousands of years. Mathematical ideas and skills arise from, and directly affect, the way of life of the people in each historical era. As a major aspect of the present technological era, mathematics should be studied by the average citizen who wishes to be literate in this twentieth century.

Elementary school teachers have a special need for understanding mathematical concepts. In a wide variety of classroom situations mathematical concepts are involved in about one out of every five hours of their teaching. Many of these hours of teaching mathematics can be fun for both the teacher and the students. This introduction contains a smorgasbord of items that have been fun for many students and teachers. The topics are not sequential and an understanding of them is not necessary for later chapters in this text, although some of the topics presented here will be presented in greater detail in later chapters. Thus this introduction is expendable in the formal structure of this text. However, it is significant for elementary school teachers as an indication of the spirit of a modern approach to the teaching of elementary school mathematics.

1 explorations with number patterns

Mathematicians love to search for patterns and generalizations in all branches of their subjects—in arithmetic, in algebra, and in geometry. A search for such patterns may not only be interesting but may also help one develop insight into mathematics as a whole.

Many patterns that often escape notice may be found in the structure of arithmetic. For example, consider the multiples of 9:

$$\begin{aligned}
1 \times 9 &= 9\\
2 \times 9 &= 18\\
3 \times 9 &= 27\\
4 \times 9 &= 36\\
5 \times 9 &= 45\\
6 \times 9 &= 54\\
7 \times 9 &= 63\\
8 \times 9 &= 72\\
9 \times 9 &= 81
\end{aligned}$$

What patterns do you notice in the column of multiples on the right? You may note that the sum of the digits in each case is always 9. You should also see that the units digit decreases (9, 8, 7, . . .), whereas the tens digit increases (1, 2, 3, . . .). What lies behind this pattern?

Consider the product

$$5 \times 9 = 45$$

To find 6×9 we need to add 9 to 45. Instead of adding 9, we may add 10 and subtract 1.

$$\begin{array}{r} 45 \\ +10 \\ \hline 55 \end{array} \qquad \begin{array}{r} 55 \\ -1 \\ \hline 54 \end{array} = 6 \times 9$$

That is, by adding 1 to the tens digit, 4, of 45, we are really adding 10 to 45. We then subtract 1 from the units digit, 5, of 45 to obtain 54 as our product.

Most people speak of "adding digits," "subtracting from the units digit," and so forth as we have done. Many elementary school teachers are more precise in their terminology and recognize that digits are *numerals*, that is, symbols for numbers, rather than numbers. Numerals can be written. Only numbers can be added or subtracted. However, unless the more precise terminology is needed to avoid major confusion, we use the commonly accepted phraseology.

The number 9, incidentally, has other fascinating properties. Of special interest is a procedure for multiplying by 9 on one's fingers. For example, to multiply 9 by 3, place both hands together as in the figure, and bend the third finger from the left. (Note that a thumb is considered to be a finger.) The result is read as 27.

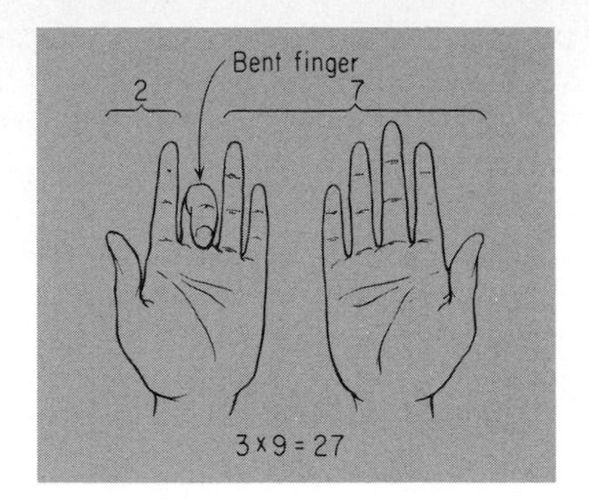

The next figure on the left shows the procedure for finding the product 7×9. Note that the seventh finger from the left is bent, and the result is read in terms of the tens digit, to the left, and the units digit to the right of the bent finger. What number fact is shown in the figure on the right?

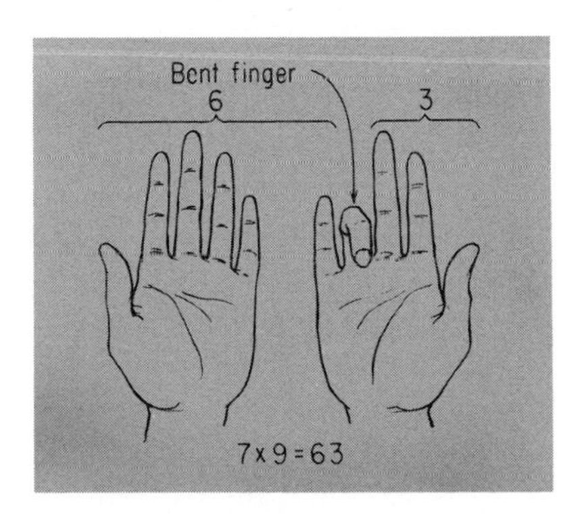

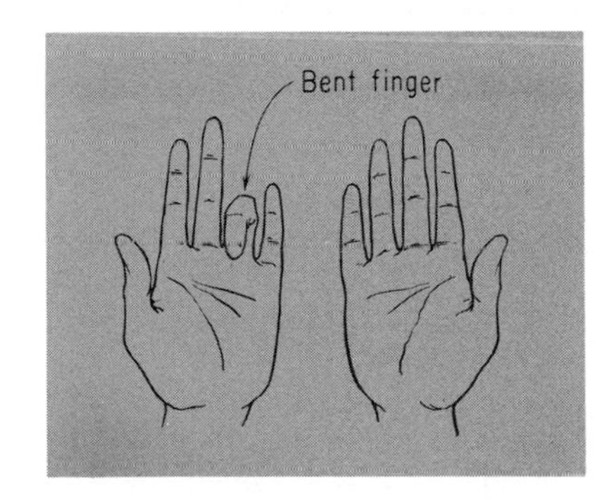

Here is one more pattern related to the number 9. You should verify that each of the following is correct:

$$
\begin{aligned}
1 \times 9 + 2 &= 11 \\
12 \times 9 + 3 &= 111 \\
123 \times 9 + 4 &= 1{,}111 \\
1{,}234 \times 9 + 5 &= 11{,}111 \\
12{,}345 \times 9 + 6 &= 111{,}111
\end{aligned}
$$

Try to find a correspondence of the number of 1's in the number symbol on the right with one of the numbers used on the left. Now see if you can supply the answers, without computation, to the following:

$$
\begin{aligned}
123{,}456 \times 9 + 7 &= ? \\
1{,}234{,}567 \times 9 + 8 &= ?
\end{aligned}
$$

Let us see *why* this pattern works. To do so we shall examine just one of the statements. A similar explanation can be offered for each of the other statements. Consider the statement:

$$12{,}345 \times 9 + 6 = 111{,}111$$

We can express 12,345 as a sum of five numbers as follows:

$$\begin{array}{r} 11{,}111 \\ 1{,}111 \\ 111 \\ 11 \\ 1 \\ \hline 12{,}345 \end{array}$$

Next we multiply each of the five numbers by 9:

$$\begin{array}{rcr} 11{,}111 \times 9 &=& 99{,}999 \\ 1{,}111 \times 9 &=& 9{,}999 \\ 111 \times 9 &=& 999 \\ 11 \times 9 &=& 99 \\ 1 \times 9 &=& 9 \end{array}$$

Finally, we add 6 by adding six ones as in the following array, and find the total sum:

$$\begin{array}{rcr} 99{,}999 + 1 &=& 100{,}000 \\ 9{,}999 + 1 &=& 10{,}000 \\ 999 + 1 &=& 1{,}000 \\ 99 + 1 &=& 100 \\ 9 + 1 &=& 10 \\ 1 &=& 1 \\ \cline{3-3} && 111{,}111 \end{array}$$

Here is another interesting pattern. After studying the pattern, see if you can add the next four lines to the table.

$$\begin{array}{rcr} 1 \times 1 &=& 1 \\ 11 \times 11 &=& 121 \\ 111 \times 111 &=& 12{,}321 \\ 1{,}111 \times 1{,}111 &=& 1{,}234{,}321 \\ 11{,}111 \times 11{,}111 &=& 123{,}454{,}321 \end{array}$$

Do you think that the pattern displayed will continue indefinitely? Compute the product $1{,}111{,}111{,}111 \times 1{,}111{,}111{,}111$ to help you answer this question.

Interesting discoveries can often be made by studying arithmetic patterns. A famous German mathematician by the name of Karl Gauss (1777–1855) is said to have been a precocious child who would often drive his teachers to despair. The story is told that on one occasion his teacher asked him to add a long column of figures, hoping to keep him suitably occupied for some time. Instead, young Gauss recognized a pattern, and gave the answer immediately. He is said to have found the sum of the first 100 counting numbers as indicated in the array shown at the top of the next page.

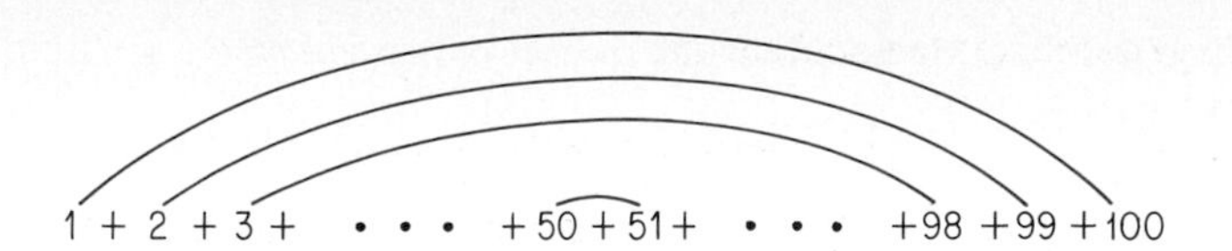

He reasoned that there would be 50 pairs of numbers, each with a sum of 101 (consider $100 + 1$, $99 + 2$, $98 + 3$, ..., $50 + 51$). Thus the sum is 50×101, that is, 5050.

Another interesting pattern emerges from the story of the man who was offered a month's employment at the rate of 1¢ for the first day, 2¢ for the second day, then 4¢, 8¢, 16¢, and so forth. He wished to determine what his total wages would be for four weeks of five days each; that is, for 20 working days.

Of course, one could list the 20 daily salaries and add, but this would be a tedious job indeed. Let us, rather, consider the big problem by exploring smaller tasks first. This is frequently a helpful technique in problem solving. Thus we consider total salaries in cents, for five, for six, for seven, and for eight days.

For 5 Days	*For 6 Days*	*For 7 Days*	*For 8 Days*
1	1	1	1
2	2	2	2
4	4	4	4
8	8	8	8
16	16	16	16
31	32	32	32
	63	64	64
		127	128
			255

Do you see a pattern emerging? Compare the total for the first five days with the salary for the sixth day; compare the total for six days with the salary for the seventh day. Notice that the total for five days is one cent less than the salary for the sixth day; the total for six days is one cent less than the salary for the seventh day; and so forth. Thus his total salary for ten days will be one cent less than the salary for the eleventh day. That is,

$$1 + 2 + 4 + 8 + 16 + 32 + 64 + 128 + 256 + 512 = 2(512) - 1, \text{ or } 1023$$

His salary for the eleventh day would be 1024 cents; in 10 days he will earn a *total* of 1023 cents, or $10.23.

To use this approach to answer the original question will require a good deal more work, but will still be easier than the addition of 20 amounts. By a doubling process we first need to find the salary for the twentieth day. The total salary for all 20 days can then be found by doubling this amount (to find the salary for the twenty-first day), and subtracting one.

To generalize, we note that the doubling process starting with one gives powers of two:

$2^0 = 1, \quad 2^1 = 2, \quad 2^2 = 4, \quad 2^3 = 8, \quad 2^4 = 16, \quad 2^5 = 32, \quad \ldots$

Then, by our discovery of a pattern, we may say

$$2^0 + 2^1 + 2^2 + 2^3 + \cdots + 2^n = 2^{n+1} - 1$$

When $n = 5$, we have

$$\begin{aligned} 2^0 + 2^1 + 2^2 + 2^3 + 2^4 + 2^5 &= 1 + 2 + 4 + 8 + 16 + 32 \\ &= 2^6 - 1 = 63 \end{aligned}$$

To find the total salary for 20 working days, we need to find the sum:

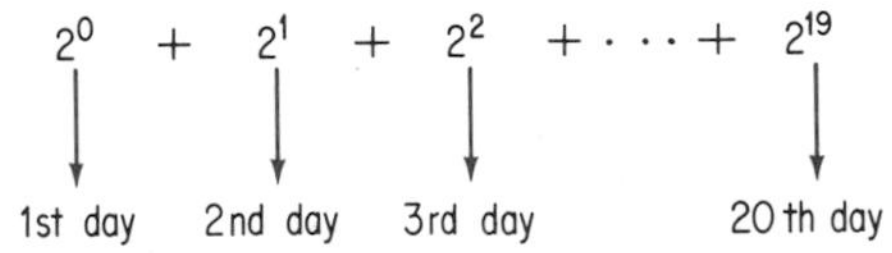

$$2^0 + 2^1 + 2^2 + \cdots + 2^{19}$$

1st day, 2nd day, 3rd day, ..., 20th day

From the preceding discussion we know that this sum is equal to $2^{20} - 1$. We can compute 2^{20} by a doubling process, or we can use a shortcut such as the following:

$$2^{10} = 2 \times 2^9 = 2 \times 512 = 1024$$

Alternatively, we may say

$$2^{10} = 2^5 \times 2^5 = 32 \times 32 = 1024$$

In a similar manner we compute

$$2^{20} = 2^{10} \times 2^{10} = 1024 \times 1024 = 1{,}048{,}576$$

and thus $2^{20} - 1 = 1{,}048{,}575$.

In 20 days, our worker would earn a total of $10,485.75.

In the following exercises the reader will have an opportunity to make other discoveries of his own through careful explorations of patterns.

EXERCISES

1. Verify that the process for finger multiplication shown in this section will work for each of the multiples of nine from 1×9 through 9×9.
2. Follow the procedure outlined in this section and show that

$$1234 \times 9 + 5 = 11{,}111$$

3. Study the following pattern and use it to express the squares of 6, 7, 8, and 9 in the same manner.

$$1^2 = 1$$
$$2^2 = 1 + 2 + 1$$
$$3^2 = 1 + 2 + 3 + 2 + 1$$
$$4^2 = 1 + 2 + 3 + 4 + 3 + 2 + 1$$
$$5^2 = 1 + 2 + 3 + 4 + 5 + 4 + 3 + 2 + 1$$

4. Study the entries that follow and use the pattern that is exhibited to complete the last four rows.

$$1 + 3 = 4 \text{ or } 2^2$$
$$1 + 3 + 5 = 9 \text{ or } 3^2$$
$$1 + 3 + 5 + 7 = 16 \text{ or } 4^2$$
$$1 + 3 + 5 + 7 + 9 = ?$$
$$1 + 3 + 5 + 7 + 9 + 11 = ?$$
$$1 + 3 + 5 + 7 + 9 + 11 + 13 = ?$$
$$1 + 3 + 5 + \cdots + (2n - 1) = ?$$

5. An addition problem can be checked by a process called **casting out nines.** To do this, you first find the sum of the digits of each of the addends (that is, numbers that are added), divide by 9, and record the remainder. Digits may be added again and again until a one-digit remainder is obtained. The sum of these remainders is then divided by 9 to find a final remainder. This should be equal to the remainder found by considering the sum of the addends (that is, the answer), adding its digits, dividing the sum of these digits by 9, and finding the remainder. Here is an example:

Addends	*Sum of Digits*	*Remainders*
4,378	22	4
2,160	9	0
3,872	20	2
1,085	14	5
11,495		11

When the sum of the remainders is divided by 9, the final remainder is 2. This corresponds to the remainder obtained by dividing the sum of the digits in the answer $(1 + 1 + 4 + 9 + 5 = 20)$ by 9.

Try this procedure for several other examples and verify that it works in each case.

6. Try to discover a procedure for checking multiplication by casting out nines. Verify that this procedure works for several cases.

7. There is a procedure for multiplying a two-digit number by 9 on one's fingers provided that the tens digit is smaller than the ones digit. The diagram at the top of the next page shows how to multiply 28 by 9.

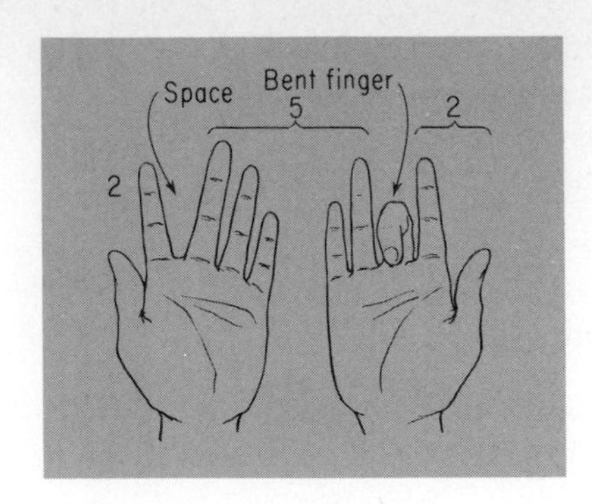

Reading from the left, put a space after the second finger and bend the eighth finger. Read the product in groups of fingers as 252.

Use this procedure to find: **(a)** 9×47; **(b)** 9×39; **(c)** 9×18; **(d)** 9×27. Check each of the answers you have obtained.

8. John offered to work for 1¢ the first day, 2¢ the second day, 4¢ the third day, and so forth, doubling the amount each day. Bill offered to work for $30 a day. Which boy would receive the most money for a job that lasted **(a)** ten days? **(b)** fifteen days? **(c)** sixteen days?

9. Using a method similar to that of Gauss, find:
(a) The sum of the first 80 counting numbers.
(b) The sum of the first 200 counting numbers.
(c) The sum of all the odd numbers from 1 through 49.
(d) The sum of all the odd numbers from 1 through 199.
(e) The sum of all the even numbers from 2 through 400.

10. See Exercise 9 and try to find a formula for the sum of:
(a) The first n counting numbers, that is,

$$1 + 2 + 3 + \cdots + (n - 1) + n.$$

(b) The first n odd numbers, that is,

$$1 + 3 + 5 + \cdots + (2n - 3) + (2n - 1).$$

11. Find the sum:

$$\frac{1}{1 \times 2} + \frac{1}{2 \times 3} + \frac{1}{3 \times 4} + \cdots + \frac{1}{9 \times 10}$$

(a) To help you discover this sum, complete the statements in the following array.

$$\frac{1}{1 \times 2} = \frac{1}{2}$$

$$\frac{1}{1 \times 2} + \frac{1}{2 \times 3} = ?$$

$$\frac{1}{1 \times 2} + \frac{1}{2 \times 3} + \frac{1}{3 \times 4} = ?$$

(b) Compare the answers on the right with the denominator of the last fraction in the given sums. Then attempt a guess at the sum of the original problem.

(c) Use the procedure for your guess to find the following sum; then confirm your answer by addition:

$$\frac{1}{1 \times 2} + \frac{1}{2 \times 3} + \frac{1}{3 \times 4} + \frac{1}{4 \times 5}$$

(d) Find the following sum:

$$\frac{1}{1 \times 2} + \frac{1}{2 \times 3} + \frac{1}{3 \times 4} + \cdots + \frac{1}{98 \times 99} + \frac{1}{99 \times 100}$$

****12.*** Note the following relationships:

$$\frac{1}{1 \times 2} = \frac{1}{1} - \frac{1}{2}, \qquad \frac{1}{2 \times 3} = \frac{1}{2} - \frac{1}{3}, \qquad \frac{1}{3 \times 4} = \frac{1}{3} - \frac{1}{4}$$

Use this pattern to confirm your answer in Exercise 11(d).

13. Find the sum:

$$\frac{1}{2} + \frac{1}{2^2} + \frac{1}{2^3} + \frac{1}{2^4} + \cdots + \frac{1}{2^n}$$

To help you discover this sum, complete these partial sums and search for a pattern.

(a) $\frac{1}{2} + \frac{1}{2^2} = \frac{1}{2} + \frac{1}{4} = ?$

(b) $\frac{1}{2} + \frac{1}{2^2} + \frac{1}{2^3} = \frac{1}{2} + \frac{1}{4} + \frac{1}{8} = ?$

(c) $\frac{1}{2} + \frac{1}{2^2} + \frac{1}{2^3} + \frac{1}{2^4} = \frac{1}{2} + \frac{1}{4} + \frac{1}{8} + \frac{1}{16} = ?$

****14.*** Try to discover the rule that is being used in each case to obtain the answer given. For example, the given information

$$2, 5 \longrightarrow 6, \qquad 3, 10 \longrightarrow 12, \qquad 7, 8 \longrightarrow 14, \qquad 5, 3 \longrightarrow 7$$

should lead you to the rule $x, y \longrightarrow x + y - 1$.

(a) $3, 4 \longrightarrow 9$; $\quad 3, 3 \longrightarrow 8$; $\quad 1, 5 \longrightarrow 8$; $\quad 2, 8 \longrightarrow 12$.
(b) $2, 4 \longrightarrow 9$; $\quad 3, 5 \longrightarrow 16$; $\quad 1, 7 \longrightarrow 8$; $\quad 3, 9 \longrightarrow 28$.
(c) $1, 5 \longrightarrow 1$; $\quad 5, 2 \longrightarrow 2$; $\quad 3, 9 \longrightarrow 3$; $\quad 6, 5 \longrightarrow 5$.
(d) $4, 8 \longrightarrow 4$; $\quad 5, 1 \longrightarrow 5$; $\quad 6, 6 \longrightarrow 6$; $\quad 8, 2 \longrightarrow 8$.
(e) $3, 4 \longrightarrow 3$; $\quad 4, 4 \longrightarrow 2$; $\quad 5, 5 \longrightarrow 0$; $\quad 3, 2 \longrightarrow 5$.

*An asterisk preceding an exercise indicates that the exercise is more difficult or challenging than the others.

2 explorations with geometric patterns

In the study of geometry we frequently form conclusions on the basis of a small number of examples, together with an exhibited pattern. Consider, for example, the problem of determining the number of triangles that can be formed from a given convex polygon by drawing diagonals from a given vertex *P*. First we draw several figures and consider the results in tabular form as follows.

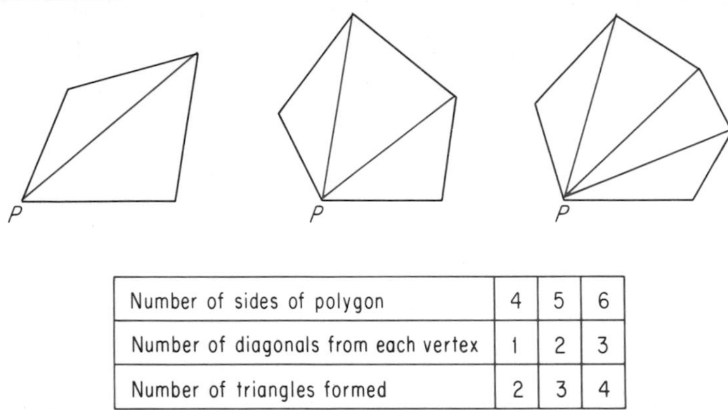

Number of sides of polygon	4	5	6
Number of diagonals from each vertex	1	2	3
Number of triangles formed	2	3	4

From the pattern of entries in the table it appears that the number of triangles formed is two less than the number of sides of the polygon. Thus we expect that we can form 10 triangles for a dodecagon, a polygon with 12 sides, by drawing diagonals from a given vertex. In general, then, for a polygon with n sides, called an *n-gon*, we can form $n - 2$ triangles.

This is reasoning by *induction*. We formed a generalization on the basis of several specific examples and an obvious pattern. This procedure does not, however, constitute a proof. In order to *prove* that $n - 2$ triangles can be formed in this way for a polygon with n sides, we must observe that two of the n sides intersect at the common point of the diagonals and that each of the other $n - 2$ sides is used to form a different triangle.

It is important to recognize that not all patterns lead to valid generalizations. Patterns offer an opportunity to make reasonable guesses, but these need to be proved before they can be accepted with certainty. Consider, for example, the maximum number of regions into which a circular region can be divided by line segments joining given points on a circle in all possible ways.

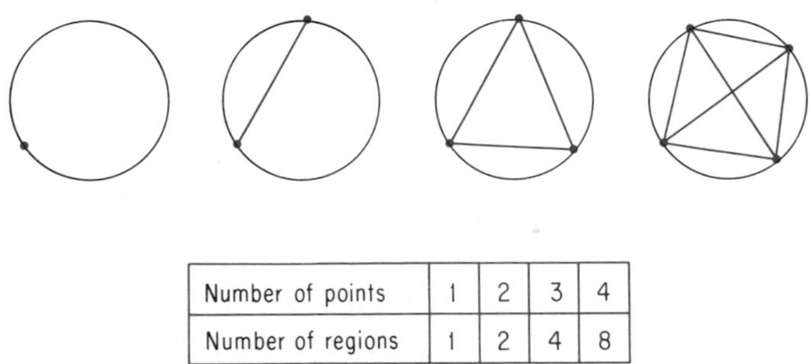

Number of points	1	2	3	4
Number of regions	1	2	4	8

Would you agree that a reasonable guess for the number of regions derived from five points is 16? Draw a figure to confirm your conjecture. What is your guess for the maximum number of regions that can be derived from six points? Again, draw a figure to confirm your conjecture. You may be in for a surprise!

EXERCISES

1. Take a piece of notebook paper and fold it in half. Then fold it in half again and cut off a corner that does not involve an edge of the original piece of paper.

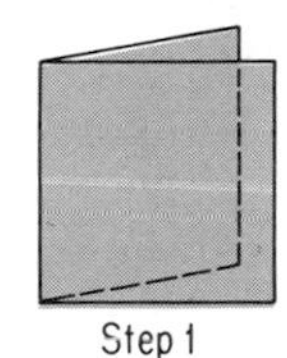

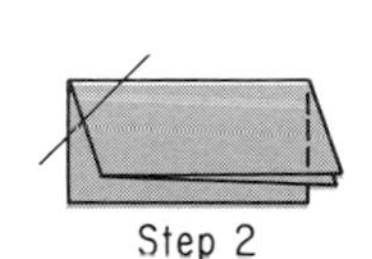

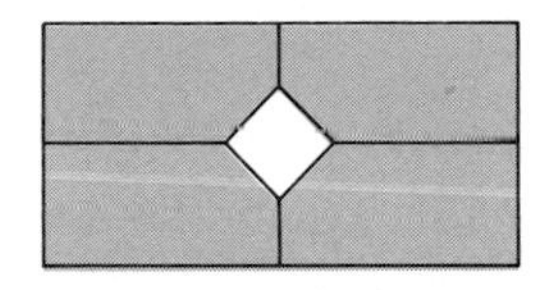

Your paper, when unfolded, should look like the preceding sketch. That is, with two folds we produced one hole. Repeat the same process but this time make three folds before cutting off an edge. Try to predict the number of holes that will be produced. How many holes will be produced with four folds? With n folds?

2. We wish to color each of the pyramids in thc accompanying figure so that no two of the faces (sides and base) that have a common edge are of the same color.

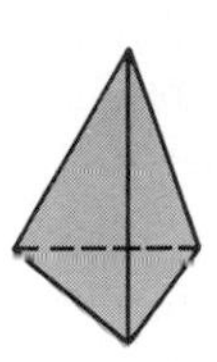

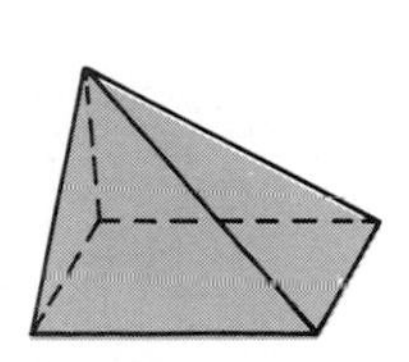

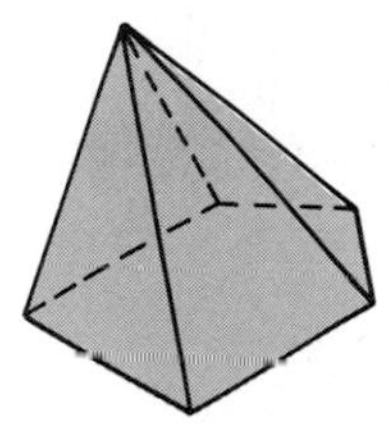

(a) What is the minimum number of colors required for each pyramid?
*(**b**) What is the relationship between the minimum number of colors required and the number of faces of a pyramid?

3. Consider the following set of figures. In each figure we count the number V of vertices, the number A of arcs, and the number R of regions into which the figure divides the plane. A square, for example, has four vertices, four arcs, and divides the plane into two regions (inside and outside of the square). See if you can discover a relationship between

V, R, and A that holds for each case. Confirm your generalization by testing it on several other figures.

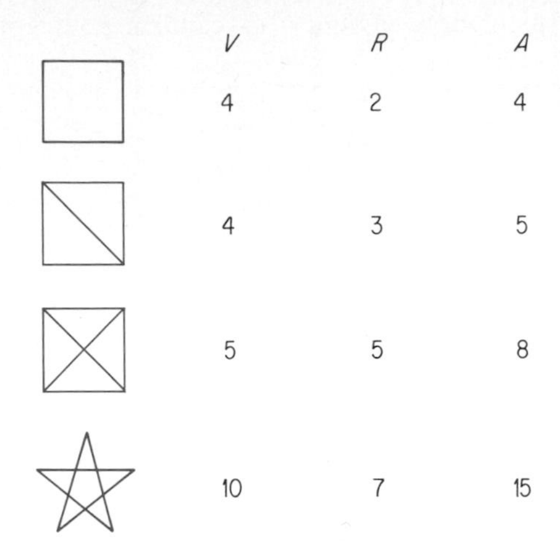

*4. What is the maximum number of pieces that can be obtained by slicing an orange with two cuts? With three cuts? With four cuts?

5. Consider the problem of arranging sets of squares that are joined along their edges. In general, these figures are referred to as **polyominoes.** A single square, a **monomino,** can be arranged in only one way. Two squares, a **domino,** can also be arranged in only one way, since the position of the 2×1 rectangle does not affect the arrangement. That is, any two such figures are congruent. There are two distinct arrangements for a **tromino,** that is, three squares.

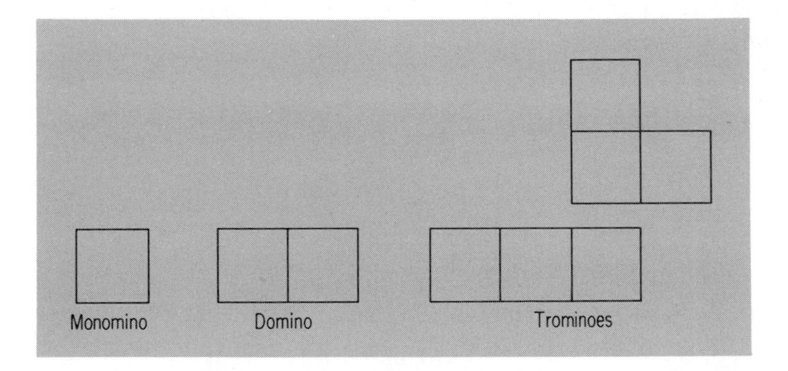

Next consider a **tetromino,** a set of four squares. Here we find that there are five possible arrangements such that no two are congruent. As in the case of the domino and other arrangements, we exclude any rearrangement that merely consists of a rotation which places the same squares in a congruent position. Here are three of the five possible tetrominoes:

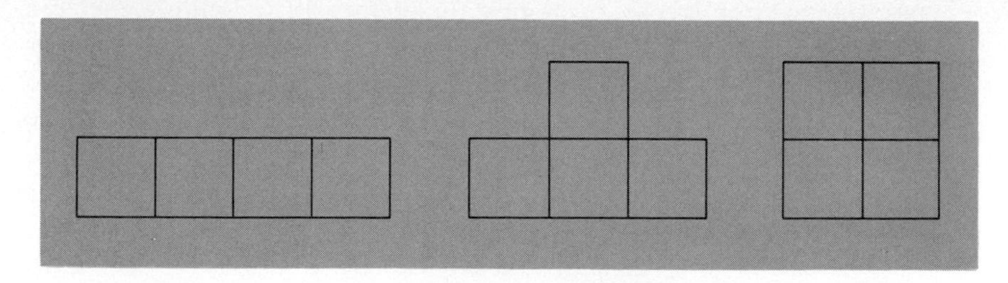

Draw the two remaining tetrominoes.

6. Consider a polyomino formed with five squares, called a **pentomino.** There are twelve such arrangements possible. Try to draw all twelve.

7. Twelve matchsticks are arranged to form the figure shown. By removing only two of the matches, form a figure that consists of two squares.

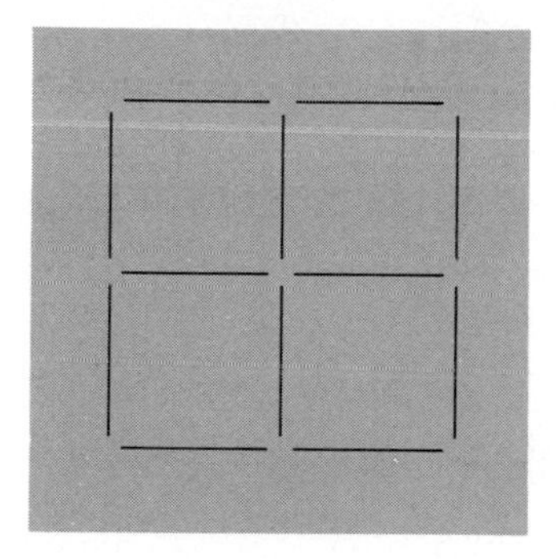

8. Twenty matchsticks are arranged to form the figure shown. By rearranging only three matchsticks, form a new figure that consists of five congruent squares.

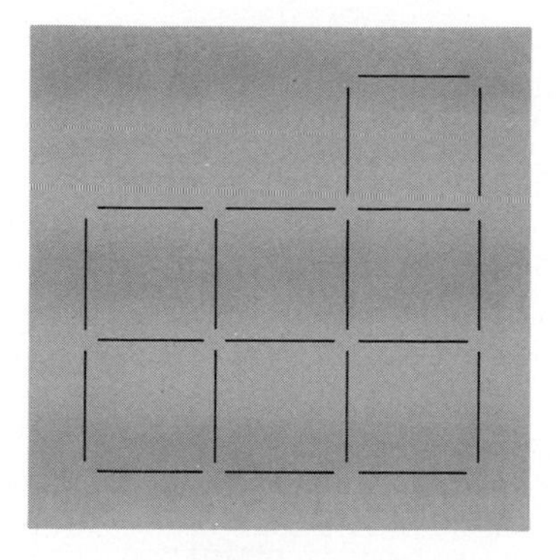

9. Arrange four 1-inch sticks and four $\frac{1}{2}$-inch sticks to form three congruent squares.

10. About the middle of the nineteenth century a scholar conjectured that all maps could be colored using at most four colors. In 1976 this conjecture was finally proved to be true. In this **four-color problem,** two countries that have common boundaries must be shown in different colors. Two countries that have only single points in common may be shown in the same color.

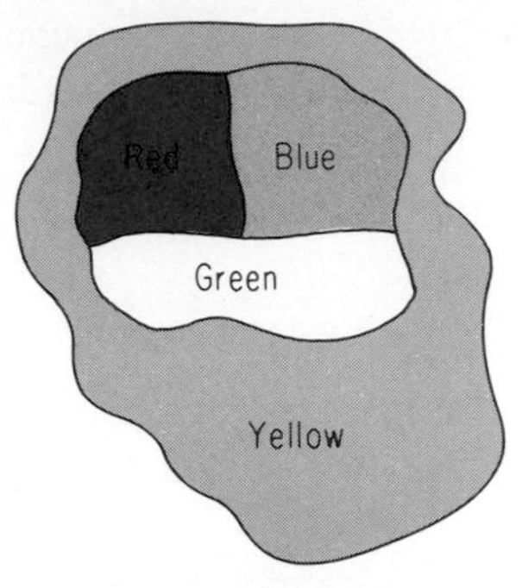

Four colors are required for this map

(a) Draw a map of five countries that requires only three colors.
(b) Can you draw a map of five countries that requires only two colors?

11. Begin with a large sheet of newspaper and fold it in half to form two layers. Then fold it in half again to form four layers. Repeat this process to form eight layers; and so on. (a) Try this experiment with different sizes of paper and determine the maximum number of times it is possible to make such folds. (b) Assume that you could continue this process for 50 times and that the original paper was 0.003 inch thick. How thick (or high) would the final pile of paper be?

3 explorations with mathematical recreations

The popularity of mathematics as a means of recreation and pleasure is evidenced by the frequency with which it is found in popular magazines and newspapers. In this section we shall explore several of these recreational aspects of mathematics.

How many of us have ever failed to be impressed by a *magic square*? The magic square shown here is arranged so that the sum of the numbers in any row, column, or diagonal is always 34.

1	12	7	14
8	13	2	11
10	3	16	5
15	6	9	4

Although there are formal methods to complete such an array, we will not go into them here. Suffice it to say that such arrangements have fascinated man for many centuries. Indeed, the first known example of a magic square is said to have been found on the back of a tortoise by the Emperor Yu in about 2200 B.C.! This was called the "lo-shu" and appeared as an array of numerals indicated by knots in strings as in the figure on the left below. Black knots were used for even numbers and white ones for odd numbers. In modern times this appears as a magic square of third order. The sum along any row, column, or diagonal is 15 as in the second figure.

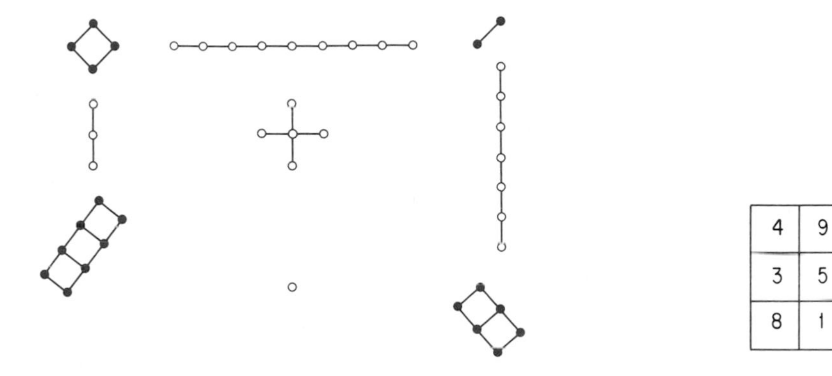

4	9	2
3	5	7
8	1	6

Closely akin to magic squares are square arrays of numbers useful in "mathemagic." Let us "build" a trick together. We begin by forming a square array and placing any six numerals in the surrounding spaces as in the figure. The numbers 3, 4, 1, 7, 2, and 5 are chosen arbitrarily. Next find the sum of each pair of numbers as in a regular addition table.

+	3	4	1
7			
2			
5			

+	3	4	1
7	10	11	8
2	5	6	3
5	8	9	6

Now we are ready to perform the trick. Have someone circle any one of the nine numerals in the box, say 10, and then cross out all the other numerals in the same row and column as 10.

+	3	4	1
7	(10)	~~11~~	~~8~~
2	~~5~~	6	3
5	~~8~~	9	6

Next circle one of the remaining numerals, say 3, and repeat the process. Circle the only remaining numeral, 9. The sum of the circled numbers is $10 + 3 + 9 = 22$.

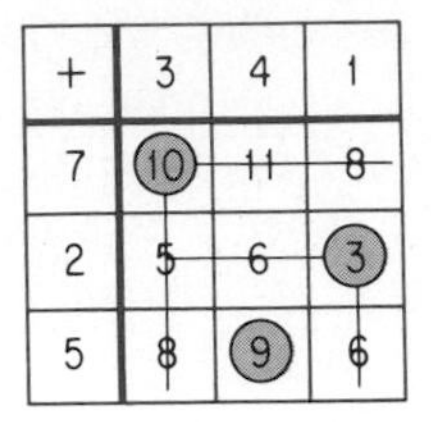

The interesting item here is that the sum of the three circled numbers will always be equal to 22, regardless of where you start! Furthermore, note that 22 is the sum of the six numbers outside the square. Try to explain why this trick works, and then build a table with 16 entries.

Another type of mathematical trick that is quite popular is the "think of a number" type. Follow these instructions:

Think of a number.
Add 3 to this number.
Multiply your answer by 2.
Subtract 4 from your answer.
Divide by 2.
Subtract the number with which you started.

If you follow these instructions carefully, your answer will always be 1, regardless of the number with which you start. We can explain why this trick works by using algebraic symbols or by drawing pictures, as shown below. Try to make up a similar trick of your own.

Think of a number:	n	□	(Number of coins in a box)
Add 3:	$n+3$	□ ○ ○ ○	(Number of original coins plus three)
Multiply by 2:	$2n+6$	□ ○ ○ ○ □ ○ ○ ○	(Two boxes of coins plus six)
Subtract 4:	$2n+2$	□ ○ □ ○	(Two boxes of coins plus two)
Divide by 2:	$n+1$	□ ○	(One box of coins plus one)
Subtract the original number, n:	$(n+1)-n=1$	○	(One coin is left)

Mathematical fallacies have always intrigued both professional and amateur mathematicians alike. Here is an arithmetic fallacy to puzzle you. You might even consider trying this on your local banker. First you need to deposit \$50 in the bank and then make withdrawals in the following manner:

	Withdraw		Balance
	Withdraw \$20,	leaving a balance of	\$30.
	Withdraw \$15,	leaving a balance of	\$15.
	Withdraw \$ 9,	leaving a balance of	\$ 6.
	Withdraw \$ 6,	leaving a balance of	\$ 0.
Adding, we have:	\$50		\$51

The total withdrawal is \$50, whereas the total of the balances is \$51. Can you therefore go to the bank to demand an extra dollar?

Here is a "proof" that $1 = 2$. Even though you may have forgotten the algebra you need to follow this, don't let it stop you; see if you can discover the fallacy.

Let $a = b$. Then

$$a^2 = b^2 = b \cdot b$$

Since $a = b$, we may write $b \cdot b$ as $a \cdot b$. Thus

$$a^2 = a \cdot b$$

Subtract b^2:

$$a^2 - b^2 = a \cdot b - b^2$$

Factor:

$$(a + b)(a - b) = b(a - b)$$

Divide by $a - b$:

$$\frac{(a + b)(a - b)}{(a - b)} = \frac{b(a - b)}{(a - b)}$$

Thus

$$a + b = b$$

Since $a = b$, we may write this as

$$b + b = b \qquad \text{or} \qquad 2b = b$$

Divide by b:

$$\frac{2b}{b} = \frac{b}{b}$$

Therefore

$$2 = 1$$

Many people enjoy solving interesting or amusing puzzles. A variety of these, many of which are old-timers, appear in the exercises that follow.

EXERCISES

1. All of the following puzzles have logical answers, but they are not strictly mathematical. See how many you can answer.
   (a) How many two-cent stamps are there in a dozen?
   (b) How many telephone poles are needed in order to reach the moon?
   (c) How far can you walk into a forest?
   (d) Two United States coins total 55¢ in value, yet one of them is not a nickel. Can you explain this?
   (e) How much dirt is there in a hole which is 3 feet wide, 4 feet long, and 2 feet deep?
   (f) There was a blind beggar who had a brother, but this brother had no brothers. What was the relationship between the two?
2. A farmer has to get a fox, a goose, and a bag of corn across a river in a boat which is only large enough for him and one of these three items. Now if he leaves the fox alone with the goose, the fox will eat the goose. If he leaves the goose alone with the corn, the goose will eat the corn. How does he get all items across the river?
3. Three cannibals and three missionaries need to cross a river in a boat big enough only for two. The cannibals are fine if they are left alone or if they are with the same number or with a larger number of missionaries. They are dangerous if they are left alone in a situation where they outnumber the missionaries. How do they all get across the river without harm?
4. A bottle and cork cost $1.50 together. The bottle costs one dollar more than the cork. How much does each cost?
5. A cat is at the bottom of a 30-foot well. Each day she climbs up 3 feet; each night she slides back 2 feet. How long will it take for the cat to get out of the well?
6. If a cat and a half eats a rat and a half in a day and a half, how many days will it take for 100 cats to eat 100 rats?
7. Ten coins are arranged to form a triangle as shown in the figure. By rearranging only three of the coins, form a new triangle that points in the opposite direction from the one shown.

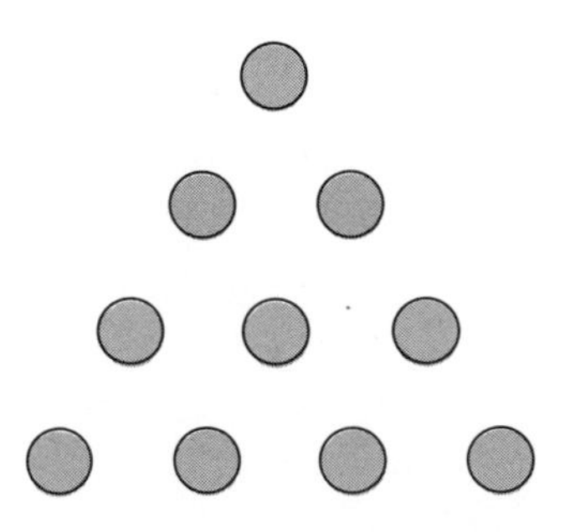

8. Use six matchsticks, all of the same size, to form four equilateral triangles. (An equilateral triangle has all three sides of the same length.)

9. A sailor lands on an island inhabited by two types of people. The A's always lie, and the B's always tell the truth. The sailor meets three inhabitants on the beach and asks the first of these: "Are you an A or a B?" The man answers, but the sailor doesn't understand him and asks the second person what he had said. The man replies: "He said that he was a B. He is, and so am I." The third inhabitant then says: "That's not true. The first man is an A and I'm a B." Can you tell who was lying and who was telling the truth?

10. A man goes to a well with three cans whose capacities are 3 gallons, 5 gallons, and 8 gallons. Explain how he can obtain exactly 4 gallons of water from the well.

11. Here is a mathematical trick you can try on a friend. Ask someone to place a penny in one of his hands, and a dime in the other. Then tell him to multiply the value of the coin in the right hand by 6, multiply the value of the coin held in the left hand by 3, and add. Ask for the result. If the number given is an even number, you then announce that the penny is in the right hand; if the result is an odd number, then the penny is in the left hand and the dime is in the right hand. Can you figure out why this trick works?

12. Consider a house with six rooms and furniture arranged as in the accompanying figure. We wish to interchange the desk and the bookcase, but in such a way that there is never more than one piece of furniture in a room at a time. The other three pieces of furniture do not need to return to their original places. Can you do this? Try it using coins or other objects to represent the furniture.

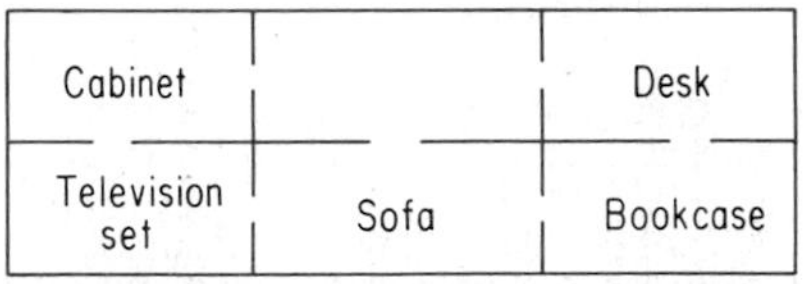

Cabinet		Desk
Television set	Sofa	Bookcase

13. Three men enter a hotel and rent a suite of rooms for \$30. After they are taken to their rooms the manager discovers he overcharged them; the suite rents for only \$25. He thereupon sends a bellhop upstairs with the \$5 change. The dishonest bellhop decides to keep \$2 and returns only \$3 to the men. Now the rooms originally cost \$30, but the men had \$3 returned to them. This means that they paid only \$27 for the room. The bellhop kept \$2. Note that $\$27 + \$2 = \$29$. What happened to the extra dollar?

14. Write the numbers from 1 through 10 using four 4's for each. Here are the first three completed for you:

$$\frac{44}{44} = 1, \qquad \frac{4}{4} + \frac{4}{4} = 2, \qquad \frac{4 + 4 + 4}{4} = 3$$

15. Arrange two pennies and two dimes as shown below. Try to interchange the coins so that the pennies are at the right and the dimes at the left. You may move only one coin at a time, you may jump over only one coin, and pennies may be moved only to the right while dimes may be moved only to the left. No two coins may occupy the same space at the same time. What is the minimum number of moves required to complete the game?

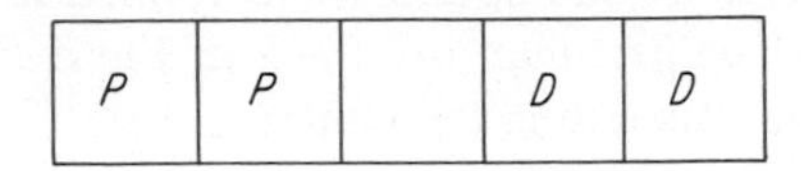

16. Repeat Exercise 15 for three pennies and three dimes, using seven blocks. What is the minimum number of moves required to complete the game?

17. What are the next two letters in the sequence:

O, T, T, F, F, S, S, . . . ?

18. What are the next two letters in the sequence:

E, F, F, N, O, S, S, . . . ?

19. Place a half-dollar, a quarter, and a nickel in one position, A, as in the figure. Then try to move these coins, one at a time, to position C. Coins may also be placed in position B. At no time may a larger coin be placed on a smaller coin. This can be accomplished in $2^3 - 1$, that is, 7 moves.

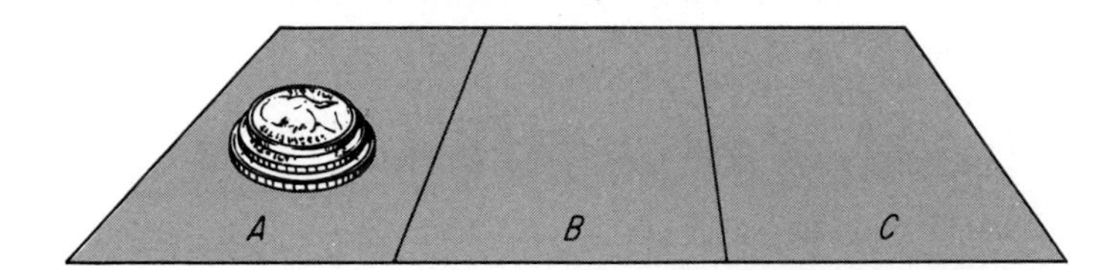

Next add a penny to the pile and try to make the change in $2^4 - 1$, that is, 15 moves.

This is an example of a famous problem called the **Tower of Hanoi.** The ancient Brahman priests were to move a pile of 64 such discs of decreasing size, after which the world would end. This would require $2^{64} - 1$ moves. Try to estimate how long this would take at the rate of one move per second.

20. Here is a game that must be played by two persons. Two players alternate in selecting one of the numbers 1, 2, 3, 4, 5, or 6. After each number is selected, it is added to the sum of those previously selected. For example, if player A selects 3 and player B selects 5, then the total is 8. If A selects 3 again, then the total is 11 and player B takes his turn. The object of the game is to be the first one to reach 50. There is a way to always win if you are permitted to go first. See if you can discover this method for winning, and then try to play the game with a classmate and with some elementary school students. Incidentally, games such as these are excellent means of providing for practice on number facts in an interesting and disguised manner.

21. Many tricks of magic have their basis in elementary mathematics and may be found in books on mathematical recreations. Here is one example of such a trick.

Have someone place three dice on top of one another while you turn your back. Then instruct him to look at and find the sum of the values shown on the two faces that touch each other for the top and middle dice, the two faces that touch each other for the middle and bottom dice, and the value of the bottom face of the bottom die. You then turn around and at a glance tell him this sum. The trick is this: You merely subtract the value showing on the top face of the top die from 21. Stack a set of three dice in the manner described and try to figure out why the trick works as it does.

22. You are given a checkerboard and a set of dominoes. The size of each domino is such that it is able to cover two squares on the board. Can you arrange the dominoes in such a way that all of the board is covered with the exception of two squares in opposite corners? (That is, you are to leave uncovered the two squares marked XX in the figure.) Try to explain why you should or should not be able to arrange the dominoes in this way.

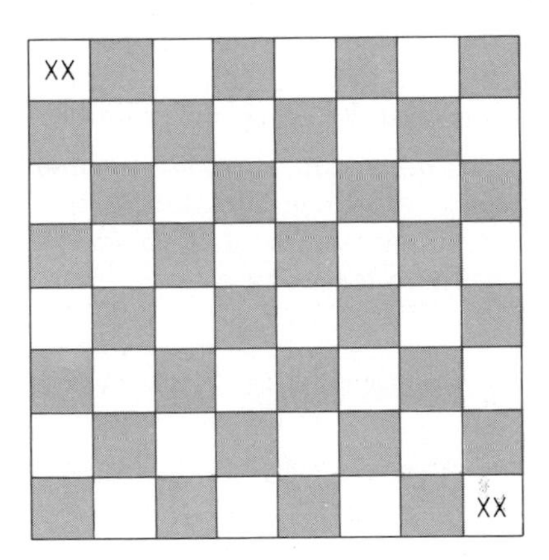

READINGS AND PROJECTS

1. Explore a recently published elementary mathematics textbook series to determine what use, if any, is made of mathematical recreations. If puzzles or games are used as an integral part of the series, bring to class a representative sample of these for different grade levels.

2. Begin your own collection of mathematical puzzles and games. It is helpful to place each individual item on an index card, with a solution and the suggested grade level for its use on the reverse side of the card. One good source of such material is Appendix A (Activities, Games, and Applications) of the thirty-fifth yearbook of the National Council of Teachers of Mathematics.
3. Read Chapter 14 (Some Puzzlers for Thinkers) of the twenty-seventh yearbook of the National Council of Teachers of Mathematics. If possible, try to use some of the suggested puzzles and games with a group of elementary school students and report on your results.
4. Martin Gardner writes a monthly column entitled "Mathematical Games" in *Scientific American.* He has assembled some of his best columns in several books, such as *Martin Gardner's Sixth Book of Mathematical Games from Scientific American*, W. H. Freeman and Co., 1971. Review at least six of the columns that have appeared within the past several years and report on any items that seem appropriate for use in an elementary mathematics class.
5. Read Chapter 3 (Recreational Activities) of *Teaching Mathematics* by Max A. Sobel and Evan M. Maletsky, Prentice-Hall, Inc., 1975. Become an "expert" on one of the suggested games and demonstrate this in class.
6. For an interesting extension of the work on finger multiplication, read the "Letter to the Editor" by Robert L. Benson on page 237 of the March 1973 issue of *School Science and Mathematics.*
7. Write a review of *A Bibliography of Recreational Mathematics*, Volume 3, a 1973 publication of the National Council of Teachers of Mathematics, prepared by William L. Schaaf. Use the references cited to prepare a report on one of the categories used in this publication.
8. Write a review of *Games and Puzzles for Elementary and Middle School Mathematics,* a 1975 publication of the National Council of Teachers of Mathematics. Learn at least five of these games and demonstrate one of them to your class.

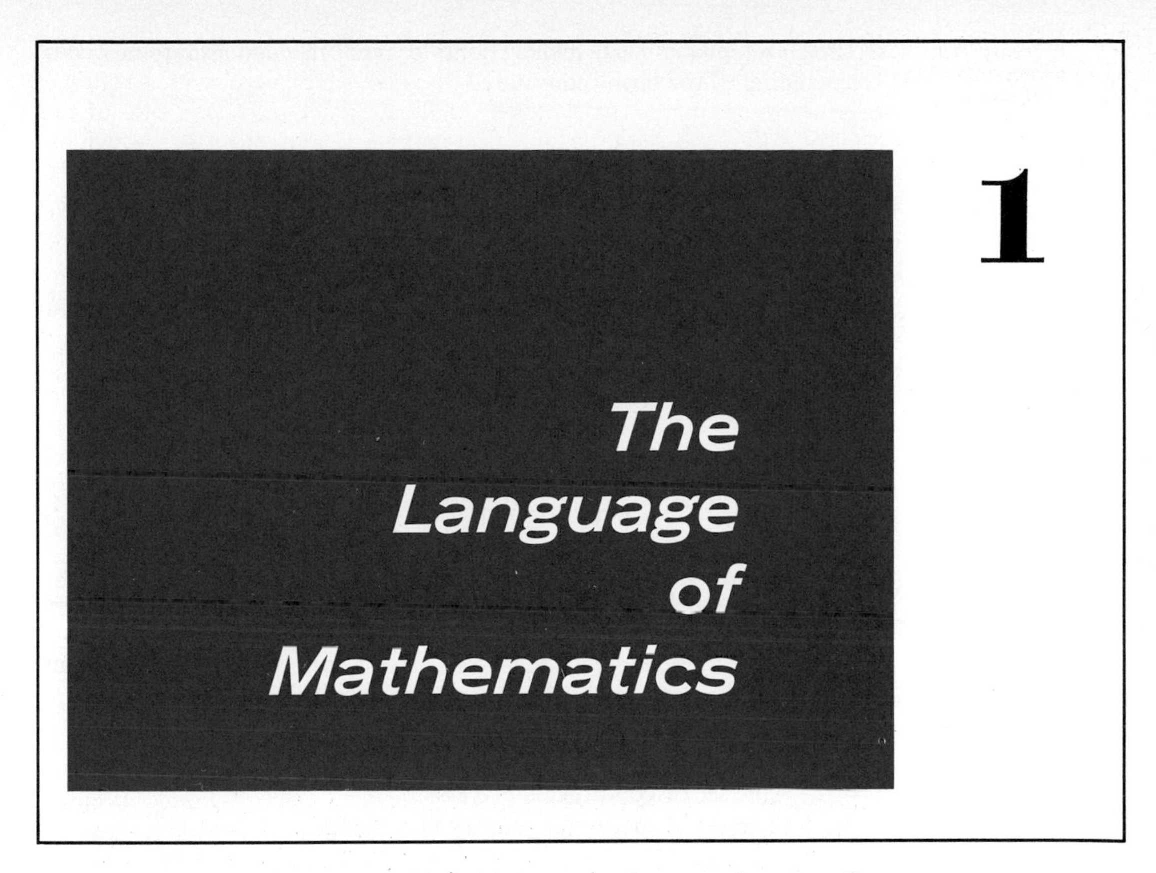

Welcome to the *set* of readers of this book. You may be a *member* of a college class for which this book is the assigned text. If so, the members of your class form a *subset* of the set of all students in such classes. Suppose that each person in your class reads this book. The set of readers of this book who are not in your class is then the *complement* of your class relative to the *universal set* of all readers of this book. If each person in your class has exactly one copy of this book of his or her own, then there is a *one-to-one correspondence* between the set of members of your class and the set of their textbooks for the class. That is, the set of members of your class and the set of their textbooks for the class are *equivalent sets.*

The study of mathematics may be considered as the study of a new language. Mathematics has its own vocabulary and its own grammer (rules for combining its terms and symbols). In this chapter we consider primarily the basic areas of the language of elementary mathematics: sets, numbers, points, and statements.

The terms set, member (or element), subset, complement, universal set, one-to-one correspondence, and equivalent sets, as used in the first paragraph of this chapter, are part of the vocabulary of the language of contemporary mathematics. Each term has a very precise meaning. We shall use examples, *counterexamples* (examples to which the term does *not* apply), exercises, and pedagogical explorations to reinforce and "sharpen" your understanding of the terms considered.

1-1 sets

There is a classic problem of which came first, the chicken or the egg. We have a similar problem in the language of mathematics—we can't define everything. Thus we must assume that you already understand some terms from your previous experiences. For example, we assume that you know what is meant by a *set* (collection) of elements and also what is meant by a particular element *being a member of* (*belonging to*) a specified set.

Consider these two sets.

> The set of letters of the English alphabet.
> The set of states of the United States of America on January 1, 1976.

You can tell whether or not any specified element belongs to either of these given sets; that is, each set is a **well-defined set.**

Whenever there can be a question as to the membership of an element in a set, the set is not a well-defined set. For example, the following sets are not well-defined sets.

> The set of good tennis players.
> The set of successful country music singers.

Contemporary mathematics is primarily concerned with well-defined sets. The elements of these sets may be numbers, points, geometric figures, blocks of various shapes (sizes, colors, materials, etc.), statements (equations, inequalities, etc.), people, or other identifiable entities.

Consider these sets of numbers.

$D = \{0, 1, 2, 3, \ldots, 9\}$, the set of **decimal digits.**
$C = \{1, 2, 3, 4, \ldots\}$, the set of **counting numbers.**
$W = \{0, 1, 2, 3, \ldots\}$, the set of **whole numbers.**

We have used braces and indicated the sets D, C, and W in **set notation.** The three dots are used to indicate elements that are not explicitly listed. The list continues in the indicated pattern either indefinitely or until the specified last element is reached. The set D has a last element and is a **finite set.** The sets C and W do not have a last element (they continue indefinitely) and are **infinite sets.**

In general, the order in which the elements of a set are listed is unimportant. However, when three dots are used to show that some elements are not explicitly listed, it is necessary to state the elements in some order so that a pattern can be observed and used to identify the missing elements. Any such ordered set of elements is a **sequence**.

Membership in, or belonging to, a set is the basis for our concept of a set. The **membership symbol** $\in$ is used as in these examples.

$$2 \in \{1, 2, 3, 4\}, \qquad 7 \notin \{1, 2, 3, 4\}$$

We read $\in$ as "is a member of" and $\notin$ as "is not a member of" the given set. Frequently capital letters are used to name sets, such as the set C of counting numbers. Then $5 \in C$; $\frac{1}{2} \notin C$.

It is interesting, at times, to seek verbal descriptions for given sets of elements that have already been listed. The next example illustrates this point.

Example 1: Write a verbal description for the set

$$Y = \{1, 3, 5, 7, 9\}$$

Solution: There are several correct responses that might be given. Two of these are: "The set of odd numbers 1 through 9"; "The set of odd numbers between 0 and 11." Note that the word "between" implies that the first and last numbers (that is, 0 and 11) are not included as members of the given set.

Two sets that have precisely the same elements are **equal sets.** We write

$$\{1, 2, 3\} = \{3, 2, 1\}$$

since order of listing does not affect membership. We also write

$$\{1, 1, 2\} = \{1, 2\}$$

since listing an element more than once does not increase the number of different members of the set. However, we shall not normally repeat elements within a listing. In general, we write $A = B$ to show that sets A and B have the same members; that is A and B are *two names for the same set*. Similarly, $A \neq B$ indicates that sets A and B do not have the same members and thus are not equal sets.

Any two equal sets may be placed in one-to-one correspondence since each element may be made to correspond to itself. For example,

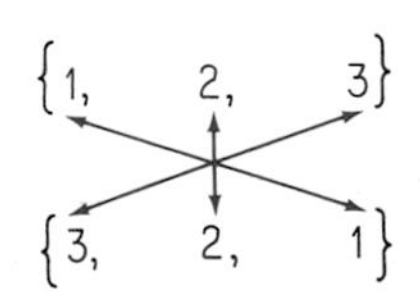

However, it may be possible to place two sets that have different elements in one-to-one correspondence. For example,

$$\begin{matrix} \{1, & 2, & 3\} \\ \updownarrow & \updownarrow & \updownarrow \\ \{a, & b, & c\} \end{matrix}$$

Any two sets that can be placed in one-to-one correspondence are **equivalent sets** in the sense that they have the same number of elements. In general, any two equal sets are equivalent sets but, as in the last example, two equivalent sets are not necessarily equal sets.

Example 2: Show that $\{c, r, a, b\}$ and $\{f, i, s, h\}$ are (a) equivalent sets of letters; (b) not equal sets.

Solution: (a)

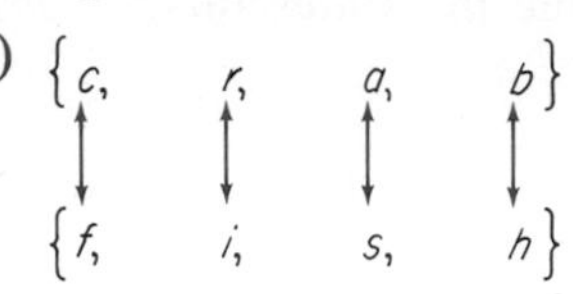

(b) The element c, for example, is a member of the first set and not a member of the second set.

Any set may be the **universal set** $\mathcal{U}$ for a particular discussion—the set of readers of this book, the set of decimal digits, the set of counting numbers, and so forth. The **complement** A' (also written $\bar{A}$) of a given set A depends upon the universal set under consideration. Thus for any given universal set $\mathcal{U}$ the set A' is the complement of A *relative to* $\mathcal{U}$. The complement of the universal set is the **empty set** (or **null set**) that contains no elements. The empty set may be denoted by $\varnothing$ or by $\{\ \}$.

Example 3: Find the complement of $\{1, 2, 3, 4, 5\}$ (a) relative to the set of decimal digits, (b) relative to the set of counting numbers.

Solution: (a) $\{0, 6, 7, 8, 9\}$; (b) $\{6, 7, 8, 9, \ldots\}$.

The language of sets is used at all levels and in all areas of mathematics. Some of the terms will be considered in the exercises; others will be developed further in future sections.

EXERCISES

State whether or not the given sets are (a) equivalent sets; (b) equal sets.

1. $\{t, o, n\}, \{n, o, t\}$

2. $\{c, a, r, t\}, \{r, a, c, k\}$

3. $\{t, a, m, p\}, \{m, a, p\}$

4. $\{l, a, z, y\}, \{f, a, s, t\}$

5. $\{d, o, n, t\}, \{d, o, n, e\}$

6. $\{h, o, t\}, \{h, e, a, t\}$

Replace the asterisk (*) by $\in$ or $\notin$ to obtain a true statement.

7. $5 * \{1, 3, 5, 7, 9\}$

8. $5 * \{2, 4, 6, 8, 0\}$

9. $\triangle * \{., |, \square\}$

10. $\triangle * \{/, \wedge, \triangle\}$

State whether or not each given set is a well-defined set.

11. The set of cities that are state capitals in the United States of America.

12. The set of states with good climates in the United States.

List the elements of each set in set notation.

13. The set of English names of the months in the year.

14. The set of counting numbers between 1 and 10.

List the elements of the complement A' of each given set A relative to the given universal set $\mathcal{U}$.

15. $\mathcal{U} = \{t, r, a, p\}$. (a) $\{a, t\}$; (b) $\{t, a, p\}$; (c) $\{p, a, r, t\}$.

16. $\mathcal{U} = \{1, 2, 3, 4, 5, 6\}$. (a) $\{1, 3, 5\}$; (b) $\{2\}$; (c) $\{\ \}$.

Identify each statement as true or false. If true, give an example; if false, give a counterexample.

17. Any two equivalent sets are equal sets.

18. Any two equal sets are equivalent sets.

Write a verbal description for each of the following sets.

19. $N = \{1, 2, 3, 4, 5\}$

20. $R = \{1, 2, 3, \ldots, 100\}$

21. $S = \{101, 102, 103, \ldots\}$

***22.** $A = \{8, 5, 4, 9, 1, 7, 6, 3, 2\}$. Give an explanation for the ordering of the elements.

PEDAGOGICAL EXPLORATIONS

The purposes of the Pedagogical Explorations throughout this book are twofold. One purpose is to supplement your background knowledge so that you will be equipped to find and use supplementary materials for students with special needs. The other purpose is to encourage you to start building a reservoir of materials for practical use in elementary school classrooms.

1. Name at least fifty words that indicate sets. For example, consider a school of fish, a swarm of bees, and a squadron of planes.

*Starred exercises either require some background that has not yet been presented in this book or need to be considered from an unusual point of view.

2. Select a word and look up its meaning in a dictionary. Then look up the meanings of the words used to define the original word. Repeat this process for the words used each time until it becomes clear that there is a set of related terms which you can understand only if you have a previous knowledge of at least one of these terms. Find several terms such that each could be used with elementary school children to illustrate the essential circularity of dictionary definitions and thus the need for experiences to supply intuitive understanding of the meanings of some words.
3. Select a daily paper or news magazine and list several of the references to sets and subsets that are made in that paper or magazine. Don't miss the committees and subcommittees.
4. Select a set such as the students of a school, members of a church, or citizens of a nation and give as many examples as you can of common significant applications of membership, of subset, and of subsets of a subset.

Students first learn the meanings of new words from examples and counterexamples of them.

5. Give some examples that could be used to help elementary school students understand what is meant by the noun *chair.* (Note that a chair is an object, a piece of furniture, not a table, and so forth.)
6. Repeat Exploration 5 for the noun *pencil.*

Suppose that you asked a student how many states there are in the United States of America and he replied 48.

7. Give a reply that could be used both to indicate a circumstance under which his answer would have been correct and to indicate politely that the stated answer is not correct for the original question.
8. Rephrase your initial question slightly to emphasize the distinction that led to his confusion.

Consider only examples that can be used with elementary school children.

9. Give several examples of well-defined sets and explain why each set is considered to be well-defined.
10. Repeat Exploration 9 for sets that are not well-defined.
11. Explain the basic distinction between well-defined sets and sets that are not well-defined.

1-2 operations with sets

The *complement* A' of a set A is the set of elements of the universal set $\mathcal{U}$ that are not elements of A. This complement of A relative to $\mathcal{U}$ may also be denoted by $\mathcal{U} - A$; $A' = \mathcal{U} - A$. For any two sets A and B the set of elements of B that are not elements of A is the **difference set** $B - A$.

Example 1: Let set $A = \{1, 2, 3, 4, 5, 7, 9\}$ and set $B = \{2, 4, 6, 8\}$. Find (a) $A - B$; (b) $B - A$; (c) $B - B$.

Solution: (a) $A - B = \{1, 3, 5, 7, 9\}$; (b) $B - A = \{6, 8\}$; (c) $B - B = \varnothing$.

The finding of all possible subsets of a given set B may be considered as an operation with the single set B. Suppose that our selected set is the set of counting numbers 1 through 9;

$$B = \{1, 2, 3, 4, 5, 6, 7, 8, 9\}$$

A set A is a *subset* of B if the set A has the property that each element of A is also an element of B. We write $A \subseteq B$ (read "A is included in B"). There are many subsets of B; a few of them are listed below:

$$A_1 = \{1, 2, 3\}$$
$$A_2 = \{1, 5, 7, 8, 9\}$$
$$A_3 = \{2\}$$
$$A_4 = \{1, 2, 3, 4, 5, 6, 7, 8, 9\}$$

Note in particular that set A_4 contains each of the elements of B and is classified as a subset of B. Any set is said to be a subset of itself. Also the empty set is a subset of every set.

A set A is a **proper subset** of a set B if A is a subset of B and there is at least one element of B that is not an element of A. We write $A \subset B$ (read "A is properly included in B"). Intuitively we speak of a proper subset as part of, but not all of, a given set. Each of the subsets A_1, A_2, A_3, and A_4 are subsets of B; the sets A_1, A_2, and A_3 are proper subsets of B; the set A_4 is not a proper subset of B. We may write in symbols:

$$A_1 \subset B, \qquad A_2 \subset B, \qquad A_3 \subset B, \qquad A_4 \subseteq B$$

Example 2: List three proper subsets of the set $\{1, 2, 3, 4\}$.

Solution: Here are three of the several possible proper subsets:

$$\{1, 2\} \qquad \{1, 3, 4\} \qquad \{1\}$$

Example 3: List all possible subsets of the set $\{1, 2\}$.

Solution: The given set has two elements. Each element may or may not be selected as a member of a particular subset. Thus, each of the elements 1, 2 may be considered as in the following array with two choices (take the element or leave it out). Since there are two elements and two choices for each element, 2×2 subsets are obtained.

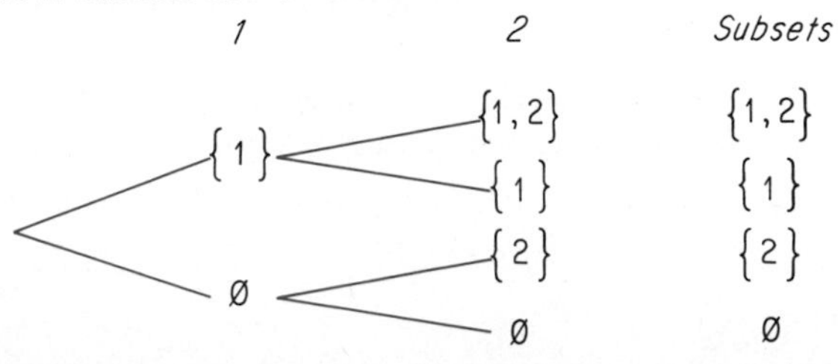

Each subset may be paired with its complement relative to the given set.

$$\begin{matrix} \{1, 2\} & \{1\} \\ \varnothing & \{2\} \end{matrix}$$

Example 4: List the subsets of {1, 2, 3} and pair each set with its complementary set.

Solution:

$$\begin{matrix} \{1, 2, 3\} & \{2, 3\} & \{1, 3\} & \{1, 2\} \\ \varnothing & \{1\} & \{2\} & \{3\} \end{matrix}$$

It seems reasonable to expect that all possible subsets should be obtained by considering the elements three at a time, two at a time, one at a time, and none at a time. As a check that we have found all possible subsets, notice that the original set had three elements; we have two choices (take it or leave it) for each element; $2 \times 2 \times 2 = 8$, and we found eight subsets.

Let us consider two sets A and B, defined as follows:

$$A = \{1, 2, 3, 4, 5, 6, 7\}$$
$$B = \{2, 5, 7, 8, 9\}$$

From these two sets let us form another set C, whose members are those elements that appear in each of the two given sets:

$$C = \{2, 5, 7\}$$

The set C consists of the elements that the sets A and B have in common and is called the **intersection** of the sets A and B. We often say that the intersection of two sets A and B (written $A \cap B$) is the set of elements that are members of both of the given sets.

Example 5: If $A = \{a, r, e\}$ and $B = \{c, a, t\}$, find $A \cap B$.

Solution: $A \cap B = \{a\}$; that is, a is the only letter that appears in each of the two given sets.

Example 6: If $X = \{1, 2, 3, 4, 5\}$ and $Y = \{4, 5, 6, 7\}$, find $X \cap Y$.

Solution:

$$X \cap Y = \{4, 5\}$$

Example 7: If $X = \{1, 2, 3\}$ and $Y = \{4, 5, 6\}$, find $X \cap Y$.

Solution: Sets X and Y have no elements in common and are said to be **disjoint sets**; their intersection is the null set. Thus we may write $X \cap Y = \varnothing$.

Example 8: If the universal set is $\{1, 2, 3, \ldots, 10\}$, $A = \{1, 2, 3, 4, 5\}$, and $B = \{3, 4, 5, 6, 7\}$, list the elements in $A' \cap B'$.

Solution: $A' = \{6, 7, 8, 9, 10\}$, $B' = \{1, 2, 8, 9, 10\}$, $A' \cap B' = \{8, 9, 10\}$.

From the two given sets A and B of this section let us next form another set D, whose members are those elements that are elements of at least one of the two given sets:

$$D = \{1, 2, 3, 4, 5, 6, 7, 8, 9\}$$

The set D is called the **union** of set A and B. We often say that the union of two sets A and B (written $A \cup B$) is the set of elements that are members of at least one of the given sets.

Example 9: Let A represent the set of names of boys on a particular committee, $A = \{\text{Bill, Bruce, Max}\}$, Let B represent the set of names of boys on another committee, $B = \{\text{Bruce, John, Max}\}$. Find $A \cup B$.

Solution: $A \cup B = \{\text{Bill, Bruce, John, Max}\}$. Here the union of the two sets is the set of names of the boys who are on at least one of the two committees. Note that the names Bruce and Max appear only once in the set $A \cup B$ even though these names are listed for both committees.

Example 10: If $A = \{1, 2, 3, 4, 5\}$ and $B = \{3, 5, 7, 9\}$, find $A \cup B$.

Solution:

$$A \cup B = \{1, 2, 3, 4, 5, 7, 9\}$$

The operations considered in this section always give rise to subsets of the universal set under consideration. Another type of operation, the *Cartesian product*, is considered in the next section.

EXERCISES

1. If $\mathcal{U} = \{1, 2, 3, 4, 5, 6, 7, 8, 9\}$, $A = \{1, 3, 5, 7, 9\}$, and $B = \{1, 4, 9\}$, find **(a)** A'; **(b)** B'.

For each of the following sets list the elements in **(a)** $A - B$; **(b)** $B - A$.

2. $A = \{1, 2, 3, 4, 5\}$, $B = \{1, 3, 5\}$
3. $A = \{1, 2, 3, 4, 5\}$, $B = \{1, 3, 5, 7\}$
4. $A = \{1, 2, 3, \ldots, 9\}$, $B = \{0, 2, 4, 6, 8\}$
5. $A = \{1, 2, 3, \ldots\}$, $B = \{1, 3, 5, \ldots\}$
6. $A = \{1, 3, 5, \ldots\}$, $B = \{2, 4, 6, \ldots\}$

For each of the following sets list the elements in **(a)** $A \cup B$; **(b)** $A \cap B$.

7. $A = \{1, 3, 4\}$, $B = \{1, 3, 5, 7\}$

8. $A = \{3, 4, 5\}, \quad B = \{4, 5, 6, 7\}$

9. $A = \{2, 4, 6, 8\}, \quad B = \{4, 6, 7, 8\}$

10. $A = \{1, 3, 5, \ldots\}, \quad B = \{2, 4, 6, \ldots\}$

11. $A = \varnothing, \quad B = \{1, 2, 3, \ldots\}$

12. $A = \{1, 2, 3, \ldots\}, \quad B = \{1, 3, 5, \ldots\}$

For each of the given universal sets list the elements in **(a)** A'; **(b)** B'; **(c)** $A' \cup B'$; **(d)** $A' \cap B'$.

13. $\mathcal{U} = \{1, 2, 3, 4, 5\}, \quad A = \{1, 2\}; \quad B = \{1, 3, 5\}$

14. $\mathcal{U} = \{1, 2, 3, \ldots, 10\}, \quad A = \{1, 3, 5, 7, 9\}; \quad B = \{2, 4, 6, 8, 10\}$

15. $\mathcal{U} = \{1, 2, 3, \ldots\}, \quad A = \{1, 3, 5, \ldots\}; \quad B = \{2, 4, 6, \ldots\}$

16. $\mathcal{U} = \{1, 2, 3, 4, 5, 6, 7\}, \quad A = \varnothing; \quad B = \{1, 2, 3, 4, 5, 6, 7\}$

17. $\mathcal{U} = \{1, 2, 3\}, \quad A = \{1\}, \quad B = \{3\}$

For Exercises 18 through 23, use $\mathcal{U} = \{1, 2, 3, 4, 5, 6, 7, 8, 9, 10\}$, *and let* $A = \{2, 3, 4, 5, 7, 9\}$, *and* $B = \{2, 4, 5, 9, 10\}$, *and list the elements in:*

18. $A' \cap B'$ *19.* $(A \cup B)'$ *20.* $A' \cup B$

21. $(A' \cap B)'$ *22.* $A \cap B$ *23.* $(A \cup B')'$

List all proper subsets of each set.

24. **(a)** $\{p\}$; **(b)** $\{p, q\}$; **(c)** $\{r, s, t, u\}$.

25. **(a)** $\{a, b, c\}$; ***(b)** $\{a, b, c, d, e\}$; ***(c)** $\varnothing$.

PEDAGOGICAL EXPLORATIONS

1. A set A is defined to be a subset of a set B if each element of A is also an element of B. Alternatively a set A may be defined as a subset of B if there is no element of A that is not also an element of B.
   (a) Do both of these definitions hold if $A = B$?
   (b) Do both of these definitions hold if A has at least one element and each element of A is an element of B?
   (c) Use the alternate definition and explain why $\varnothing \subseteq B$ for any set B.
   (d) Use the original definition and explain why $\varnothing \subseteq B$ for any set B.
2. Consider a set of 25 members of a class that is studying this book. Think of at least ten subsets of this set. Be sure to include several types of subsets that could be used in a discussion with an elementary school class.
3. Consider only examples that could be used with primary school children. Give at least five examples of:
   (a) A pair of complementary sets.

(b) An intersection of sets.
(c) A union of sets.

4. Make an array as in Example 3 of this section for the subsets of each of these sets.
(a) $\{a\}$ (b) $\{a, b, c\}$
(c) $\{a, b, c, d\}$ (d) $\{a, b, c, d, e\}$

5. Observe the numbers of subsets obtained in Exploration 4 and complete the following array.

Number of elements	0	1	2	3	4	5	6	10	50	n
Number of subsets	1									
Number of proper subsets										

1-3 relations among sets

We have already considered several relations among sets:

$A = B$, A and B are two names for the same set.
$A \subseteq B$, A is a subset of B.
$A \subset B$, A is a proper subset of B.
A and B are complementary sets.
A and B are disjoint sets.

We have also mentioned one-to-one correspondence as the basis for equivalent sets. We now consider one-to-one correspondences in further detail.

Two sets, $X = \{x_1, x_2, \ldots\}$ and $Y = \{y_1, y_2, \ldots\}$, are in **one-to-one correspondence** if we can find a pairing of the x's and y's such that each x corresponds to one and only one y and each y corresponds to one and only one x. Consider the sets $R = \{a, b, c\}$ and $S = \{\$, +, \%\}$. We can match the element $a \in R$ with any one of the elements in S (three choices); we can then match $b \in R$ with any one of the other two elements of S (two choices), and then match $c \in R$ with the remaining element of S (one possibility). Thus a one-to-one correspondence of the sets R and S can be shown in $3 \times 2 \times 1$ (that is, 6) ways:

{a, b, c} ↕ {\$, +, %}   {a, b, c} ↕ {\$, %, +}   {a, b, c} ↕ {+, %, \$}

{a, b, c} ↕ {+, \$, %}   {a, b, c} ↕ {%, \$, +}   {a, b, c} ↕ {%, +, \$}

Two sets A and B that can be placed in a one-to-one correspondence are said to be **equivalent sets** (written $A \longleftrightarrow B$ and read as "A is equivalent to B"). Any two equivalent sets have the same number of elements, that is, the same **cardinality.**

You may use the concept of a one-to-one correspondence to show whether or not any two sets of elements have the same number of members. You also use this concept of a one-to-one correspondence when you count the elements of a set.

Example: Count the elements of the set $\{a, b, c, d, e\}$ by using a one-to-one correspondence with a subset of the set of counting numbers.

Solution:

$$\begin{matrix} \{a, & b, & c, & d, & e\} \\ \updownarrow & \updownarrow & \updownarrow & \updownarrow & \updownarrow \\ \{1, & 2, & 3, & 4, & 5\} \end{matrix}$$

The set $\{a, b, c, d, e\}$ has five elements.

Any set A of elements that may be placed in one-to-one correspondence with the set of elements $\{1, 2, 3, 4, 5\}$ is said to have five elements; we write $n(A) = 5$ to show that the number of elements in the set A is 5. Any set B of elements that may be placed in one-to-one correspondence with the set of elements $\{1, 2, 3, 4, \ldots, k-1, k\}$ is said to have k elements; $\mathbf{n(B) = k}$. Notice this use of the set of counting numbers and one-to-one correspondence to determine how many elements there are in a set; that is, to determine the **cardinal number** of a set. When the counting numbers are taken in order, the last number used in the one-to-one correspondence is the cardinal number of the set. We define the cardinality of the empty set to be zero; that is, $n(\varnothing) = 0$. Thus the whole numbers may be used as, and introduced as, the cardinal numbers of finite sets.

A one-to-one correspondence between two sets is actually a very special type of correspondence between two sets. As an example of a more general correspondence, consider any classification or sorting procedure. For example, suppose that eight kindergarten children were given sets of 3, 2, 2, 1, 3, 1, 2, and 3 crayons, respectively. They might recognize the number of crayons that each was given. In this way they would create a correspondence, or *mapping*, of the sets of crayons into the set of numbers $\{1, 2, 3\}$.

A slightly different example of a mapping is obtained when we think of a village postal clerk sorting letters into boxes. There is one box for each family in the village. The mapping of the letters into the boxes may not involve all of the boxes since some of the families may not get any letters. Also some families may get more than one letter.

As in our last two examples a correspondence of a set A with a set B may not be one-to-one. If each element of A corresponds to exactly one element of B, then the set A is mapped into the set B. A mapping of a set A into a set B such that each element of B is the *image* of exactly one element of A is a *one-to-one correspondence*.

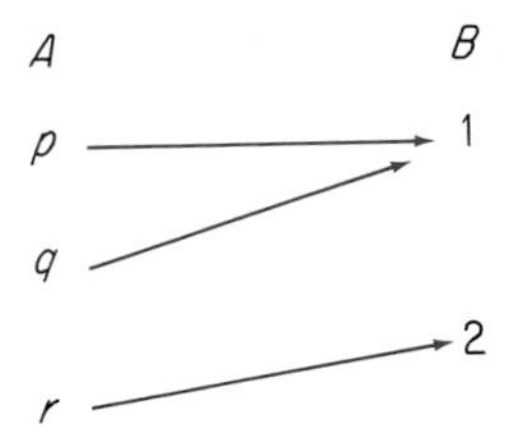

Consider the sets $A = \{p, q, r\}$ and $B = \{1, 2\}$ with a mapping of A into B as indicated in the figure. Note that p is mapped into 1, q is mapped into 1, and r is mapped into 2. We may indicate this mapping of A into B by the set of ordered pairs

$$\{(p, 1), (q, 1), (r, 2)\}$$

In each ordered pair the first element is a member of the set A and the second element is a member of the set B. Any pair (s, t) with one element s considered as the first element and the other element t considered as the second element is an **ordered pair** of elements.

For any two given sets A and B the set of all possible ordered pairs with an element of A as the first element and an element of B as the second element is the **Cartesian product** of A and B, written $A \times B$ and read as "A cross B." For $A = \{p, q, r\}$ and $B = \{1, 2\}$,

$$A \times B = \{(p, 1), (p, 2), (q, 1), (q, 2), (r, 1), (r, 2)\}$$

Similarly,

$$B \times A = \{(1, p), (1, q), (1, r), (2, p), (2, q), (2, r)\}$$

The elements of a Cartesian product are often represented in an array with the elements of A identifying the columns and the elements of B identifying the rows.

2	$(p, 2)$	$(q, 2)$	$(r, 2)$
1	$(p, 1)$	$(q, 1)$	$(r, 1)$
	p	q	r

Each of the ordered pairs of the mapping $\{(p, 1), (q, 1), (r, 2)\}$ is an element of the array for $A \times B$. Thus the set of ordered pairs of the mapping is a subset of $A \times B$ as indicated in the next array.

2	$(p, 2)$	$(q, 2)$	$(r, 2)$
1	$(p, 1)$	$(q, 1)$	$(r, 1)$
	p	q	r

For any sets A and B, any mapping of A into B is a subset of $A \times B$ and is called a **function.** Any subset of $A \times B$ is called a **relation.** Then a function (mapping) is a relation in which each element of A corresponds to exactly one element of B. Mappings, relations, and functions are very useful terms in the language of mathematics.

EXERCISES

Show all possible one-to-one correspondences of the two given sets.

1. $\{1, 2\}$ and $\{s, t\}$
2. $\{1, 2, 3\}$ and $\{a, b, c\}$

Create a set with the given number of elements. Then use a one-to-one correspondence with a subset of the counting numbers to show that the cardinal number of the set is the given number.

3. Three.
4. Six.
5. Ten.

Find the cardinal number of each set.

6. $\{p\}$
7. $\{11, 12, \ldots, 18\}$
8. $\{100, 101, \ldots, 110\}$
9. $\varnothing$
10. $\{1, 3, 5, 7, \ldots, 19\}$
11. $\{2, 4, 6, \ldots, 102\}$

Represent each relation by **(a)** *a set of ordered pairs;* **(b)** *an indicated portion of a Cartesian product.*

12.

Alan → Joan
Joe → Doris
Pete → Betty

13.

a → 2
b → 1
c → 3

14.

a → q
b → q
c → r
(p)

15.

1 → x
2 → x
2 → y
3 → z

Identify by exercise number the relations in Exercises 12 through 15 that are:

16. Functions.
17. One-to-one.

State whether you would expect each of the following statements to be always true or not always true. If the statement is not always true, then give at least one counterexample.

18. (a) $(A \cap B) \subseteq A$ (b) $(A \cap B) \subset B$

19. (a) $B \subset (A \cup B)$ (b) $A \subseteq (A \cup B)$

20. (a) $(A - B) \subset A$ (b) $(A - B) \subseteq A$

Describe, if possible, the conditions on the sets A and B such that each statement will always be **(a)** *true;* **(b)** *false.*

21. $(A - B) = A$

22. $(A \cap B) = A$

23. $(A \cup B) = A$

24. $(A \cap B) = (B \cap A)$

25. $(A \cup B) = (B \cup A)$

26. $[A \cap (A \cup B)] = A$

27. $[A \cap (A \cup B)] = B$

28. $(A \cup B) = (A \cap B)$

PEDAGOGICAL EXPLORATIONS

The concept of infinity as used in mathematics is a very difficult one for most people to acquire. It is helpful to contrast the idea of very large sets with that of infinite sets.

For each of the following finite situations, first make an educated guess. Then find the correct answer.

1. To the nearest day, how long would it take to count to one billion at the rate of one number per second?
2. Estimate how many pennies it would take to make a stack one inch high. Approximately how high would a stack of one million pennies be?
3. One million one-dollar bills are placed end-to-end along the ground. To the nearest ten miles, how long would this strip of bills be?
4. To the nearest billion, estimate the number of seconds that elapse in a century.
5. To the nearest day how long would it take to spend $1,000,000 at the rate of $1.00 per minute?

One of the largest numbers ever named is a **googol**, which has been defined as 1 followed by 100 zeros:

1000000000000000000000000000000000000000000000000000
0000000000000000000000000000000000000000000000000

This number is larger than what is considered to be the total number of protons or electrons in the universe! A googol can be expressed, using exponents, as 10^{100}.

Even larger than a googol is a **googolplex**, defined as 1 followed by a googol of zeros. One famous mathematician claimed that there would not even be room between the earth and the moon to write all the zeros in a googolplex!

*6. How many zeros are there in the number represented as a googol times a googol? Express this number using exponents. Is this number smaller or larger than a googolplex?

The use of one-to-one correspondences may be extended to infinite sets. We now use n to represent any counting number and explore this extension. For example, the set of odd counting numbers may be placed in one-to-one correspondence with the set of all counting numbers.

$$\begin{array}{ccccccccc} \{1, & 2, & 3, & 4, & 5, & \ldots, & n, & \ldots\} \\ \updownarrow & \updownarrow & \updownarrow & \updownarrow & \updownarrow & & \updownarrow & \\ \{1, & 3, & 5, & 7, & 9, & \ldots, & 2n-1, & \ldots\} \end{array}$$

Any set that can be placed in one-to-one correspondence with the set of counting numbers has $\aleph_0$ (read as "aleph null") as its **transfinite cardinal number**. Then, even though the set of counting numbers is an infinite set, it is correct to say that there are $\aleph_0$ counting numbers. There are also $\aleph_0$ odd counting numbers.

7. Show a one-to-one correspondence between the set of counting numbers and the set of even counting numbers

$$\{2, 4, 6, \ldots, 2n, \ldots\}$$

8. Use the correspondences obtained preceding Exploration 7 and in Exploration 7 to give a justification for the statement:

(a) $\aleph_0 + \aleph_0 = \aleph_0$ (b) $\aleph_0 = 2\aleph_0$

Show a one-to-one correspondence between the set of counting numbers and the given set. Then explain the statement of equality listed with the set.

9. $\{101, 102, 103, 104, \ldots, 100 + n, \ldots\}$, $\aleph_0 = \aleph_0 - 100$
10. $\{4, 5, 6, 7, \ldots, n + 3, \ldots\}$, $\aleph_0 = \aleph_0 - 3$
11. $\{0, 1, 2, 3, 4, \ldots, n - 1, \ldots\}$, $\aleph_0 = \aleph_0 + 1$
12. $\{a, b, c, 0, 1, 2, 3, \ldots, n - 4, \ldots\}$, $\aleph_0 = \aleph_0 + 4$

Note that the arithmetic involving $\aleph_0$ follows a different set of rules than ordinary arithmetic.

1-4 sets of numbers

Three very different aspects are considered as we think about numbers:

Symbols or names for numbers.
Uses of numbers.
Different types of numbers.

The symbol 5 is one name for the number five. Any symbol for a number is a **numeral.** Since

$$𝍸, \quad V, \quad 2 + 3, \quad 10 \div 2, \quad \sqrt{25}$$

are all names for the same number, we write

$$𝍸 = V = 2 + 3 = 10 \div 2 = \sqrt{25}$$

Thus, as for sets, we use the **equality symbol** = for two names (symbols) for the same number. Then $\neq$ means that the symbols are names for different numbers.

A few textbooks insist that the reader consciously distinguish between number and numeral each time that either is used. Throughout this text we usually make the assumption that the reader understands from the context of a statement whether the numeral (symbol) or the number (concept, idea) is meant. Thus we use statements such as "the number 5" instead of "the number represented by the numeral 5." In a few instances, as illustrated in the explorations, it is necessary to distinguish between a number and one of its symbols. Hence our vocabulary should include both *number* and *numeral*.

Any whole number is the *cardinal number* of one or more sets of numbers. Also any finite set has a whole number as its cardinal number. Thus one use of numbers is as cardinal numbers of sets, that is, to tell us how many elements there are in a set.

If we say that Jack is 183 centimeters (about six feet) tall, we are using 183 as a measure of Jack's height in centimeters, that is, as the cardinal number of the set of units (each one centimeter long) that are needed to match Jack's height. Any number that is used as a *measure* is a cardinal number. Other types of numbers and many other types of measures are commonly used. For example,

> Jane's new car cost $4,000.
> Bill is ten dollars in debt; -10 dollars.
> Ruth is one meter 63 centimeters tall; 1.63 meters.
> The distance from home plate to second base on a baseball diamond is $90\sqrt{2}$ feet, that is, about 38.7 meters.

Note that, as in the case of Ruth's height (to the nearest centimeter) and the distance from home plate to second base (to the nearest tenth of a meter), measures are often **approximate numbers,** that is, numbers *rounded off* to the nearest unit. The measure of 38.7 in meters is 387 in decimeters (tenths of a meter). Any measure may be considered as a whole number of the smallest unit used. In the case of Bill's debt of ten dollars, the unit is one dollar of debt and the measure is ten. Many texts call a number of units, such as 38.7 meters, a **measurement.** Then a *measure* is a number; a *measurement* is a number of units.

Several other types of numbers are considered later in detail but mentioned here in recognition of the vocabulary of mathematics. Consider:

$I = \{\ldots, -3, -2, -1, 0, 1, 2, 3, \ldots\}$, **integers**
$N = \{\ldots, -3, -2, -1\}$, **negative integers**
$P = \{1, 2, 3, \ldots\}$, **positive integers**
fractional numbers such as $\frac{1}{2}, \frac{2}{3}, \frac{24}{7}$
mixed numbers such as $2\frac{1}{3}, 5\frac{3}{4}, 3\frac{3}{7}$
rational numbers such as $2, -2, \frac{5}{8}, 3\frac{1}{8}, -\frac{2}{3}$
real numbers such as $3, -3, -\frac{1}{2}, 5\sqrt{2}, \sqrt[3]{6}$

A symbol for a fractional number is often called a **fraction.** Formal definitions will be given as the various sets of numbers are considered in detail. However, note that the set of positive integers appears to be the same (for all practical purposes is the same) as the set of counting numbers, also called the the set of **natural numbers.** The set of whole numbers may be called the set of **nonnegative integers,** that is, the set consisting of zero and the positive integers. Different texts may use different terms. Different people (students or teachers) may think of sets of number in different ways—often simply at different times or in different situations.

There are two other common uses of numbers. For example, you expect page 35 of this book to follow page 34. Counting numbers are often used to assign an order to the elements of a finite set. Often slightly different words such as first, second, third, and fourth are used to indicate this use of numbers. Numbers that are used to assign an order to the elements of a set are called **ordinal numbers.** Finally, numbers may be used primarily for **identification** where cardinality, order, and measure serve secondary roles. For example, consider the number of your driver's license, your telephone number, or your social security number. Each of these numbers is used primarily for identification. However, a number that is primarily used for identification may also be looked up in an ordered list to trace a telephone call or to identify the owner of a car. The three uses of numbers—as cardinal numbers (how many, including how many units), as ordinal numbers, and for identification—are widely used in the increasingly numerous and elementary areas of applications of mathematics.

The ordinality of the counting numbers may be extended to obtain **order relations** for all real numbers. These relations involve the symbols:

$<$ is less than.
$\leq$ is less than or equal to.
$>$ is greater than.
$\geq$ is greater than or equal to.

Each of the following is a correct statement using one of these relations:

$$2 < 5, \quad -2 < 1, \quad -3 < -1$$
$$3 \leq 3, \quad 2 \leq 3, \quad -1 > -3$$
$$5 > 2, \quad 3 \geq 2, \quad 3 \geq 3$$

If any of these statements do not appear to be correct, the difficulty may be with the use of "or", which we shall consider in detail shortly. These statements may also be considered with reference to a number line as in the next section.

A symbol, usually a letter, that may be replaced by any member of a specified set is called a **variable.** For example:

For any counting number n, $n < n + 1$.

This general statement is made possible by the use of n as a variable that has the set of counting numbers as its set of possible replacements, its **replacement set.**

Variables are also used, for example, in the definitions of multiples of numbers. Any number that can be expressed in the form

$$2n, \qquad n \text{ an integer}$$

is a **multiple of 2** and is called an **even number.** Any number that can be expressed in the form

$$2n + 1, \qquad n \text{ an integer}$$

is an **odd number.** When any integer t is divided by 2 either the *division is exact* (has remainder 0) and $t = 2n$ for some integer n, or there is a remainder of 1 and $t = 2n + 1$ for some integer n. Thus every integer is either an even number or an odd number. Note that $0 = 2 \times 0$ and thus 0 is an even number.

In the general language of mathematics we frequently speak of multiples of any number. For example,

$3n$, n an integer—multiples of 3;
$10n$, n an integer—multiples of 10;
$n\sqrt{2}$, n an integer—multiples of $\sqrt{2}$.

The replacement set for n specifies the type of multiples under consideration. For example, we use

$$5n, \qquad n \text{ a positive integer}$$

for *positive integral multiples* of 5, that is, $\{5, 10, 15, 20, \ldots\}$. We use

$$7n, \qquad n \text{ an integer}$$

for *integral multiples* of 7, that is, $\{\ldots, -21, -14, -7, 0, 7, 14, 21, \ldots\}$. Thus the set $\{s, 2s, 3s, 4s, \ldots\}$ may be described as the set of positive integral multiples of s for any number s.

Since $24 = 3 \times 8$, we say that 24 is a *multiple* of 8. We may also say that 8 is a *factor* of 24. In general, if m is a multiple of k, then k is a **factor** of m. Notice also that 0 is an integral multiple of any number n and any number n is a factor of 0.

EXERCISES

Tell whether each number is used primarily as a cardinal (including measure) number, an ordinal number, or for identification.

1. Doris is in the eighth row.
2. There are 36 students in the class.
3. It takes 9 people to field a baseball team.
4. This is the second time he is taking the course.
5. I am listening to 93 on the FM dial.
6. The card catalog lists the book as 376.33.
7. This is the second problem on the list.
8. There are thirty volumes in this set of books.
9. The tenth volume includes geometry.
10. My license plate number is 7136.

Represent each set in set notation.

11. The positive integral multiples of 3.
12. The positive integral multiples of 4.
13. The integral multiples of 5.
14. The integral multiples of 6.
15. The negative integral multiples of 7.
16. The negative integral multiples of 10.

Represent the intersection of the sets in the specified exercises (**a**) *in set notation and* (**b**) *as simply as possible in words.*

17. Exercises 11 and 13.
18. Exercises 12 and 14.
19. Exercises 11 and 14.
20. Exercises 13 and 16.

Find the set of positive integral factors of each number.

*21. 6

*22. 13

*23. 17

*24. 12

*25. 27

*26. 36

PEDAGOGICAL EXPLORATIONS

1. Select an elementary school grade level and make a set of ten or more problems designed for student explorations and practice to help the students recognize the wide variety of common uses of numbers.

2. Repeat Exploration 1 to establish firmly the dependence of finite cardinal numbers upon one-to-one correspondences.
3. Repeat Exploration 1 to show the existence of at least one cardinal number that is not a finite cardinal number.
4. Addition is presented to first grade children using disjoint sets. Give examples indicating why it is essential for the sets to be disjoint.
5. Make appropriate visual aids and describe at least one way of showing first grade students that (a) $3 + 2 = 5$; (b) $1 + 4 = 5 = 4 + 1$.
6. As in Exploration 5, but for second or third grade, explain the statement: If $3 + 2 = 5$, then $2 + 3 = 5$, $5 - 2 = 3$, and $5 - 3 = 2$.
7. Let A be the set of all positive integral factors of 36 and B the set of all positive integral factors of 24.
   (a) Describe $A \cap B$.
   (b) Describe a usc of members of $A \cap B$ in reducing the fraction $\frac{24}{36}$.
8. Let C be the set of all positive integral multiples of 15 and D the set of all positive integral multiples of 12.
   (a) Describe $C \cap D$.
   (b) Describe a use of a member of $C \cap D$ in finding $\frac{2}{15} + \frac{7}{12}$.

1-5 sets of points

Any geometric figure is a set of points. In this textbook, as in most textbooks for the early grades, we consider how geometric figures look. Preschool children form circles in their games without any conscious analysis of the properties of a circle. Yet a child who gets out of position relative to the circle concept of another child is usually told by that child to get into place.

We identify geometric figures by name but we do not try to verbalize their formal definitions. Suppose that several children are in a circle with a ball at the center. One child may insist that another child not be closer to the center than he is. The intuitive concept of equidistant from the center is clearly present but no formal definitions are appropriate or needed. Thus we are concerned with the development and the recognition of patterns that should precede a formal study of geometric figures.

Here are some pictures of a few of the figures that school children name and use in their games and other activities.

Points

Lines

The arrowheads indicate that even though the pictures are restricted to the page, the lines extend indefinitely in the indicated directions.

Circles

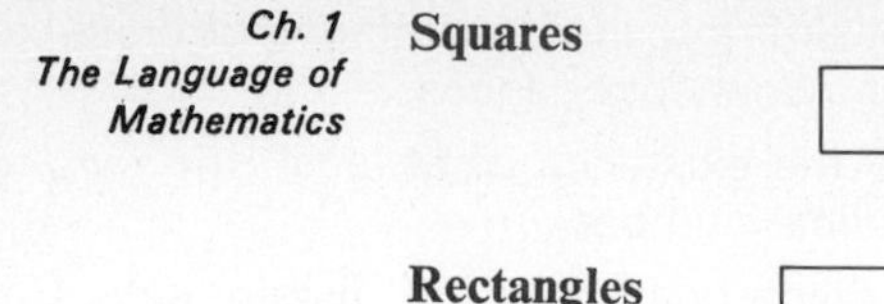

Squares

Rectangles

Triangles

Gradually relationships among figures are observed. For example, in most textbooks, some rectangles are squares; all squares are rectangles. The qualification "in most textbooks" emphasizes the lack of uniformity in books. Thus when any particular book is read or used the reader must make a special effort to recognize the precise definitions used *in that book*. Such ambiguities are not a unique phenomenon of the "new math." Nearly a century ago Lewis Carroll wrote in Chapter VI of *Through the Looking Glass:*

> "When I use a word," Humpty Dumpty said, in a rather scornful tone, "it means just what I chose it to mean—neither more nor less."
> "The question is," said Alice, "whether you *can* make words mean so many different things."
> "The question is," said Humpty Dumpty, "which is to be master—that's all."

Descriptions of geometric figures are refined as relationships are observed. Usually after several refinements we are content to call the description a definition. In the language of mathematics symbols as well as words are carefully defined. As in the following examples a small change in a symbol can substantially change its meaning.

Rays such as $\overrightarrow{AB}$

A B C

have exactly one **endpoint** and extend in only one direction. A ray may be named by its endpoint and any one of its other points. Thus $\overrightarrow{AB}$ and $\overrightarrow{AC}$ are two names for the same ray.

Lines such as $\overleftrightarrow{PQ}$

P Q Z

extend in two opposite directions and do not have any endpoints. In identifying a line any two of its distinct points may be used. Thus

$$\overleftrightarrow{PQ}, \quad \overleftrightarrow{PZ}, \quad \overleftrightarrow{QZ}, \quad \overleftrightarrow{QP}, \quad \overleftrightarrow{ZP}, \quad \overleftrightarrow{ZQ}$$

are all names for the same line, $\overleftrightarrow{PQ}$.

Line segments such as $\overline{RS}$ R ——— S

do not extend in any direction but have two endpoints, R and S. Note that $\overline{SR}$ and $\overline{RS}$ are two names for the same line segment.

Example 1: Consider the given line and identify each set of points.

(a) $\overline{PQ} \cup \overline{QS}$ (b) $\overline{PR} \cap \overline{QS}$
(c) $\overrightarrow{QR} \cup \overrightarrow{SR}$ (d) $\overrightarrow{PQ} \cap \overrightarrow{QS}$

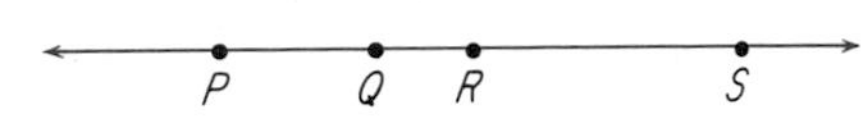

Solution: (a) $\overline{PS}$; (b) $\overline{QR}$; (c) $\overleftrightarrow{PQ}$ (any one of several names for the line may be used); (d) $\overrightarrow{QS}$ (this ray may also be named $\overrightarrow{QR}$).

Any line may be used as a number line. A number line is one of the most common and useful geometric figures. To obtain a **number line,** draw any line, select any point of that line as the **origin** with *coordinate* 0 (zero), and select any other point of the line as the **unit point** with coordinate 1 (one). Usually the number line is considered in a horizontal position with the unit point on the right of the origin.

The length of the line segment with the origin and the unit point as endpoints is the **unit distance,** or **unit of length,** for marking off a scale on the line. The points representing any given counting numbers may be obtained by marking off successive units to the right of the origin. The numbers are the **coordinates** of the points; the points are the **graphs** of the numbers.

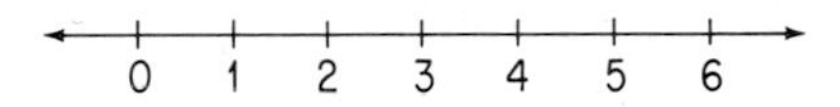

We can now **graph** different sets of numbers on a number line as in the following example. We shall use the verb "graph" to mean "draw the graph of."

Example 2: On a number line graph the set of whole numbers less than 3.

Solution: Draw a number line and place solid dots at the points that correspond to 0, 1, and 2.

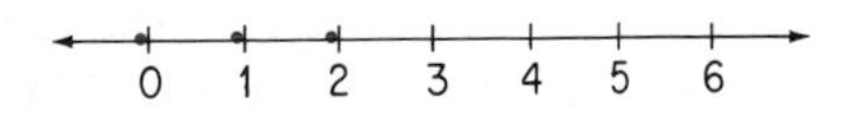

Graphs of sets of integers may be represented on a number line by marking off units to the left as well as to the right of the origin.

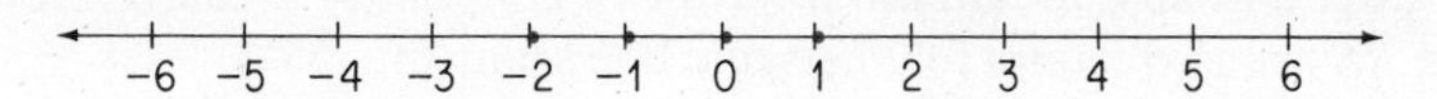

Graphs of sets of real numbers may also be represented on a number line. Any number line has the property that every point of the line has a real number as its coordinate and every real number has a point of the line as its graph.

Order relations among integers (and indeed any real numbers) may be represented on a number line. If the unit point is selected on the right of the origin:

$a < b$ if the point with coordinate a is on the left of the point with coordinate b. For example, $2 < 5$, $-1 < 2$, $-3 < -1$.

$a \leq b$ if the point with coordinate a is at or on the left of the point with coordinate b; that is, the point a is not on the right of b, $(a \not> b)$. For example, $3 \leq 3$, $2 \leq 3$.

$b > a$ if $a < b$. For example, $5 > 2$ since $2 < 5$; $-1 > -3$ since $-3 < -1$.

$b \geq a$ if $a \leq b$. For example, $3 \geq 2$ since $2 \leq 3$.

These concepts are considered in detail in the discussions of various sets of numbers.

Example 3: Graph on a real number line:
(a) $x > 2$ (b) $x < -1$ (c) $-2 \leq x \leq 1$

Solution:

(a)

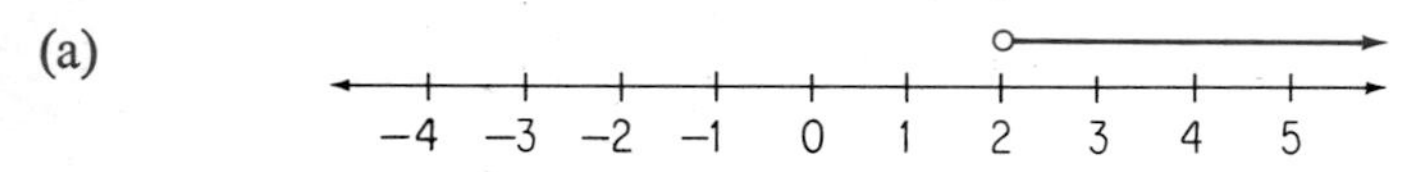

(b)

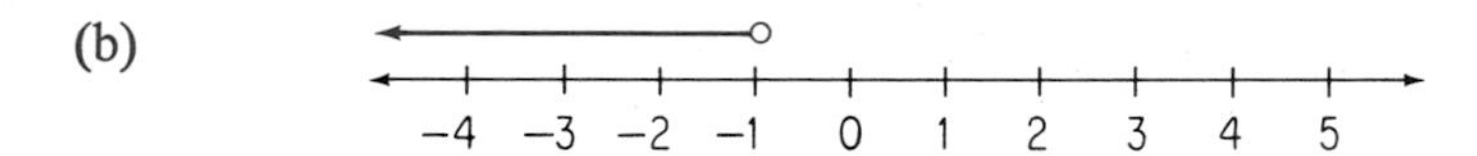

(c)

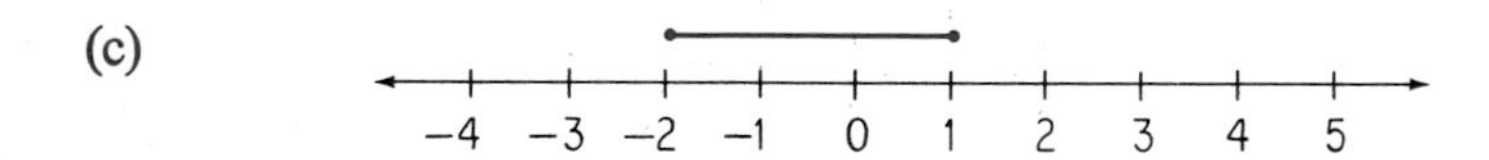

Note that *hollow dots* are used when points such as $x = 2$ in part (a) and $x = -1$ in (b) are *not* included in the graph; *solid dots* are used as in part (c) when the endpoints are included in the graph. The graphs may be drawn either *on* or *above* the number line, whichever is clearer to the reader.

Many elementary school texts consider any geometric figure that can be drawn on a sheet of paper without lifting the pencil from the paper to be a continuous **plane curve.**

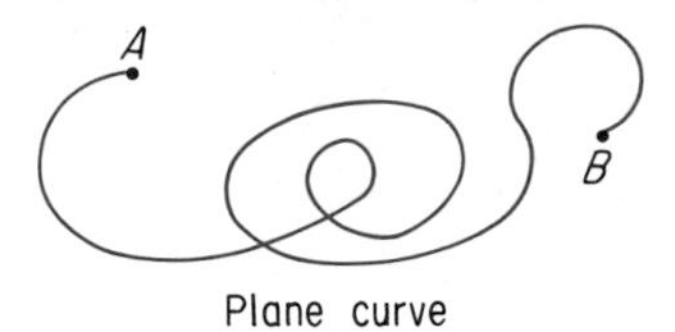

Plane curve

The plane curve in the figure was drawn by starting at the point A and ending at B. If A and B are two names for the same point ($A = B$), then the curve is a **closed curve.** If $A \neq B$, then the curve is not a closed curve and is called an **open curve.**

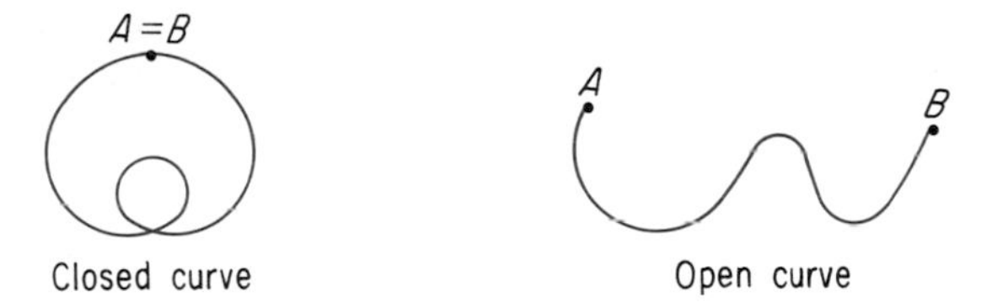

Closed curve Open curve

If, other than possibly ending at the starting point, no point of a plane curve is used more than once as the curve is drawn, then the curve is a **simple plane curve.** The following curves are simple curves. Note that in geometry curves do not have to be "crooked" or "smooth."

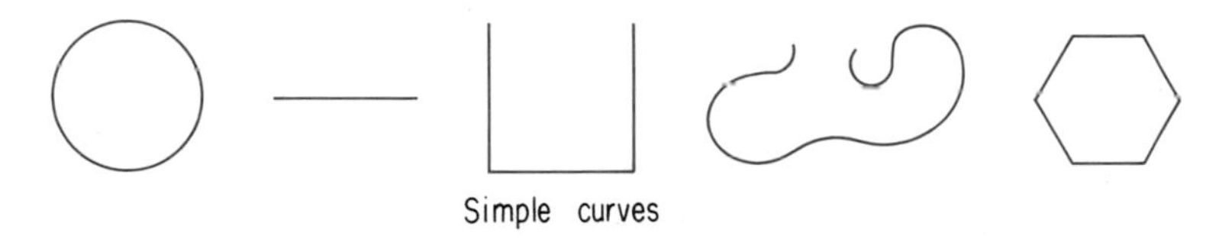

Simple curves

The next set of curves are not simple curves. Note in the examples that a simple curve may be either open or closed. Also a curve that is not simple may be either open or closed.

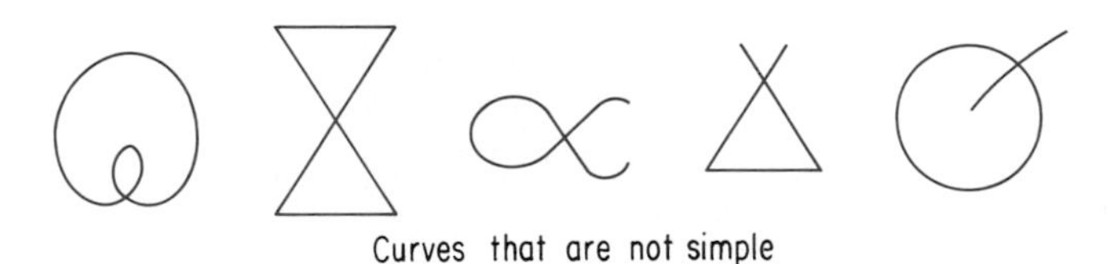

Curves that are not simple

A simple closed curve that is made up of three line segments is a *triangle*. As in the following figure, any triangle ABC, symbolized by $\triangle ABC$, has the points A, B, and C as **vertices** (singular: vertex); has the line segments $\overline{AB}$, $\overline{BC}$, $\overline{AC}$ as **sides**; has an **interior**; and has an **exterior.**

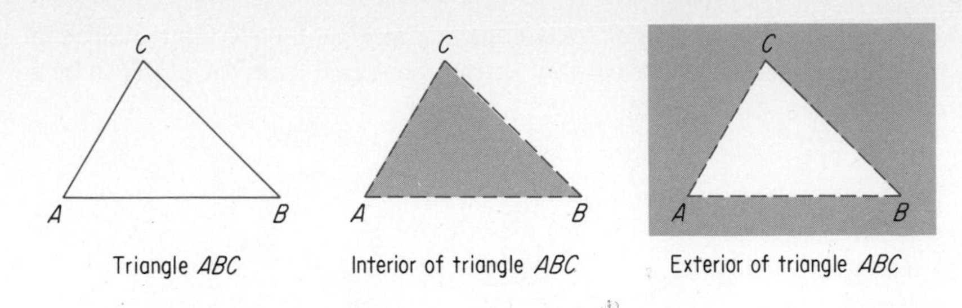

Triangle ABC Interior of triangle ABC Exterior of triangle ABC

The union of the points of the sides of any triangle *is* the triangle, that is, the set of points that form the triangle. The union of a triangle and its interior is a **triangular region.** Similarly, there are **circular regions, square regions,** and **rectangular regions.**

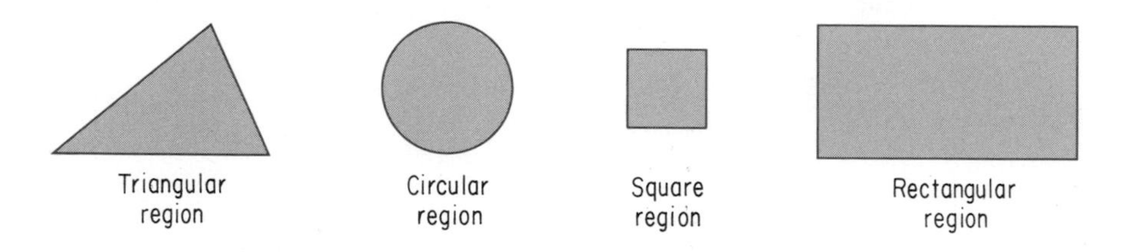

Triangular region Circular region Square region Rectangular region

The union of any two rays, such as $\overrightarrow{AB}$ and $\overrightarrow{AC}$, with a common endpoint A is a **plane angle** BAC, $\angle BAC$, with the point A as its **vertex** and the rays $\overrightarrow{AB}$ and $\overrightarrow{AC}$ as its **sides.** Note that when the angle is named by three points, the vertex is the second point named.

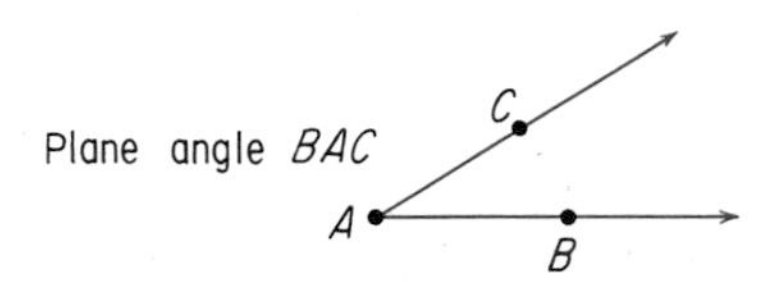

Plane angle BAC

In some texts the sides of an angle may be on the same line (as often happens in secondary school mathematics). In other texts there is an arbitrary requirement that the rays of an angle be on different lines. Any angle with its sides on different lines has an **interior** and an **exterior** as in the figure.

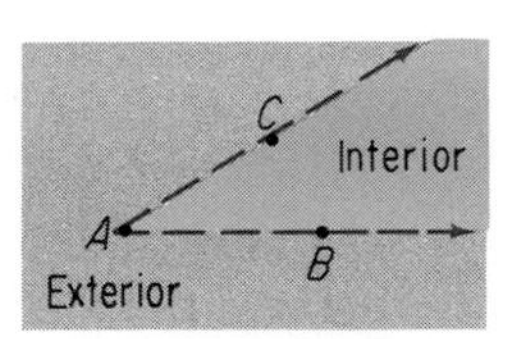

Example 4: Consider the figure at the top of the next page and identify each set of points: (a) $\overrightarrow{AB} \cap \overrightarrow{ED}$; (b) $\angle BAE \cap \angle BCD$; (c) (interior $\triangle ACE$) $\cap$ (exterior $\angle EFD$); (d) three angles such that the intersection of their interiors is the interior of $\triangle ABF$.

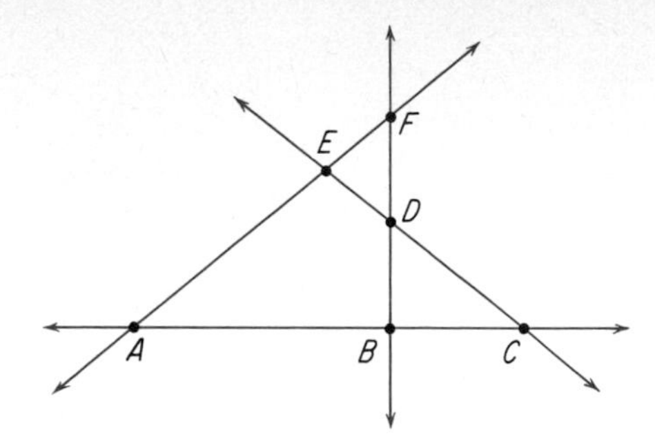

Solution: (a) $\{C\}$; (b) $\overline{AC} \cup E$; (c) interior $\triangle BCD$; (d) $\angle ABF$, $\angle BFA$, and $\angle FAB$.

Example 5: Consider the two given intersecting lines and identify each set of points: (a) $\overleftrightarrow{AB} \cap \overleftrightarrow{DE}$; (b) $\overrightarrow{BC} \cup \overrightarrow{BD}$; (c) $\angle CBD \cap \overrightarrow{ED}$; (d) (interior $\angle CBE$) $\cap \overleftrightarrow{AC}$.

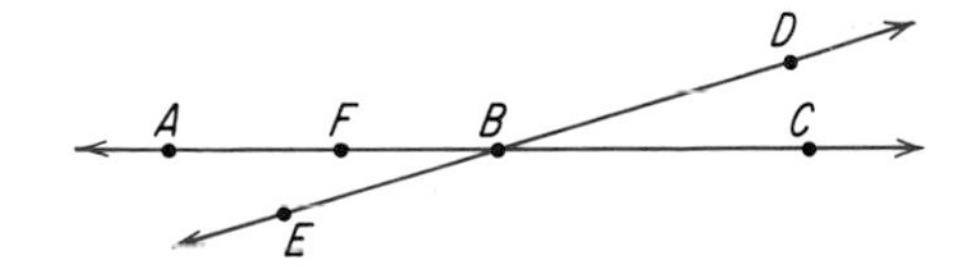

Solution: (a) The two lines intersect in the single point B; (b) $\angle CBD$; (c) $\overrightarrow{BD}$; (d) $\varnothing$.

The elementary language of mathematics includes space as well as plane figures. Consider, for example, the **sphere, cube, cylinder,** and **pyramid** shown in the figure.

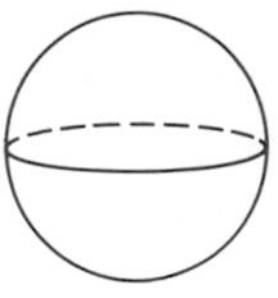

Sphere

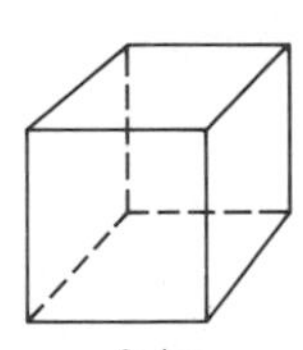

Cube

Cylinder

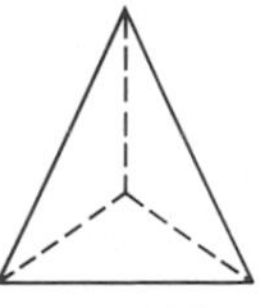

Pyramid

The pyramid shown in the figure is a **triangular pyramid** because it has a triangular base. There are also square pyramids, rectangular pyramids, and others.

Space curves are often most easily visualized as parts of a common space figure. For example, consider the space curve $ABCDE$ composed of the indicated edges of the cube in the figure shown at the top of the next page.

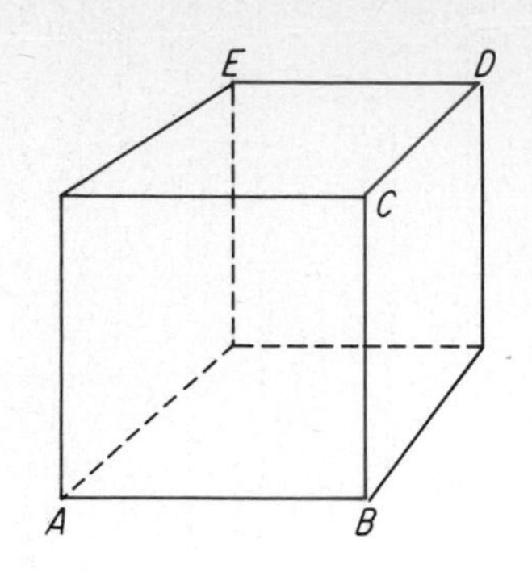

EXERCISES

Graph on a real number line.

1. $x > -1$
2. $x < 3$
3. $x \leq 0$
4. $x \geq -2$
5. $-1 \leq x \leq 3$
6. $-2 \leq x \leq 2$

Sketch a plane curve that is:

7. Simple but not closed.
8. Closed but not simple.
9. Simple and closed.
10. Neither simple nor closed.

Consider $\overleftrightarrow{AD}$ and $\overrightarrow{CE}$ in the given figure and identify each set of points.

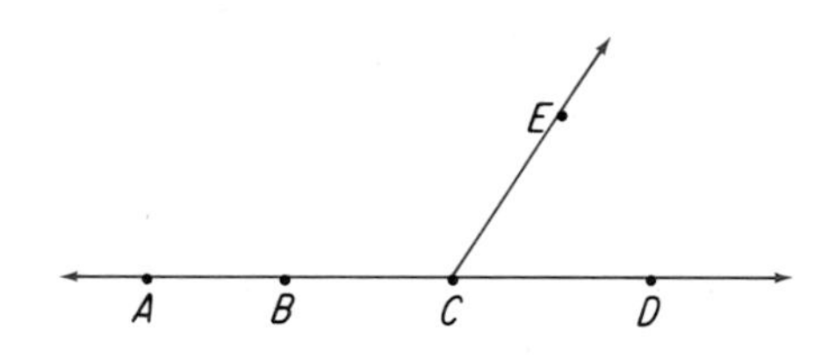

11. $\overline{AB} \cup \overline{BD}$
12. $\overline{AB} \cap \overline{CD}$
13. $\overline{AC} \cap \overline{BD}$
14. $\overline{AB} \cap \overline{BC}$
15. $\overline{AD} \cap \overline{BC}$
16. $\overline{AD} \cup \overline{BC}$
17. $\overleftrightarrow{AB} \cap \overline{BC}$
18. $\overleftrightarrow{AB} \cap \overrightarrow{BC}$
19. $\overrightarrow{BC} \cup \overrightarrow{CD}$
20. $\overrightarrow{BC} \cap \overrightarrow{CD}$
21. $\overrightarrow{BC} \cap \overrightarrow{BA}$
22. $\overrightarrow{DC} \cup \overrightarrow{CA}$
23. $\angle BCE \cap \overline{BD}$
24. $\overrightarrow{CD} \cup \overrightarrow{CE}$
25. $\angle BCE \cap \angle DCE$
26. $\angle BCE \cup \angle ACE$
27. $\angle BCE \cup \angle DCE$
28. $\angle BCE \cap \angle ACE$

Consider the given four lines and identify each set of points.

sec. 1-5
sets of
points

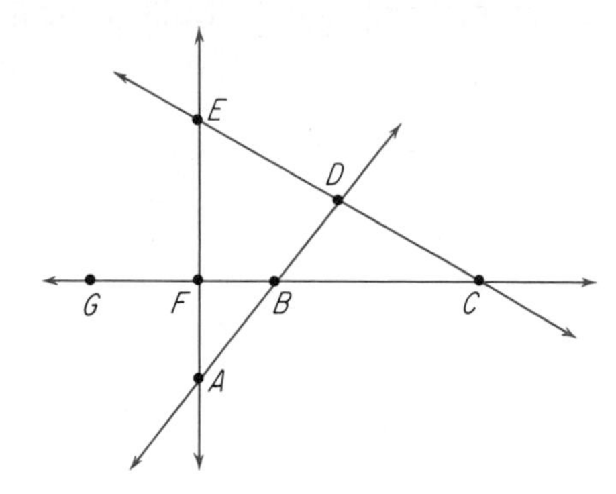

29. $\overrightarrow{GF} \cap \overrightarrow{DE}$

30. $\overrightarrow{CG} \cup \overrightarrow{CE}$

31. $\triangle CFE \cap \overrightarrow{AB}$

32. $\triangle ABF \cap \triangle BCD$

33. $\triangle ABF \cap \overrightarrow{CG}$

34. $\triangle CBD \cap \triangle CFE$

35. (Interior $\triangle CBD$) $\cap$ (interior $\triangle CFE$)

36. (Interior $\triangle ABF$) $\cap$ (exterior $\triangle CBD$)

37. (Exterior $\triangle ABF$) $\cap$ $\overleftrightarrow{CD}$

PEDAGOGICAL EXPLORATIONS

1. Make a collection of ways in which names of geometric figures are used in colloquial speech.

Games may be used to present many mathematical concepts very effectively. Frequently slight modifications of the rules for a particular game lead to another very different game. The following games are described in terms of points and line segments on a piece of paper. However other procedures such as pegs and elastics on a peg board may be used instead.

2. Make several dots for points on a piece of paper as in the figure. Two players take turns drawing line segments with the given dots as endpoints. No line segment may cross another or contain more than two of the given points. The last person able to play is the winner.

3. As in Exploration 2 but with the added restriction that no point may be the endpoint of more than two line segments.

4. As in Exploration 2 but with the added restriction that no point may be the endpoint of more than three line segments.

5. As in Exploration 2, if you begin with three dots and your opponent has the first move, can you win?

A simple, closed, plane curve that is composed of line segments is a **polygon** with the line segments as **sides** and their endpoints as **vertices**. For each type of polygon (**a**) sketch at least two figures of different shapes; (**b**) state the number of sides of each polygon; and (**c**) state the number of vertices of each polygon. Consult a dictionary if necessary.

6. Triangle.

7. Quadrilateral

8. Pentagon.

9. Hexagon.

10. Heptagon.

11. Octagon.

12. Nonagon.

13. Decagon.

14. Dodecagon.

***15.** Icosagon.

Paradoxes may be used to provide elementary school students with insight into the properties of infinite sets. Here is an example from geometry. Consider the fact that a line segment contains an infinite number of points. From this one may demonstrate the fact that two line segments of unequal length nevertheless contain the same number of points. Thus consider segments $\overline{AB}$ and $\overline{CD}$ where the lines $\overleftrightarrow{CA}$ and $\overleftrightarrow{DB}$ meet at a point P, as shown in the figure at the top of the next page. We can show that there is a one-to-one correspondence between the points of the line segments $\overline{AB}$ and $\overline{CD}$.

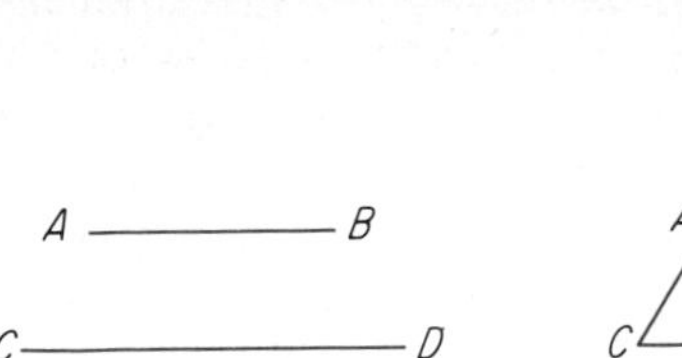

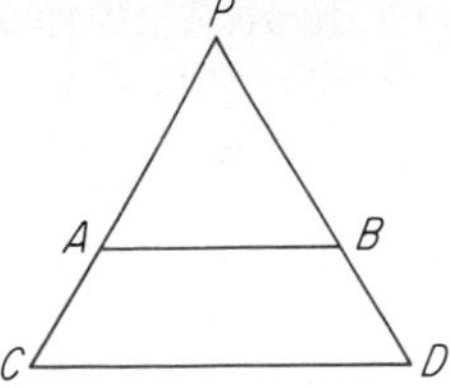

16. Copy the above figure. Select a point M on $\overline{CD}$ and draw the line $\overleftrightarrow{PM}$. Note that this line crosses $\overline{AB}$ at a point M'. Is there a point M' for each point M? Could two distinct points M_1 and M_2 have the same corresponding point M'? (Remember that there is exactly one line on two distinct points and at most one point on two distinct lines.) Are there at least as many points on $\overline{AB}$ as on $\overline{CD}$?

17. Repeat Exploration 16 selecting points N on $\overline{AB}$ and obtaining points N' on $\overline{CD}$ to determine that there are at least as many points on $\overline{CD}$ as on $\overline{AB}$.

18. Explain why there are exactly as many points on $\overline{CD}$ as on $\overline{AB}$; indeed, there is a one-to-one correspondence between these sets of points.

1-6 Euler diagrams and Venn diagrams

Relationships among sets are often represented by sets of points. We shall use a rectangular region to represent the universal set. Then a particular subset A will be represented by a circular or other convenient region. As in the case of A', the complement of A in the given figure, we use a dashed line for the part of the region's boundary that does not belong to the region.

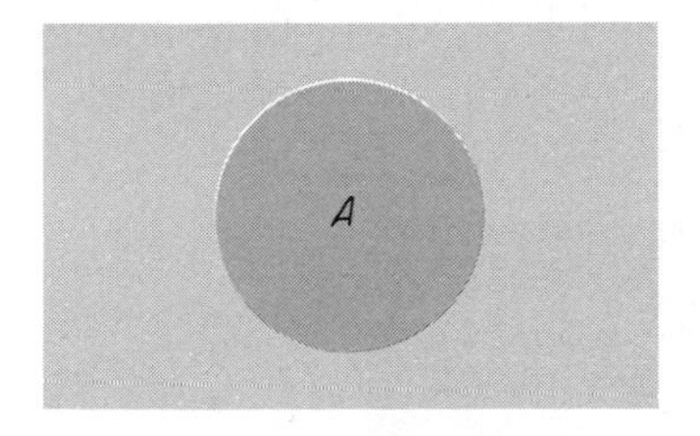

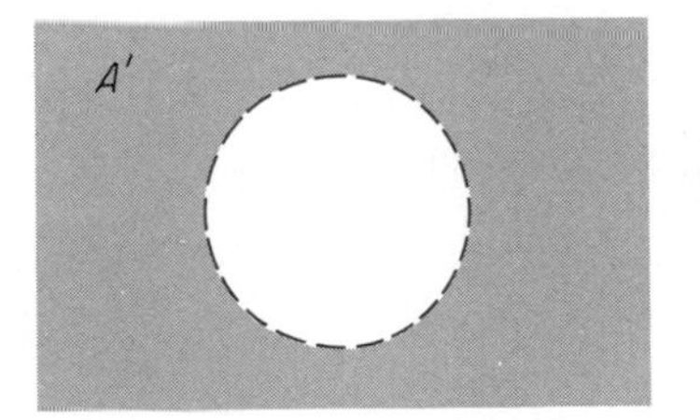

The shading is often omitted when the meaning is clear without the shading. With or without the shading, the figures are called **Euler diagrams.** We may use Euler diagrams to show the intersection of two sets.

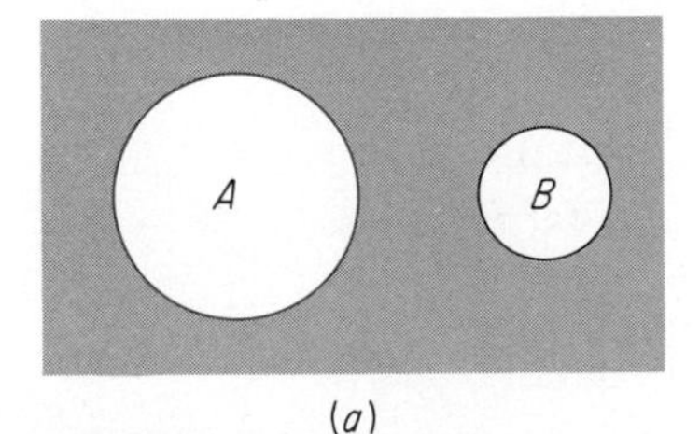

(a)

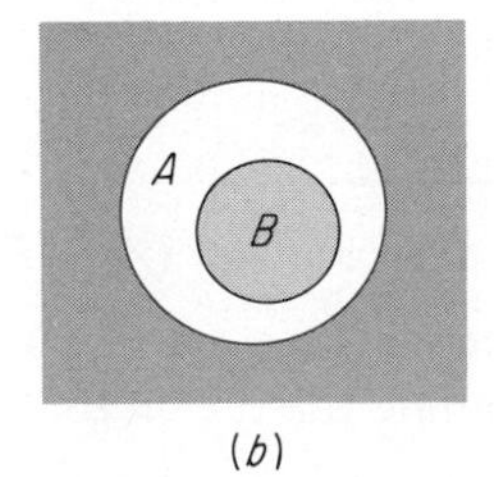

(b)

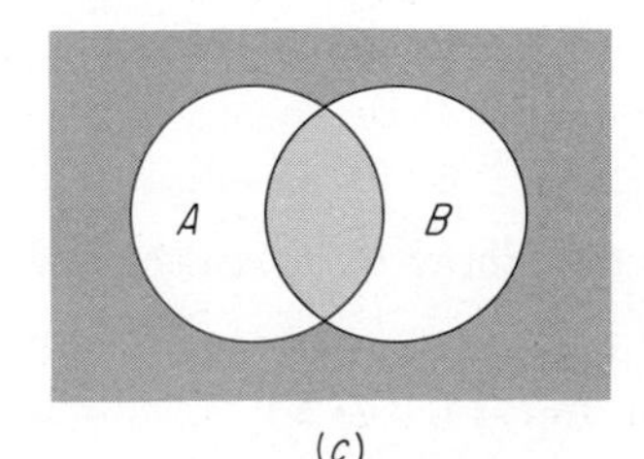

(c)

Note that in part (a) of the preceding figure, $A \cap B$ is the empty set; in part (b), $A \cap B = B$. We may also use Euler diagrams to show the union of two sets. Note that $A \cup B = A$ is shown in part (b) of the figure below.

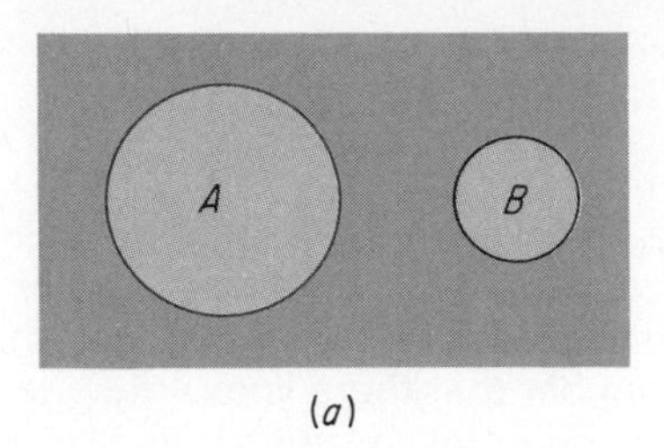

(a)

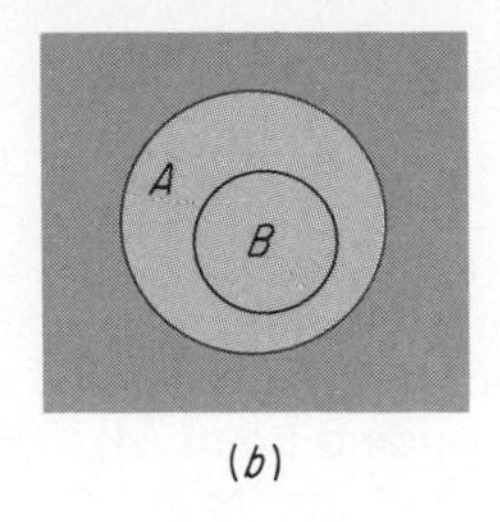

(b)

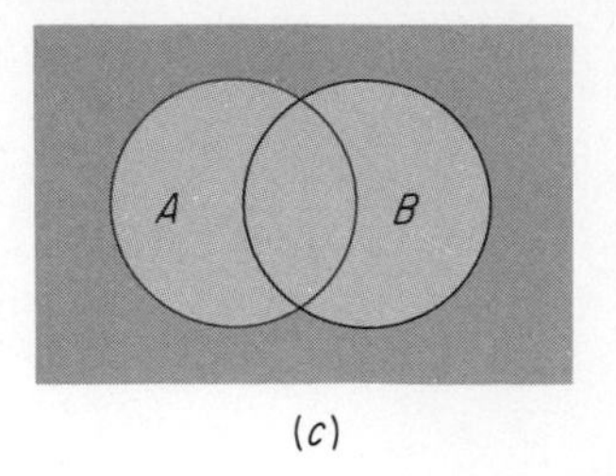

(c)

The figures for intersection and union may also be used to illustrate the following properties of any two sets A and B:

$$(A \cap B) \subseteq A, \qquad (A \cap B) \subseteq B$$

$$A \subseteq (A \cup B), \qquad B \subseteq (A \cup B)$$

We consider only well-defined sets (§1-1), and thus each element of the universal set $\mathcal{U}$ is a member of exactly one of the sets A and A'. When two sets A and B are considered, an element of $\mathcal{U}$ must belong to exactly one of the four sets:

$A \cap B$, i.e., both A and B

$A \cap B'$, i.e., A and not B

$A' \cap B$, i.e., B and not A

$A' \cap B'$, i.e., neither A nor B

An Euler diagram in which each of these four regions is represented is often called a **Venn diagram.** Notice that in the two preceding illustrations of Euler diagrams, the figures in the parts (c) were of the type known as Venn diagrams.

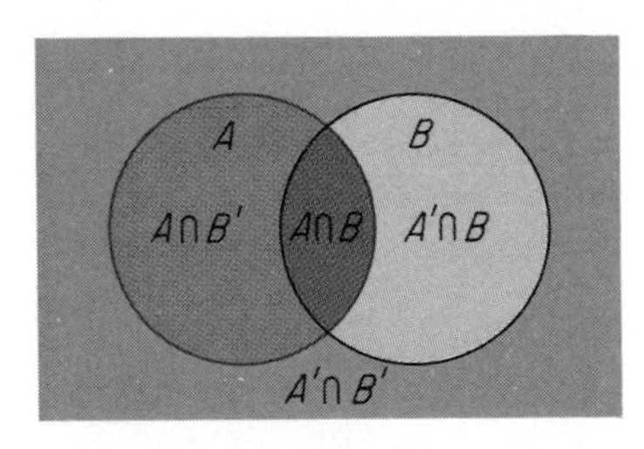

Venn diagrams may be used to show that two sets refer to the same set of points, that is, are equal.

Example 1: Show by means of a Venn diagram that $(A \cup B)' = A' \cap B'$.

Solution: We make separate Venn diagrams for $(A \cup B)'$ and $A' \cap B'$. The diagram for $(A \cup B)'$ is made by shading the region for A horizontally, shading the region for B vertically, identifying the region for $A \cup B$ as consisting of the points in regions that are shaded in any way (horizontally, vertically, or both horizontally and vertically), and identifying the region for $(A \cup B)'$ as consisting of the points in the region without horizontal or vertical shading.

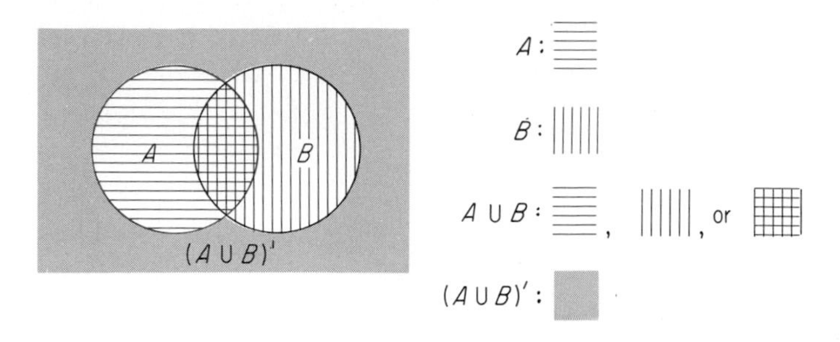

The diagram for $A' \cap B'$ is made by shading the region for A' horizontally, shading the region for B' vertically, and identifying the region for $A' \cap B'$ as consisting of all points in regions that are shaded both horizontally and vertically.

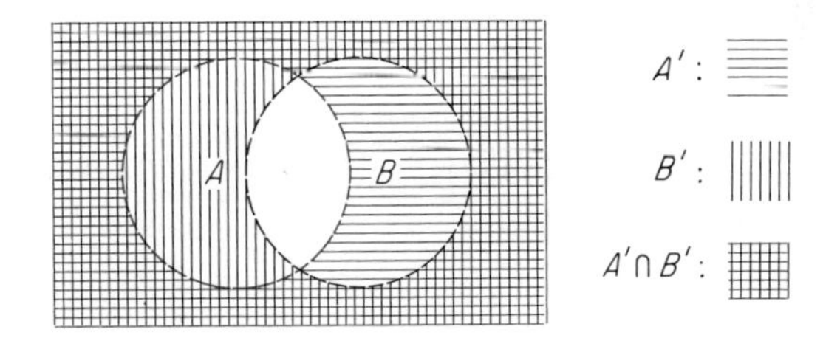

The solution is completed by observing that the region for $(A \cup B)'$ in the first Venn diagram is identical with the region for $A' \cap B'$ in the second Venn diagram; the two sets have the same elements and therefore are equal.

We may also use Venn diagrams for three sets. In this case there are eight regions that must be included, and the figure is usually drawn as follows.

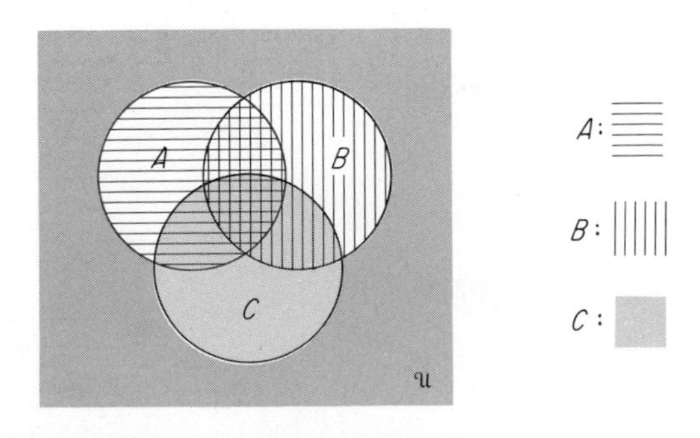

Venn diagrams for three sets are considered in the next two examples and in Exercises 19 through 29.

Example 2: Show that $A \cap (B \cup C) = (A \cap B) \cup (A \cap C)$.

Solution: Set A is shaded with vertical lines; $B \cup C$ is shaded with horizontal lines. The intersection of these sets, $A \cap (B \cup C)$, is the region that has both vertical and horizontal shading.

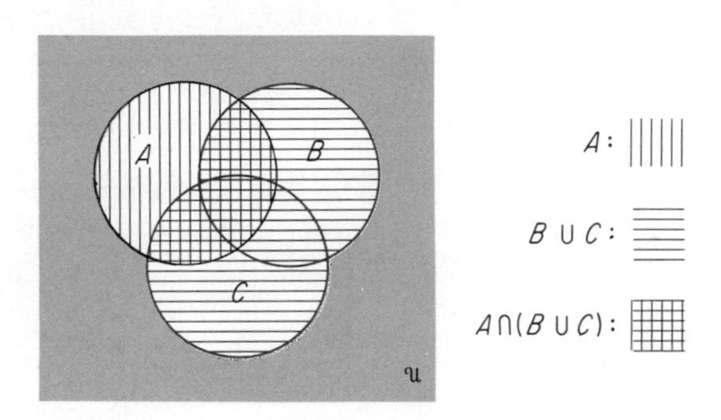

The set $A \cap B$ is shaded with horizontal lines; $A \cap C$ is shaded with vertical lines. The union of these sets is the subset of $\mathcal{U}$ that is shaded with lines in either or in both directions.

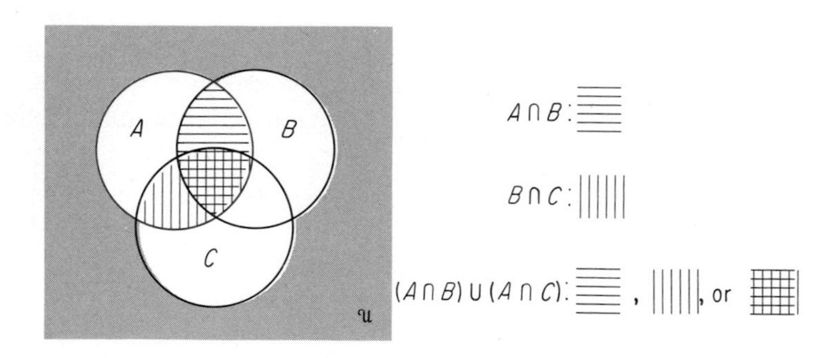

Note that the final results in the two diagrams are the same, thus showing the equivalence of the following sets:

$$A \cap (B \cup C) \quad \text{and} \quad (A \cap B) \cup (A \cap C).$$

Example 3: In the figure at the top of the next page, find
(a) $n(A \cap B \cap C)$;
(b) $n(A \cap B' \cap C)$.

Solution: (a) There is one element in the intersection of all three sets. Thus $n(A \cap B \cap C) = 1$. (b) There are four elements that are in both sets A and C, but not in set B. Thus $n(A \cap B' \cap C) = 4$.

Example 4: In a group of 35 students, 15 are studying algebra, 22 are studying geometry, 14 are studying trigonometry, 11 are studying both algebra and geometry, 8 are studying geometry and trigonometry, 5 are studying algebra and trigonometry, and 3 are studying all three subjects. How many of these students are not taking any of these subjects? How many are taking only geometry?

Solution: This problem can easily be solved by means of a Venn diagram with three circles to represent the set of students in each of the listed subject-matter areas. It is helpful to start with the information that there are 3 students taking all three subjects. We write the number 3 in the region that is the intersection of all three circles. Then we work backwards; since 5 are taking algebra and trigonometry, and 3 of these have already been identified as also taking geometry, there must be exactly 2 of them taking only algebra and trigonometry; that is, there must be 2 in the region representing algebra and trigonometry but *not* geometry. Continuing in this manner, we enter the given data in the figure.

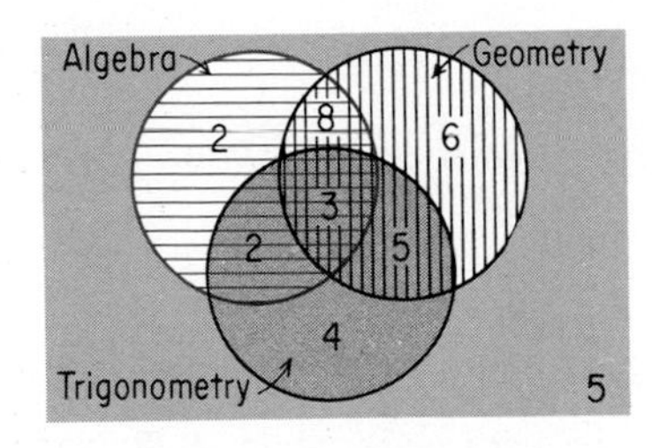

Since the total of the numbers in the various areas is 30, there must be 5 students not in any of the classes listed in the various regions. Also, reading directly from the figure, we find that there are 6 students taking geometry only.

The terms Euler diagram, Euler circle, and Venn diagram are used synonymously in some elementary books. Their use in this book is the historical one and should provide a basis for comprehension of any other special uses encountered.

EXERCISES

Consider the given diagram and find each number.

1. **(a)** $n(A \cap B)$ **(b)** $n(A)$
(c) $n(B \cap A')$ **(d)** $n(B \cup A)$

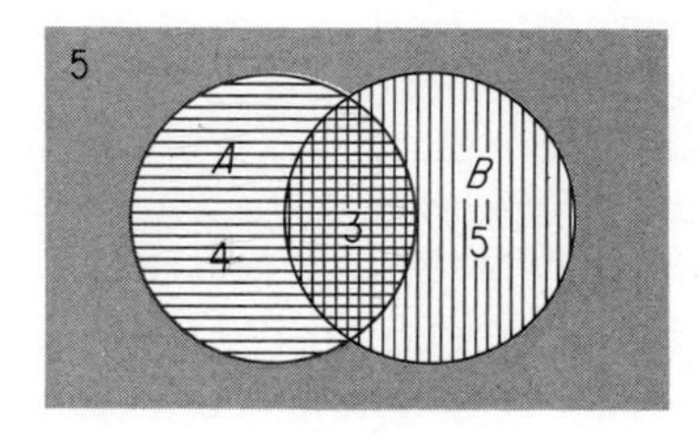

2. **(a)** $n(P \cup Q)$ **(b)** $n(P' \cap Q')$
(c) $n(P' \cup Q)$ **(d)** $n(P \cup Q')$

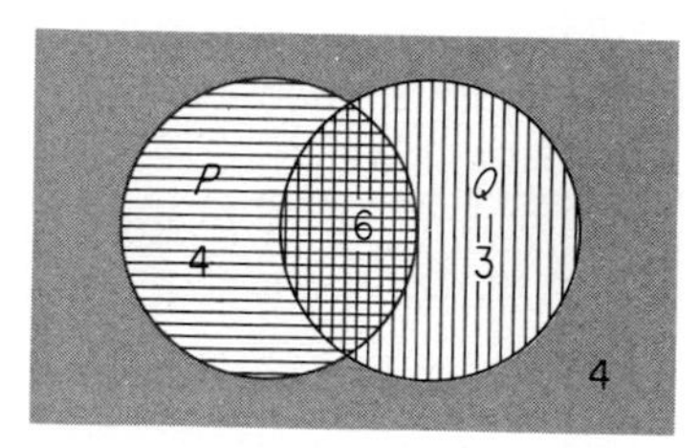

Represent each of the following by an Euler diagram.

3. $A \subset B$

4. A and B are disjoint sets.

5. $A \cup B$ when $A \cap B = \varnothing$.

6. $A \subseteq B$ and $B \subseteq A$.

7. $(A \cup B)'$ when A and B are disjoint sets.

8. $(A \cup B \cup C)'$ when A, B, and C are disjoint sets.

Represent each set by a Venn diagram.

9. $A' \cup B$

10. $A' \cap B$

11. $A \cap B'$

12. $A \cup B'$

In Exercises 13 and 14 show each relation by Venn diagrams.

13. $(A \cap B)' = A' \cup B'$ **14.** $A \cup B' = (A' \cap B)'$

Consider the given diagram and find each number.

15. (a) $n(A \cap B \cap C)$ (b) $n(A \cap B \cap C')$
(c) $n(A \cap B' \cap C')$ (d) $n(A)$
(e) $n(A \cup B)$ (f) $n(B \cup C)$

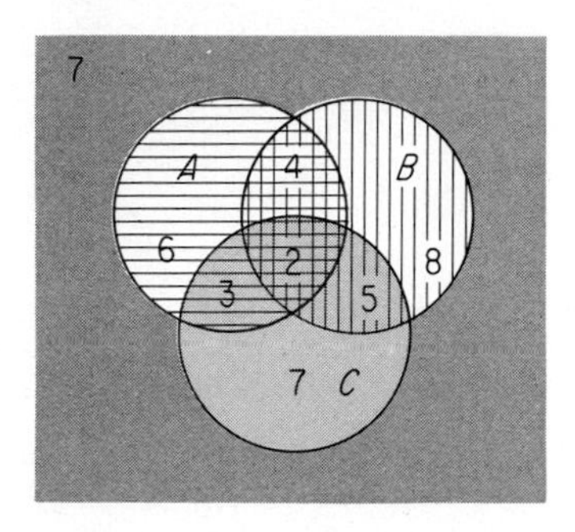

16. (a) $n(R' \cap S \cap T)$ (b) $n(R')$
(c) $n(R' \cup S)$ (d) $n(S' \cup T')$
(e) $n(R' \cup S' \cup T')$ (f) $n(R \cup S' \cup T)$

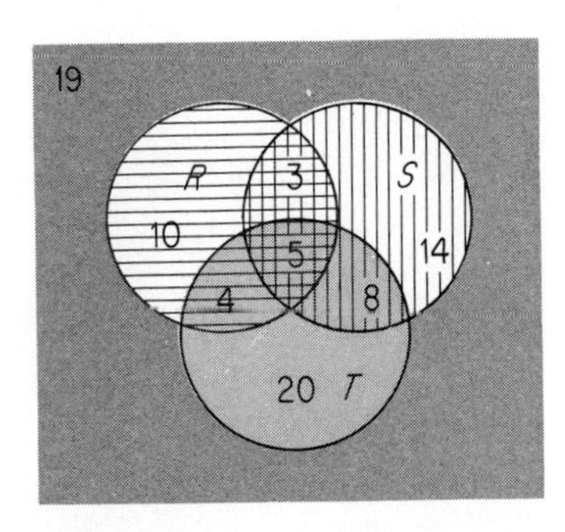

17. (a) $n(A \cup B)$ (b) $n(B \cap C)$
(c) $n(A \cap B')$ (d) $n(A \cup B \cup C)$
(e) $n(A \cup B' \cup C')$ (f) $n(A \cap B' \cap C')$

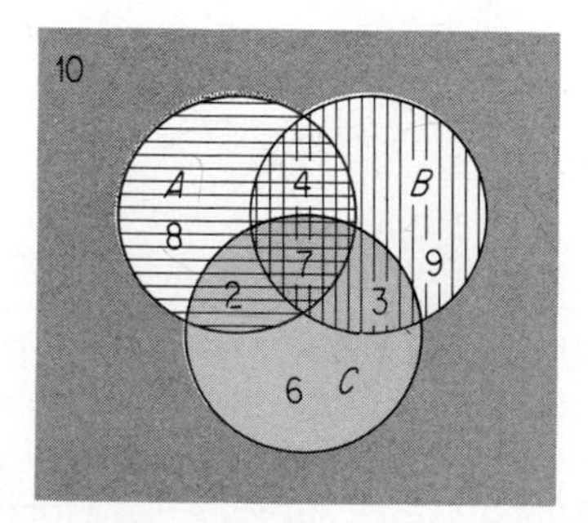

18. (a) $n(X \cup Y)$ (b) $n(X \cup Z)$
(c) $n(X')$ (d) $n(X \cup Y')$
(e) $n(X' \cap Y \cap Z)$ (f) $n(X \cap Y' \cap Z')$

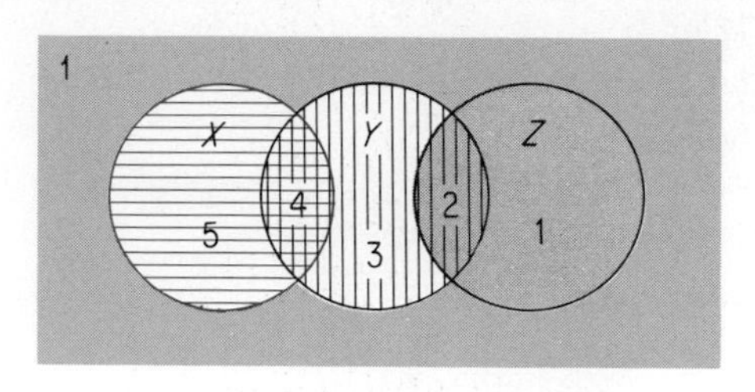

Represent each of the following by a Venn diagram.

19. (a) $A \cap B \cap C$ (b) $A \cap B \cap C'$
(c) $A \cap B' \cap C$ (d) $A \cap B' \cap C'$

20. (a) $A' \cap B \cap C$ (b) $A' \cap B \cap C'$
(c) $A' \cap B' \cap C$ (d) $A' \cap B' \cap C'$

Identify each shaded region by set notation.

21.

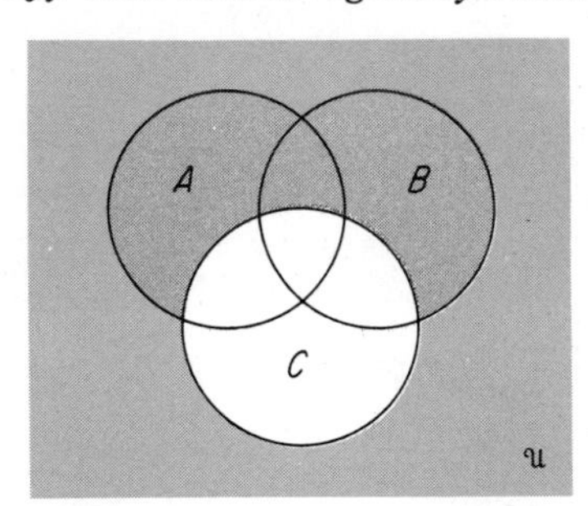

22.

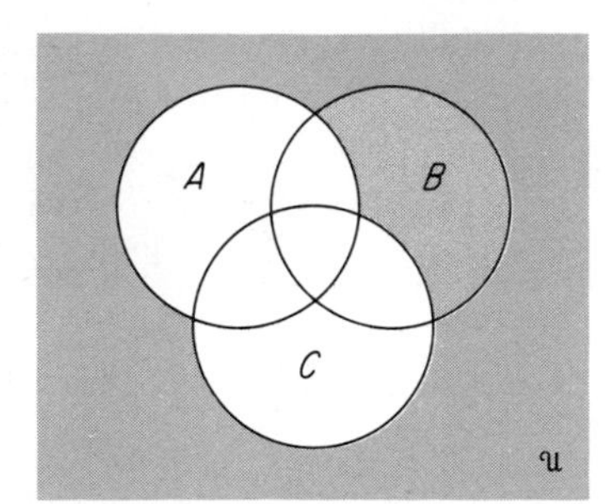

23.

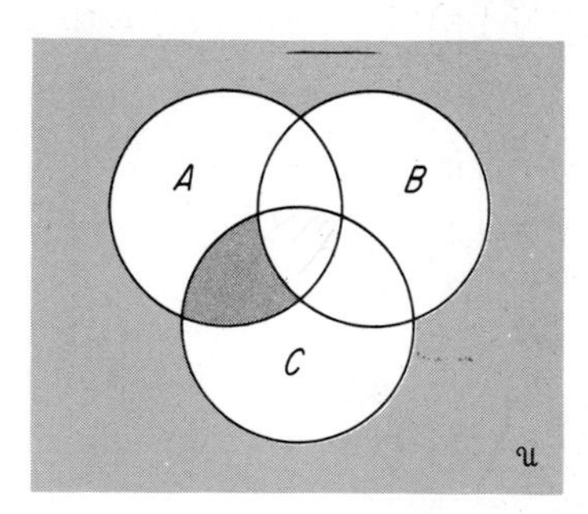

24.

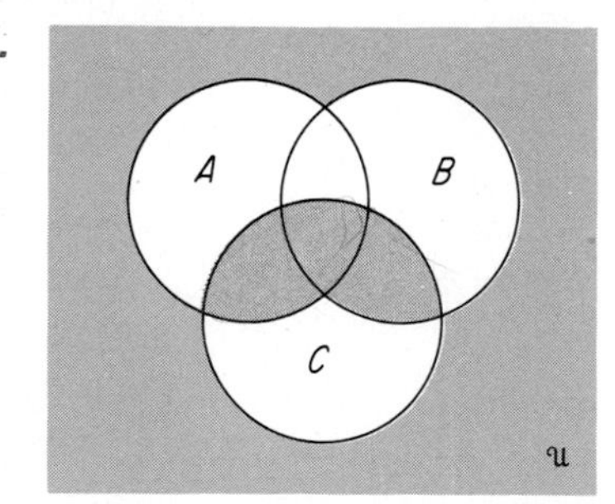

Use Venn diagrams to solve each problem.

25. In a survey of 50 students, the following data were collected: There were 19 taking biology, 20 taking chemistry, 19 taking physics, 7 taking physics and chemistry, 8 taking biology and chemistry, 9 taking biology and physics, 4 taking all three subjects. **(a)** How many of the group are

not taking any of the three subjects? **(b)** How many are taking only chemistry? **(c)** How many are taking physics and chemistry but not biology?

26. Fifty cars belong to students in a certain dormitory. Of these cars 42 have radios, 15 have tape decks, 10 have air conditioners, 2 have all three, 6 have a radio and an air conditioner, 5 have a tape deck and an air conditioner, and 10 have a radio and a tape deck. **(a)** How many of these cars have an air conditioner but no radio? **(b)** How many have no radio, no tape deck, and no air conditioner?

27. Repeat Exercise 26 for 4 cars having all three items and all other data as before.

28. Suppose that the student who collected the data for Exercise 26 stated that 5 of the cars had all three items and gave the other data as before. **(a)** Did he make a careful survey? **(b)** Explain your answer to part (a).

29. A survey was taken of 30 students enrolled in three different clubs, A, B, and C. Show that the following data that were collected are inconsistent: 18 in A, 10 in B, 9 in C, 5 in B and C, 6 in A and B, 9 in A and C, 3 in A, B, and C.

PEDAGOGICAL EXPLORATIONS

Show each relation by Venn diagrams.

1. $A \cup (B \cup C) = (A \cup B) \cup C$

2. $A \cap (B \cap C) = (A \cap B) \cap C$

Elementary school children may be asked to copy given numerals into specified regions of a given Venn diagram.

3. Copy the given diagram and then copy the numbers in the appropriate regions. {8, 11, 15, 17, 7, 6, 5, 2, 1}

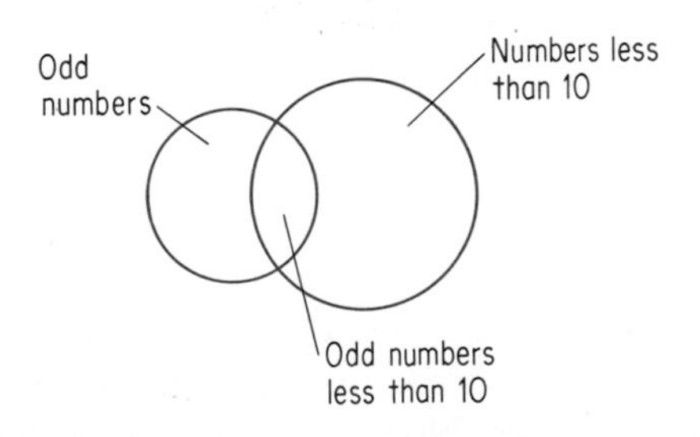

4. Select any sets A and B with $n(A) = 6$, $n(B) = 4$, and such that:

(a) $n(A \cup B) = 7$ **(b)** $n(A \cup B) = 6$

(c) $n(A \cup B) = 9$ **(d)** $n(A \cap B) = 2$

5. In Exploration 4 find:
   (a) $n(A \cap B)$ if $n(A \cup B) = 7$
   (b) $n(A \cap B)$ if $n(A \cup B) = 6$
   (c) $n(A \cap B)$ if $n(A \cup B) = 10$
   (d) $n(A \cup B)$ if $n(A \cap B) = 2$
   (e) A formula for $n(A \cup B)$ in terms of $n(A)$, $n(B)$, and $n(A \cap B)$.

Here is a Venn diagram for four sets A, B, C, and D, with region D shaded.

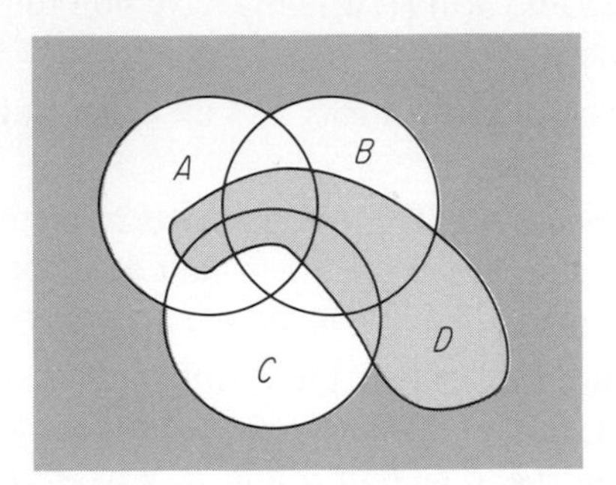

Use such diagrams to determine whether each statement given in Explorations 6 and 7 is true or false.

6. $A \cap (B \cup C \cup D) = (A \cap B) \cup (A \cap C) \cup (A \cap D)$
7. $A \cup (B \cap C \cap D) = (A \cup B) \cap (A \cup C) \cap (A \cup D)$

8. If $A \subseteq B$, describe
   (a) $A \cap B$;
   (b) $A \cup B$;
   (c) $A \cap B'$;
   (d) $A' \cup B$.
9. If $A \subseteq B$ and $B \subseteq A$, describe
   (a) $A \cup B$;
   (b) $A \cap B$;
   (c) $A \cap B'$;
   (d) $A' \cup B$.

1-7 flow charts

At one time or another, each of us has to give detailed instructions of some kind. We may give someone directions for getting to a specific store; we may give a recipe for making a certain kind of cake. Our instructions may be either oral or written. Now that people have to give instructions not only to other people but also to computers, which lack the human ability to bridge even minor gaps in the instructions, flow charts have been found useful as a means of organizing and conveying precise instructions. Here are several examples of very simple flow charts. Each flow chart consists of an input of one or more elements, one or more rules, and an output.

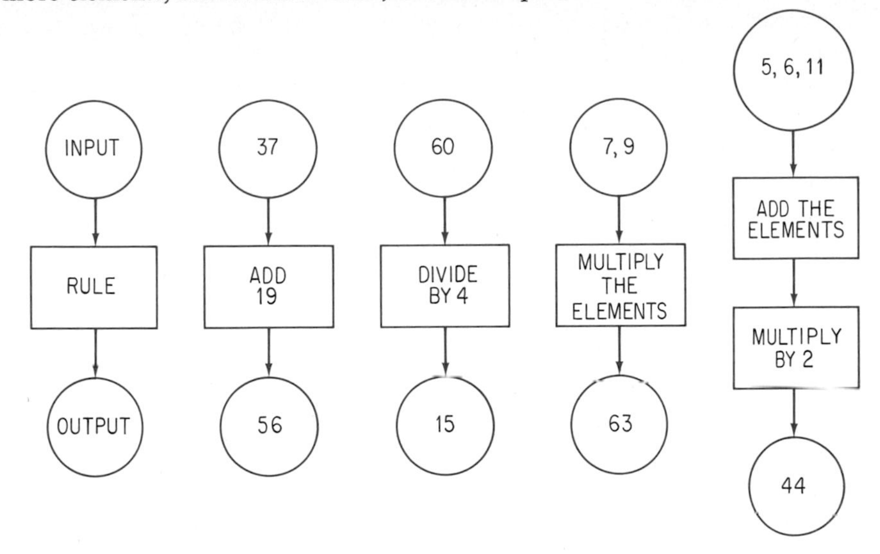

Note that circular regions are used for inputs and outputs. Rules are usually given within rectangular boxes. We shall use circular regions also for intermediate results as in the next two examples. If these operations were performed on a computer, then the intermediate results would be stored until used in the next step. Such storage cells are shown in the next figures.

We can also use variables for the input numbers. In the following two flow charts, the two rules have been interchanged. Notice that the two outputs are different. Each of the two flow charts is accompanied by one that illustrates the steps for the replacement $n = 7$ and the intermediate results for the operations involved.

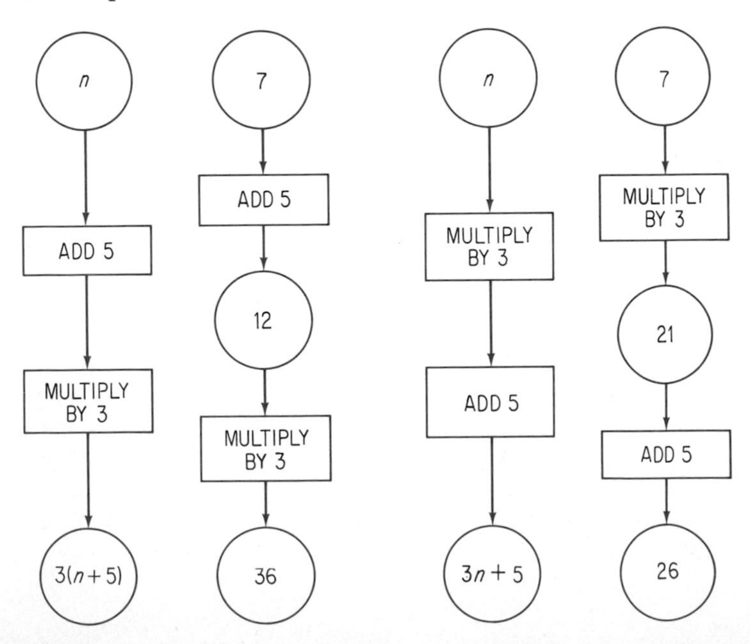

Sometimes a flow chart will contain a decision box that asks a question. In such cases you must answer the question as "YES" or "NO" and follow the arrow out of the box in the direction indicated by your answer. Such boxes are usually diamond shaped.

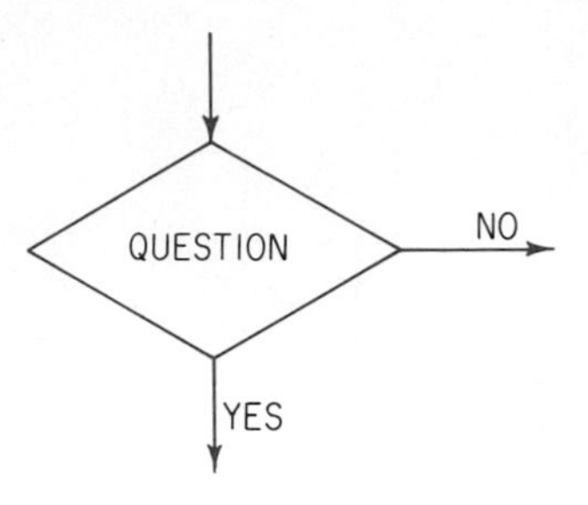

Here is a flow chart that contains a decision box. If your answer to the question is affirmative, you follow the arrow down; if your answer is negative, you branch to the right.

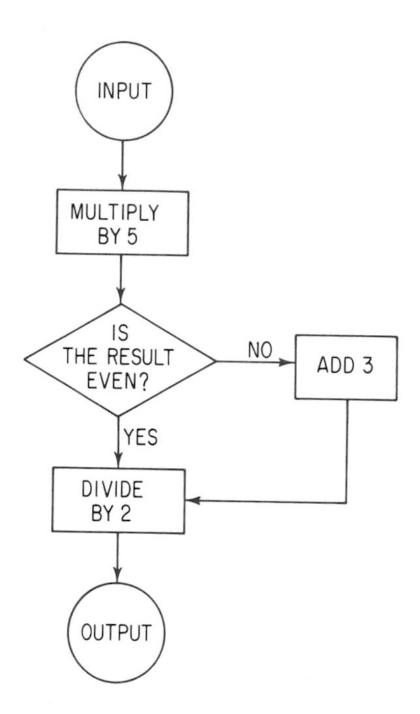

Suppose the input number is 4. Then in the first step, $5 \times 4 = 20$. The result is even, so move down in the flow chart and divide by 2 to obtain the output number 10. Next suppose that the input number is 3. Then in the first step, $5 \times 3 = 15$. The result is not even, so move to the right in the flow chart and add 3 to obtain 18. In the next step, divide by 2 to obtain the output number 9. Now find the output for an input of 6. The output should be 15.

Find the output of the given flow chart for the indicated value of n.

1.

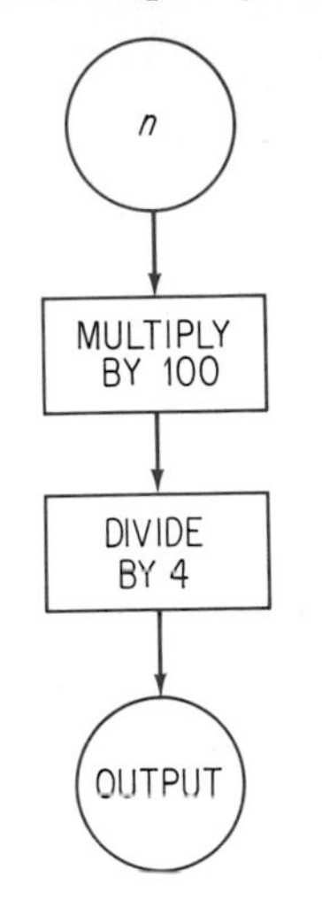

(a) $n = 83$
(b) $n = 211$

2.

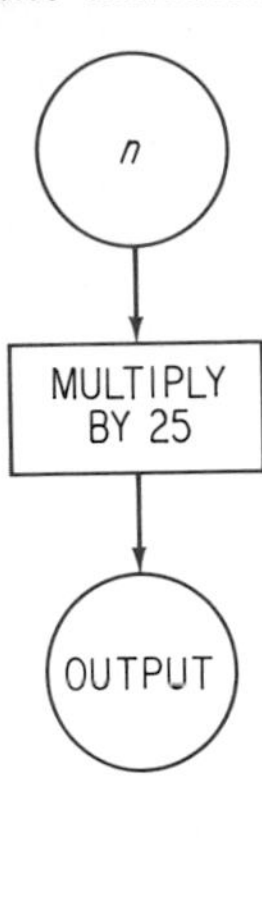

(a) $n = 83$
(b) $n = 211$

(*Note:* Compare your outputs for Exercises 1 and 2.)

For each of the following the rule and the output are given. Find the input.

3. (a)

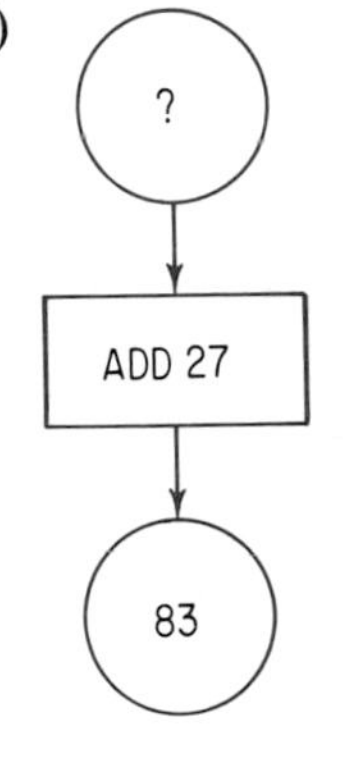

(b)

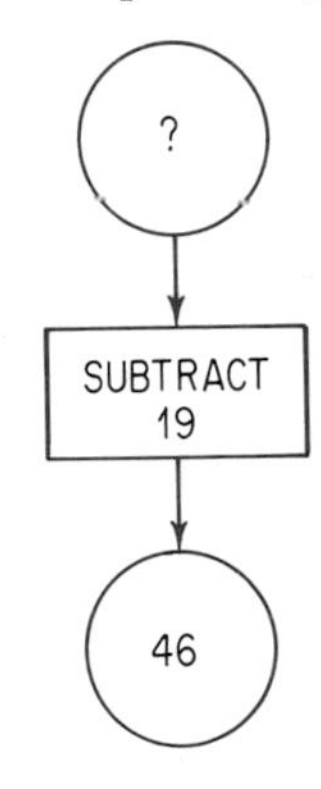

4. (a)

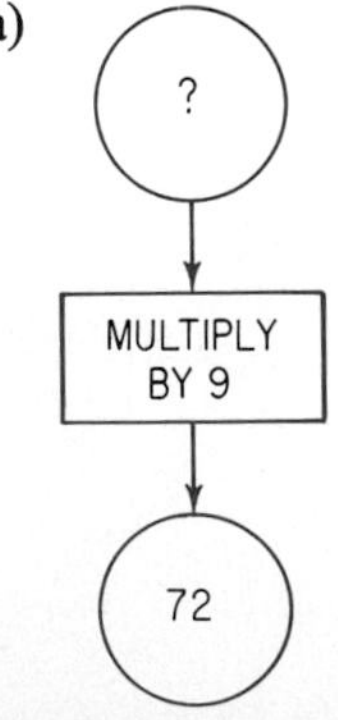

(b)

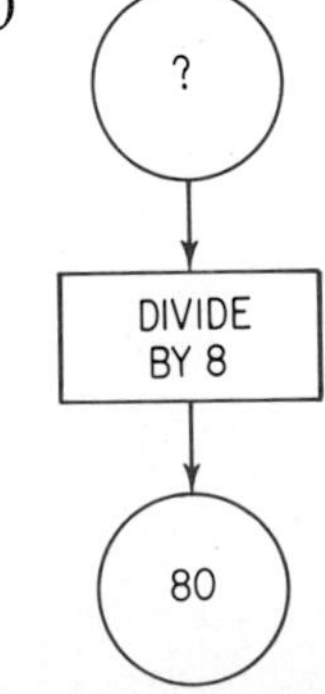

Use the flow chart shown, and the input numbers given in Exercises 5 *through* 10, *to find the outputs.*

5. $n = 3$
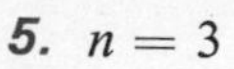
6. $n = 5$
7. $n = 7$
8. $n = 12$
9. $n = 13$
10. $n = 17$
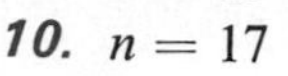

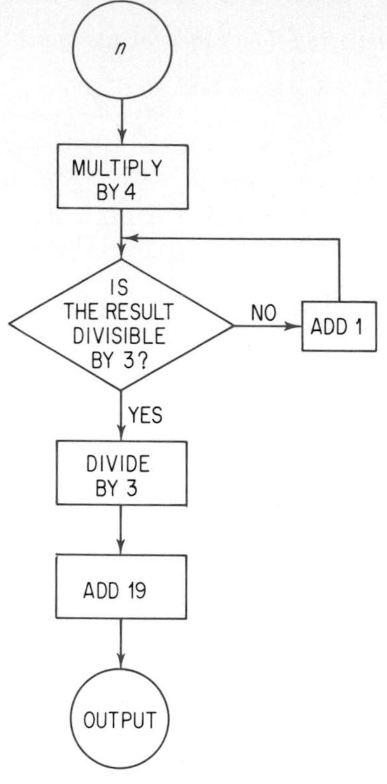

Flow charts can be used to generate a sequence of terms, as shown in the following two exercises.

11. **12.**

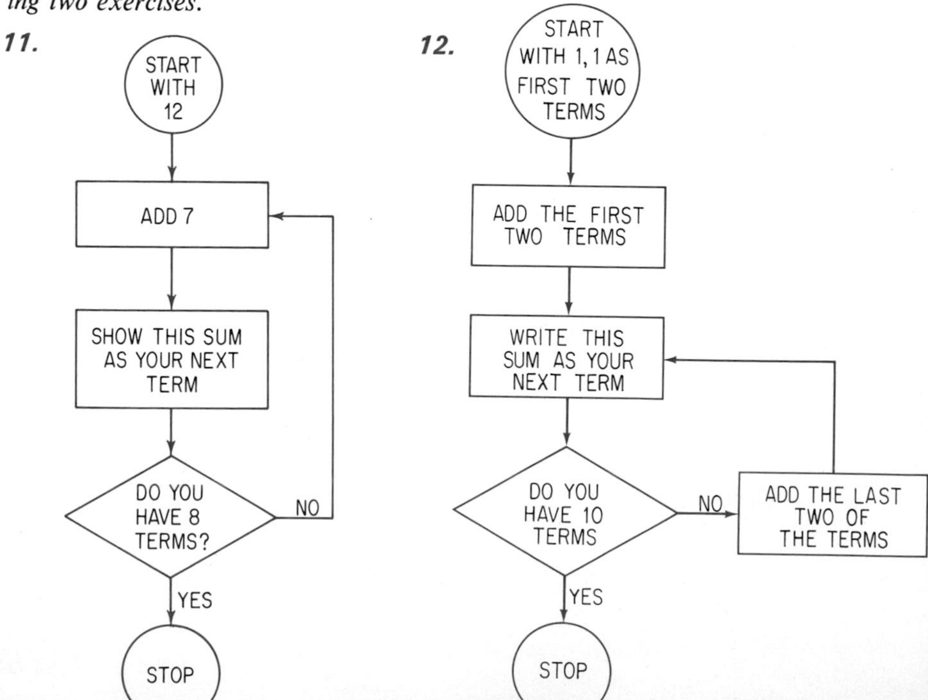

PEDAGOGICAL EXPLORATIONS

1. Select a common activity such as frying an egg (once over lightly, please), starting an automobile engine, buying an ice cream cone of your favorite flavor, etc., and make a flow chart of the related activities.
2. Make a flow chart of the essential steps for you to complete this course to your satisfaction.
3. Prepare worksheets for introducing flow charts using arithmetic operations on numbers. Note the opportunities for extensive practice (drill).
4. Repeat Exploration 3 using algebraic expressions.

1-8 sets of statements

In the language of mathematics as in ordinary language we use several forms of statements. For example, the following are **simple sentences.**

p: 5 is less than 7.
q: $7 \times 8 = 65$.
r: n is a whole number.
s: n is a negative integer.
t: n is zero.

It is convenient to use letters to identify sentences. The sentence p is *true*; q is *false*. Sentences that can be identified as true or false are often called *statements*. Notice that a command such as "Stand up and be counted" is neither true nor false and is not considered to be a statement as we are using the word here.

If n is replaced by 5, then r is true but s and t are false. If n is replaced by zero, then r and t are true but s is false. If n represents any counting number, then r is true but s and t are false. If n may be replaced by any one of the numbers $\{-7, -5, -3, -1\}$, then s is true but r and t are false. Note that without some restriction upon the possible values of n the sentences r, s, and t cannot be identified as true or false. Such a restriction may be provided by specifying a replacement set for n as in the previous examples. Such a restriction may also be provided by the introduction of one or the other of these **quantifiers:**

There exists at least one number n such that (*existential quantifier*).

For all real numbers n it is true that (*universal quantifier*).

Any sentence that involves a variable is meaningless without two items of information:

What replacement set is to be used for the variable?
What quantifier is to be used?

In each case the information may be clear from the context of the sentence. For example, in mathematics it is often assumed that any variable represents a real number unless otherwise specified. Also any sentence that is not always true is assumed to be false as a general statement. In other words, it is assumed for all statements that the universal quantifier is intended if no quantifier is specified.

The special symbols for quantifiers are easily remembered; $\exists$ is read as "there exists" and $\forall$ is read as "for all." Some examples of their use are:

$\exists x\colon\ x + 2 = 5$, x a whole number	(true)
$\forall x\colon\ x + 2 = 2 + x$	(true)
$\forall x\colon\ x^2 > 0$	(false for $x = 0$)
$\exists x\colon\ x^2 < 0$	(false)

The sentence $x + 2 = 5$ is true if $x = 3$ and false if $x \neq 3$. Any sentence whose truth value depends upon the replacement selected for the variable is an **open sentence.** To **solve** an open sentence we find the set of replacements for which the statement is true.

Example 1: Solve: $x + 3 = 5$.

Solution: If $x = 2$, then $x + 3 = 5$. If $x \neq 2$, then $x + 3 \neq 5$. The **solution set** of the open sentence $x + 3 = 5$ is $\{2\}$.

The statement (equation) given in Example 1 can be thought of as a **set-selector;** it selects from the replacement set of x just those numbers that make the sentence true when used as replacements for x. The selected set is the solution set of the equation. The solution set $\{2\}$ is "the set of all x such that $x + 3 = 5$." We may designate this solution set in **set-builder notation** as

$$\{x \mid x + 3 = 5\}$$

The replacement set of the variable may be indicated in the set-builder notation as follows:

$$\{x \mid x + 3 = 5,\ x \text{ an integer}\}$$

We may also write

$$\{x \mid x + 3 = 5\} = \{2\}$$

That is, the set of all x such that $x + 3 = 5$ is the set consisting of the element 2.

In contemporary mathematics, as we have already noted, we are primarily concerned with well-defined sets. We are also primarily concerned with sentences that are either statements (true or false) or can be made into statements by using a quantifier. Frequently we use the term *statement* whether the necessary quantifier is explicitly stated or not. However, it should always be possible to read aloud any mathematical sentence without grammatical embarrassment.

Consider the previous statements r, s, and t for integral values of n.

r: n is a whole number.
s: n is a negative integer.
t: n is zero.

The statements r and t are **consistent statements** since they can both be true, in particular, if $n = 0$. The statements s and t are **contrary statements** since they cannot both be true. The statements r and s are also contrary statements. However, there is an important distinction between the two situations. The statements s and t cannot both be true but may both be false, for example, if $n = 7$. The statements r and s cannot both be true and also cannot both be false for any integral value of n; that is, exactly one of the statements r and s must be true and the other must be false. Any two statements such that the truth value of one statement assures the opposite truth value for the other statement are **contradictory statements.**

Example 2: Relative to the statement "n is an integer equal to 3," select from the given statements
(a) the statements that are consistent;
(b) the statements that are contrary;
(c) the statements that are contradictory.

(i) $2n = 6$ (ii) $n + 2 = 6$
(iii) $n \neq 3$ (iv) $n^2 = 9$

Solution: (a) (i), (iv); (b) (ii), (iii); (c) (iii).

A **compound statement** is formed by combining two or more simple statements. An example is the following:

Today is Friday in Chicago and this month is October.

In this illustration the two simple statements are combined by the connective "and." Other connectives could also have been used. Consider the same simple statements using the connective "or":

Today is Friday in Chicago or this month is October.

We shall consider such compound statements and determine the conditions under which they are true or false, assuming that the simple statements are true. In doing this we use letters or variables to represent statements and symbols to represent connectives. For example, we may use p and q to represent these simple statements:

p: Today is Friday in Chicago.
q: This month is October.

The following *connectives* are commonly used:

$\wedge$: and
$\vee$: or
$\sim$: not

We may use p and q as previously defined and write these statements in symbolic form together with their translations in words:

$p \wedge q$: Today is Friday in Chicago and this month is October.
$p \vee q$: Today is Friday in Chicago or this month is October.
$\sim p$: Today is not Friday in Chicago.

Example 3: Translate $p \wedge (\sim q)$, where p and q are as given in this section.

Solution: Today is Friday in Chicago and this month is not October.

Example 4: Write, in symbolic form: Today is not Friday in Chicago or this month is not October.

Solution: Use p and q as given in this section.

$$(\sim p) \vee (\sim q)$$

The **truth values** (T, F) of compound statements depend upon the truth values of the simple statements that are used. In general:

$\sim p$ is true if p is false, false if p is true; that is, $\sim p$ and p are contradictory statements.
$p \wedge q$ is true if *both* p and q are true, false in all other cases.
$p \vee q$ is true if at least one of the statements p, q is true; false if both p and q are false.

This is the *inclusive use* of "or" since if p or q or both p and q are true, then $p \vee q$ is true. As in §1-4 each of the following statements is a true statement:

$$3 \leq 3, \quad 2 \leq 3, \quad 3 \geq 3, \quad 3 \geq 2$$

The word "or" is widely used in everyday speech:

At nine o'clock tomorrow morning I'll either be in a conference in New York or in my dentist's office in Boston.
Jane looks either very tired or ill.

The first of these two examples illustrates the *exclusive use* of "or" since it is not possible for both statements to be true. The **exclusive or** (often denoted by $\underline{\vee}$, called "vel") is true when, and only when, exactly one of the given statements is true. The second example illustrates the *inclusive use* of "or" since at least one, possibly both, statements may be true.

Solve each statement for integers n.

1. $3n = 12$

2. $n + 3 = 12$

3. $n - 2 = 12$

4. $n \div 2 = 12$

5. $2n + 2 = 12$

6. $3n - 6 = 12$

Let n be any whole number and consider these statements.

(i) $n = 5$
(ii) $5n = 5$
(iii) $2n \neq 10$
(iv) $2n \neq 2$
(v) $n + 2 = n$
(vi) $n \times 2 = n$

7. Select the statements that are **(a)** consistent with, **(b)** contrary to, **(c)** contradictory to, the statement $n = 1$.

8. Repeat Exercise 7 for $n = 5$.

9. Repeat Exercise 7 for $n = 0$.

10. Repeat Exercise 7 for $n \neq 1$.

Think of "short" as "not tall" and use

p: Jim is tall;
q: Bill is short.

11. Write each of these statements in symbolic form.
(a) Jim is short and Bill is tall.
(b) Neither Jim nor Bill is tall.
(c) Jim is not tall and Bill is short.
(d) It is not true that Jim and Bill are both tall.
(e) Either Jim or Bill is tall.

12. Assume that Bill and Jim are both tall. Which of the statements in Exercise 11 are true?

Think of "sad" as "not happy" and use

p: Joan is happy;
q: Mary is sad.

13. Write each of these statements in symbolic form.
(a) Joan and Mary are both happy.
(b) Either Joan is happy or Mary is happy.
(c) Neither Joan nor Mary is happy.
(d) It is not true that Joan and Mary are both sad.
(e) It is not true that neither Joan nor Mary is happy.

14. Assume that Joan and Mary are both happy. Which of the statements in Exercise 13 are true?

Give each of these statements in words. Use

> p: *I like this book;*
> q: *I like mathematics.*

15. (a) $p \wedge q$ (b) $\sim q$
(c) $\sim p$ (d) $(\sim p) \wedge (\sim q)$

16. (a) $(\sim p) \wedge q$ (b) $p \vee q$
(c) $\sim(p \wedge q)$ (d) $\sim[(\sim p) \wedge q]$

17. Assume that you like this book and that you like mathematics. Which of the statements in Exercises 15 and 16 are true for you?

18. Assume that you like this book but that you do not like mathematics. Which of the statements in Exercises 15 and 16 are true for you?

***19.** Assume that two given statements p and q are both true and indicate whether or not you would expect each of the following statements to be true:
(a) $p \wedge q$ (b) $p \vee q$
(c) $p \vee (\sim q)$ (d) $(\sim p) \vee q$

***20.** Repeat Exercise 19 under the assumption that p is true and q may be true or false.

PEDAGOGICAL EXPLORATIONS

Elementary school teachers find many ways to help students understand *and* or *or.* This is necessary, for example, to enable students to follow instructions.

> Put the blocks in the box *and* the paper in the basket.
> Bring a sandwich *or* a piece of fruit for recess.
> List the nouns on page 4 *or* page 5.
> Please sharpen the pencil *and* give it to me.
> Read the story *and* write a report on it by Friday.

Students also need to understand compound sentences as they extend their vocabulary.

> This chain is large *and* heavy.
> Each rain hat on the pegs is all black *or* all yellow.
> A polygon that is a rhombus *and* also a rectangle is a square.

Teachers need to understand the use of compound statements so that they can avoid giving ambiguous instructions and also so that they can determine which students follow instructions. For example, one teacher told the class to draw a picture and read a book or write a story. Sally read a book and wrote a story. Linda drew a picture and read a book. Jack drew a picture and wrote a story. Joe wrote a story. Bill played the piano. Which students completed the assignment? Actually the instructions are subject to two different interpretations. Under one interpretation everyone is to draw a picture and then read a book or write a story or do both. Under the other interpretation writing a story is an alternative for both drawing a picture and reading a book. If the first interpretation is intended, the instructions should be restated for that interpretation. If the second interpretation is intended, the instructions should be restated, possibly in two parts, for that interpretation. As stated the teacher's instructions are ambiguous.

1. Give five examples of instructions to primary school students that require the students to understand *and* and *or*.
2. Repeat Exploration 1 for fifth or sixth graders.
3. For elementary school children give two examples of ambiguous instructions and restate each instruction for each of its interpretations.

CHAPTER TEST

1. Let $\mathcal{U} = \{m, o, r, a, b, l, e\}$ and list the elements of A' when A is defined as **(a)** $\{m, o, r, e\}$; **(b)** $\{l, a, b, o, r\}$.
2. Show all possible one-to-one correspondences of the two sets $\{t, o, p\}$ and $\{r, a, t\}$.
3. List all possible subsets of $\{h, a, t\}$.

Identify as true or as not always true (false).

4. The sets $\{b, a, t\}$ and $\{l, a, b\}$ are
   (a) equal sets **(b)** equivalent sets
5. **(a)** $(\mathcal{U} \cup B) \subseteq \mathcal{U}$ **(b)** $\mathcal{U} \cap \varnothing \subset \varnothing$
6. **(a)** $(A \cup B) \subset (B \cup A)$ **(b)** $(A \cap B) \subset (A \cup B)$
7. **(a)** $75 \in \{2, 4, 6, \ldots\}$ **(b)** $75 \in \{1, 3, 5, \ldots\}$
8. The relation

   1 → b, a
   2 → b
   3 → b, c

   is
   (a) one-to-one **(b)** a function
9. **(a)** $\forall x: x \geq x - 1$ **(b)** $\exists x: x^2 = x$
10. A number that is a multiple of 6 and also a multiple of 5 is a multiple of
    (a) 15 **(b)** 30

11. (a) There is one and only one line that contains all of the points of a given ray.
(b) Any square may be considered as a rectangle.

12. The statements $x < 6$ and $x \geq 7$ for integers x are
(a) contrary (b) contradictory

13. If statements p and q are both true, then
(a) $(\sim p) \vee q$ is true (b) $\sim(p \wedge \sim q)$ is true

For $\mathcal{U} = \{0, 1, 2, 3, 4, 5, 6, 7, 8, 9\}$, $A = \{2, 4, 6\}$, *and* $B = \{0, 3, 6, 9\}$, *find:*

14. $A' \cap B$ **15.** $(A \cup B)'$

Use the given diagram and find:

16. $n(P' \cap Q \cap R)$ **17.** $n(Q \cup R)$

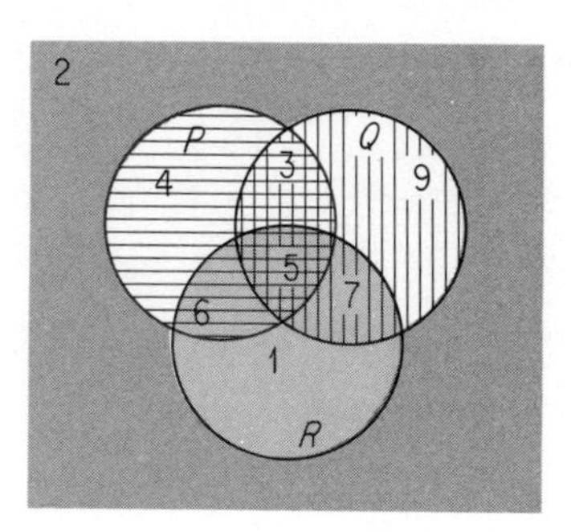

18. Suppose that you operate a newsstand and sell three newspapers: the Morning Star, the Evening Gazette, and the Daily Bulletin. You expect 650 people to buy at least one paper, 50 to buy all three papers, 130 to buy only the Star, 300 to buy the Bulletin, 150 to buy the Bulletin and the Gazette, 175 to buy the Bulletin and the Star, and 75 to buy only the Gazette. How many do you expect to buy both the Gazette and the Star?

19. Graph on a real number line:
(a) $x \leq 3$ (b) $-1 < x \leq 0$

Draw a flow chart to represent:

20. $2(n + 5)$ **21.** $2n + 5$

Use p: Wendy is fast;
q: Jan is slow.

Think of "slow" as "not fast," and write each of these statements in symbolic form.

22. (a) Wendy is fast and Jan is slow.
(b) Neither Wendy nor Jan is slow.

23. (a) Wendy is not fast but Jan is fast.
(b) Either Wendy or Jan is fast.

24. It is not true that Wendy and Jan are both fast.

25. Assume that Wendy and Jan are both fast. Which of the statements in Exercises 22 through 24 are then true?

READINGS AND PROJECTS

1. Look through the arithmetic materials for grades 1 and 2 in at least one recent series of textbooks and prepare a report on the development of the language of mathematics.

2. *The Arithmetic Teacher* is the journal that is designed especially for the teachers of mathematical concepts in elementary schools. People who wish to receive this journal regularly (eight issues per year) join the National Council of Teachers of Mathematics (often abbreviated NCTM) and specify *The Arithmetic Teacher* as their preferred journal. The dues of full-time undergraduate students are one-half the dues of regular members.

   Select a recent issue of *The Arithmetic Teacher* and:
   (a) Describe its general content noting especially items that might be useful to you in your future teaching.
   (b) Prepare a report on an article or feature that particularly appeals to you.

3. Read "A Mathematics Fair in the Lower Grades" by William W. K. Freeman and Diana Lambdin Kroll on pages 624 through 628 of the November 1974 issue of *The Arithmetic Teacher.*

4. *The Mathematics Teacher* is published eight times each year by the National Council of Teachers of Mathematics and is designed primarily for teachers of mathematical concepts in secondary schools. However, there are often items that are of interest to elementary school teachers. For example, see "Mathematics in Use, as Seen on Postage Stamps" by William L. Schaaf on pages 16 through 24 of the January 1974 issue.

   A collection of postage stamps that are related to mathematics could make an excellent school display. Indeed, enlarged reprints of several of the pictures from the article just cited are included in poster format in the January 1975 (pages 39 through 42) and subsequent issues of *The Mathematics Teacher.*

5. Read "What Does 'Everyman' Really Need from School Mathematics?" by Max S. Bell on pages 196 through 202 of the March 1974 issue of *The Mathematics Teacher.*

6. Read "Master of Tessellations: N. C. Escher, 1898–1972" by Ernest R. Ranucci on pages 299 through 306 of the April 1974 issue of *The Mathematics Teacher.* See also "Designs with Tesselations" by Evan M. Maletsky on pages 335 through 338 of the same issue.

7. Read "Where Do You Stand? Computational Skill is Passé" on pages 485 through 488 of the October 1974 issue of *The Mathematics Teacher.*

8. The game of "life" is described in an article "Population Explosion: An Activity Lesson" by Donald T. Piele on pages 496 through 502 of the October 1974 issue of *The Mathematics Teacher*. This game can be played by children in the upper elementary school grades and shows one of the many relationships of mathematical concepts to the critical problems of our time. There is a bibliography for those who would like further details. Learn to play the game and teach it to a friend.

9. Simply for your own pleasure read *The Dot and the Line, A Romance in Lower Mathematics* by Norton Juster, Random House, Inc., 1963. The book is very light reading and universally enjoyed by those who take a few minutes to look through it. Also try to find an opportunity to view the film (available on 16 mm) of the same name.

10. Read Chapters 2 (Arithmetic in Today's Culture), 5 (Arithmetic in Kindergarten and Grades 1 and 2), and 11 (Definitions in Arithmetic) of the twenty-fifth yearbook of the National Council of Teachers of Mathematics.

11. Read Booklet No. 1 (Sets) of the twenty-ninth yearbook of the National Council of Teachers of Mathematics. Complete at least three of the exercise sets given there.

12. Read "Let's Do It! Developing the Concept of Grouping" by James V. Bruni and Helen Silverman on pages 474 through 479 of the October 1974 issue of *The Arithmetic Teacher.*

13. Read Chapter 13 (Arithmetic for the Fast Learner in English Schools) of the twenty-seventh yearbook of the National Council of Teachers of Mathematics. Do at least three of the suggested activities.

14. Read Chapters 3 (The Textbook as an Instructional Aid), 4 (Other Printed Materials), and 11 (A Systems Approach to Mathematics Instruction) of the thirty-fourth yearbook of the National Council of Teachers of Mathematics. Start a collection of references and materials for the grades in which you are most interested.

15. Read "Using Creativity in Elementary School Mathematics" by Linda R. Jensen on pages 210 through 215 of the March 1976 issue of *The Arithmetic Teacher.* Add as many items as you can to her list of ways to use a box of sugar cubes and a pack of 3-by-5 cards to generate creativity in a specified elementary school grade level.

16. Explore the consumer-related puzzles given in "Sam Loyd, America's Greatest Puzzlist" by Raymond E. Spaulding on pages 201 through 211 of the March 1976 issue of *The Mathematics Teacher.* Try to solve as many of these as you can before referring to the solution provided.

2 Logic

A minimal introduction of logical concepts was provided in §1-8. Further details are included in this chapter. The primary goal of the chapter is for the reader to learn sufficient logical principles:

> To comprehend the structure of most statements that one encounters.
> To avoid making ambiguous or easily misunderstood statements.
> To recognize whether or not an argument is logically sound (valid).
> To be able to present a valid argument for any defensible position.

These goals are sought through an explicit recognition of the ambiguities of ordinary language (written or oral) and the basic ingredients of a valid argument.

Consider such a common type of statement as

> If you want to be strong, eat TAG.

The advertiser wants you to conclude that if you eat TAG, then you will be strong. However such an inference does not necessarily follow, as we shall try to make clear in this chapter. The recognition of ambiguities in ordinary language is achieved through various interpretations of sample statements and through the identification of several of the many ways of making the same statement but using different words. The basic ingredients of a valid argument are considered under the section headings of implications and mathematical proofs.

2-1 language

In studying or applying mathematics we need to use our accustomed language, either spoken or written, together with our specialized mathematical language. Also, in order to communicate effectively, we must find ways of being precise in our use of language. In other words, we need to learn how to say exactly what we mean.

Suppose that you made the following assignment:

Do Exercises 1 through 5 and 7 or 10 and 12.

Then assume that six of your students did the exercises indicated after their names.

Joan:	1 through 10
Jack:	10 and 12
Bill:	1 through 5 and 7
Tom:	1 through 5, 10, and 12
Alice:	1 through 12
Joe:	1 through 5, 7, and 12

Which of these students completed your assignment? Just what did you mean by that assignment?

Alice apparently recognized the fact that she could not clearly identify the assignment. She took no risks and made sure that she completed the assignment—regardless of its interpretation—by doing all of the exercises so that the assignment would have to be a subset of the exercises that she actually did.

Joan also did extra exercises but could not be sure that she completed the assignment. For example, the assignment could have been intended as

Do Exercises 1 through 5, and 7 or 10, and 12; that is,
do Exercises (1 through 5) and (7 or 10) and 12.

We often use punctuation such as commas and semicolons to imply grouping of written words. When these same words are spoken, brief pauses represent the punctuation and thus imply the groupings. In the interpretation that we just considered the groupings are shown first by ordinary punctuation; second by the use of parentheses.

The assignment could also have been interpreted as

Do Exercises 1 through 5, and 7; or 10 and 12.

Under this interpretation each of the six students could claim to have completed the assignment.

Note that you could have stated the first interpretation precisely in either of these forms:

Do Exercises 1, 2, 3, 4, 5, 12, and either 7 or 10.
Do Exercises 1 through 5, 7, and 12. You may substitute Exercise 10 for Exercise 7.

Similarly you could have stated the second interpretation precisely in the form:

Either do Exercises 1 through 5 and 7 or do Exercises 10 and 12.

The previous examples involved an ambiguity based upon unspecified groupings of words. Another ambiguity is based upon the inclusive use of *or* and the exclusive use of *or*. Consider these statements:

Jane is in Chicago or London at this moment.
Jack or Bill has at least $5 with him.

The first statement implies that exactly one (**exclusive or**) of the situations holds. The second statement is satisfied if either one or both (**inclusive or**) of the people have $5.

Ambiguities arise in statements such as the following:

Tom or Don should carry the chair into my office.

In mathematics we always use the inclusive or (one or the other or both) unless the statement is specifically restricted. For example, the last cited statement could be restricted to the exclusive or as follows:

Tom or Don, but not both, should carry the chair into my office.

Statements that involve quantifiers (§1-8) are particularly troublesome as sources of confusion in the use of the English language. For example, each of these statements has the same meaning as the others:

All of my students are not dishonest.
No one of my students is dishonest.
All of my students are honest.

Suppose that we wish to deny (**negate**) a statement such as:

Some books are worth reading.

This given statement actually means that:

There is at least one book that is worth reading.

The negation of this statement may be expressed in any one of these four equivalent ways:

It is not true that some books are worth reading.
There aren't any books worth reading.
No book is worth reading.
All books are not worth reading.

Note that in order to negate a "some" (existentially quantified) statement, we use an "all" (universally quantified) statement.

We can also negate universally quantified statements. Consider the statement:

All of my students are honest.

This statement is false if at least one student is dishonest. The negation of the given statement may be expressed in any one of these forms:

Not all of my students are honest.
At least one of my students is dishonest.
There is at least one dishonest student in my class.

Thus to negate a universally quantified statement we use an existentially quantified statement.

One of the most common sources of confusion in our language involves the use of "all are not" when the intended meaning is "not all are." Many supposedly well-educated people display this confusion. For example, a teacher might say:

All of you will not pass this course.

However, the intended meaning of the statement is:

Not all of you will pass this course.

In other words, at least one of you will not pass the course.

Our examples of confusions in the use of spoken or written language have been illustrated by relatively simple statements. Legal statements, advertisements, political promises, and numerous other common uses of language sometimes deliberately play upon such confusions. The study of logic can be very worthwhile solely for the purpose of increasing your facility with ordinary language.

1. Consider the assignment:

   Do Exercises 5 or 7 and 10 or 12.

   Suppose that

   Alice does Exercise 5.
   Bill does Exercises 7 and 12.
   Charles does Exercises 7 and 10.
   Don does Exercises 5 and 10.
   Ed does Exercise 12.

   For each student use parentheses to indicate at least one interpretation of the assignment under which that student could properly claim that the assignment had been completed.
2. Which of the students identified in Exercise 1 could not properly claim to have completed the assignment if it had been given as:

   Do Exercises 5 and 7 or 10 or 12.

3. Repeat Exercise 2 for the assignment:

   Do Exercise 1, 3, 5, 7, 10, or 12.

State in words, and in at least two different forms, the denial (negation) of each given statement.

4. All apples are pieces of fruit.
5. All numbers are whole numbers.
6. Any whole number is positive.
7. Any fraction represents a rational number.
8. There exist rational numbers that are not integers.
9. There exist real numbers that are not rational numbers.
10. Some complex numbers are real numbers.
11. Some rational numbers are integers.

PEDAGOGICAL EXPLORATIONS

There exist situations in which statements are neither true nor false and our usual rules of logic cannot be used.

1. Obtain a plain 3 × 5 card or similar piece of paper. On one side write

   The statement on the other side of this card is true.

   Then on the other side of the card write

   The statement on the other side of this card is false.

   Discuss the possible sets of truth values for the preceding statements.

2. There is reported to be a town in which the barber is a man who shaves all men in the town who do not shave themselves. Who shaves the barber?
3. Discuss the truth values of this statement:

   The sentence you are reading is false.

2-2 truth values of statements

How do you "prove" a statement to a friend? Undoubtedly there are several ways, including these three:

1. Find the statement in an encyclopedia or other reference book that he will accept without further proof.
2. Prove to him that the statement is a necessary consequence of some statement that he accepts.
3. Prove to him that the statement cannot be false.

In mathematics there are also several ways of "proving" statements. In essence each proof is based upon:

1. Statements that are accepted as true (assumed).
2. Sequences of statements (arguments) such that each statement is either assumed or is a *logical consequence* of the preceding statements, and the statement to be proved is included in the sequence.
3. Proofs that statements cannot be false.

Thus a proof is concerned with the truth of the statement under consideration.

A statement may be always true.

All tigers are members of the cat family.

Other statements may not always be true.

All balls are spherical.

In this case we may think of a basketball as spherical and a football as not spherical. In the language of mathematics, as in §1-8, we use T for true and F

for false. A statement is *true* if it is always true (cannot be false); a statement is considered to be *false* if it is not always true.

Truth tables (tables of truth values) or Venn diagrams may be used to summarize (or define) the truth values of statements. For example, consider $\sim p$, the **negation** of p. In the Venn diagram the points of each region indicate the instances under which the particular statement represented by that region is true. Thus p is true for the points of the circular region; $\sim p$ is true for the points outside the circular region of p. We consider only statements p that have exactly one of the truth values T, F. This assumption is indicated in the Venn diagram by the fact that every element of the universal set is a member of exactly one of the regions p, $\sim p$.

p	$\sim p$
T	F
F	T

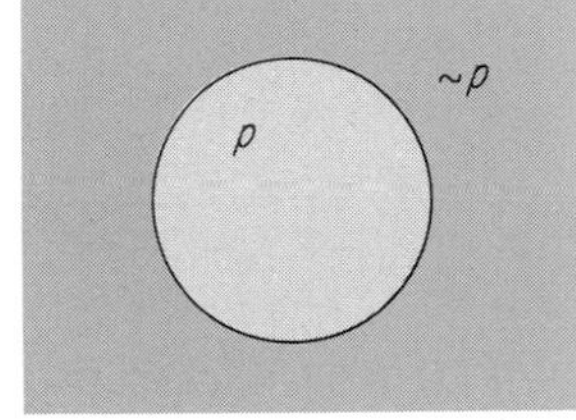

For any two statements p and q four possibilities arise as shown by the four regions of the next Venn diagram.

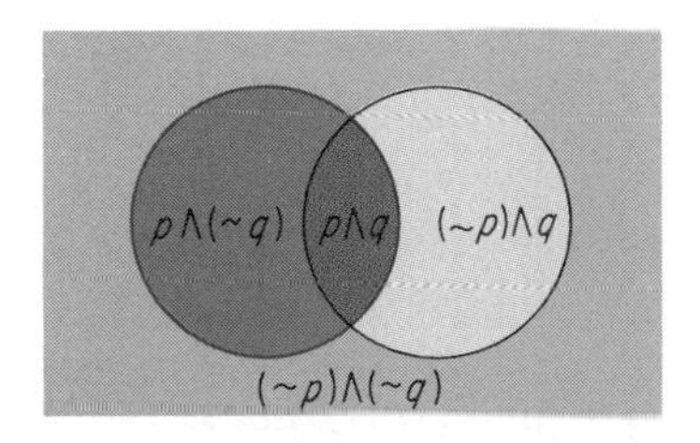

The four possibilities for p and q may also be readily recognized from the fact that provision must be made for each of the statements to be either true or false. Thus if p is true, q may be true or false.

		p	q
p true	q true	T	T
p true	q false	T	F

Also if p is false, q may be true or false.

		p	q
p false	q true	F	T
p false	q false	F	F

We use truth tables in which these four possible combinations of truth values provide a key upon which we base the definitions for the truth values of compound statements. Two compound statements were considered in §1-8 and their truth values described in words:

$p \wedge q$	p and q	the **conjunction** of p and q
$p \vee q$	p or q	the **disjunction** of p and q

The definitions of the truth values of these statements are given in the following truth tables. The key used here in presenting truth values for p and for q is the conventional one. However, the order in which the pairs of truth values is presented is entirely arbitrary; other orders may be found in other textbooks.

p	q	$p \wedge q$
T	T	T
T	F	F
F	T	F
F	F	F

p	q	$p \vee q$
T	T	T
T	F	T
F	T	T
F	F	F

As stated in §1-8, $p \wedge q$ is true if both p and q are true, false otherwise; $p \vee q$ is true if at least one of the statements p, q is true, false otherwise. Thus the compound statement $p \vee q$ is true unless both p and q are false.

Venn diagrams for $p \wedge q$ and $p \vee q$ may be indicated as in the following figures.

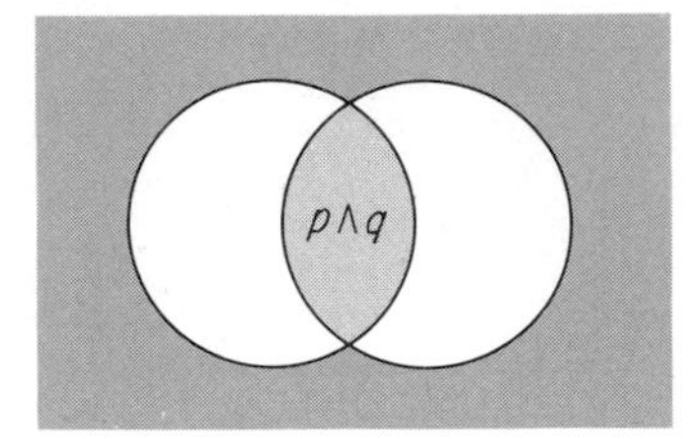

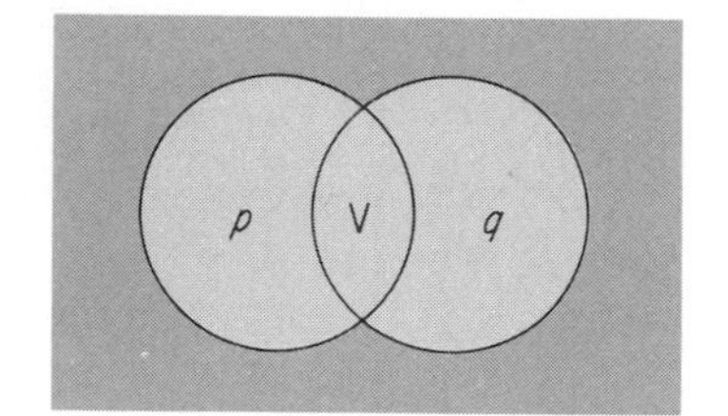

We can use either truth tables or Venn diagrams to summarize the truth values of a wide variety of compound statements. We shall emphasize truth tables since the steps used in forming a truth table are easily identified and are also precisely the steps that must be used to form the Venn diagram. To illustrate this procedure we shall construct a truth table for the statement $p \wedge (\sim q)$.

First set up a table with the appropriate headings as follows:

p	q	p	$\wedge$	$(\sim q)$
T	T			
T	F			
F	T			
F	F			

Now complete the column headed "p" by using the truth values that appear under p in the first column. In the column headed "$\sim q$" write the negation of the values given under q in the second column. (Why?) Our table now appears as follows:

p	q	p	$\wedge$	$(\sim q)$
T	T	T		F
T	F	T		T
F	T	F		F
F	F	F		T

Finally we find the conjunctions of the values given in the third and fifth columns. The completed table appears as follows, with the order in which the columns were considered indicated by (a), (b), and (c) and the final results in bold print:

p	q	p	$\wedge$	$(\sim q)$
T	T	T	**F**	F
T	F	T	**T**	T
F	T	F	**F**	F
F	F	F	**F**	T
		(a)	(c)	(b)

We can summarize this table by saying that the statement $p \wedge (\sim q)$ is true only in the case when p is true and q is false. Consider, for example,

p: This class has 28 members.
q: Some member of this class is absent today.

Then the statement $p \wedge (\sim q)$ may be stated as:

This class has 28 members and no one is absent today.

This statement, as in the general case, is true only if p is true and q is false.

EXERCISES

Construct truth tables for:

1. $(\sim p) \wedge q$ **2.** $(\sim p) \vee q$ **3.** $(\sim p) \vee (\sim q)$
4. $(\sim p) \wedge (\sim q)$ **5.** $\sim(p \wedge q)$ **6.** $p \vee (\sim q)$

In Exercises 7 through 10 copy and complete each truth table.

7.

p	q	$\sim$	$[p$	$\vee$	$(\sim q)]$
T	T				
T	F				
F	T				
F	F				
		(d)	(a)	(c)	(b)

8.

p	q	$\sim$	$[(\sim p)$	$\vee$	$q]$
T	T				
T	F				
F	T				
F	F				
		(d)	(a)	(c)	(b)

9.

p	q	$\sim$	$[(\sim p)$	$\wedge$	$(\sim q)]$
T	T				
T	F				
F	T				
F	F				
		(d)	(a)	(c)	(b)

10.

p	q	$\sim$	$[(\sim p)$	$\vee$	$(\sim q)]$
T	T				
T	F				
F	T				
F	F				
		(d)	(a)	(c)	(b)

11. Define $p \underline{\vee} q$ to mean "p or q but not both" and construct a truth table for this connective.

12. Construct a truth table for $p|q$, which we define to be true when p and q are not both true and to be false otherwise.

13. In Exercise 12 express $p|q$ in terms of other connectives that we have previously defined.

14. Use p: I like this book; q: I like mathematics. Tell the conditions under which each of the statements in Exercises 1 and 4 is true.

PEDAGOGICAL EXPLORATIONS

Elementary school students need to learn to identify statements as true or as false. Suppose that there are thirty chairs in this classroom and there are exactly twenty-six people in the room. Identify each statement as true or false.

1. There are thirty chairs in this classroom.
2. There are thirty chairs in this classroom and there are twenty-seven people in the room.
3. There are thirty chairs in this classroom or there are thirty people in this room.

4. Either there are thirty chairs in this classroom or there are not twenty-six people in this room.
5. There are at least twenty-eight chairs in this classroom.
6. There are at most thirty chairs in this classroom.

Compound statements are frequently used in elementary school classrooms.

7. Describe a situation as in the paragraph above Explorations 1 through 6. Then make up at least five compound statements that could be used for elementary school students to identify as true or as false.
8. Prepare a set of transparencies for use with an overhead projector to show $p \wedge q$ and $p \vee q$. Use a rectangular region for the universe of discussion. Prepare two overlays, each with a circular region to represent the truth value of a simple statement. Then place these overlays in various positions to show the conjunction and disjunction of statements.

2-3 conditional statements

Many of the statements that we make in everyday conversation are based upon a condition. For example, consider the following:

> If the telephone rings, then Jane will answer it.
> If I have no homework, then I shall go bowling.
> If I bribe the instructor, then I shall pass the course.

Each of these statements is expressed in the **if-then** form:

> If p, then q.

Any if-then statement can be expressed in symbols as $p \longrightarrow q$, which is read as "if p, then q." The **conditional symbol** $\longrightarrow$ is another connective. It is used to form a compound statement $p \longrightarrow q$, a **conditional statement.**

Our first task is to consider the various possibilities for p and q in order to define $p \longrightarrow q$ for each of these cases. One way to do this would be to present a completed truth table and to accept this as our definition of $p \longrightarrow q$. Let us, however, attempt to justify the entries in such a table. Consider again the conditional statement:

> If the telephone rings, then Jane will answer it.

If the telephone rings and Jane answers it, then the given statement is obviously true. On the other hand, the statement is false if the telephone rings and Jane does not answer it. Assume now that the telephone does not ring; then the given statement is true whether Jane uses the phone or not. In other words, if the telephone does not ring, then she has no responsibility for answering it. We can summarize these assertions of the truth values of a conditional statement as in the following truth table.

p	q	$p \rightarrow q$
T	T	T
T	F	F
F	T	T
F	F	T

Consider the statement:

If it rains, then I shall give you a ride home.

Have I lied to you:

1. If it rains and I give you a ride home?

2. If it rains and I do not give you a ride home?

3. If it does not rain and I give you a ride home?

4. If it does not rain and I do not give you a ride home?

According to the accepted meanings of the words used, you have a right to feel that I lied to you only if it rains and I do not give you a ride home. In other words, the conditional statement is true unless the premise is true and the conclusion is false.

If we are to communicate effectively with each other, we must accept definitions of the meanings of the words and of the sentence structures that we use. Thus, if we are to define the meaning of $p \longrightarrow q$, we must have an explicit truth value for each of the four possible combinations of truth values of p and q. This definition is shown in the given truth table. Notice that this definition is in perfect agreement with our accepted interpretations of the related sentences of the preceding illustration.

According to the truth table for $p \longrightarrow q$, any statement of the form "if p, then q" is false only when p is true and q is false. On the other hand, if p is false, then the statement $p \longrightarrow q$ is accepted as true regardless of the truth value of q. Thus each of the following statements is true by this definition.

If $2 + 3 = 7$, then George Washington is now the President of the United States.
If $2 + 3 = 7$, then the moon is made of green cheese.
If $2 + 3 = 7$, then July follows June.

If you have difficulty accepting any of these statements as true, then you should review the definition of the truth values of if-then statements. Remember also that there need be no relationship between p and q in an if-then statement, although we tend to use related statements in this way in everyday life.

Example 1: Give the truth value of each statement:

(a) If $5 + 7 = 12$, then $6 + 7 = 13$.
(b) If $5 \times 7 = 35$, then $6 \times 7 = 36$.
(c) If $5 + 7 = 35$, then $6 + 7 = 13$.
(d) If $5 + 7 = 35$, then $6 \times 7 = 36$.

Solution: Think of each statement in the form $p \longrightarrow q$.

(a) For p true and q true, the statement $p \longrightarrow q$ is *true*.
(b) For p true and q false, the statement $p \longrightarrow q$ is *false*.
(c) For p false and q true, the statement $p \longrightarrow q$ is *true*.
(d) For p false and q false, the statement $p \longrightarrow q$ is *true*.

Example 2: Find all possible replacements for integers x such that the following statement is true:

If $3 \times 4 = 34$, then $x - 3 = 5$.

Solution: Think of the statement in the form $p \longrightarrow q$ for

$$p: \quad 3 \times 4 = 34$$
$$q: \quad x - 3 = 5$$

Since p is false, the given statement $p \longrightarrow q$ is true whether q is true or false. That is, regardless of the value assigned to x, the following statement is always true:

If $3 \times 4 = 34$, then $x - 3 = 5$.

Given any conditional statement, $p \longrightarrow q$, three other related statements may be identified:

Statement:	$p \longrightarrow q$	If p, then q.
Converse:	$q \longrightarrow p$	If q, then p.
Inverse:	$(\sim p) \longrightarrow (\sim q)$	If not p, then not q.
Contrapositive:	$(\sim q) \longrightarrow (\sim p)$	If not q, then not p.

Here are five examples of conditional statements with their converses, inverses, and contrapositives:

1. *Statement:* If it is snowing, I leave my car in the garage.
Converse: If I leave my car in the garage, it is snowing.
Inverse: If it is not snowing, I do not leave my car in the garage.
Contrapositive: If I do not leave my car in the garage, it is not snowing.

2. *Statement:* If $\triangle ABC \cong \triangle XYZ$, then $\triangle ABC \sim \triangle XYZ$.
Converse: If $\triangle ABC \sim \triangle XYZ$, then $\triangle ABC \cong \triangle XYZ$.
Inverse: If $\triangle ABC$ is not congruent to $\triangle XYZ$, then $\triangle ABC$ is not similar to $\triangle XYZ$.
Contrapositive: If $\triangle ABC$ is not similar to $\triangle XYZ$, then $\triangle ABC$ is not congruent to $\triangle XYZ$.

3. *Statement:* If x is negative, then $x \neq 0$.
   Converse: If $x \neq 0$, then x is negative.
   Inverse: If x is not negative, then $x = 0$.
   Contrapositive: If $x = 0$, then x is not negative.
4. *Statement:* If $x + 2 = 5$, then $x = 3$.
   Converse: If $x = 3$, then $x + 2 = 5$.
   Inverse: If $x + 2 \neq 5$, then $x \neq 3$.
   Contrapositive: If $x \neq 3$, then $x + 2 \neq 5$.
5. *Statement:* $p \longrightarrow (\sim q)$.
   Converse: $(\sim q) \longrightarrow p$.
   Inverse: $(\sim p) \longrightarrow \sim (\sim q)$, which can be simplified as $(\sim p) \longrightarrow q$.
   Contrapositive: $\sim (\sim q) \longrightarrow (\sim p)$, or simply $q \longrightarrow (\sim p)$.

The following truth tables for these variants of a conditional statement $p \longrightarrow q$ summarize this discussion and specify the truth values for each statement:

		Statement			Converse			Inverse			Contrapositive		
p	q	$p \to q$			$q \to p$			$(\sim p) \to (\sim q)$			$(\sim q) \to (\sim p)$		
T	T	T	**T**	T	T	**T**	T	F	**T**	F	F	**T**	F
T	F	T	**F**	F	F	**T**	T	F	**T**	T	T	**F**	F
F	T	F	**T**	T	T	**F**	F	T	**F**	F	F	**T**	T
F	F	F	**T**	F	F	**T**	F	T	**T**	T	T	**T**	T

From the table you should see again that the contrapositive of a conditional statement has the same truth values as the statement. Also note that the converse and the inverse of any conditional statement have the same truth values.

Example 3: Write the contrapositive of the following statement:

If I work hard, then I will pass the course.

Solution:

If I do not pass the course, then I have not worked hard.

This statement has the same truth values as the original statement. If you fail the course, then you have not worked hard, because if you had worked hard, then you would have passed the course.

In Exercises 1 through 4 consider the statements:

p: *You pass the examination.*
q: *You pass the course.*

Then translate each symbolic statement into an English sentence.

1. $p \rightarrow q$ **2.** $q \rightarrow p$ **3.** $(\sim p) \rightarrow (\sim q)$ **4.** $(\sim q) \rightarrow (\sim p)$

5. Repeat Exercises 1 through 4 for the statements:
p: John drives a red car.
q: John lives in a white house.

6. Give the truth value of each statement:
(a) If $2 \times 3 = 5$, then $2 + 3 = 6$.
(b) If $2 \times 3 = 5$, then $2 + 3 = 5$.
(c) If $2 + 3 = 5$, then $2 \times 3 = 5$.

7. Give the truth value of each statement:
(a) If $5 \times 6 = 56$, then $5 + 6 = 11$.
(b) If $5 \times 6 = 42$, then $5 + 6 = 10$.
(c) If $5 \times 6 = 42$, then $5 + 5 = 10$.

8. Assume that $a \times b = c$, $b \times c = d$, and $c \neq d$. Then give the truth value of each statement:
(a) If $a \times b = c$, then $b \times c = d$.
(b) If $a \times b = d$, then $b \times c = c$.
(c) If $a \times b = d$, then $b \times c = d$.
(d) If $a \times b = c$, then $b \times c = c$.

Write the converse, inverse, and contrapositive of each statement.

9. If we can afford it, then we buy a new car.

10. If we play pingpong, then you win the game.

11. If two sides and the included angle of one triangle are congruent to two sides and the included angle of another triangle, then the triangles are congruent.

12. If $x > 2$, then $x \neq 0$.

13. If $x(x - 1) = 0$, then $x = 1$.

14. $(\sim p) \rightarrow q$.

Exercises 15, 16, 17, and 18, refer to the statements given in Exercises 11, 12, and 13. Tell whether or not you accept as always true:

15. The given statement.

16. The converse of the given statement.

17. The inverse of the given statement.

18. The contrapositive of the given statement.

Find all possible replacements for the real variable x for which each sentence is a true statement.

19. If $2 + 3 = 5$, then $x + 1 = 8$.

20. If $2 + 3 = 6$, then $x + 1 = 8$.

**21.* If $x + 1 = 8$, then $2 + 3 = 5$.

**22.* If $x + 1 = 8$, then $2 + 3 = 6$.

**23.* If $3 - x = 1$, then $2 \times 5 = 13$.

**24.* If $3 - x = 1$, then $2 \times 5 = 10$.

PEDAGOGICAL EXPLORATIONS

1. Conditional statements are used extensively in elementary school classrooms. Usually the student needs to determine first whether the "if part" (premise) applies to him. If the premise does not apply to him (is not true for him), then he may disregard the conclusion without disobeying the instruction. Note that the failure of the premise to apply to a particular student corresponds to the last two rows of the truth table for implications.

   Raise your hand if you are six years old or in the first grade.
   Bring your paper to the desk if you have done all the problems and checked them all.

   Give five conditional statements that could be used with first graders.

2. From newspapers and magazines prepare a collection of statements that are written in if-then form.

2-4 conditional and biconditional statements

The words *necessary* and *sufficient* are often used in conditional statements. For example, consider the statement:

Working hard is a sufficient condition for passing the course.

Let us use p to mean "work hard" and q to represent "pass the course." We need to decide whether the given statement means "if p, then q" or "if q, then p." The word sufficient can be interpreted to mean that working hard is adequate or enough—but not necessary—for passing. That is, there may be other ways to pass the course, but working hard will do it. Thus we interpret the statement to mean:

If you work hard, then you will pass the course.

The symbolic statement $p \longrightarrow q$ may thus be used for each of these statements:

If p, then q.
p is a sufficient condition for q.

Next consider the statement:

Working hard is a necessary condition for passing the course.

Here you are told that working hard is necessary or essential in order to pass. That is, regardless of what else you do, you had better work hard if you wish to pass. However there is no assurance that working hard alone will do the trick. It is necessary, but not sufficient. (You may also have to get good grades.) Therefore, we interpret the statement to mean:

If you pass the course, then you have worked hard.

The symbolic statement $q \longrightarrow p$ may thus be used for each of these statements:

If q, then p.
p is a necessary condition for q.

Still another form to consider is the statement "q, only if p." In terms of the example used in this section we may write this as:

You will pass the course only if you work hard.

Note that this does *not* say that working hard will insure a passing grade. It does mean that if you have passed, then you have worked hard. That is, "q, only if p" is equivalent to the statement "if q, then p."

We can also interpret this in another way. This statement "q, only if p" means "if not p, then not q." The contrapositive of this last statement, however, is "if q, then p." In terms of our illustration this means that if you do not work hard, then you will not pass. Therefore, if you pass, then you have worked hard.

To summarize our discussion, each of the following statements represents a form of the conditional statement $p \longrightarrow q$:

If p, then q.
q, if p.
p is a sufficient condition for q.
q is a necessary condition for p.
p, only if q.

The last form is probably the hardest to use. It should appear reasonable when considered as a restatement of the fact that if $p \longrightarrow q$, then it is impossible to have p without having q. Many people find it helpful to remember these facts:

> The "if part" is sufficient.
> The "then part" is necessary.
> If you can identify one part of a conditional statement as necessary, then the other part is sufficient.
> If you can identify one part of a conditional statement as sufficient, then the other is necessary.
> If you can identify one part of a conditional statement as the "if part," then the other part is the "only-if part."
> If you can identify one part of a conditional statement as the "only-if part," then the other part is the "if part."

The many distinct ways of expressing a conditional statement illustrate the difficulty of understanding the English language. We shall endeavor to reduce the confusion by expressing conditional statements in the form

If p, then q. Symbolically, $p \longrightarrow q$.

Example 1: Write each statement in if-then form:
(a) Right angles are congruent.
(b) $x > 5$, only if $x \geq 0$.

Solution: (a) If two angles are right angles, then they are congruent. (b) If $x > 5$, then $x \geq 0$.

Example 2: Translate into symbolic form, using

p: I shall work hard.
q: I shall get an A.

(a) I shall get an A only if I work hard.
(b) Working hard will be a sufficient condition for me to get an A.
(c) If I work hard, then I shall get an A, and if I get an A, then I shall have worked hard.

Solution: (a) $q \longrightarrow p$; (b) $p \longrightarrow q$; (c) $(p \longrightarrow q) \wedge (q \longrightarrow p)$.

The statement $(p \longrightarrow q) \wedge (q \longrightarrow p)$ in Example 2(c) is one form of the **biconditional statement** $p \longleftrightarrow q$ (read as "p if and only if q"); the symbol $\longleftrightarrow$ is the **biconditional symbol.**

Any biconditional statement

$$p \leftrightarrow q$$

is a statement that p is a sufficient condition for q and also p is a necessary condition for q. We may condense this by saying that p is a **necessary and sufficient conditon** for q. The biconditional statement may be stated in either of these forms:

p if and only if q; that is, p **iff** q.
p is a necessary and sufficient condition for q.

Example 3: Complete a truth table for the statement:

$$(p \rightarrow q) \wedge (q \rightarrow p)$$

Solution:

p	q	$(p \rightarrow q)$	$\wedge$	$(q \rightarrow p)$
T	T	T	T	T
T	F	F	F	T
F	T	T	F	F
F	F	T	T	T
		(a)	(c)	(b)

In Example 3 we constructed a truth table for $(p \rightarrow q) \wedge (q \rightarrow p)$. However, we have previously agreed that this conjunction of statements is a form of $p \leftrightarrow q$. This enables us to construct a truth table for $p \leftrightarrow q$. Note that $p \leftrightarrow q$ is true if and only if p and q have the same truth values.

p	q	$p \leftrightarrow q$
T	T	T
T	F	F
F	T	F
F	F	T

From the truth table we see that $p \leftrightarrow q$ is true when p and q are both true or are both false. Thus each of these biconditional statements is true:

$2 \times 2 = 4$ if and only if $7 - 5 = 2$ (Both parts are true.)
$2 \times 2 = 5$ if and only if $7 - 5 = 3$ (Both parts are false.)

Each of the following biconditional statements is false because exactly one part of each statement is false:

$$2 \times 2 = 4 \text{ if and only if } 7 - 5 = 3$$
$$2 \times 2 = 5 \text{ if and only if } 7 - 5 = 2$$

Example 4: Under what conditions is the following statement true?

I shall get an A if and only if I work hard.

Solution: The statement has the form $p \leftrightarrow q$. Such a statement is true only if either p and q are both true or p and q are both false. Thus the given statement is true in the following two cases.

(a) You get an A and you work hard.
(b) You do not get an A and you do not work hard.

EXERCISES

Write each statement in if-then form.

1. All ducks are birds.
2. Vertical angles are congruent.
3. Complements of the same angle are congruent.
4. Supplements of congruent angles are congruent.
5. Any two parallel lines are coplanar.
6. All triangles are polygons.
7. All circles are round.
8. All mathematics books are dull.
9. All teachers are boring.
10. All p are q.
11. You will like this book only if you like mathematics.
12. A necessary condition for liking this book is that you like mathematics.
13. To like this book it is sufficient that you like mathematics.
14. A sufficient condition for liking this book is that you like mathematics.
15. Liking this book is a necessary condition for liking mathematics.

Write each statement in symbolic form using:

p: *I feel chilly.*
q: *I put on a sweater.*

16. If I feel chilly, then I put on a sweater.

17. I shall put on a sweater only if I feel chilly.

18. If I do not feel chilly, then I do not put on a sweater.

19. Feeling chilly is a necessary condition for me to put on a sweater.

20. I put on a sweater if and only if I feel chilly.

21. If I feel chilly, then I do not put on a sweater.

22. For me to put on a sweater it is sufficient that I feel chilly.

23. A necessary and sufficient condition for me to put on a sweater is that I feel chilly.

Write each statement in symbolic form using:

p: *I miss my breakfast.*
q: *I get up late.*

24. I miss my breakfast if and only if I get up late.

25. A necessary condition for me to miss my breakfast is that I get up late.

26. For me to miss my breakfast it is sufficient that I get up late.

27. A necessary and sufficient condition for me to miss my breakfast is that I get up late.

28. For me not to miss my breakfast it is necessary that I do not get up late.

Express each statement in if-then form and classify as true or false.

29. $11 - 3 > 8$ if $9 + 3 < 10$.

30. A necessary condition for 2×2 to be equal to 5 is that $8 - 5 = 3$.

31. For 7×4 to be equal to 25 it is sufficient that $5 + 3 = 8$.

32. $7 \times 6 = 40$ only if $8 \times 5 \neq 40$.

33. $7 \times 6 = 42$ only if $8 \times 5 \neq 40$.

Decide relative to each given assertion whether or not a young man who makes the assertion, receives the money, and fails to marry the young lady can be sued for breach of promise.

34. I shall marry your daughter only if you give me $10,000.

35. A sufficient condition for me to marry your daughter is that you give me $10,000.

PEDAGOGICAL EXPLORATIONS

1. Difficulties often arise with the words "only if." For example, the statement "I shall get an A, only if I work hard" means "If I get an A, then I

have worked hard." However, in everyday language most people tend to interpret the first statement (incorrectly) as "If I work hard, then I shall get an A." Find several other common examples of such confusions involving "only if."

2. A statement such as "You may leave early if you finish the quiz" logically means "If you finish the quiz, then you may leave early." However many tend to interpret this permissive use of *if* as meaning *only if.* Thus they would then incorrectly think of the given statement as meaning "If you leave early, then you have finished the quiz."

   The use of an if-then statement to give permission frequently has the special meaning of "don't . . . unless. . . ." With this usage, the statement

   You may leave early if you finish the quiz

   is interpreted as

   Don't leave early unless you have finished the quiz.

   Prepare a set of three "if-then" statements that give permission, and rewrite each in the form "don't . . . unless. . . ."

2-5 implications

Consider the statement

p: Jane is in Chicago or London at this moment.

There are three *logical possibilities*:

Jane is in Chicago at this moment.
Jane is in London at this moment.
Jane is not in either Chicago or London at this moment.

Note that the possibility of one person (Jane) being in two distinct (and far apart) places at the same time is not a logical possibility.

Any statement that is true for all of the logical possibilities (circumstances) in which it may be considered is a **logically true statement.** The statement p that we have just considered is true in the first two of its listed logical possibilities but may be false in the third, for example, if Jane is in New York at this moment. Accordingly, this statement p is not a logically true statement.

A statement such as "$2 \times 3 = 6$" is a logically true statement (cannot be false) in our number system. Another example of a logically true statement is given in Example 1.

Any statement has a set of logical possibilities, the *universal set* of that statement. The set of its logical possibilities for which the given statement is true is its **truth set.** If its truth set is the same as its universal set, then the statement is a logically true statement. The Venn diagram representation of a statement is a representation of its truth set.

Example 1. State all logical possibilities and indicate whether or not the statement is logically true.

$x > 0$ or $x < 2$ for real numbers x.

Solution: Think of the given statement as $p \vee q$ for

$$p:\ x > 0$$
$$q:\ x < 2$$

These statements may be represented on a number line as in the figure.

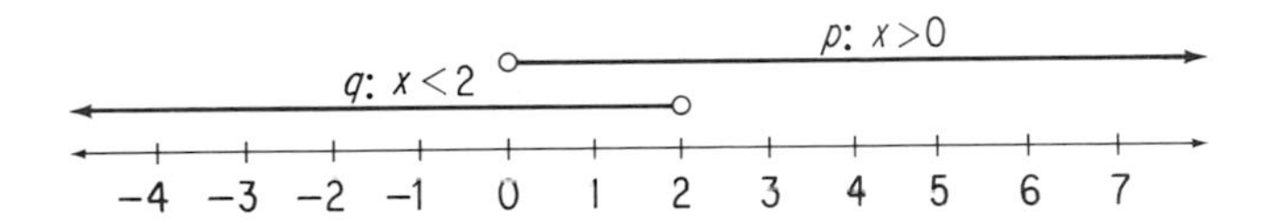

Then consider the **expanded truth table**:

p	q	Logically possible ?	$p \vee q$
T	T	Yes, for $0 < x < 2$	T
T	F	Yes, for $x > 2$	T
F	T	Yes, for $x < 0$	T
F	F	No	—

As may be observed from the number line, the circumstance of having both p and q false is not a logical possibility. That is, it is not possible to find a real number that does not belong to at least one of the sets p, q. Accordingly, we have not indicated a truth value for the given statement in this circumstance. The given statement is true for all of its logical possibilities and therefore is a logically true statement.

Example 2. State all logical possibilities and indicate whether or not the statement is logically true.

Bob is in college and doing good work.

Solution: Think of the given statement as $p \wedge q$ for

p: Bob is in college.
q: Bob is doing good work.

Then consider the expanded truth table:

p	q	Logically possible?	$p \wedge q$
T	T	Yes	T
T	F	Yes	F
F	T	Yes	F
F	F	Yes	F

Since the given statement is false for at least one of its logical possibilties, the given statement is not a logically true statement.

Statements that cannot be false (logically true statements) are particularly important since they provide essentially no basis for misunderstandings. If we use only logically true statements, we should be able to communicate completely with all people who understand such statements. Accordingly, we try to use only logically true statements throughout contemporary mathematics and especially whenever we are concerned with proofs of statements.

A conditional statement $p \longrightarrow q$ that is also a logically true statement is an **implication** $p \Rightarrow q$ (read as "p implies q"). The statement p is the **premise** of the implication; the statement q is the **conclusion** of the implication.

Example 3. Determine whether or not the given conditional statement is an implication.

If n is a counting number, then $n^2 > 0$.

Solution: Think of the given statement as $p \longrightarrow q$ for

p: n is a counting number.
q: $n^2 > 0$.

Then we have the following expanded truth table:

p	q	Logically possible?	$p \rightarrow q$
T	T	Yes	T
T	F	No	—
F	T	Yes; suppose $n = -1$	T
F	F	Yes, for $n = 0$	T

Since $p \longrightarrow q$ is true for each of its logical possibilities, $p \Rightarrow q$; that is, the given statement is an implication. In other words, it is always true that if n is a counting number, then $n^2 > 0$.

Determine whether or not the given conditional statement is an implication.

If n is a whole number, then $n^2 > 0$.

Solution: Think of the given statement as $p \longrightarrow q$ for

p: n is a whole number.
q: $n^2 > 0$.

Then we have the expanded truth table:

p	q	Logically possible ?	$p \rightarrow q$
T	T	Yes	T
T	F	Yes, for $n = 0$	F
F	T	Yes	T
F	F	Yes	T

Since $p \longrightarrow q$ is false for at least one of its logical possibilities, the given conditional statement is not an implication. In other words, the given statement is not an implication because it is possible to find at least one whole number, in this case 0, whose square is not greater than 0.

Example 5. Assume that all cases are logically possible and determine whether or not the given conditional statement is an implication.

$$[(p \longrightarrow q) \wedge p] \longrightarrow q$$

Solution: We make a truth table for the given conditional statement.

p	q	$[(p \longrightarrow q)$	$\wedge$	$p]$	$\longrightarrow$	q
T	T	T	T	T	T	T
T	F	F	F	T	T	F
F	T	T	F	F	T	T
F	F	T	F	F	T	F
		(a)	(c)	(b)	(e)	(d)

Since the given conditional statement is true for all of its logical possibilities, as shown in column (e) of the truth table, the statement is an implication.

A biconditional statement $p \longleftrightarrow q$ that is also a logically true statement is a **logical equivalence** $p \Longleftrightarrow q$ (read as "p is logically equivalent to q"). The procedures for identifying logically equivalent statements are similar to those for implications and are developed in the exercises. Remember that $p \longrightarrow q$ and $q \longrightarrow p$ are both true for the TT and the FF rows of the truth table and only for those rows. Thus two logically equivalent statements must have the same logical possibilities and for each of their logical possibilities must have the same truth values.

Example 6. Assume that all cases are logically possible and show by a truth table that any conditional statement is logically equivalent to its contrapositive statement.

Solution:

p	q	$[p \rightarrow q]$	$\longleftrightarrow$	$[(\sim q)$	$\longrightarrow$	$(\sim p)]$
T	T	T	T	F	T	F
T	F	F	T	T	F	F
F	T	T	T	F	T	T
F	F	T	T	T	T	T
		(a)	(e)	(b)	(d)	(c)

As shown in columns (a) and (d) the two conditional statements have the same truth values and therefore are logically equivalent.

EXERCISES

State the logical possibilities for each of these statements and indicate whether or not the given statement is a logically true statement.

1. Bill is driving a Ford.
2. Don has a textbook for this course.
3. $7 \times 8 = 56$ or $5 \times 6 = 56$
4. $10 \div 5 = 2$ or $10 = 2 \times 5$
5. $x \leq 0$ or $x > 0$ for real numbers x.
6. $x < 7$ or $x > 5$ for real numbers x.
7. Ginny is over 21 years old and she isn't 16 years old.
8. If Ginny is over 21 years old, then she isn't 16 years old.
9. If my house number is 41, then it isn't 74.
10. If n is an integer, then either $n > 5$ or $n \leq 4$.

In Exercises 11 through 18 assume that each variable x represents a real number and determine whether or not each statement is an implication.

11. If $2x = 6$, then $x = 3$.
12. If $x + 5 = 7$, then $x = 2$.
13. If $x \leq 3$, then $x = 3$.
14. If $x = -2$, then $x^2 = 4$.
15. If $x^2 = 4$, then $x = -2$.
16. If $x^3 = -8$, then $x = -2$.
17. If Eve ate an apple, then she ate a piece of fruit.
18. If Adam did not eat an apple, then he did not eat a piece of fruit.

Assume that all cases are logically possible and determine whether or not each statement is an implication.

19. $[\sim(\sim p)] \rightarrow p$
20. $[(p \rightarrow q) \wedge (\sim q)] \rightarrow (\sim p)$
21. $[(p \rightarrow q) \wedge q] \rightarrow p$
22. $[(p \rightarrow q) \wedge (\sim p)] \rightarrow (\sim q)$
23. $[(p \vee q) \wedge (\sim p)] \rightarrow q$
24. $[(p \rightarrow q) \wedge (q \rightarrow r)] \rightarrow (p \rightarrow r)$

Assume that all cases are logically possible and determine whether or not each statement is a logical equivalence.

25. $[\sim(\sim p)] \leftrightarrow p$
26. $(p \wedge q) \leftrightarrow (q \wedge p)$
27. $(p \vee q) \leftrightarrow (q \vee p)$
28. $(p \rightarrow q) \leftrightarrow [(q \vee (\sim p)]$
29. $(p \rightarrow q) \leftrightarrow [\sim(p \wedge (\sim q))]$
30. $[(p \vee q) \wedge (\sim q)]] \leftrightarrow p$

PEDAGOGICAL EXPLORATIONS

Refreshments for the rest period are brought to a kindergarten class of six boys and five girls. In each exploration an assumption and two or more possible situations are stated. For each situation indicate whether or not all logical possibilities have been included and, if not, describe or give the unlisted possibilities.

1. Assume that one of the members of the class brings the refreshments.
   - (a) A girl brought the refreshments.
   - (b) A boy brought the refreshments.
   - (c) A girl or a boy brought the refreshments.
   - (d) Judy, a member of the class, brought the refreshments, and Bill did not bring the refreshments.
2. Assume that two of the members of the class brought the refreshments.
   - (a) Two boys brought the refreshments, or two girls brought the refreshments.
   - (b) John and Barbara, members of the class, brought the refreshments.

For each type of statement give at least five examples that could be used with elementary school children.

3. Logically true statements about toys. For example, the blocks on my desk are either all red or not all red.
4. Logically true statements about numbers. For example, $n < 5$ or $n > 2$.
5. Logically true statements about people. For example,

   Ann is older than Jane, or
   Ann and Jane are the same age, or
   Jane is older than Ann.

6. Statements that are not logically true but may be either true or false. For example, Ann is older than Jane or Jane is older than Ann.
7. Statements that are not logically true and are false for all of their logical possibilities. Such statements are called **self-contradictory statements.** For example, Ann is older than Jane and Jane is older than Ann.
8. Implications.
9. Logically equivalent statements.

2-6 mathematical proofs

Any proof includes, at least informally, some given statements that are assumed and one or more statements that are to be proved. The assumed statements are the "**given**" (often called the **premises**) of the proof. The statements that are to be proved are the **conclusions** of the proof. A correct mathematical proof is based upon a *valid argument*. Specifically, any argument, mathematical or otherwise, is a **valid argument** if the conjunction of the premises implies the conclusions. In other words, an argument is valid, if, under the assumption of the premises, the conclusions cannot fail to be true.

Whenever we get involved in an argument, we should check the truth of the premises, note the symbolic form of each step of the argument, consider a symbolic statement of the entire argument, and accept the argument as valid if and only if its symbolic statement is logically true.

Example 1: Determine whether or not the following argument is valid:
Given: If Mary is a junior, she is taking algebra.
Given: Mary is a junior.
Conclusion: Mary is taking algebra.

Solution: Use

p: Mary is a junior.
q: Mary is taking algebra.

and think of the argument as:
Given: $p \longrightarrow q$.
Given: p.
Conclusion: q.
The argument is valid if and only if the statement

$$[(p \longrightarrow q) \land p] \longrightarrow q$$

is logically true. As in Example 5 of §2-5 this symbolic statement of the argument is an implication. Thus the argument is valid.

The argument used in Example 1 illustrates one of the basic rules of inference, the **law of detachment**, also called **modus ponens**

> If a statement of the form "if p, then q" is assumed to be true, and if p is known to be true, then q must be true.

Example 2: Determine whether or not the following argument is valid:
Given: If $3x - 1 = 8$, then $3x = 9$.
Given: $3x - 1 = 8$.
Conclusion: $3x - 9$.

Solution: For

$$p: \quad 3x - 1 = 8$$
$$q: \quad 3x = 9$$

the argument is valid. It has the form of the implication

$$[(p \longrightarrow q) \land p] \Rightarrow q$$

Example 3: Determine whether or not the following argument is valid:
Given: If you worked hard, then you passed the course.
Given: You passed the course.
Conclusion: You worked hard.

Solution: For

p: You worked hard.
q: You passed the course.

the argument has the form

$$[(p \longrightarrow q) \land q] \longrightarrow p$$

This statement is not an implication (§2-5, Exercise 21). Thus the argument is not valid.

Notice that anyone who uses the argument in Example 3 to convince his friends that he has worked hard is hoping that his friends will think that the converse (§2-3) of any implication is also an implication; that is,

If $p \Rightarrow q$, then $q \Rightarrow p$.

We know from our comparison of truth values of $p \rightarrow q$ and $q \rightarrow p$ that a conditional statement does not necessarily imply its converse. As another example of this type of reasoning consider the advertisement:

If you want to be healthy, eat KORNIES.

The advertiser hopes that the consumer will incorrectly assume the converse statement:

If you eat KORNIES, then you will be healthy.

The argument is not valid and is called a **fallacy**.

Example 4: Determine whether or not the following argument is valid:
Given: If you worked hard, then you passed the course.
Given: You did not pass the course.
Conclusion: You did not work hard.

Solution: For

p: You worked hard.
q: You passed the course.

the argument has the form

$$[(p \rightarrow q) \wedge (\sim q)] \Rightarrow (\sim p)$$

This statement is an implication (§2-5, Exercise 20); the argument is valid.

Notice that the argument in Example 4 is based upon the equivalence of any given conditional statement $p \rightarrow q$ and its contrapositive statement $(\sim q) \rightarrow (\sim p)$ as we observed in Example 6 of §2-5.

Example 5: Determine whether or not the following argument is valid:
Given: If you worked hard, then you passed the course.
Given: You did not work hard.
Conclusion: You did not pass the course.

Solution: For

p: You worked hard.
q: You passed the course.

the argument has the form

$$[(p \longrightarrow q) \wedge (\sim p)] \longrightarrow (\sim q)$$

This statement is not logically true (§2-5, Exercise 22), and the argument is not valid. It is another example of a fallacy.

Notice that anyone who uses the argument in Example 5 expects his listeners to assume that the inverse $(\sim p) \longrightarrow (\sim q)$ of any true statement $p \longrightarrow q$ must be true. As in the case of a conditional statement and its converse, we know that a statement does not necessarily imply its inverse. As another example of this type of reasoning, consider the advertisement:

> If you brush your teeth with SCRUB, then you will have no cavities.

The advertiser would like you to assume, fallaciously, the inverse statement:

> If you do not brush your teeth with SCRUB, then you will have cavities.

The final form of valid reasoning that we shall consider here is based upon the implication

$$[(p \longrightarrow q) \wedge (q \longrightarrow r)] \Rightarrow (p \longrightarrow r)$$

(§2-5, Exercise 24). Consider the following example:

> If you like this book, then you like mathematics.
> If you like mathematics, then you are intelligent.
> Therefore, if you like this book, then you are intelligent.

The first two statements are premises (assumed to be logically true) and the third statement is the conclusion. In this and in other cases in which the argument is valid, the conclusion is often called a **valid conclusion** of the premises.

EXERCISES

In each exercise assume that the premises are logically true and determine whether or not the argument is valid.

1. *Given:* If Elliot is a freshman, then Elliot takes mathematics.
Given: Elliot is a freshman.
Conclusion: Elliot takes mathematics.

2. *Given:* If you like dogs, then you will live to be 120 years old.
   Given: You like dogs.
   Conclusion: You will live to be 120 years old.
3. If the Yanks win the game, then they win the pennant.
   They do not win the pennant.
   Therefore, they did not win the game.
4. If you like mathematics, then you like this book.
   You do not like mathematics.
   Therefore, you do not like this book.
5. If you work hard, then you are a success.
   You are not a success.
   Therefore, you do not work hard.
6. If you are reading this book, then you like mathematics.
   You like mathematics.
   Therefore, you are reading this book.
7. If you are reading this book, then you like mathematics.
   You are not reading this book.
   Therefore, you do not like mathematics.
8. If you work hard, then you will pass the course.
   If you pass the course, then your teacher will praise you.
   Therefore, if you work hard, then your teacher will praise you.
9. If you like this book, then you like mathematics.
   If you like mathematics, then you are intelligent.
   Therefore, if you are intelligent, then you like this book.
10. If you are happy, then you are lucky.
    If you are lucky, then you will be rich.
    Therefore, if you do not become rich, then you are not happy.

In each exercise use all of the given premises and supply a valid conclusion for them.

11. If you drink milk, then you will be healthy.
    You are not healthy.
    Therefore, ...
12. If you eat a lot, then you will gain weight.
    You eat a lot.
    Therefore, ...
13. If you like to fish, then you enjoy swimming.
    If you enjoy swimming, then you are a mathematician.
    Therefore, ...
14. If you do not work hard, then you will not get an A.
    If you do not get an A, then you will have to repeat the course.
    Therefore, ...

15. If you like this book, then you are not lazy,
If you are not lazy, then you will become a mathematician.
Therefore, ...

sec. 2-6
mathematical proofs

PEDAGOGICAL EXPLORATIONS

1. Begin a collection of fallacies in reasoning that you find in newspapers or magazines, or that you hear on radio or television.
2. Examine a newspaper or magazine for if-then statements that are used in advertising. Then analyze the statement to see whether the converse or inverse of the statement is used.
3. Give five valid arguments that could be used with elementary school children.
4. Give five arguments that could be used with elementary school children to illustrate arguments that are not valid.
5. Mathematical proofs are ultimately of the form $p \Rightarrow q$, where p and q may be compound statements. Accordingly, a valid argument may be of any form that is logically equivalent to an implication. Assume that all cases are logically possible and give at least five different possible forms for a valid argument $p \Rightarrow q$.

The following explorations are based upon sets of premises written by Charles Lutwidge Dodgson, who used the name of Lewis Carroll as author of *Alice's Adventures in Wonderland* and *Through the Looking Glass.* Supply a conclusion so that each argument will be valid.

6. Babies are illogical.
   Nobody is despised who can manage a crocodile.
   Illogical persons are despised.
   Therefore, ...
7. No ducks waltz.
   No officers ever decline to waltz.
   All my poultry are ducks.
   Therefore, ...
8. No terriers wander among the signs of the zodiac.
   Nothing that does not wander among the signs of the zodiac is a comet.
   Nothing but a terrier has a curly tail.
   Therefore, ...

2-7 what was actually said?

Often we read a sentence or hear a speaker make a statement to which we react with a question such as "What does that actually mean?" Sometimes the confusion is purposeful; at all times logical concepts should be helpful in determining what was actually stated—whether or not that is what was meant.

In a sense the ultimate usefulness of any elementary consideration of logic is the improvement of one's ability (as a speaker or writer and as a listener or reader) to participate effectively in communication. Numerous confusions arise from advertisements in which abuses of logical principles are purposely used to mislead the public without breaking laws regarding truth in advertising. Confusions arise from colloquialisms and fads in the use of words. The tendency of some people to talk for ten minutes to convey the content of one simple sentence can also cause confusion.

Each of us can profit from saying what we mean and understanding what others say. A knowledge of the languages of statements (logic) and sets can help us do this. Examples 1, 2, and 3 illustrate the uses of principles of the language of statements; Examples 4, 5, and 6 illustrate the use of the language of sets.

Example 1. Find a simpler logically equivalent statement for the given statement and express the equivalence of the two statements in symbolic form.

I ain't going nowhere.

Solution: I'm going somewhere. For p: I'm going somewhere, the symbolic form is $[\sim(\sim p)] \leftrightarrow p$, §2-5 Exercise 25, the "double negative."

Example 2. Repeat Example 1 for the following statement.

Either John or Bill will take Judy to the dance but Bill will be in the hospital all that week.

Solution: John will take Judy to the dance. For

p: John will take Judy to the dance.
q: Bill will take Judy to the dance.

the symbolic form is $[(p \vee q) \wedge \sim q] \leftrightarrow p$, §2-5 Exercise 30.

Example 3. Repeat Example 1 for the statement $x^2 = 9$ and $x \neq 3$.

Solution: If $x^2 = 9$, then $x = 3$ or $x = -3$. Since $x \neq 3$, the solution is $x = -3$. For

p: $x = -3$
q: $x = 3$

the symbolic form is $[(p \vee q) \wedge \sim q] \leftrightarrow p$, §2-5, Exercise 30.

In the eighteenth century the Swiss mathematician Leonhard Euler used diagrams to present a visual approach to the study of the validity of arguments. The following diagrams are an aid to the reasoning that is in progress but are not essential to the reasoning process. We use a region, usually a circular region, P to represent the situations in which a statement p is true. Then p is false in all situations represented by points of P'. Similarly, we use a region Q for a statement q, and so forth. Then $p \longrightarrow q$ if and only if $P \subseteq Q$. In the language of sets Euler diagrams may be used to test the validity of an argument. We draw one or more Euler diagrams, making use of the premises and avoiding the introduction of additional premises. The structure of the diagram indicates whether or not the argument is valid. As usual we assume for each example that its premises are true, whether or not we believe them.

sec. 2-7 what was actually said?

Example 4. Test the validity of this argument:

All undergraduates are sophomores.
All sophomores are attractive.
Therefore, all undergraduates are attractive.

Solution: A diagram of this argument follows:

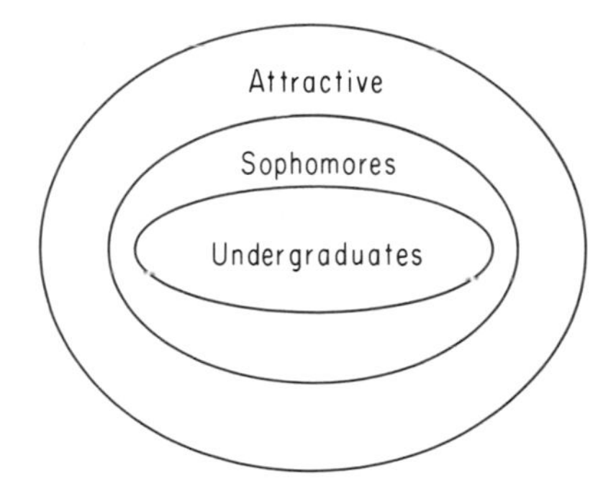

The argument is valid.

Notice in Example 4 that you may or may not agree with the conclusion or with the premises. The important thing is that you are *forced* to draw the diagram in this manner. The conclusion is an inescapable consequence of the given premises. You should not allow the everyday meaning of words to alter your thinking. Thus the preceding argument can be stated abstractly as in these three statements:

All u's are s's.
All s's are a's.
Therefore, all u's are a's.

Here we rely on logic alone rather than any preconceived notions about the meaning of such words as "undergraduate," "sophomore," and "attractive."

Example 5: Draw a diagram and test the validity of this argument:

All freshmen are clever.
All attractive people are clever.
Therefore, all freshmen are attractive.

Solution: The first premise tells us that the set of freshmen is a subset of the set of clever individuals. This is drawn as follows:

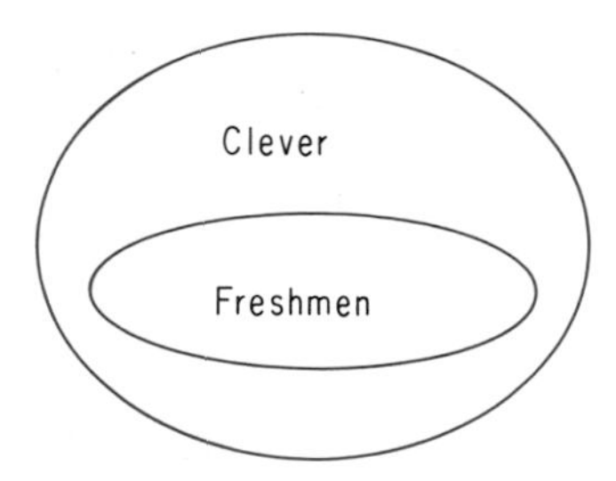

Next we need to draw a circle to represent the set of attractive people as a subset of the set of clever individuals. However, there are several possibilities here:

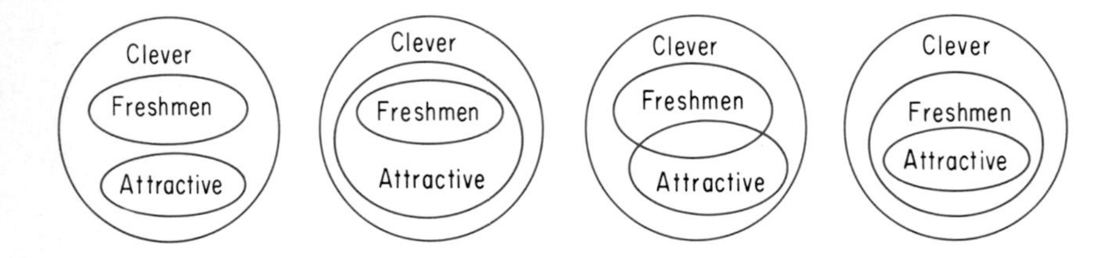

Each of the figures drawn represents a distinct possibility, but only one shows that all freshmen are attractive. Therefore, since we are not *forced* to arrive at this conclusion, the argument is said to be *not valid*. Each of the figures, on the other hand, forces you to arrive at the following conclusions:

Some clever people are attractive.
Some clever people are freshmen.

Example 6: Test the validity of the following argument:

All a's are b's.
Some a's are c's.
Therefore, some c's are b's.

Solution: Note that since some a's are c's, then there are also some c's that are a's. However, since all a's are b's, it follows that some c's must also be b's as in these diagrams:

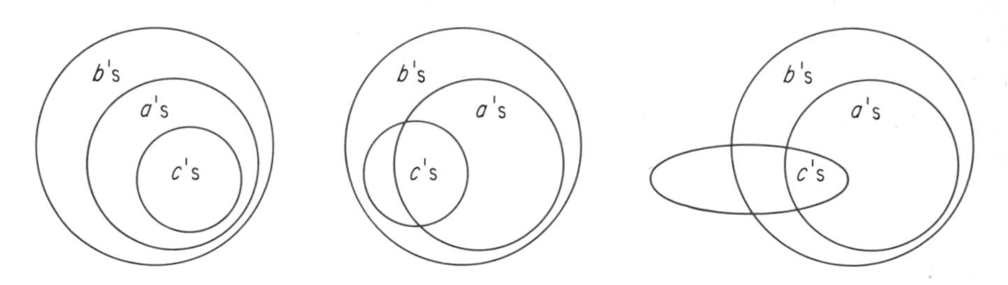

The argument is valid.

EXERCISES

Find a simpler logically equivalent statement for each given statement.

1. Ain't nobody home.
2. Not all of my sisters are blondes.
3. I am not failing all of my courses.
4. Tonight I'll go to the library or study; however, I'm not going to study.
5. $x \neq 5, x^2 = 25, x < 0$
6. $x^2 = 1, x > 0, x^3 = 1$

Use Euler diagrams to test the validity of each argument.

7. All students love mathematics.
   Harry is a student.
   Therefore, Harry loves mathematics.
8. All juniors are brilliant.
   All brilliant people love mathematics.
   Therefore, if you are a junior, then you love mathematics.
9. All young people are beautiful.
   All beautiful people like this book.
   Therefore, if you like this book, then you are a young person.
10. All mathematics teachers are dull.
    Some Ph.D.'s are dull.
    Therefore:
    (a) Some mathematics teachers are Ph.D.'s.
    (b) Some dull people have Ph.D.'s.

11. All juniors are clever.
Some juniors are males.
Therefore:
(a) Some males are clever.
(b) Some males are juniors.
(c) Some clever people are males.
(d) All males are juniors.

12. All mathematics teachers are interesting.
All attractive individuals are interesting.
Some mathematics teachers are kind.
Therefore:
(a) Some interesting people are kind.
(b) Some mathematics teachers are attractive.
(c) All mathematics teachers are attractive.
(d) All mathematics teachers are kind.
(e) Some kind individuals are attractive.
(f) No mathematics teachers are attractive.
(g) No attractive individuals are interesting.

PEDAGOGICAL EXPLORATIONS

1. One elementary textbook uses Euler circles (diagrams) to show the use of the words "all," "some," and "no." Here is an example:

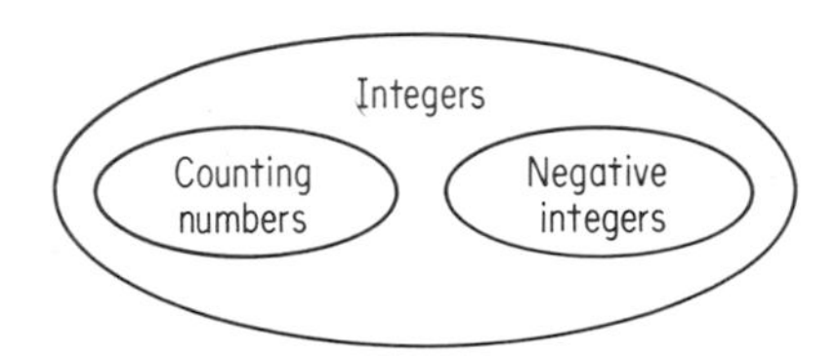

All counting numbers are integers.
Some integers are counting numbers.
No counting numbers are negative integers.

Draw a similar diagram to show this set of statements:

All triangles are polygons.
Some polygons are squares.
No square is a triangle.

Then prepare at least five other similar figures using arithmetic or geometric concepts that are found at the elementary school level.

2. Another interesting exercise for elementary school youngsters is to present an Euler diagram and have them draw inferences from it. In the following you are to complete each blank space with one of the words "all," "some," or "no."

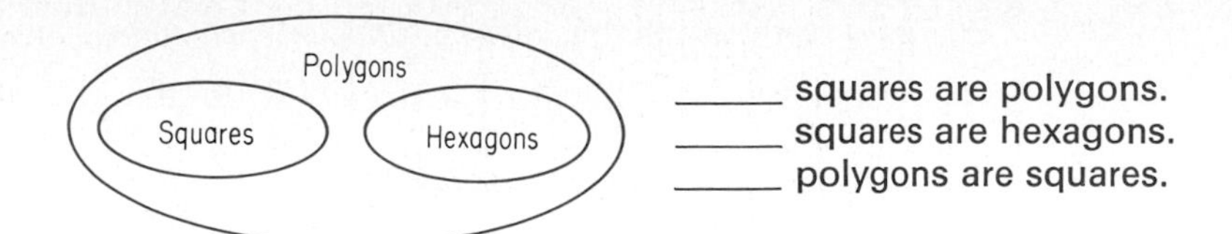

______ squares are polygons.
______ squares are hexagons.
______ polygons are squares.

Prepare a collection of at least five other similar exercises, using concepts found at various grades in the elementary mathematics curriculum.

3. Discuss the validity of the arguments used in the following situation.

A man approached the girl at the check-out counter of a local market and asked the price of a box of blueberries.
Clerk: "65 cents a box."
Customer: "What! They're selling them for 55 cents a box across the street."
Clerk: "Why don't you buy them there?"
Customer: "Because they're all sold out."
Clerk: "Oh! If we were all sold out, our price would be 45 cents a box."

4. Try to find other descriptions of situations, such as the above, which involve logical concepts.

CHAPTER TEST

1. Use the given symbols and write each statement in symbolic form.

p: John studies hard.
q: Jim does not study hard.

(a) Neither John nor Jim studies hard.
(b) It is not true that John and Jim both do not study hard.

State in words, and in at least three different forms, the denial (negation) of each of the given statements.

2. All geometric figures are polygons.

3. Some integers are counting numbers.

4. There exists a whole number that is not a counting number.

Assume that all cases are logically possible and construct truth tables for:

5. $\sim[(\sim p) \vee q]$ ***6.*** $(p \wedge q) \rightarrow (\sim p)$ ***7.*** $\sim[p \leftrightarrow (\sim q)]$

Write the converse and the contrapositive of each statement.

8. If apples are red, then they are ripe. ***9.*** If $x < 3$, then $x \neq 4$.

Give the truth value of each statement.

10. **(a)** If $7 \times 11 = 78$, then $11 \times 7 = 88$.
(b) If $5 + 2 = 5$, then $2 \times 3 = 6$.

11. **(a)** If $5 \times 7 = 57$, then $5 + 7 = 12$.
(b) If $5 + 7 = 12$, then $5 \times 7 = 57$.

Find all possible replacements for the real variable x such that each sentence is a true statement.

12. **(a)** If $3 + 4 = 34$, then $x + 2 = 7$.
(b) If $3 + 4 = 7$, then $x + 2 = 34$.

13. **(a)** If $x + 2 = 7$, then $3 + 4 = 7$.
(b) If $x + 2 = 7$, then $3 + 4 = 34$.

Write each statement in if-then form.

14. All horses are quadrupeds.

15. Knowing Judy is a sufficient reason for liking her.

Write each statement in symbolic form.

16. **(a)** p is sufficient for q.
(b) p is necessary and sufficient for $(\sim q)$.

17. Use

p: I wear a coat.
q: It is snowing.

(a) For me to wear a coat it is necessary that it be snowing.
(b) I wear a coat only if it is snowing.

Determine whether or not the given conditional statement is an implication.

18. $[(p \rightarrow q) \wedge (\sim q)] \rightarrow (\sim p)$ ***19.*** $\{[(\sim p) \rightarrow q] \wedge p\} \rightarrow q$

Identify the logical possibilities for each of these statements and indicate whether or not the given statement is a logically true statement.

20. Jim has a foreign car.

21. If n is an integer, then n is a negative number or n is a whole number.

Find a logically equivalent statement for each given statement and express the equivalence of the two statements in symbolic form.

22. Not all of my bills are overdue.

23. Ain't nobody out on this rainy night.

Use Euler diagrams to test the validity of each argument.

24. All a's are b's. All b's are c's. Some d's are not c's.
Therefore: **(a)** All a's are c's. **(b)** Some d's are not b's.

25. **(a)** All dogs love their masters.
If an animal loves its master, the animal should be well cared for.
Therefore, all dogs should be well cared for.
(b) If x is a positive integer, then $x^2 > 0$.
$x^2 \ngtr 0$.
Therefore, x is not a positive integer.

READINGS AND PROJECTS

1. Look through one or more recent series of elementary school textbooks and report on the readiness for, and use of, logical concepts such as and, or, and the converse of a statement.

2. Proofs in elementary school classes are usually at the level of giving reasons or explanations. The explanations by the teacher should be logically correct. Most students are in the process of learning to provide logical explanations.

Start a collection of "striking" examples of youthful reasoning such as these two gathered by the Swiss psychologist Jean Piaget:

(a) When you're walking in the moonlight, the moon moves along with you. (b) Trees make the wind blow by waving their branches.

3. Start a collection of newspaper, magazine, etc. clippings that appear to be particularly useful in the introduction of logical concepts to elementary school children.

4. Elementary school children need to experience ideas in order to understand them. For example, many of the confusions in our ordinary language can be removed by using appropriate representations. The following examples are based upon straws in cups. Numerous other situations could be used. For our example there are exactly three possible cases:

I. All cups have straws.

II. Some but not all cups have straws.

III. No cups have straws.

No other cases are possible. Therefore, any two statements that hold for the same cases must be equivalent statements.

(a) List the cases for which each of the following statements is true.

- **(i)** All cups have straws.
- **(ii)** No cups have straws.
- **(iii)** Some cups have straws.
- **(iv)** All cups are without straws.
- **(v)** Some cups are without straws.
- **(vi)** There is no cup with a straw.
- **(vii)** There is no cup without a straw.
- **(viii)** Every cup is without a straw.
- **(ix)** Not every cup has a straw.
- **(x)** Not every cup is without a straw.

(b) Collect the listed statements in subsets of equivalent statements so that no statement in one subset is equivalent to a statement in another subset.

5. Read "Sorting, Classifying, and Logic" by Douglas E. Cruikshank on pages 588 through 598 of the November 1974 issue of *The Arithmetic Teacher.*

6. Read, or reread, Lewis Carroll's *Alice in Wonderland* and *Through the Looking Glass* for numerous amusing examples of logical principles and his "light touch" in handling them.

7. Lewis Carroll, the author of *Alice in Wonderland,* was actually the mathematician C. L. Dodgson. His *Symbolic Logic* (fourth edition, 1896) and *The Game of Logic* (1886) were published as a single paperback volume in 1958 by Dover Publications, Inc.

Symbolic Logic includes multitudes of intriguing illustrative examples, solutions, and an "Appendix addressed to teachers." The appendix includes comparisons of Euler's method of diagrams, Venn's method of diagrams, and Dodgson's method of diagrams as well as symbolic and other methods.

Look over *Symbolic Logic,* don't miss the introduction "To Learners," and report on the aspects of the book that you find most interesting. Dodgson claimed that he used all of these materials with children at most 14 years old.

8. Start a collection of logical paradoxes; that is, sets of statements that do not seem to "make sense" under our usual laws of logic. For example, consider the statements given in Pedagogical Explorations 1, 2, and 3 of §2-1.

9. Read Booklet No. 12 (Logic) of the thirtieth yearbook of the National Council of Teachers of Mathematics. Complete at least three of the exercise sets given there.

3 Numeration and Mathematical Systems

Some time ago an editorial in a leading newspaper stated the following startling news in its headlines:

$$1 + 1 \text{ is no longer equal to } 2!$$

Of course this headline was misleading. Indeed, the editorial included the statement that our schools were teaching elementary school children that $1 + 1 = 10$. Actually, the writer of the editorial had taken addition in base 2 out of context and had misrepresented its proper notation. In order to convey the true meaning of the symbols used, he should have stated that

$$1 + 1 = 10_{\text{two}}$$

which means that $1 + 1$ can be thought of as one set of two units and no single units. In other words, $10_{\text{two}} = (1 \times 2) + (0 \times 1) = 2 + 0 = 2$. Therefore, $1 + 1$ is still equal to 2!

Critics of new approaches to mathematics may have some valid arguments to their claims that modern youngsters have fallen off in their ability to perform arithmetic computations. However, it is seldom fair to criticize an idea out of its proper context. Newer programs of any sort should be evaluated only after the objectives and content are well understood. The purpose of this chapter is to clarify several aspects of contemporary mathematics programs. When the reader has completed this chapter, he should be able to understand and appreciate such statements as these:

$$8 + 7 = 3 \qquad \text{(on a 12-hour clock)}$$
$$4 \times 3 = 22_{\text{five}} \qquad \text{(in base 5)}$$

3-1 Egyptian numeration

We habitually take for granted the use of our system of numeration, as well as our computational procedures. However, these represent the creative work of man through the ages. We can gain a better appreciation of our system of numeration and methods of computation by examining other systems.

In this section we shall travel through the years and first explore the system of numeration used by the ancient Egyptians as shown in their hieroglyphics. We shall continue our study of numeration systems in later chapters, culminating (in Chapter 14) with a study of binary notation, the mathematical basis for the modern computer.

Throughout these explorations the reader should note the growth that has taken place in mankind's mathematical thinking from early times to the present day, a growth that matches his creative developments in other areas. Systems of numeration form an important part of each educated person's knowledge and provide the basis for many topics that are currently emphasized in elementary and secondary school mathematics curricula.

Many early people used strokes to represent numbers:

| || ||| |||| ||||| |||||| ||||||| |||||||| . . .

1   2   3   4   5   6   7   8   . . .

The number of strokes that a person can recognize as a distinct number varies from person to person. Somewhere around 5, 6, 7, or 8 we start seeing "many" strokes instead of a specific number of strokes. Several primitive tribes had number languages that consisted only of

one   two   many

or perhaps

one   two   three   many

The early Egyptians used groupings of strokes so that they could recognize the number of strokes for 1 through 9.

Other patterns were also used, often determined by the writing space available. Since strokes were either present or not present, they could be arranged horizontally, vertically, or in both rows and columns. For example, consider the groupings in the following figure.

|, ||, |||, ||||, $\begin{smallmatrix}||\\|||\end{smallmatrix}$, $\begin{smallmatrix}|||\\|||\end{smallmatrix}$, $\begin{smallmatrix}|||\\||||\end{smallmatrix}$, $\begin{smallmatrix}||||\\||||\end{smallmatrix}$, $\begin{smallmatrix}|||\\|||\\|||\end{smallmatrix}$

Probably because man has ten fingers a new symbol is often introduced for ten. We would write this symbol in our notation as 10. The ancient Egyptians used the symbol ∩. The ancient Babylonians used the symbol <. The Romans used X. All these symbols—and there are others—are merely numerals; that is, they are different ways of representing (providing a name for) the same number.

Let us explore one system, the one used by the ancient Egyptians, in greater detail. They used a new symbol for each power of ten. Here are some of their symbols, a description of the physical objects they are supposed to represent, and the number they represent as expressed in our notation:

Symbol	Object	Number
|	Vertical staff	1
∩	Heel-bone	10
𓍢	Scroll	100
𓆼	Lotus flower	1000
𓂭	Pointing finger	10,000

The scroll is often called a coil of rope and was sometimes "coiled" counterclockwise instead of clockwise.

This Egyptian system is said to have a **base** of ten, but has no place value. The "base" ten is due to the use of powers of ten. We call our system of numeration a **decimal system** to emphasize our use of powers of ten. The absence of a place value means that the position of the symbol does not affect the number represented. For example, in our decimal system of numeration, 23 and 32 represent different numbers. In the Egyptian system ∩| and |∩ are different ways of writing eleven, that is, different names or numerals for the same number. (The former notation is the one normally found in their hieroglyphics.) Here are some other comparisons of decimal and ancient Egyptian number symbols.

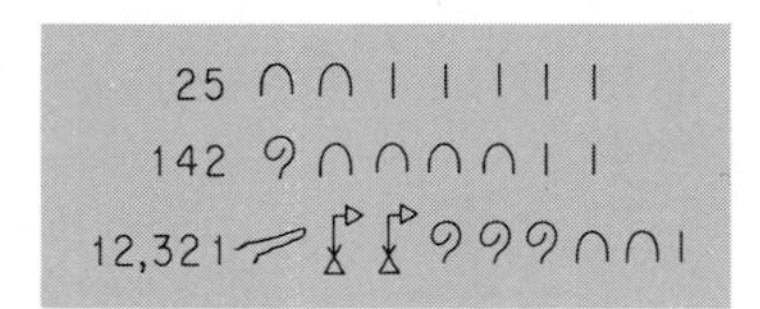

Computation in the ancient Egyptian system is possible, although tedious. For example, the figure at the top of the next page shows the steps used to add 27 and 35.

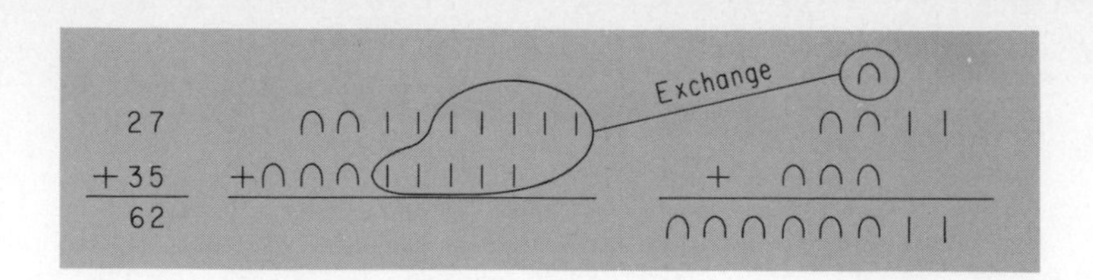

Observe that, in this Egyptian system, an indicated collection of ten ones was replaced by a symbol for ten before the final computation took place. In our decimal system we mentally perform a similar exchange of ten ones for a ten when we express (7 + 5) as one ten and two ones. We exchange kinds of units in a similar manner in subtraction.

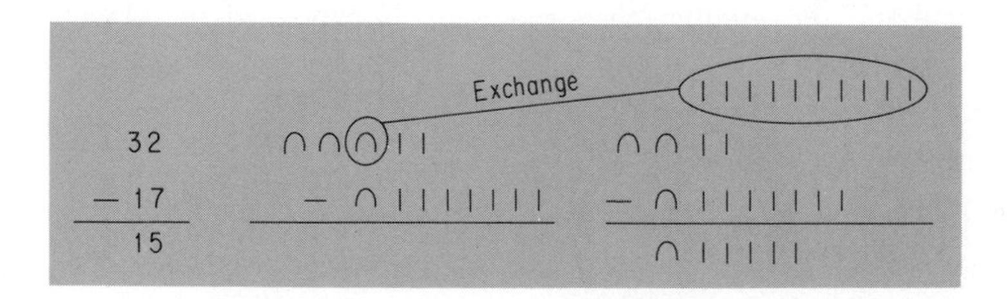

The ancient Egyptians multiplied by a process of doubling. This process is based on the fact that any number may be represented as a sum of powers of two. For example, $19 = 1 + 2 + 16$. To find the product 19×25 we first double 25; we then continue to double and keep a record of the multiples of 25 as follows:

(1)	x	25	=	(25)
(2)	x	25	=	(50)
4	x	25	=	100
8	x	25	=	200
(16)	x	25	=	(400)

Then we find the product 19×25 by adding the multiples of 25 that correspond to 1, 2, and 16:

$$19 = 1 + 2 + 16$$
$$19 \times 25 = (1 + 2 + 16) \times 25$$
$$= 25 + 50 + 400 = 475$$

Example 1: Use the Egyptian method of doubling to find the product 23×41.

Solution:

(1) × 41 = (41)		
(2) × 41 = (82)		$23 = 1 + 2 + 4 + 16$
(4) × 41 = (164)		$23 \times 41 = (1 + 2 + 4 + 16) \times 41$
8 × 41 = 328		$= 41 + 82 + 164 + 656 = 943$
(16) × 41 = (656)		

Later the Egyptians adopted a more refined and automatic procedure for multiplication known as **duplation and mediation** that involves doubling one factor and halving the other. For example, to find the product 19×25 we may successively halve 19, discarding remainders at each step, and successively double 25:

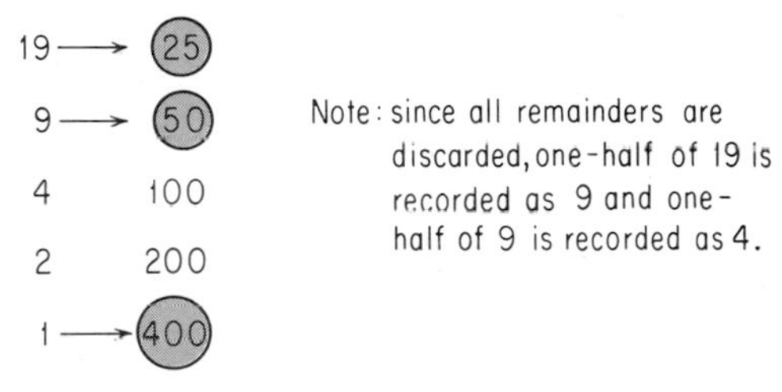

This process is complete when a 1 appears in the column of numbers that are being halved. Opposite each number in this column of halves there is a corresponding number in the column of numbers being doubled. The product 19×25 is found as the sum of the numbers that are opposite the odd numbers in the column of halves:

$$19 \times 25 = 25 + 50 + 400 = 475$$

Note that this process automatically selects the addends to be used in determining the product; one need not search for the appropriate powers of two to be used. Also note that this process drastically reduces the number of multiplication and division facts that must be learned in order to compute.

Example 2: Use the process of duplation and mediation to find the product 23×41.

Solution:

23 → (41)		
11 → (82)		
5 → (164)		$23 \times 41 = 41 + 82 + 164 + 656 = 943$
2 328		
1 → (656)		

EXERCISES

Write in ancient Egyptian notation.

1. 25   **2.** 246   **3.** 3417

4. 60   **5.** 12,407   **6.** 5723

Write in decimal notation.

7. ∩ ∩ | | **8.** 𓍢 𓍢 ∩ |

9. 𓆼 𓍢 | | **10.** 𓆼 𓆼 𓍢 𓍢 ∩ |

11. 𓆼 𓍢 𓍢 𓍢 ∩ ∩ | | | | **12.** 𓆼 𓍢 𓍢 ∩ | |

Write in ancient Egyptian notation and perform the indicated operation in that system.

13. $\begin{array}{r} 83 \\ +32 \\ \hline \end{array}$ **14.** $\begin{array}{r} 153 \\ +62 \\ \hline \end{array}$ **15.** $\begin{array}{r} 238 \\ +135 \\ \hline \end{array}$

16. $\begin{array}{r} 541 \\ -215 \\ \hline \end{array}$ **17.** $\begin{array}{r} 2651 \\ -528 \\ \hline \end{array}$ **18.** $\begin{array}{r} 507 \\ -124 \\ \hline \end{array}$

Use the Egyptian method of doubling to find these products.

19. 17×45 **20.** 15×35

21. 29×41 **22.** 43×29

Use the Egyptian method of duplation and mediation to find these products.

23. 29×44 **24.** 34×61

25. 31×22 **26.** 27×70

PEDAGOGICAL EXPLORATIONS

1. The early Egyptians often used their numerals to form patterns, or pictures. For example, they were able to write 25 in such ways as these:

||∩|∩|| ∩|||||∩ |∩|||∩|

Why can't we do likewise with our numerals? What are the basic advantages of our decimal system of notation that makes it desirable to give up such opportunities for "artistic effect?"

2. Elementary school textbooks vary greatly in their emphasis upon the distinction between numbers and numerals. The following questions may be used to identify some of the difficulties and ways of handling them. For example, a common practice is to use single quotes to refer to a numeral. Thus '2' refers to the numeral rather than the number, whereas 2 without quotes refers to the number.

   (a) Which is larger, 3 or 7 ?

   (b) Which is larger, '3' or '7' ?

   (c) Comment upon this statement: One half of 22 is 11, but one half of '22' is '2'.

3. Roman numerals are often used to state dates of construction of buildings on the cornerstones and in other places as well. Often the pages that precede the introduction to a book are given in Roman numerals. Find as many examples as you can of the use of such numerals, and then develop a short unit on Roman numerals for use at a specified elementary school grade level.
4. Neither the Egyptians nor the Romans had a symbol for zero. Why do we need such a symbol in our system of numeration? Why did they find it unnecessary to invent such a symbol in their systems? (*Hint:* Compare our method for writing 7, 70, and 700 with the Egyptian and Roman numerals for these numbers.)
5. Select an elementary school grade level and for students in that grade describe another early (before 400 A.D.) system of numeration.

3-2 early methods of computation

There exist numerous examples of the ways in which early civilizations dealt with computations. Some of these may prove of interest to the reader. The method of multiplication that appeared in one of the first published arithmetic texts in Italy, the *Treviso Arithmetic* (1478), is an interesting one. Although evidence exists of the use of this method by the early Hindus and Chinese, it later became widely used by the Arabians who passed it on to the Europeans during the Middle Ages. We shall refer to this process here as "galley" multiplication, although it was called "Gelosia" multiplication in the original text. Let us use the process to find the product of 457 and 382.

First prepare a "galley" with three rows and three columns, and draw the diagonals as in the figure. Our choice for the number of rows and columns is based on the fact that we are to multiply two three-digit numbers.

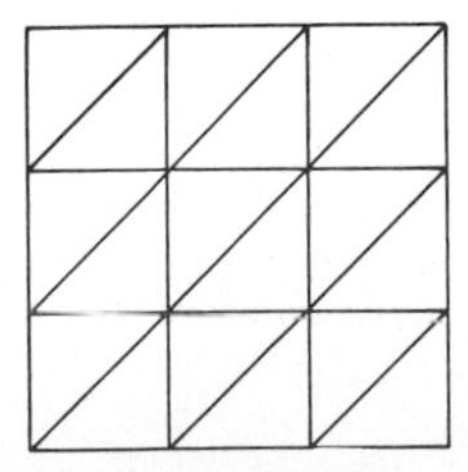

Place the digits 4, 5, and 7 in order from left to right at the top of the columns. Place the digits 3, 8, and 2 in order from top to bottom at the right of the rows. Then each product of a digit of 457 and a digit of 382 is called a **partial product** and is placed at the intersection of the column and row of the digits. The diagonal separates the digits of the partial product (tens digit above units digit). For example, $3 \times 7 = 21$, and this partial product is placed in the upper right-hand corner of the galley; $5 \times 8 = 40$, and this partial product is placed in the center of the galley; $4 \times 2 = 8$, and this partial product is entered as 08 in the lower left-hand corner of the galley. See if you can justify each of the entries in the completed array shown below.

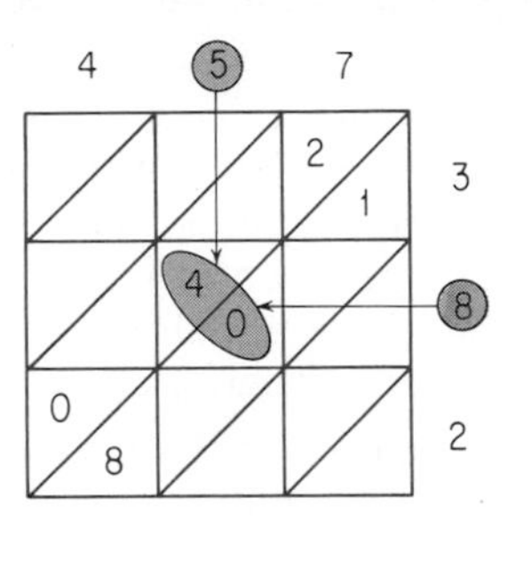

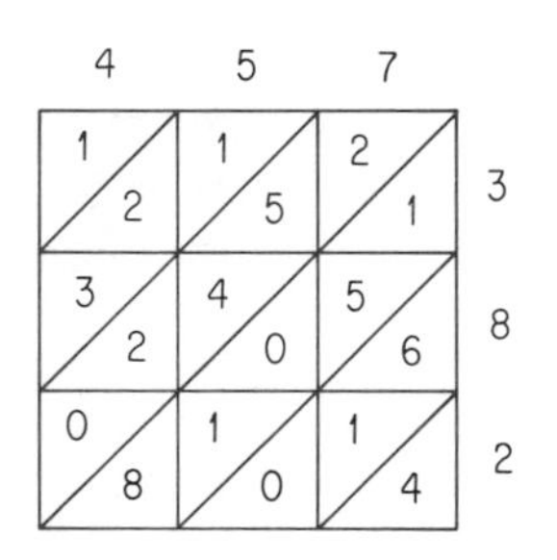

After all partial products have been entered in the galley, we add along diagonals, starting in the lower right-hand corner and carrying to the next diagonal sum where necessary. The figure on the left indicates this pattern. The completed problem appears in the figure on the right. We read the final answer, as indicated by the arrow in the figure, as 174,574. Note that we read the digits in the opposite order to that in which they were obtained.

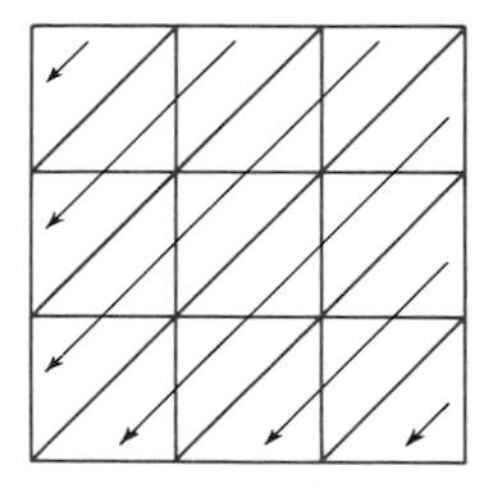

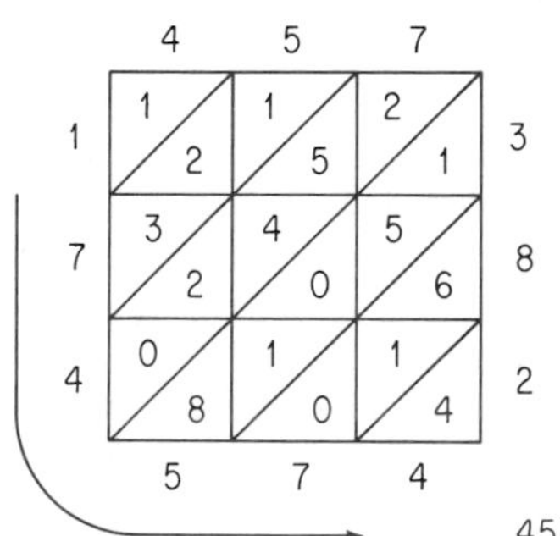

$457 \times 382 = 174{,}574$

Example: Use galley multiplication and multiply 372 by 47.

Solution:

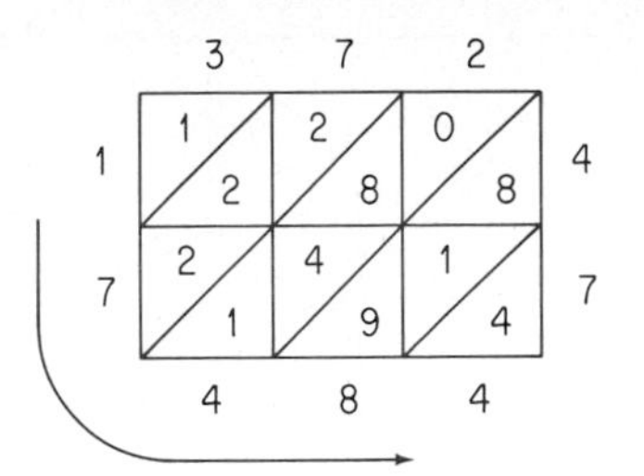

This procedure works because we are really listing all partial products before we add. Compare the following two computations.

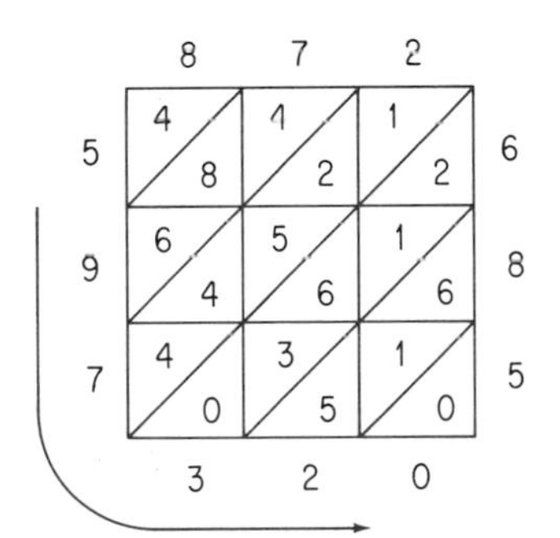

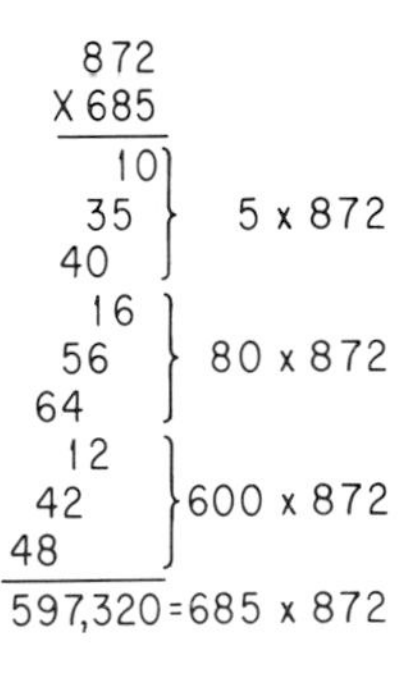

Note that the numerals along the diagonals correspond to those in the columns at the right.

The Scottish mathematician, John Napier (1550–1617), made use of this system as he developed what proved to be one of the forerunners of the modern computing machines. His device is referred to as **Napier's rods,** or **Napier's bones,** named after the material on which he had numerals printed.

To make a set of these rods we need to prepare a collection of strips of paper, or other material, with multiples of each of the digits listed. Study the set of rods shown in part (A) of the figure at the top of the next page.

Note, for example, that the rod headed by the numeral 9 lists the multiples of 9:

9, 18, 27, 36, 45, 54, 63, 72, and 81

We can use these rods to multiply two numbers. To multiply 7×483, place the rods headed by numerals 4, 8, and 3 alongside the index as shown in part (B) of the same figure.

Index	0	1	2	3	4	5	6	7	8	9
1	0/0	0/1	0/2	0/3	0/4	0/5	0/6	0/7	0/8	0/9
2	0/0	0/2	0/4	0/6	0/8	1/0	1/2	1/4	1/6	1/8
3	0/0	0/3	0/6	0/9	1/2	1/5	1/8	2/1	2/4	2/7
4	0/0	0/4	0/8	1/2	1/6	2/0	2/4	2/8	3/2	3/6
5	0/0	0/5	1/0	1/5	2/0	2/5	3/0	3/5	4/0	4/5
6	0/0	0/6	1/2	1/8	2/4	3/0	3/6	4/2	4/8	5/4
7	0/0	0/7	1/4	2/1	2/8	3/5	4/2	4/9	5/6	6/3
8	0/0	0/8	1/6	2/4	3/2	4/0	4/8	5/6	6/4	7/2
9	0/0	0/9	1/8	2/7	3/6	4/5	5/4	6/3	7/2	8/1

(A)

Index	4	8	3
1	0/4	0/8	0/3
2	0/8	1/6	0/6
3	1/2	2/4	0/9
4	1/6	3/2	1/2
5	2/0	4/0	1/5
6	2/4	4/8	1/8
7	2/8	5/6	2/1
8	3/2	6/4	2/4
9	3/6	7/2	2/7

(B)

Consider the row of numerals alongside 7 on the index.

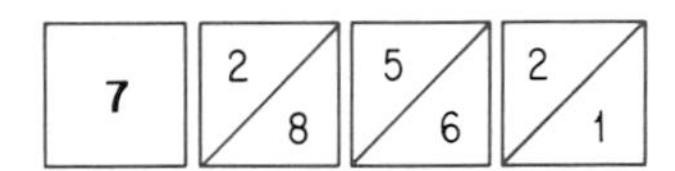

Add along the diagonals, as in "galley multiplication."

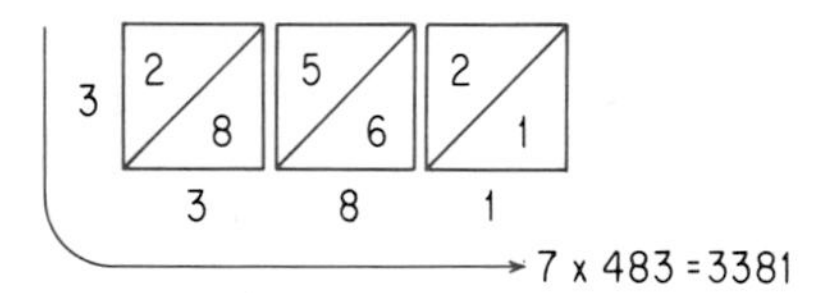

The same arrangement of rods may be used to read immediately the product of 483 and any other one-digit number. With practice one can develop skill in using these rods for rapid computation. For example, $8 \times 483 = 3864$ as follows:

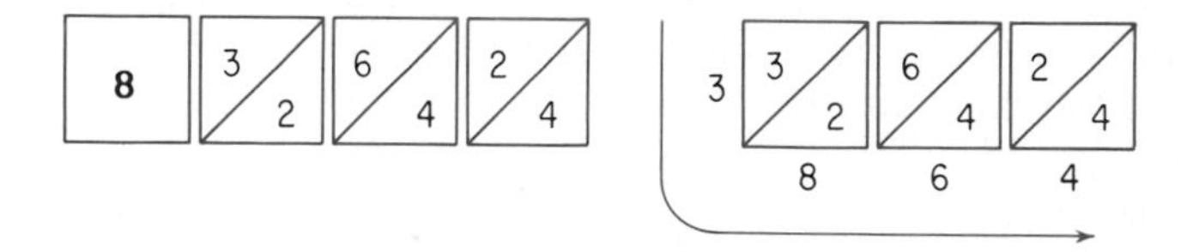

Combining the two previous results, we can find the product 87×483:

$$\begin{array}{r} 483 \\ \times 87 \\ \hline \end{array} \qquad \begin{array}{rcr} 7 \times 483 &=& 3{,}381 \\ 80 \times 483 &=& 38{,}640 \\ \hline 87 \times 483 &=& 42{,}021 \end{array}$$

Note that we are able to read only products with a one-digit multiplier directly from the rods.

EXERCISES

Each diagram shows a use of Napier's rods. State each product in terms of the numbers that are multiplied.

1. 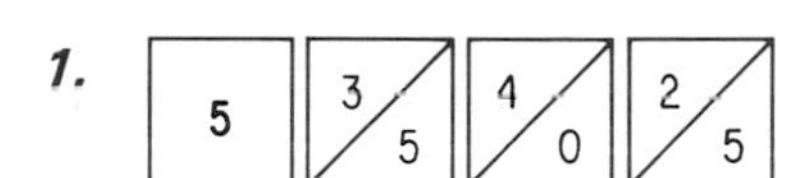

2. 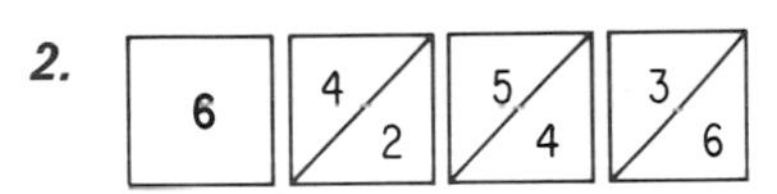

3. 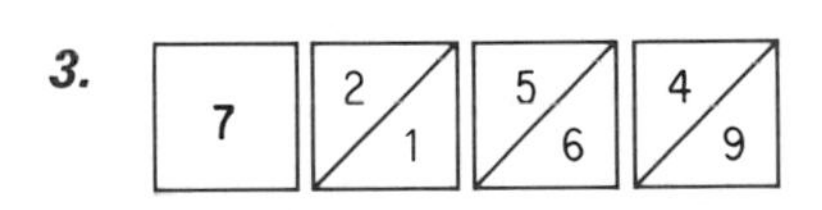

4. 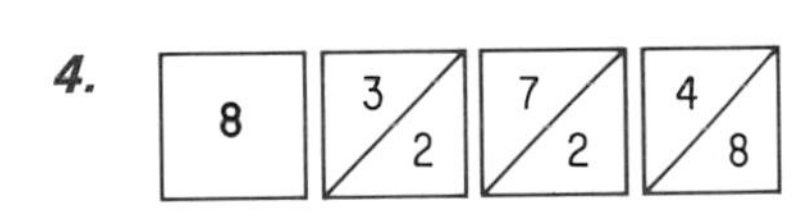

5. 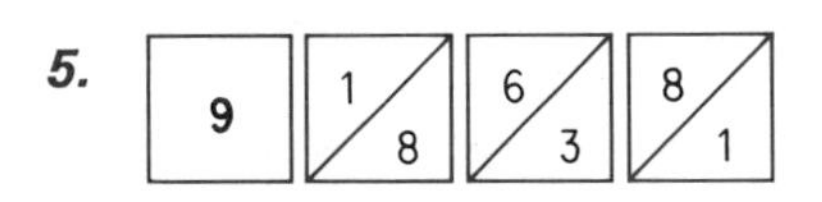

6. 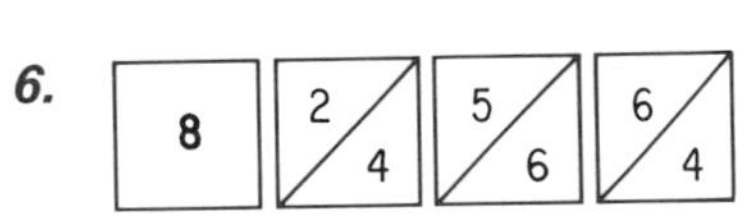

Multiply, using the "galley" method.

7. $\begin{array}{r} 942 \\ \times 73 \\ \hline \end{array}$ **8.** $\begin{array}{r} 568 \\ \times 429 \\ \hline \end{array}$ **9.** $\begin{array}{r} 432 \\ \times 276 \\ \hline \end{array}$

10. $\begin{array}{r} 8764 \\ \times 37 \\ \hline \end{array}$ **11.** $\begin{array}{r} 8035 \\ \times 289 \\ \hline \end{array}$ **12.** $\begin{array}{r} 5081 \\ \times 2376 \\ \hline \end{array}$

Construct a set of Napier's rods and use them to find each of the following products.

13. $\begin{array}{r} 365 \\ \times 7 \\ \hline \end{array}$ **14.** $\begin{array}{r} 427 \\ \times 9 \\ \hline \end{array}$ **15.** $\begin{array}{r} 387 \\ \times 5 \\ \hline \end{array}$

16. $\begin{array}{r} 259 \\ \times 8 \\ \hline \end{array}$ **17.** $\begin{array}{r} 3765 \\ \times 6 \\ \hline \end{array}$ **18.** $\begin{array}{r} 427 \\ \times 36 \\ \hline \end{array}$

PEDAGOGICAL EXPLORATIONS

1. Write up a lesson plan on galley multiplication for use in an elementary school class (specify the grade that you are considering).
2. Repeat Exploration 1 for multiplication using Napier's rods.
3. Prepare a demonstration set of Napier's rods for use in front of a class or prepare a set for use on an overhead projector.
4. Find the product 39 × 756 by using the "galley" method and by using Napier's rods. Which process is easier to use? Which process do you think would be the better to use in an elementary mathematics class in order to show the comparison with our usual multiplication process?

3-3 clock arithmetic

A **mathematical system** involves a set of elements (such as the set of counting numbers), one or more operations (such as addition, subtraction, multiplication, and division), one or more relations (such as equality), and some axioms (rules) which the elements, operations, and relations satisfy. For example, we assume that every system involves the relation of equality and the axiom that $a = a$; that is, any quantity is equal to itself. Actually, you have been working with mathematical systems ever since you started school. We shall now study a specific mathematical system, and later will relate the properties of this system to those of familiar sets of numbers.

If it is now 9 P.M. as you begin to read this section, what time will it be in 5 hours? (We hope it won't take you that long to complete your reading!) Do you see that the statement

$$9 + 5 = 2$$

is a correct statement if we are talking about positions on a 12-hour clock? It represents one of the fundamental addition facts of **12-hour clock arithmetic.**

Let us consider the numerals 1 through 12 on a 12-hour clock as the elements of a set T and consider addition on this clock to be based upon counting in a clockwise direction. Thus to find the sum $9 + 5$ we start at 9 and count 5 units in a clockwise direction to obtain the result 2.

Verify that each of the following is correct:

$$8 + 7 = 3 \quad \text{(on a 12-hour clock)}$$
$$5 + 12 = 5 \quad \text{(on a 12-hour clock)}$$
$$3 + 11 = 2 \quad \text{(on a 12-hour clock)}$$

We may make a table of addition facts on a 12-hour clock as shown at the top of the next page.

+	1	2	3	4	5	6	7	8	9	10	11	12
1	2	3	4	5	6	7	8	9	10	11	12	1
2	3	4	5	6	7	8	9	10	11	12	1	2
3	4	5	6	7	8	9	10	11	12	1	2	3
4	5	6	7	8	9	10	11	12	1	2	3	4
5	6	7	8	9	10	11	12	1	2	3	4	5
6	7	8	9	10	11	12	1	2	3	4	5	6
7	8	9	10	11	12	1	2	3	4	5	6	7
8	9	10	11	12	1	2	3	4	5	6	7	8
9	10	11	12	1	2	3	4	5	6	7	8	9
10	11	12	1	2	3	4	5	6	7	8	9	10
11	12	1	2	3	4	5	6	7	8	9	10	11
12	1	2	3	4	5	6	7	8	9	10	11	12

Note that regardless of where we start on the clock, we shall always be at the same place 12 hours later. Thus for any element t of the set T we have

$$t + 12 = t \qquad \text{(on a 12-hour clock).}$$

Let us attempt to define several other operations for this arithmetic on a 12-hour clock. What does multiplication mean? Multiplication by an integer may be considered as repeated addition. For example, 3×5 on a 12-hour clock is equivalent to $5 + 5 + 5$. Since $5 + 5 = 10$ and $10 + 5 = 3$, we know that $3 \times 5 = 3$ on a 12-hour clock.
Verify that each of the following is correct:

$$4 \times 5 = 8 \qquad \text{(on a 12-hour clock)}$$
$$3 \times 9 = 3 \qquad \text{(on a 12-hour clock)}$$
$$3 \times 7 = 9 \qquad \text{(on a 12-hour clock)}$$

The following examples provide further illustrations of clock arithmetic.

Example 1: Solve the equation $t + 6 = 2$ for t, where t may be replaced by any one of the numerals on a 12-hour clock.

Solution: Observe from row 6 of the table of addition facts that on a 12-hour clock $8 + 6 = 2$; therefore $t = 8$. Note that we may also solve for t by starting at 6 on the clock and then count hours in a clockwise direction until we reach 2. The number of hours counted is the desired value of t since if $6 + t = 2$, then $t + 6 = 2$.

Example 2: Using the numerals on a 12-hour clock, find a replacement for t such that $9 \div 7 = t$, i.e., $\frac{9}{7} = t$.

Solution: We can think of the statement $\frac{9}{7} = t$ as one of the four equivalent statements:

$$9 = 7 \times t, \qquad 9 = t \times 7$$

$$\frac{9}{7} = t, \qquad \frac{9}{t} = 7$$

Then we use the sentence $9 = 7 \times t$ to solve the problem. How many groups of 7 must be added to obtain 9 on a 12-hour clock? We can count by 7's clockwise from 12 as many times as necessary to obtain 9.

$7 \times 1 = 7$

$7 \times 2 = 14 = 2$ (on a 12-hour clock)

$7 \times 3 = 21 = 9$ (on a 12-hour clock)

We can also use row 7 of a multiplication table, or trial and error, to find that $7 \times 3 = 9$ (on a 12-hour clock); thus $t = 3$.

Example 3: What is $3 - 7$ on a 12-hour clock?

Solution: Let t represent a numeral on a 12-hour clock. The statement "$3 - 7 = t$" is equivalent to "$t + 7 = 3$." From the table of addition facts $8 + 7 = 3$ (on a 12-hour clock); thus $t = 8$.

From a slightly different point of view we may solve Example 3 by counting in a clockwise direction from 12 to 3, and then counting 7 units in a counterclockwise direction. We complete this process at 8. Thus $3 - 7 = 8$ (on a 12-hour clock).

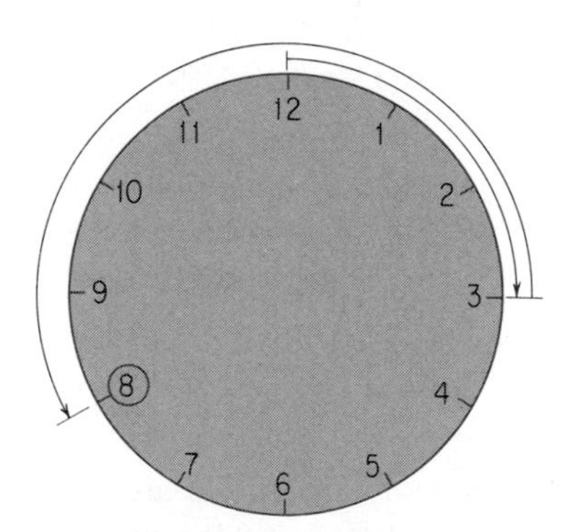

Note that clock arithmetic may be considered as an arithmetic of remainders. Multiples of 12 are discarded so that each answer may be expressed as a number 1 through 12. For example, consider the product $5 \times 8 = 40$:

$$5 \times 8 = 40 = (3 \times 12) + 4$$

In clock arithmetic (3×12) represents three rotations and is disregarded; this is analogous to dividing 40 by 12 and considering the remainder only.

$$5 \times 8 = 4 \quad \text{(on a 12-hour clock)}$$

EXERCISES

Solve each problem as on a 12-hour clock.

1. $9 + 8$	**2.** $7 + 11$	**3.** $7 - 12$
4. $6 - 9$	**5.** 4×9	**6.** 7×7
7. $11 \div 5$	**8.** $1 \div 7$	**9.** $8 + 8$
10. $6 + 10$	**11.** 3×8	**12.** 9×9
13. $2 - 10$	**14.** $5 - 11$	**15.** $4 \div 7$
16. $4 \div 5$	**17.** $12 \div 5$	***18.** $6 \div 10$

Find all possible replacements for t for which each sentence is a true statement for the numerals on a 12-hour clock.

19. $t + 8 = 5$	**20.** $t - 5 = 11$	**21.** $8 + t = 3$
22. $4 - t = 10$	**23.** $3 \times t = 3$	**24.** $7 \times t = 11$
25. $\frac{t}{5} = 7$	**26.** $\frac{t}{7} = 8$	***27.** $\frac{2}{t} = 3$
***28.** $2 + t = 2 - t$	***29.** $t + 12 = t$	***30.** $3 - t = 5 + t$

31. Make a complete table of multiplication facts for the numbers on a 12-hour clock.

32. Use the results of Exercise 31 to list as many interesting patterns as you can discover about multiplication on a 12-hour clock.

PEDAGOGICAL EXPLORATIONS

1. Elementary school students are often taught to consider number facts by families of related facts. For example, here is a family of related number facts:

$$5 - 2 = 3 \qquad 5 = 2 + 3 \qquad 5 = 3 + 2 \qquad 5 - 3 = 2$$

Here is another family of related number facts:

$$6 \div 2 = 3 \qquad 6 = 2 \times 3 \qquad 6 = 3 \times 2 \qquad 6 \div 3 = 2$$

Any two members of a family of related number facts are equivalent statements. Any statement may be used to "solve" or to "check" another statement in the same family.

Use families of number facts to explain your answers to Exercises 7 and 14 of this section.

2. Refer to a general textbook on methods of teaching in elementary schools, and/or speak with several elementary school teachers, and list some of the necessary ingredients for a successful lesson plan. For example, some educators claim that the four essential aspects of a lesson plan include objectives, materials, methodology, and evaluation of student learning. If possible, obtain a sample lesson plan from an elementary school teacher.
3. Use the information gathered in Exploration 2 and prepare a 20-minute lesson plan that introduces a unit on clock arithmetic at a specified elementary grade level. Make use of at least one visual aid in your lesson. If possible, present this lesson to a class and report on your results.

3-4 modular arithmetic

Let us next consider a mathematical system based on a clock for five hours, numbered 0, 1, 2, 3, and 4 as in the figure below.

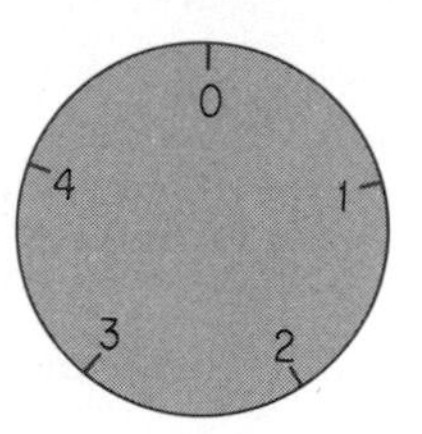

Addition on this clock may be performed by counting as on an ordinary clock. However, it seems easier to think of addition as rotations in a clockwise direction. Thus $3 + 4$ indicates that we are to start at 0 and move 3 units, then 4 more. The result is 2.

We interpret 0 to mean no rotation as well as to designate a position on the clock. Then we have such facts as:

$$3 + 0 = 3 \quad \text{(on a five-hour clock)}$$
$$4 + 1 = 0 \quad \text{(on a five-hour clock)}$$

Verify that the following table of addition facts for the numbers on a five-hour clock is correct:

+	0	1	2	3	4
0	0	1	2	3	4
1	1	2	3	4	0
2	2	3	4	0	1
3	3	4	0	1	2
4	4	0	1	2	3

Multiplication on a five-hour clock is defined as repeated addition, as in §3-3. For example, 3×4 on a five-hour clock is equivalent to $4 + 4 + 4$. Since $4 + 4 = 3$, and $3 + 4 = 2$, we see that $3 \times 4 = 2$ on a five-hour clock.

We define subtraction and division using equivalent statements as follows:

Subtraction: Since the statements "$a - b = x$" and "$a = b + x$" are equivalent, we define $a - b = x$ *if and only if* $b + x = a$.

Division: Since the statements "$a \div b = x$" and "$a = b \times x$" are equivalent, we define $a \div b = x$ *if and only if* $b \times x = a$.

Regardless of where we start on this five-hour clock, we shall always be at the same place five hours later. Usually the symbols of a mathematical system based on clock arithmetic begin with 0, and the system is called a **modular arithmetic.** In the illustration of this section we say that we have an arithmetic modulo 5, and we use the symbols 0, 1, 2, 3, 4. Specific facts in this mathematical system are written as follows:

$3 + 4 \equiv 2 \pmod 5$, read "$3 + 4$ is congruent to 2, modulo 5"
$4 \times 2 \equiv 3 \pmod 5$, read "$4 \times 2$ is congruent to 3, modulo 5"

In general, two numbers are **congruent modulo 5** if and only if they differ by a multiple of 5. Thus 3, 8, 13, and 18 are all congruent to each other modulo 5. We may write, for example

$$18 \equiv 13 \pmod 5$$
$$8 \equiv 3 \pmod 5$$

If any whole number is divided by 5, then the remainder must be 0, 1, 2, 3, or 4. Thus each whole number is congruent modulo 5 to exactly one element of the set F, where $F = \{0, 1, 2, 3, 4\}$. The elements of F are the elements of arithmetic modulo 5.

Example 1: Solve for x where x may be replaced by any element in arithmetic modulo 5: $2 - 3 = x$.

Solution: The statement "$2 - 3 = x$" is equivalent to "$3 + x = 2$." Since $3 + 4 = 2$, $x = 4$.

Example 2: Solve for x in arithmetic modulo 5: $\frac{2}{3} = x$.

Solution: Note that the statement "$\frac{2}{3} = x$" is equivalent to "$3 \times x = 2$." Since $3 \times 4 = 2$, $x = 4$.

Let us explore the properties of the mathematical system based upon the set $F = \{0, 1, 2, 3, 4\}$ and the operation addition as defined in this section. This operation is a **binary operation;** it associates a single element with any *two* elements of the given set. The following properties are of special interest.

1. *The set F of elements in arithmetic modulo 5 is closed with respect to addition.* That is, for any pair of elements, there is a unique element which represents their sum and which is also a member of the original set. We note, for example, that there is one and only one entry in each place of the table given in this section, and that each entry is an element of the set F.

2. *The set F of elements in arithmetic modulo 5 satisfies the commutative property for addition.* That is,

$$a + b = b + a$$

where a and b are any elements of the set F. Specifically, we see that $3 + 4 = 4 + 3$, $2 + 3 = 3 + 2$, etc.

3. *The set F of elements in arithmetic modulo 5 satisfies the associative property for addition.* That is,

$$(a + b) + c = a + (b + c)$$

for all elements a, b, and c of the set F.

As a specific example we evaluate $3 + 4 + 2$ in two ways:

$$(3 + 4) + 2 \equiv 2 + 2 \text{ or } 4, \text{ modulo } 5$$
$$3 + (4 + 2) \equiv 3 + 1 \text{ or } 4, \text{ modulo } 5$$

4. *The set F of elements in arithmetic modulo 5 includes an identity element for addition.* That is, the set contains an element 0 such that the sum of any given element and 0 is the given element.

$$0 + 0 = 0, \quad 1 + 0 = 1, \quad 2 + 0 = 2$$
$$3 + 0 = 3, \quad 4 + 0 = 4$$

5. *Each element in arithmetic modulo 5 has an inverse with respect to addition.* That is, for each element a of set F there exists an element a' of F such that $a + a' = 0$, the identity element. The element a' is said to be the inverse of a. Specifically, we have the following:

The inverse of 0 is 0; $0 + 0 \equiv 0 \pmod 5$.
The inverse of 1 is 4; $1 + 4 \equiv 0 \pmod 5$.
The inverse of 2 is 3; $2 + 3 \equiv 0 \pmod 5$.
The inverse of 3 is 2; $3 + 2 \equiv 0 \pmod 5$.
The inverse of 4 is 1; $4 + 1 \equiv 0 \pmod 5$.

We may summarize these five properties by saying that the set F of elements in arithmetic modulo 5 forms a **commutative group under addition.**

EXERCISES

Each of the following is based upon the set of elements in arithmetic modulo 5.

1. Make a complete table of multiplication facts modulo 5.

2. Verify that the commutative property for multiplication holds for at least two specific instances.
3. Verify that the associative property for multiplication holds for at least two specific instances.
4. What is the identity element with respect to multiplication?
5. Find the inverse of each element with respect to multiplication.
6. Verify for at least two specific instances that in arithmetic modulo 5

$$a(b + c) = ab + ac$$

Solve each of the following for x.

7. $3 + x \equiv 1 \pmod 5$
8. $x + 4 \equiv 1 \pmod 5$
9. $x + 3 \equiv 2 \pmod 5$
10. $4 + x \equiv 3 \pmod 5$
11. $2 - 3 \equiv x \pmod 5$
12. $2 - 4 \equiv x \pmod 5$
13. $1 - 4 \equiv x \pmod 5$
14. $1 - x \equiv 4 \pmod 5$
15. $2 \times x \equiv 1 \pmod 5$
16. $4 \times x \equiv 3 \pmod 5$
17. $4 \times x \equiv 1 \pmod 5$
18. $3 \times x \equiv 1 \pmod 5$
19. $\frac{3}{x} \equiv 4 \pmod 5$
20. $\frac{2}{x} \equiv 4 \pmod 5$
21. $\frac{1}{2} \equiv x \pmod 5$
22. $\frac{4}{3} \equiv x \pmod 5$
23. $x + 3 \equiv x \pmod 5$
24. $x + 1 \equiv 3 - x \pmod 5$

*25. We describe the property illustrated by the equation $3 \times 4 \equiv 0 \pmod{12}$, where the product is zero but neither number is zero, by saying that 3 and 4 are zero divisors. Are there other zero divisors in arithmetic modulo 12? If so, list them.

26. Consider a modulo 7 system where the days of the week correspond to numbers as follows: Monday—0; Tuesday—1; Wednesday—2; Thursday—3; Friday—4; Saturday—5; Sunday—6. Memorial Day, May 30, is the 150th day of the year and falls on a Thursday. In that same year, on what day of the week does July 4, the 185th day of the year, fall? On what day does Christmas, the 359th day of the year fall?

PEDAGOGICAL EXPLORATIONS

Operations in which each input consists of a single number are **unary operations.** The following unary operations are indicated with an arrow to show the correspondence or mapping of the given number (input) onto the new number (output). Each operation in the exercises at the top of the next page has been given a name to aid class discussions.

1. This is how we "star" a number.
$6^* \longrightarrow 8 \quad 7^* \longrightarrow 9 \quad 0^* \longrightarrow 2$

Can you "star" these numbers?
$5^* \longrightarrow ? \quad 9^* \longrightarrow ? \quad 12^* \longrightarrow ?$

2. This is how we "quote" a number.
"5" ⟶ 26 "3" ⟶ 10 "7" ⟶ 50

Can you "quote" these numbers?
"6" ⟶ ? "10" ⟶ ? "9" ⟶ ?

3. This is how we "bar" a number.
$\overline{2} \longrightarrow 7 \quad \overline{5} \longrightarrow 16 \quad \overline{11} \longrightarrow 34$

Can you "bar" these numbers?
$\overline{4} \longrightarrow ? \quad \overline{6} \longrightarrow ? \quad \overline{12} \longrightarrow ?$

Arrows are also often used for binary operations. Try to discover the meaning of the operation # from the examples given. The binary operation # has a different meaning in each exercise.

4. $3 \# 4 \longrightarrow 7,\quad 2 \# 0 \longrightarrow 2,\quad 5 \# 1 \longrightarrow 6,\quad 2 \# 2 \longrightarrow 4.$
5. $2 \# 4 \longrightarrow 3,\quad 1 \# 7 \longrightarrow 4,\quad 8 \# 10 \longrightarrow 9,\quad 0 \# 2 \longrightarrow 1.$
6. $5 \# 3 \longrightarrow 9,\quad 2 \# 4 \longrightarrow 7,\quad 0 \# 3 \longrightarrow 4,\quad 1 \# 5 \longrightarrow 7.$
7. $2 \# 3 \longrightarrow 7,\quad 3 \# 4 \longrightarrow 13,\quad 1 \# 5 \longrightarrow 6,\quad 0 \# 3 \longrightarrow 1.$

*8. $3 \# 4 \longrightarrow 2,\quad 4 \# 3 \longrightarrow 5,\quad 5 \# 1 \longrightarrow 9,\quad 1 \# 1 \longrightarrow 1.$

*9. $3 \# 5 \longrightarrow 2,\quad 4 \# 5 \longrightarrow 1,\quad 2 \# 8 \longrightarrow 0,\quad 1 \# 6 \longrightarrow 3.$

10. Consider a soldier facing in a given direction. He is then given various commands such as "right face," "left face," "about face," and "as you were." The last command tells him to retain whatever position he may be in at the time.

A few specific examples of his movements should clarify matters. If he is facing to the north and is given the command "about face," then in his new position he is facing to the south. If we let R represent the command "right face," A represent the command "about face," and L represent the command "left face," then R followed by A is equivalent to the single command L.

Use the symbol $\circ$ to represent the operation "followed by" and verify that each of the following is correct:

$$L \circ L = A$$
$$L \circ A = R$$
$$A \circ R = L$$

Now suppose that our soldier makes a left turn and then we wish him to remain in that position. We make use of the command "as you were," represented by E. Thus $L \circ E = L$; also $E \circ L = L$.

(a) Make a table summarizing all possible movements, using the headings E, R, L, and A.
(b) Find $R \circ R$, $A \circ A$, $L \circ L$.
(c) Does $R \circ L = L \circ R$?
(d) Does $R \circ (L \circ A) = (R \circ L) \circ A$?
(e) Does the set of commands appear to be commutative and associative with respect to the operation $\circ$?

(f) Find the identity element with respect to $\circ$?
(g) List each element of the set with its inverse with respect to $\circ$.
(h) Is the set of commands closed with respect to $\circ$?

3-5 decimal notation and divisibility

We now turn our attention to the **decimal system of numeration.** It is a decimal system in that it is based on powers or groups of tens. Furthermore, it has place value in that the value of any digit used depends upon the position which it occupies. Thus, the two numerals 4 in 484 have quite different values.

To illustrate this latter concept we will write this number, 484, in what is known as **expanded notation:**

$$\begin{aligned} 484 &= 4 \text{ hundreds} + 8 \text{ tens} + 4 \text{ ones} \\ &= (4 \times 100) + (8 \times 10) + (4 \times 1) \end{aligned}$$

Note that one of the numerals 4 represents 4 hundreds, whereas the other 4 represents 4 units, that is, 4 ones.

It is convenient to use exponents when one is writing a number in expanded notation. An **exponent** is a number that tells how many times another number, called the **base,** is used as a factor in a product. For example, in the expression 7^2, the numeral 2 is the exponent and 7 is the base. Note that $7^2 = 7 \times 7 = 49$; also $7^3 = 7 \times 7 \times 7 = 343$.

Using exponents to write 484 in expanded notation, we have

$$\begin{aligned} 484 &= (4 \times 100) + (8 \times 10) + (4 \times 1) \\ &= (4 \times 10^2) + (8 \times 10) + (4 \times 1) \end{aligned}$$

Zero plays an important role in place value notation. Indeed, it is the use of the symbol 0 that allows us to write numbers as large as we wish in decimal notation, using only its ten digits. Thus we may write 10, 100, 1000, etc., using only the digits 0 and 1, whereas in the Egyptian system each new power of 10 required a new symbol. In Egyptian numeration, ∩ represented 10, 𓍢 represented 100, 𓆼 represented 1000, etc.

Example: Write 2306 in expanded notation.

Solution:

$$2306 = (2 \times 10^3) + (3 \times 10^2) + (0 \times 10) + (6 \times 1)$$

We may use the exponent 1 to indicate that a number is to be used as a factor only once; thus $10^1 = 10$. We define $10^0 = 1$. Using these exponents, we have

$$2306 = (2 \times 10^3) + (3 \times 10^2) + (0 \times 10^1) + (6 \times 10^0)$$

Either displayed form may be considered as in expanded notation.

Any whole number N less than 10,000 can be expressed in the form

$$N = 1000T + 100h + 10t + u$$

where T, h, t, and u are elements, not necessarily distinct, of the set {0, 1, 2, 3, 4, 5, 6, 7, 8, 9}. Then

$$\frac{N}{2} = \frac{1000T}{2} + \frac{100h}{2} + \frac{10t}{2} + \frac{u}{2} = 500T + 50h + 5t + \frac{u}{2}$$

Thus $N/2$ is a whole number if and only if $u/2$ is a whole number; that is, N is divisible by 2 if and only if u is divisible by 2.

$$\frac{N}{3} = \frac{(999 + 1)T}{3} + \frac{(99 + 1)h}{3} + \frac{(9 + 1)t}{3} + \frac{u}{3}$$

$$= 333T + 33h + 3t + \frac{T + h + t + u}{3}$$

Thus $N/3$ is a whole number if and only if $(T + h + t + u)/3$ is a whole number; that is, N is divisible by 3 if and only if the sum of its decimal digits is divisible by 3.

$$\frac{N}{4} = \frac{1000T}{4} + \frac{100h}{4} + \frac{10t}{4} + \frac{u}{4} = 250T + 25h + \frac{10t + u}{4}$$

Thus $N/4$ is a whole number if and only if $(10t + u)/4$ is a whole number; that is, N is divisible by 4 if and only if the number represented by its tens and units digits is divisible by 4.

According to these rules 1976 is divisible by 2 since its units digit 6 is divisible by 2; 1976 is not divisible by 3 since the sum of its digits is 23 and 23 is not divisible by 3; 1976 is divisible by 4 since 76 is divisible by 4. In the case of divisibility by 3 the process may be continued as follows: 1976 is not divisible by 3 since the sum of its digits is 23, the sum of the digits of 23 is 5, and 5 is not divisible by 3.

Other explanations of divisibility rules are considered in the exercises.

EXERCISES

Write in expanded notation.

1. 257 ***2.*** 372 ***3.*** 3504 ***4.*** 5240

5. 235,100 ***6.*** 304,065 ***7.*** 500,200 ***8.*** 100,090

Write in decimal notation.

9. $(8 \times 10^3) + (1 \times 10^2) + (6 \times 10^1) + (5 \times 10^0)$

10. $(4 \times 10^5) + (3 \times 10^4) + (0 \times 10^3) + (4 \times 10^2)$
$+ (2 \times 10^1) + (8 \times 10^0)$

11. $(6 \times 10^5) + (9 \times 10^3) + (5 \times 10^2) + (2 \times 10^0)$

12. $(8 \times 10^6) + (6 \times 10^5) + (4 \times 10^4)$ **13.** $(3 \times 10^6) + (2 \times 10^0)$

14. (6×10^7) **15.** (8×10^9)

16. (7×10^8)

Test each number for divisibility by **(a)** 2, **(b)** 3, **(c)** 4, **(d)** 5, **(e)** 6, **(f)** 8, **(g)** 9, **(h)** 25, **(i)** 100.

17. 5280 **18.** 225 **19.** 1728 **20.** 16,275 **21.** 17,540

22. 19,678 **23.** 36,000 **24.** 27,600 **25.** 45,460

For decimal notation try to find a rule for divisibility by:

26. 5 **27.** 6 **28.** 8

29. 9 **30.** 25 **31.** 100

PEDAGOGICAL EXPLORATIONS

1. Construct a table of the counting numbers through 100, suitable for a bulletin board display. Use underlines in different colors to represent divisibility by specific counting numbers. For example, you might use a blue underline to show that a number is divisible by 2, a red underline for divisibility by 3, and so on. Then the counting number 6 would have both a blue and a red underline. Indicate the divisibility of each number in the table by the numbers 2, 3, 4, 5, 6, and 7.

For numbers expressed in decimal notation, construct a flow chart for testing divisibility by:

2. 2 **3.** 3 **4.** 4 **5.** 5

6. 6 **7.** 8 **8.** 10 **9.** 15

3-6 other systems of numeration

Let us now turn our attention to various systems of numeration. In our decimal system objects are grouped and counted in tens and powers of ten. For example, the diagram shows how one might group and count 134 items.

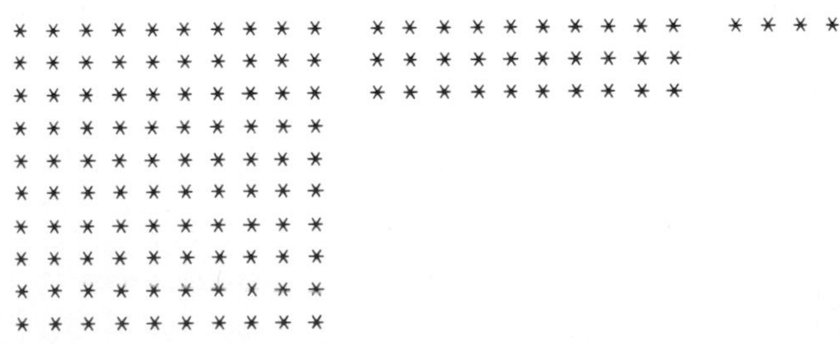

$134 = (1 \times 10^2) + (3 \times 10) + (4 \times 1)$

In the next figure we see 23 asterisks grouped in three different ways.

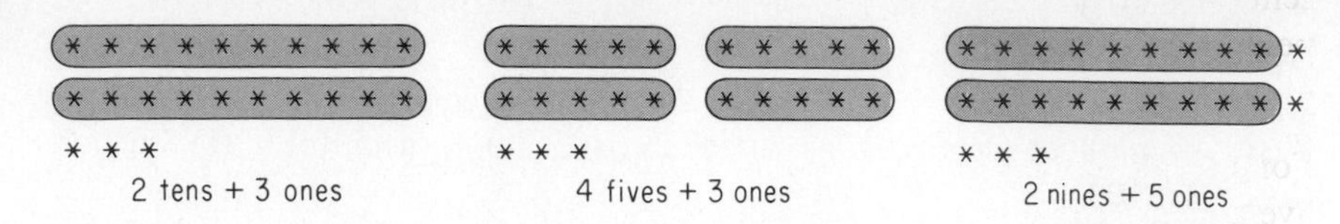

If we use a subscript to indicate our manner of grouping, we may write many different numerals (names) for the number of items in the same collection:

$$23_{\text{ten}} = 43_{\text{five}} = 25_{\text{nine}}$$

(2 *tens* + 3 ones) (4 *fives* + 3 ones) (2 *nines* + 5 ones)

43_{five} is read "four three, base five."
25_{nine} is read "two five, base nine."

Each of these numerals represents the number of asterisks in the same set of asterisks. Still another numeral for this number is 35_{six}:

$$35_{\text{six}} = 3 \textit{ sixes} + 5 \text{ ones} = 18 + 5 = 23$$

We call our decimal system of numeration a **base ten** system; when no subscript is used, the numeral is understood to be expressed in base ten. When we group by fives, we have a **base five** system of numeration; that is, we name our system of numeration by the manner in which the grouping is accomplished.

Example 1: Draw a diagram for 18 objects and write the corresponding numeral (a) in base five and (b) in base eight notation.

Solution:

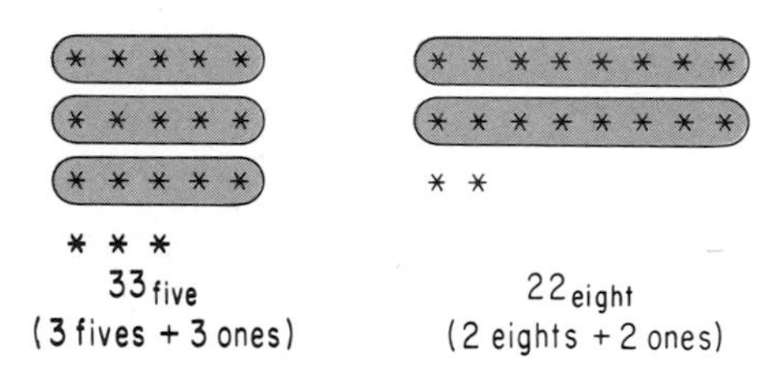

The use of dollar bills and coins is a helpful aid to see the relationship between numbers expressed in base five and base ten. Consider, for example, a sum of 123¢. Using a dollar, dimes, and pennies we may think of this sum as follows:

$$123¢ = 1 \text{ dollar} + 2 \text{ dimes} + 3 \text{ pennies}$$

Compare this expression with this expanded notation:

$$123 = (1 \times 10^2) + (2 \times 10) + (3 \times 1)$$

Now let us consider the base five numeral 123_{five}. A concrete representation of this numeral is to think of it in terms of quarters, nickels, and pennies in this manner:

$$123_{\text{five}} = 1 \text{ quarter} + 2 \text{ nickels} + 3 \text{ pennies}$$

For expanded base five notation we write numbers in terms of powers of five and use the digits 0, 1, 2, 3, and 4. For example:

$$123_{\text{five}} = (1 \times 5^2) + (2 \times 5) + (3 \times 1)$$

Using the monetary expansion as an aid, or expanded base five notation, we see that $123_{\text{five}} = 38_{\text{ten}}$. We can also show this by means of the following diagram that includes one group of 25, two groups of five, and three ones.

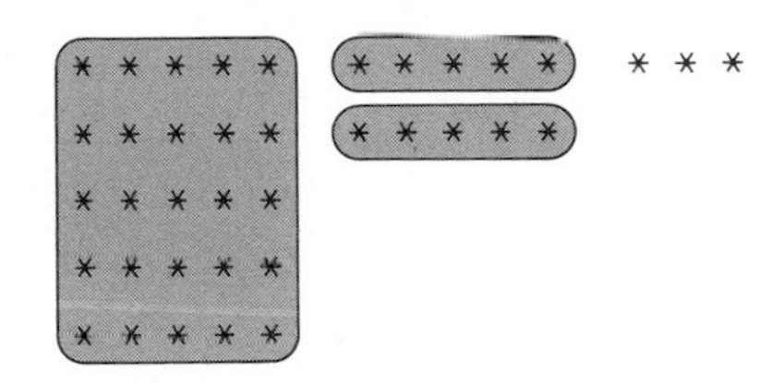

Throughout the remainder of this section we shall work exclusively with base five notation, although the principles developed apply to all other bases as well. For convenience we shall write all numbers in base five notation using the numeral 5 rather than the word "five," as a subscript. Thus we shall write 123_5 although there is no numeral 5 in this system of notation. We begin with this table that contains the numbers 1 to 30 written in base five notation.

Base Ten	*Base Five*	*Base Ten*	*Base Five*	*Base Ten*	*Base Five*
1	1	11	21_5	21	41_5
2	2	12	22_5	22	42_5
3	3	13	23_5	23	43_5
4	4	14	24_5	24	44_5
5	10_5	15	30_5	25	100_5
6	11_5	16	31_5	26	101_5
7	12_5	17	32_5	27	102_5
8	13_5	18	33_5	28	103_5
9	14_5	19	34_5	29	104_5
10	20_5	20	40_5	30	110_5

To translate a number that is not listed in the table from base five notation to base ten notation, express the number in terms of powers of five and simplify.

Example 2: Write 432_5 in base ten notation.

Solution: If we think of this expression in terms of quarters, nickels, and pennies, we have the following:

$$432_5 = 4 \text{ quarters} + 3 \text{ nickels} + 2 \text{ pennies} = 117¢ = 117_{10}$$

Using expanded notation instead we have this solution:

$$\begin{aligned} 432_5 &= (4 \times 5^2) + (3 \times 5^1) + (2 \times 5^0) \\ &= (4 \times 25) + (3 \times 5) + (2 \times 1) \qquad (\textit{Note: } 5^0 = 1.) \\ &= 117 \end{aligned}$$

The use of coins in base five fails us when we have a numeral with more than three digits. Thus we may think of $5^0 = 1$ as a penny, $5^1 = 5$ as a nickel that is worth 5 pennies, and $5^2 = 25$ as a quarter that is worth 25 pennies. However, $5^3 = 125$ and we do not have a coin worth 125 pennies. For such a number we must make use of expanded notation, as in Example 3.

Example 3: Write 3214_5 in base ten notation.

Solution:

$$\begin{aligned} 3214_5 &= (3 \times 5^3) + (2 \times 5^2) + (1 \times 5^1) + (4 \times 5^0) \\ &= (3 \times 125) + (2 \times 25) + (1 \times 5) + (4 \times 1) \\ &= 434 \end{aligned}$$

To translate from base ten to base five, any one of several procedures may be adopted. Consider the problem

$$339 = (\quad)_5$$

When a number is expressed to the base five, it is written in terms of powers of five:

$$5^0 = 1, \quad 5^1 = 5, \quad 5^2 = 25, \quad 5^3 = 125, \quad 5^4 = 625, \quad \ldots$$

The highest power of 5 that is not greater than the given number is 5^3. This power of 5, namely $5^3 = 125$, can be subtracted from 339 twice. Then the remainder 89 is positive and less than 125.

$$\begin{array}{r} 339 \\ -125 \\ \hline 214 \\ -125 \\ \hline 89 \end{array}$$

Thus, we write 2×5^3 in the expansion of 339 to the base five.

The next power of 5 is 5^2. This number can be subtracted from 89 three times to obtain a nonnegative remainder less than 25.

$$\begin{array}{r} 89 \\ -25 \\ \hline 64 \\ -25 \\ \hline 39 \\ -25 \\ \hline 14 \end{array}$$

Thus, we write 3×5^2 in the expansion.

Finally, we subtract 5 from 14 twice, write 2×5 in the expansion, and obtain 4 as a remainder.

$$\begin{array}{r} 14 \\ -5 \\ \hline 9 \\ -5 \\ \hline 4 \end{array}$$

$$\begin{aligned} 339 &= 2(125) + 3(25) + 2(5) + 4 \\ &= (2 \times 5^3) + (3 \times 5^2) + (2 \times 5^1) + (4 \times 5^0) = 2324_5 \end{aligned}$$

A group of 339 elements can be considered as two groups of 125 elements, three groups of 25 elements, two groups of 5 elements, and 4 elements.

An alternative procedure for changing 339 to the base five depends upon successive division by 5:

$$\begin{aligned} 339 &= 67 \times 5 + 4 \\ 67 &= 13 \times 5 + 2 \\ 13 &= \ \ 2 \times 5 + 3 \end{aligned}$$

Next, substitute from the third equation into the second. Then substitute from the second equation to the first, and simplify as follows:

$$\begin{aligned} 13 &= 2 \times 5 + 3 \\ 67 &= 13 \times 5 + 2 = (2 \times 5 + 3) \times 5 + 2 \\ 339 &= 67 \times 5 + 4 = [(2 \times 5 + 3) \times 5 + 2] \times 5 + 4 \\ &= (2 \times 5^3) + (3 \times 5^2) + (2 \times 5^1) + (4 \times 5^0) = 2324_5 \end{aligned}$$

The arithmetical steps involved in these computations can be performed as shown in the following array (often called an *algorithm*).

$$\begin{array}{rl} 5\,\underline{)\,339} & \\ 5\,\underline{)\,67} & \text{—— } 4 \uparrow \\ 5\,\underline{)\,13} & \text{—— } 2 \\ 5\,\underline{)\,2} & \text{—— } 3 \\ 0 & \text{—— } 2 \end{array}$$

Read upward as 2324_5.

Note that the remainder is written after each division by 5. Then the remainders are read in reverse order to obtain the expression for the number to the base five. This procedure works for integers only, not for fractional parts of a number.

Example 4: Write 423 in base five notation.

Solution:

```
5 ) 423
  5 ) 84 — 3 ↑
  5 ) 16 — 4 |        Answer: 3143₅
   5 ) 3 — 1 |
       0 — 3 |
```

Answer: 3143_5

Check:

$$3143_5 = (3 \times 5^3) + (1 \times 5^2) + (4 \times 5^1) + (3 \times 5^0)$$
$$= 375 + 25 + 20 + 3 = 423$$

After computations in other bases have been considered, the method of successive division by the new base may be used in changing from one base to another. For example, we may use this procedure to change from base five to base ten, successively dividing by 20_5. The computation must be done in base five notation. For the present we may change from base five to base ten by expressing the number in terms of powers of 5.

EXERCISES

Write numerals for each of the following collections in the bases indicated by the manner of grouping.

1. (* * * * *) (* * * * *) * *

2.

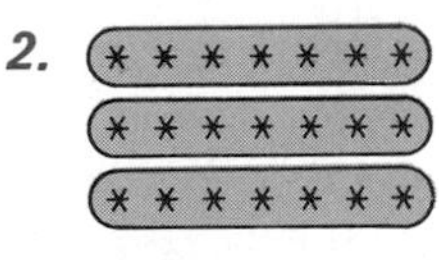

3.

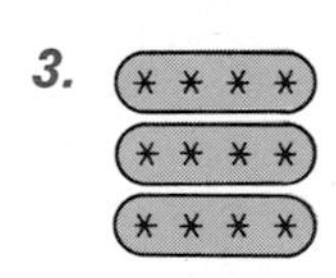

4.

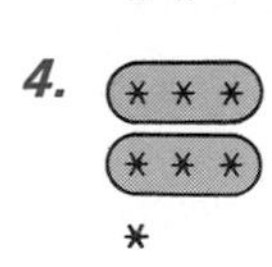

Draw a diagram to show the meaning of each of the following.

5. 34_{five} **6.** 25_{six} **7.** 25_{seven}

8. 42_{nine} **9.** 13_{four} **10.** 22_{three}

Change to base ten notation.

11. 34_{five} **12.** 25_{seven} **13.** 43_{eight}

14. 54_{six} **15.** 43_{five} **16.** 102_{four}

Write each number in decimal notation.

17. 413_5
18. 330_5
19. 444_5
20. 3210_5
21. 421_5
22. 2413_5
23. 4032_5
24. 243_5
25. 4341_5
26. 1423_5

Write each number in base five notation.

27. 182
28. 493
29. 982
30. 596
31. 625
32. 816
33. 337
34. 756
35. 607
36. 1000

Extend the concepts of this section and write each number in decimal notation.

***37.** 337_8
***38.** 3013_4
***39.** 11011_2
***40.** 421_{12}
***41.** 232_{20}
***42.** 512_6
***43.** 314_{15}
***44.** 4352_8

PEDAGOGICAL EXPLORATIONS

1. For base five notation try to find rules for divisibility by (**a**) 10_5, that is 5; (**b**) 4; (**c**) 2.
2. Prepare a brief introduction to base five notation for a specific elementary grade level. Use as many ways as possible to create interest in the topic.
3. Prepare a 10-minute talk appropriate for a P.T.A. meeting that explains the meaning of base five notation, and that also attempts to convince the parents of the value of such a unit of study in the elementary curriculum.
4. Some first grade classes have been introduced to the basic idea of base five notation through the use of *hand numerals.* Thus 32_H may be used to represent 3 hands and 2 fingers, that is, 17. Prepare a table that shows how to represent the numbers from 1 through 24 using hand numerals.
5. Does the use of coins (pennies, nickels, quarters) for base five numerals illustrate a place-value system? Explain your answer and then discuss both the advantages and limitations of the use of these coins to teach the concepts of base five notation.

6. Try to find some other concrete representations for writing numbers in bases other than five and ten.
7. Explain why we cannot use 1 as a base in a system of numeration.

3-7 computation in base five notation

Coins can be used effectively to help develop skills of addition and subtraction in base five. Consider, for example, this problem:

In adding, the 6 pennies can be exchanged for 1 nickel and 1 penny. In the "nickel's column" we then have $1 + 3 + 4$ or 8 nickels. This can then be exchanged for 1 quarter and 3 nickels. Finally, in the "quarter's column" we have 3 quarters.

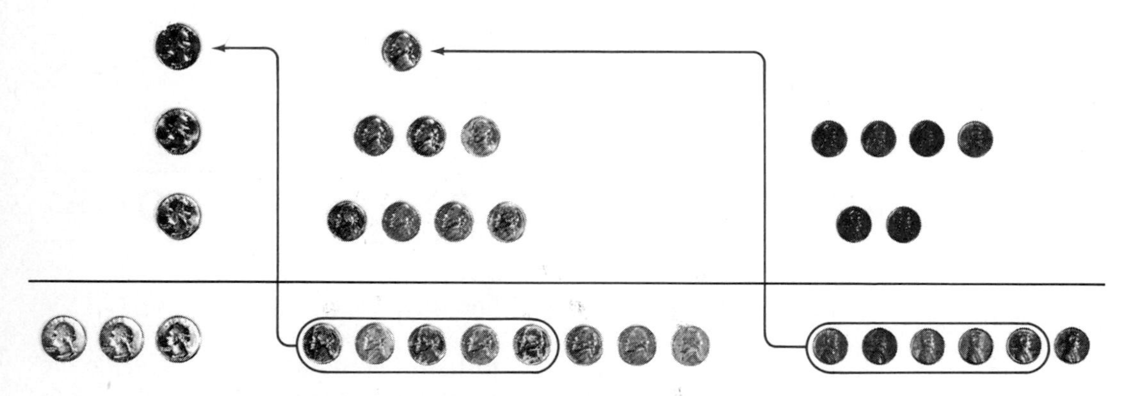

The final result gives a total of 3 quarters, 3 nickels, and 1 penny. In base five notation this may be written as 331_5. To compute without such aids as coins we need the basic addition facts in base five notation. For example, we need to be able to find such sums as $4_5 + 3_5$. Mentally we can represent this sum as

* * * *   +   * * *

which can then be regrouped as

and which can then be written as 12_5. That is, $4_5 + 3_5 = 12_5$. In base ten language, we may also think of this as $4 + 3 = 7$, which is then equivalent to *one* group of five and *two* ones.

sec. 3-7
computation in base
five notation

Shown below is a table of the number facts needed for addition problems in base five. (You should verify each entry.)

+	0	1	2	3	4
0	0	1	2	3	4
1	1	2	3	4	10_5
2	2	3	4	10_5	11_5
3	3	4	10_5	11_5	12_5
4	4	10_5	11_5	12_5	13_5

The facts in this table may be used in finding sums of numbers, as illustrated in the following example.

Example 1: Find the sum of 432_5 and 243_5. Then check in base ten notation.

Solution:

$$\begin{array}{r} 432_5 \\ +243_5 \\ \hline 1230_5 \end{array}$$

Check:

$$\begin{array}{rll} 432_5 = & (4 \times 5^2) + (3 \times 5^1) + (2 \times 5^0) & = 117 \\ +243_5 = & (2 \times 5^2) + (4 \times 5^1) + (3 \times 5^0) & = \ \ 73 \\ \hline 1230_5 = & (1 \times 5^3) + (2 \times 5^2) + (3 \times 5^1) + (0 \times 5^0) & = 190 \end{array}$$

Here are the steps used in Example 1. In each case the familiar symbol 5 has been used in place of 10_{five} to help the reader recognize that powers of 5 are involved. This convention will be followed throughout this chapter.

First add the column of ones.

$$\begin{array}{r} (4 \times 5^2) + (3 \times 5^1) + (2 \times 5^0) \\ (2 \times 5^2) + (4 \times 5^1) + (3 \times 5^0) \\ \hline 10_5 \end{array}$$

Next write the sum 10_5 of the ones as 1×5^1 in the fives' column, and add the column of fives.

$$\begin{array}{r} 1 \times 5^1 \qquad\qquad \\ (4 \times 5^2) + (3 \times 5^1) + (2 \times 5^0) \\ (2 \times 5^2) + (4 \times 5^1) + (3 \times 5^0) \\ \hline (13_5 \times 5^1) + (0 \times 5^0) \end{array}$$

Then write $10_5 \times 5^1$ from the previous sum in the 5^2 column and add the column of 5^2 entries.

$$\begin{array}{r} 1 \times 5^2 \qquad\qquad\qquad\qquad\qquad\quad \\ (4 \times 5^2) + (3 \times 5^1) + (2 \times 5^0) \\ (2 \times 5^2) + (4 \times 5^1) + (3 \times 5^0) \\ \hline (12_5 \times 5^2) + (3 \times 5^1) + (0 \times 5^0) = 1230_5 \end{array}$$

Note that we "carry" groups of five in base five computation, just as we "carry" groups of ten in decimal computation.

Subtraction is not difficult if it is thought of as the inverse of addition. The table of addition facts in base five may again be used.

Example 2: Subtract in base five and check in base ten: $211_5 - 142_5$.

Solution: Think of 211_5 as

$$(2 \times 5^2) + (1 \times 5^1) + (1 \times 5^0)$$

then as

$$(1 \times 5^2) + (11_5 \times 5^1) + (1 \times 5^0)$$

then as

$$(1 \times 5^2) + (10_5 \times 5^1) + (11_5 \times 5^0).$$

Thus:

$$\begin{array}{rrl} & 211_5 = & (1 \times 5^2) + (10_5 \times 5^1) + (11_5 \times 5^0) \\ (-) & 142_5 = & (1 \times 5^2) + \quad (4 \times 5^1) + \quad (2 \times 5^0) \\ \hline & 14_5 & \qquad\qquad\quad (1 \times 5^1) + \quad (4 \times 5^0) \end{array}$$

Check:

$$\begin{array}{rrr} & 211_5 = & 56 \\ (-) & 142_5 = & 47 \\ \hline & 14_5 = & 9 \end{array}$$

This problem can be solved by exchanging and thinking in base five in the following steps:

$$\text{(a)} \begin{array}{r} 211_5 \\ -142_5 \\ \hline \end{array} \qquad \text{(b)} \begin{array}{r} 20^11_5 \\ -14\ 2_5 \\ \hline 4_5 \end{array} \qquad \text{(c)} \begin{array}{r} 1^10^11_5 \\ -1\ 4\ 2_5 \\ \hline 1\ 4_5 \end{array}$$

Each raised "1" can be interpreted to show that the number 211_5 has not been changed but rather expressed in a different form:

$$211_5 = (2 \times 5^2) + (1 \times 5^1) + (1 \times 5^0)$$

$$20^11_5 = (2 \times 5^2) + (0 \times 5^1) + (1 \times 5^1) + (1 \times 5^0) = 211_5$$

$$1^10^11_5 = (1 + 5^2) + (1 \times 5^2) + (0 \times 5^1) + (1 \times 5^1) + (1 \times 5^0) = 211_5$$

Next we consider multiplication in base five, as in Example 3.

Example 3: Find the product of 243_5 and 4_5.

Solution: We need the following facts:

$$4 \times 3 = 12_{10} = (2 \times 5^1) + (2 \times 5^0) = 22_5$$
$$4 \times 4 = 16_{10} = (3 \times 5^1) + (1 \times 5^0) = 31_5$$
$$4 \times 2 = 8_{10} = (1 \times 5^1) + (3 \times 5^0) = 13_5$$

The pattern for the computation when multiplying and "carrying" in base five notation is precisely the same as that used in base ten computation. A long form may be used to avoid "carrying" or a condensed form may be used. In the long form zeros may be added when needed for each partial product, or indentation of partial products may be used to indicate the powers of 5 involved.

$$\begin{array}{rl} 243_5 & \\ \times 4_5 & \\ \hline 22_5 & = 4_5 \times 3_5 \\ 310_5 & = 4_5 \times 40_5 \\ 1300_5 & = 4_5 \times 200_5 \\ \hline 2132_5 & \end{array}$$

Condensed forms:

$$\begin{array}{r} 243_5 \\ \times 4_5 \\ \hline 22_5 \\ 31_5 \\ 13_5 \\ \hline 2132_5 \end{array} \qquad \begin{array}{r} 243_5 \\ \times 4_5 \\ \hline 2132_5 \end{array}$$

Multiplication by more than a one-digit multiplier is possible in other bases. Again, the pattern for computation is similar to that used in base ten computation. Consider, for example, the product $34_5 \times 243_5$:

$$\begin{array}{rl} 243_5 & \\ \times 34_5 & \\ \hline 2132 & = 4_5 \times 243_5 \\ 13340 & = 30_5 \times 243_5 \\ \hline 21022_5 & \end{array}$$

As in base ten computation, the product $30_5 \times 243_5$ can be written without the final numeral 0 since the place value is taken care of by indenting.

The pattern used for division in base ten can also be used for division in base five. Consider, for example, the problem $121_5 \div 4_5$. As a first step it is helpful to make a table of multiples of 4_5 in base five:

$$1_5 \times 4_5 = 4_5$$
$$2_5 \times 4_5 = 13_5$$
$$3_5 \times 4_5 = 22_5$$
$$4_5 \times 4_5 = 31_5$$

Since 12_5 is greater than 4_5 and less than 13_5, the first numeral in the quotient is 1.

$$\begin{array}{r} 1 \\ 4_5\overline{)121_5} \\ 4 \\ \hline 31 \end{array}$$

As in base ten computation, we multiply $1_5 \times 4_5$ and subtract this from 12_5 in the dividend. The next digit is then brought down. Note that $12_5 = 7$ and thus $12_5 - 4 = 3$.

Next we must divide 31_5 by 4_5. Note from the table of multipliers of 4_5 that $4_5 \times 4_5 = 31_5$. Thus we can complete the division, and there is no remainder.

$$\begin{array}{r} 14_5 \\ 4_5\overline{)121_5} \\ \underline{4} \\ 31 \\ \underline{31} \end{array} \qquad \textit{Check:} \quad \begin{array}{r} 14_5 \\ \underline{\times 4_5} \\ 121_5 \end{array}$$

Example 4: Divide 232_5 by 3_5 and check by multiplication in base five.

Solution:

$$\begin{array}{l} 1_5 \times 3_5 = 3_5 \\ 2_5 \times 3_5 = 11_5 \\ 3_5 \times 3_5 = 14_5 \\ 4_5 \times 3_5 = 22_5 \end{array} \qquad \begin{array}{r} 42_5 \\ 3_5\overline{)232_5} \\ \underline{22} \\ 12 \\ \underline{11} \\ 1 \end{array} \qquad \textit{Check:} \quad \begin{array}{r} 42_5 \\ \underline{\times 3_5} \\ 231_5 \\ \underline{+1_5} \\ 232_5 \end{array}$$

The quotient is 42_5 and the remainder is 1_5. Note that in the check we multiply the quotient by the divisor, and add the remainder to obtain the dividend.

EXERCISES

Add in base five and check in base ten.

1. $\begin{array}{r} 42_5 \\ \underline{+34_5} \end{array}$ **2.** $\begin{array}{r} 44_5 \\ \underline{+14_5} \end{array}$

3. $\begin{array}{r} 243_5 \\ \underline{+334_5} \end{array}$ **4.** $\begin{array}{r} 433_5 \\ \underline{+443_5} \end{array}$

5. $\begin{array}{r} 3343_5 \\ \underline{+3422_5} \end{array}$ **6.** $\begin{array}{r} 4124_5 \\ \underline{+4442_5} \end{array}$

Subtract in base five and check in base ten.

7. $\begin{array}{r} 243_5 \\ \underline{-31_5} \end{array}$ **8.** $\begin{array}{r} 41_5 \\ \underline{-24_5} \end{array}$

9. $322_5 - 123_5$

10. $422_5 - 244_5$

11. $1021_5 - 403_5$

12. $3004_5 - 1312_5$

13. $400_5 - 133_5$

14. $2003_5 - 1014_5$

Multiply in base five and check in base ten.

15. $342_5 \times 3_5$

16. $243_5 \times 4_5$

17. $1424_5 \times 3_5$

18. $2104_5 \times 2_5$

19. $44_5 \times 32_5$

20. $33_5 \times 43_5$

21. $342_5 \times 34_5$

***22.** $3244_5 \times 324_5$

Divide in base five and check in base ten.

23. $4_5\overline{)134_5}$

24. $3_5\overline{)231_5}$

25. $3_5\overline{)1234_5}$

26. $4_5\overline{)3042_5}$

27. $22_5\overline{)143_5}$

28. $32_5\overline{)3031_5}$

29. $23_5\overline{)1341_5}$

30. $41_5\overline{)43201_5}$

31. Construct a table showing the basic multiplication facts for base five.

***32.** Complete the following tables of addition and multiplication facts for base four.

+	0	1	2	3
0				
1				
2				
3				

x	0	1	2	3
0				
1				
2				
3				

PEDAGOGICAL EXPLORATIONS

1. Pencils are normally ordered by the dozen (12) or ordered by the gross ($12^2 = 144$). Using this representation, we may think of 123_{12} as showing one gross, 2 dozen, and 3 pencils. Use an appropriate sketch to illustrate each of the following addition and subtraction problems in terms of numbers of pencils:

(a) $\begin{array}{r} 248_{12} \\ +\ 376_{12} \\ \hline \end{array}$ (b) $\begin{array}{r} 525_{12} \\ -\ 349_{12} \\ \hline \end{array}$

2. Draw a set of pictures of hands (five fingers) and fingers that might be used for an early elementary class to explain the solution to these problems:

(a) $\begin{array}{r} 24_{5} \\ +\ 13_{5} \\ \hline \end{array}$ (b) $\begin{array}{r} 41_{5} \\ -\ 23_{5} \\ \hline \end{array}$

3. Prepare a worksheet suitable for a specific elementary grade level with problems translating from base five to base ten. Begin the sheet with an illustrative concrete example to aid the students in their work.

4. Repeat Exploration 3 for translation from base ten to base five.

5. Explore a set of recently published elementary school mathematics textbooks to determine what consideration, if any, is given to work on computation in other number bases.

CHAPTER TEST

1. Write 342 in ancient Egyptian notation.

2. Use the Egyptian method of doubling to find the product 21×37.

3. Use the Egyptian method of duplation and mediation to find the product 19×37.

4. Multiply, using the "galley" method: 437×572.

5. Write in expanded notation: 23,457.

Solve as on a 12-hour clock.

6. $8 + 7$

7. 7×11

8. $3 - 10$

9. $2 \div 5$

Find all possible replacements for x for which each sentence is a true statement.

10. $4 + x \equiv 1 \pmod{5}$

11. $1 - x \equiv 4 \pmod{5}$

12. $2x \equiv 3 \pmod{5}$

13. $x + 2 \equiv x \pmod{5}$

14. $\frac{x}{3} \equiv 2 \pmod{5}$

15. $\frac{3}{x} \equiv 4 \pmod{5}$

Change to base ten.

16. 324_{five}

17. 53_{eight}

Write each number in base five notation.

18. 432

19. 739

Compute in base five.

20. $\begin{array}{r} 423_5 \\ +342_5 \\ \hline \end{array}$

21. $\begin{array}{r} 4102_5 \\ -1324_5 \\ \hline \end{array}$

22. $\begin{array}{r} 2432_5 \\ \times 4_5 \\ \hline \end{array}$

23. $\begin{array}{r} 3142_5 \\ \times 23_5 \\ \hline \end{array}$

24. $3_5 \overline{)1334_5}$

25. $14_5 \overline{)4334_5}$

READINGS AND PROJECTS

1. Examine several recently published elementary school mathematics textbooks and observe the uses that are made of nondecimal systems of notation.
2. Read Chapter 16 (On Divisibility Rules) of the twenty-seventh yearbook of the National Council of Teachers of Mathematics.
3. There is a great deal of current controversy concerning the so-called "new math" that was introduced into the elementary schools in the 1960's. Prepare a short talk on this topic appropriate for a P.T.A. meeting, including both advantages and disadvantages of new programs. Debate the issue with a classmate who has opposing points of view from your own. Several sources to consider are:
   - (a) "Is the 'New Math' Really Better?" by Irving M. Cowle on pages 68 through 73 of the January 1974 issue of *The Arithmetic Teacher.*
   - (b) *Why Johnny Can't Add, the Failure of the New Math* by Morris Kline, published by St. Martin's Press, Inc., 1973.
4. For a further discussion and extension of work with Napier's rods, read the column by Martin Gardner in the March 1973 issue of *Scientific American.*

5. Discuss the advantages of teaching the history of early numeration systems in elementary school mathematics classes. For an interesting article on this topic see "A Discovery Approach with Ancient Numeration Systems" by Robert W. Keller on pages 543 and 544 of the November 1972 issue of *The Arithmetic Teacher.*
6. Numerous suggestions can be found in the literature for teaching techniques suitable for the slow learner. For example, it is claimed that they learn best through an activity-oriented curriculum centered about laboratory experiences. Review several professional journals; then read and report on at least three different articles dealing with teaching the slow learner. Prepare a list of ten specific suggestions for teaching mathematics to the slow learner in the elementary school.
7. Repeat Exploration 6 for the gifted student.
8. Read Chapter 3 (Structuring Arithmetic) and Chapter 4 (Guiding the Learner to Discover and Generalize) of the twenty-fifth yearbook of the National Council of Teachers of Mathematics. Using this material as a basis, prepare a 15–20-minute lesson plan that leads a specified elementary mathematics class to formulate a generalization through a discovery approach.
9. Read Booklet No. 3 (Numeration Systems for the Whole Numbers) of the twenty-ninth yearbook of the National Council of Teachers of Mathematics. Complete at least three of the exercise sets given there.
10. Read Chapter 5 [Adjustment of Instruction (Elementary School)] of the thirty-fifth yearbook of the National Council of Teachers of Mathematics. Report on the author's point of view on within-class grouping as a means of adjusting instruction to individual needs.
11. Read "Analysis of Questions Teachers Ask About Teaching Modern Mathematics in the Elementary School" by Robert W. Green on pages 98 through 102 of the February 1976 issue of *The Arithmetic Teacher.* Summarize the recommendations cited by the author as they apply to elementary school teacher training programs.
12. Read "'New Math' in the Gay Nineties" by Gary A. Deatsman on pages 165 and 166 of the March 1976 issue of *The Arithmetic Teacher.* Try to locate and review an early arithmetic textbook, published before 1900.
13. Obtain a copy of "Overview and Analysis of School Mathematics, Grades K–12," a 1975 report of the National Advisory Committee on Mathematical Education (NACOME) of the Conference Board of the Mathematical Sciences. Read and report on their findings and recommendations concerning "new math" programs in the elementary schools. In particular, see pages 1 through 4, 10 through 14, and 21 through 22.

Whole Numbers: Properties and Operations

How many of you can find the quotient $9065 \div 37$? Undoubtedly every reader of this book will have little trouble completing this long division:

```
     245
37)9065
   74
   166
   148
    185
    185
```

On the other hand, how many of you can answer such questions as these:

> Why did you subtract 74 from 90?
> Why did you "bring down" the 6?
> What does 166 have to do with the original problem?

The only honest answer that most people can give to questions such as these is, "That's the way I was taught to do it. I don't know why."

Among the important objectives of this chapter is that of providing the rationale behind the fundamental operations of addition, subtraction, multiplication, and division. Therefore we shall explore *why* our methods of computation work as they do.

The famous mathematician Leopold Kronecker (1823–1891) said "God created the natural numbers; everything else is man's handiwork." He was referring to what we generally call the set of counting numbers:

$$\{1, 2, 3, 4, 5, \ldots\}$$

Another objective of this chapter is to explore the properties of this set of numbers and to begin to unfold "man's handiwork" as we examine extensions to other sets of numbers.

4-1 counting numbers

A number is associated with a set of elements whenever the elements of the set are counted. The set of counting numbers is often represented as the set C:

$$C = \{1, 2, 3, 4, 5, \ldots\}$$

Sums and products of counting numbers may be introduced by using cardinal numbers of sets. Consider, for example, the sum $3 + 4$. Then consider two sets A and B with three and four elements respectively, and with no elements in common:

Let $A = \{a, b, c\}$.
Let $B = \{k, l, m, n\}$.
Then $A \cup B = \{a, b, c, k, l, m, n\}$.

The sum $3 + 4$ is then found to be the number of elements in $A \cup B$, namely 7.

In general, if A and B are two sets such that $A \cap B = \varnothing$, then

$$n(A) + n(B) = n(A \cup B)$$

Multiplication may be introduced in terms of successive addition. For example:

$$3 \times 2 = 2 + 2 + 2$$

Thus multiplication may be introduced by using equivalent sets that have no elements in common, that is, disjoint equivalent sets. For example, consider three disjoint sets A, B, and C each containing two elements and with no common elements:

Let $A = \{a, b\}$, $B = \{p, q\}$, and $C = \{x, y\}$.
Also, $A \cap B = \varnothing$, $A \cap C = \varnothing$, and $B \cap C = \varnothing$.
Then $A \cup B \cup C = \{a, b, p, q, x, y\}$.

Thus the product 3×2 is the number of elements in $A \cup B \cup C$, namely 6.

As in §1-3 we may also show multiplication through the use of a *Cartesian product*.

Example 1: Use a Cartesian product to illustrate $3 \times 2 = 6$.

Solution: Let $P = \{a, b, c\}$ and $Q = \{m, n\}$. Then $n(P) = 3$, $n(Q) = 2$, and $P \times Q$ is this set of ordered pairs:

$$\{(a, m), (b, m), (c, m), (a, n), (b, n), (c, n)\}$$

Thus $3 \times 2 = n(P) \times n(Q) = n(P \times Q) = 6$.

Each of the operations of addition and multiplication is a *binary operation.* The properties of the set of counting numbers under addition and multiplication may be described by using the terminology in the listing that follows. Thus, for any counting numbers a, b, and c we have:

Closure, +: There is one and only one counting number $a + b$. Whenever two counting numbers are added, the sum is a unique element of the set of counting numbers. That is, the set of counting numbers is closed with respect to addition.

Closure, ×: There is one and only one counting number $a \times b$. Whenever two counting numbers are multiplied, the product is a unique element of the set of counting numbers. That is, the set of counting numbers is closed with respect to multiplication.

Commutative, +: $a + b = b + a$. The sum of any two counting numbers is the same regardless of the order in which the numbers are added.

Commutative, ×: $a \times b = b \times a$. The product of any two counting numbers is the same regardless of the order in which the numbers are multiplied.

Associative, +: $(a + b) + c = a + (b + c)$. The sum of three counting numbers is the same regardless of the order in which they are associated.

Associative, ×: $(a \times b) \times c = a \times (b \times c)$. The product of three counting numbers is the same regardless of the order in which they are associated.

Identity, ×: $a \times 1 = 1 \times a = a$. The counting number 1 is the **multiplicative identity** and is also called the **identity element for multiplication.** This property is often referred to as the **multiplication property of one.**

Distributive Property, × with respect to +: $a \times (b + c) = (a \times b) + (a \times c)$.

Consider, for example, the expression $3(5 + 8)$. According to the distributive property, we may evaluate this expression in two different ways and obtain the same answer either way. Thus

$$3(5 + 8) - 3(13) = 39$$
$$(3)(5) + (3)(8) = 15 + 24 = 39$$

so that

$$3(5 + 8) = (3)(5) + (3)(8)$$

This example illustrates the fact that, because of the distributive property, we may either add first and then multiply, or find the two products first and then add. In general, we say that for all replacements of a, b, and c,

$$a(b + c) = ab + ac$$

Formally, we call this the **distributive property for multiplication with respect to addition,** or simply the *distributive property*. Note that addition is *not* distributive with respect to multiplication since, for example,

$$3 + (5 \times 8) \neq (3 + 5) \times (3 + 8)$$

that is,

$$3 + 40 \neq 8 \times 11$$

It is the distributive property that allows us, in algebra, to make such statements as:

$$2(a + b) = 2a + 2b$$
$$3(x - y) = 3x - 3y$$

It is the distributive property that elementary school youngsters use in multiplication. Consider the problem 7×43. The distributive property is used by thinking of 7×43 as:

$$7 \times (40 + 3) = (7 \times 40) + (7 \times 3) = 280 + 21 = 301$$

In our usual algorithm we have:

$$\begin{array}{rl} 43 & \\ \times 7 & \\ \hline 21 & = 7 \times 3 \\ 280 & = 7 \times 40 \\ \hline 301 & = (7 \times 3) + (7 \times 40) = (7 \times 40) + (7 \times 3) \end{array}$$

The distributive property can also be used in developing shortcuts in multiplication. Thus the product 8×99 can be found quickly as follows:

$$8 \times 99 = 8(100 - 1) = 800 - 8 = 792$$

Note that the set of counting numbers does not include an identity for addition since zero is not a counting number. Moreover, the set of counting numbers is *not* closed under subtraction or under division. For example, $2 - 5$ and $12 \div 5$ are meaningless if you are restricted to using counting numbers.

Example 2: Apply properties of counting numbers to the left member of the equation $25 \times (11 \times 4) = (25 \times 4) \times 11$ to obtain the right member of the equation. Show each step and name each property.

Solution:

$$\begin{aligned} 25 \times (11 \times 4) &= 25 \times (4 \times 11) && \text{Commutative, } \times \\ &= (25 \times 4) \times 11 && \text{Associative, } \times \end{aligned}$$

EXERCISES

1. Find $A \times B$ for $A = \{1, 2\}$ and $B = \{1, 2, 3, 4\}$, thereby showing that $2 \times 4 = 8$.
2. Find $P \times Q$ for $P = \{1, 2, 3\}$ and $Q = \{1, 2, 3\}$, thereby showing that $3 \times 3 = 9$.
3. Show that subtraction of counting numbers is not commutative.
4. Show that division of counting numbers is not commutative.
5. Does $8 - (3 - 2) = (8 - 3) - 2$? Is subtraction of counting numbers associative?
6. Does $12 \div (6 \div 2) = (12 \div 6) \div 2$? Is division of counting numbers associative?
7. Show that addition of counting numbers is not distributive with respect to multiplication.

For each arithmetic statement, name the property of counting numbers that is illustrated.

8. $3 \times 2 = 2 \times 3$
9. $6 + 5 = 5 + 6$
10. $2 + (3 \times 5) = 2 + (5 \times 3)$
11. $2 \times (3 \times 5) = (2 \times 3) \times 5$
12. $4580 = 1 \times 4580$
13. $25 \times (14 + 26) = (25 \times 14) + (25 \times 26)$
14. $17 \times (15 + 21) = 17 \times (21 + 15)$
15. $48 + (19 + 7) = (19 + 7) + 48$
16. $27 \times (5 + 3) = (5 + 3) \times 27$
17. $17 \times (10 + 1) = (17 \times 10) + (17 \times 1)$
18. $5280 \times (7 + 13) = 5280 \times (13 + 7)$

Apply properties of counting numbers to the left member of each equation to obtain the right member. Show each step and name each property.

19. $92 + (50 + 8) = (92 + 8) + 50$

20. $(25 \times 17) \times 4 = 17 \times 100$

21. $37 \times (1 + 100) = 3700 + 37$

22. $(2 + 3) + (8 + 7) = (2 + 8) + (3 + 7)$

23. $(73 + 19) + (7 + 1) = (73 + 7) + (19 + 1)$

24. $(26 \times 1) \times (10 \times 2) = 10 \times (26 \times 2)$

25. In ordinary arithmetic, is addition distributive with respect to addition? That is, does $a + (b + c) = (a + b) + (a + c)$ for all possible replacements of a, b, and c?

26. In ordinary arithmetic, is multiplication distributive with respect to multiplication? That is, does $a \times (b \times c) = (a \times b) \times (a \times c)$ for all possible replacements of a, b, and c?

Use the distributive property to find each product by means of a shortcut.

27. 7×79 **28.** 6×58 **29.** 8×92 **30.** 9×63

For each sentence find the replacement set for the counting number n.

31. $8(2 + 5) = (8)(2) + (8)(n)$

32. $5(8 + 4) = (5)(8) + (n)(4)$

33. $7(9 + n) = (7)(9) + (7)(5)$

34. $n(5 + 8) = (3)(5) + (3)(8)$

35. $(6 + 3)5 = (6)(5) + (n)(5)$

36. $(5 + n)3 = (5)(3) + (7)(3)$

37. $(3)(5) + (3)(6) = 3(5 + n)$

38. $(4)(6) + (4)(n) = 4(6 + 2)$

39. $(3)(n) + (7)(n) = (3 + 7)n$

40. $3(4 + n) = (3)(4) + (3)(n)$

PEDAGOGICAL EXPLORATIONS

1. Elementary school students use the distributive property of multiplication with respect to addition in many ways. For example:

$$30 + 40 = (3 \times 10) + (4 \times 10) = (3 + 4) \times 10 = 7 \times 10 = 70$$
$$(17 \times 3) + (13 \times 3) = (17 + 13) \times 3 = 30 \times 3 = 90$$

Make a worksheet of 20 problems to help elementary school students recognize the usefulness of this distributive property.

2. The properties of counting numbers are used to simplify arithmetic computations throughout the elementary school grades. For example, the number fact $7 \times 10 = 70$ is easily learned. However, the problem 10×7 often leads an elementary school student to computations 1×7, 2×7, 3×7, . . . , 10×7 and is rather hard *until* the student realizes that 7×10 and 10×7 are two ways of stating the same problem. Many educators feel that students *should not be told* but rather *should be helped to discover* for themselves such useful properties of numbers. In other words, many matched pairs of problems (7×10 and 10×7, 2×15 and 15×2, and so forth) should be given until the student observes the pattern of the commutativity of multiplication. Such experiences of discovery and the recognition of the labor-saving effect of the pattern that is discovered should be developed whenever possible. Make a worksheet of 20 problems to help elementary school students recognize and make effective use of the commutative property for multiplication.

3. Repeat Exploration 2 for the commutative property for addition.

4. Two rows of three blocks may be used to represent 2×3. Describe how you could represent 3×2 and use cubical blocks to illustrate $2 \times 3 = 3 \times 2$.

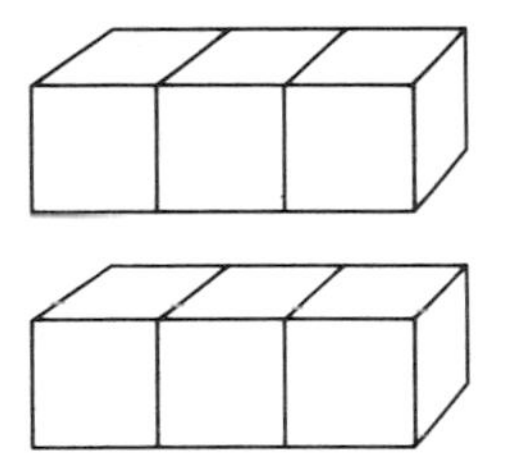

5. Repeat Exploration 4 to show that $(2 \times 3) \times 4 = 2 \times (3 \times 4)$.

6. Repeat Exploration 4 to show that $2 \times (3 + 4) = (2 \times 3) + (2 \times 4)$.

7. Operations may be considered in terms of a **function machine** with three basic parts:

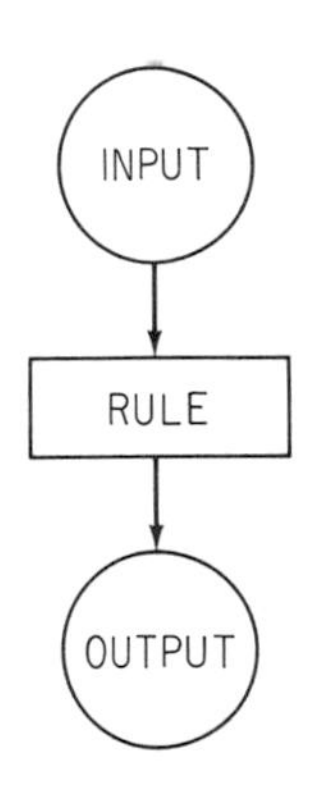

For the rule "add 5" complete this array:

Input	3	7	___	___	21	___
Output	___	___	7	11	___	21

8. Make a worksheet for elementary school students for the rule: (**a**) subtract 1; (**b**) multiply by 5; (**c**) square the number; (**d**) square the number and add 2; (**e**) square the number and subtract 1.

9. In the introduction to this chapter there appears a quotation from the German mathematician Kronecker. Start a collection of such famous sayings that might be of interest in mathematics classes. For example, Archimedes is supposed to have said, "Give me a place to stand and a lever long enough and I will move the earth."

4-2 prime numbers

Before extending the set of counting numbers to other sets we shall explore an interesting and useful classification of the counting numbers according to certain of its subsets. First, however, some definitions are needed.

The counting number 6 is divisible by 2, since there is a counting number 3 such that $6 = 2 \times 3$; the counting number 7 is not divisible by 2, since there exists no counting number b such that $7 = 2 \times b$. In general, a counting number n is **divisible by** a counting number t if and only if there is a counting number k such that $n = t \times k$. If n is divisible by t, then n is a **multiple of** t and t is a **factor of** n. For example, 6 is a multiple of 2 and 2 is a factor of 6.

The counting numbers are often considered in terms of the numbers by which they are divisible. The set

$$A = \{2, 4, 6, 8, 10, 12, \ldots\}$$

consists of the numbers that are divisible by 2; that is, the numbers expressible in the form $2k$, where k stands for a counting number. The set

$$B = \{3, 6, 9, 12, 15, 18, \ldots\}$$

consists of the numbers divisible by 3; the set

$$C = \{4, 8, 12, 16, 20, 24, \ldots\}$$

consists of the numbers divisible by 4; the set

$$D = \{5, 10, 15, 20, 25, 30, \ldots\}$$

consists of the numbers divisible by 5; the set

$$E = \{6, 12, 18, 24, 30, 36, \ldots\}$$

consists of the numbers divisible by 6; and so forth.

Notice that $C \subset A$; in other words, any number that is divisible by 4 is also divisible by 2. Notice also that $A \cap B = E$; in other words, the set of numbers that are divisible by both 2 and 3 is the set of numbers that are divisible by 6.

The number 1 divides every counting number since $k = 1 \times k$ for every counting number k. Accordingly, the number 1 is called a **unit**. Since the set A does not include all counting numbers, divisibility by 2 is a special property of the elements of the set A.

The number 2 is a member of set A and is not a member of any other one of the sets B, C, D, or E. The number 2 is not divisible by any counting number except itself and 1. Thus the number 2 has exactly two distinct counting numbers as factors, itself and 1. Any counting number that has exactly two distinct counting numbers as factors is a **prime number.** Thus 2 is prime, 3 is prime, 4 is not prime since 4 is divisible by 2 (that is, 4 is a member of the set A as well as the set C), 5 is prime, and 6 is not prime since 6 is divisible by both 2 and 3. The counting numbers that are greater than 1 and are not prime are **composite numbers.** Note that every counting number greater than 1 is either prime or composite. The number 1 is neither prime nor composite.

We could extend the list of sets A, B, C, D, E, F to identify other prime numbers, that is, elements that belong to one and only one of these sets. However, the method used to select these sets may be applied to the entire set of counting numbers. We shall illustrate this for the set $\{1, 2, 3, \ldots, 100\}$ and thereby illustrate a method for finding prime numbers that was discovered by a Greek mathematician named Eratosthenes over two thousand years ago. The method is known as the **Sieve of Eratosthenes.** First we prepare a table of the counting numbers through 100. Then selected members of this set are excluded as follows.

1	2	3	4	5	6	7	8	9	10
11	12	13	14	15	16	17	18	19	20
21	22	23	24	25	26	27	28	29	30
31	32	33	34	35	36	37	38	39	40
41	42	43	44	45	46	47	48	49	50
51	52	53	54	55	56	57	58	59	60
61	62	63	64	65	66	67	68	69	70
71	72	73	74	75	76	77	78	79	80
81	82	83	84	85	86	87	88	89	90
91	92	93	94	95	96	97	98	99	100

Cross out 1, since we know that it is not classified as a prime number. Draw a circle around 2, the smallest prime number. Then cross out every following multiple of 2, since each one is divisible by 2 and thus is not prime. That is, cross out the numbers in the set $\{4, 6, 8, \ldots, 100\}$.

Draw a circle around 3, the next prime number in our list. Then cross out each succeeding multiple of 3. Some of these numbers, such as 6 and 12, will already have been crossed out because they are multiples of 2. That is, they are members of both sets A and B.

The number 5 is prime and is circled, and we exclude each fifth number after 5. The next prime number is 7, and we exclude each seventh number after 7. Note that 49 is the first multiple of 7 that has not already been excluded as being a member of another set of multiples. The next prime number is 11. Since all multiples of 11 in this set have already been excluded, the remaining numbers that have not been excluded are prime numbers and may be circled.

Notice that 49 is the first number that is divisible by 7 and is not also divisible by a prime number less that 7. In other words, each composite number less than 7^2 has at least one of its factors less than 7. Similarly, we might have observed that each composite number less than 5^2 has at least one factor less than 5. In general, *for any prime number p each composite number less than p^2 has a prime number less than p as a factor.*

We use this property to tell us when we have excluded all composite numbers from a set. In the set of numbers {1, 2, . . . , 100} we have considered the primes 2, 3, 5, and 7. The next prime is 11. Thus by our method we have already excluded all composite numbers up to but not including 11^2, that is, 121. In particular, we have identified the set of prime numbers less than or equal to 100.

Example 1: List the set of prime numbers less than 70.

Solution: From the chart we identify this set as

{2, 3, 5, 7, 11, 13, 17, 19, 23, 29, 31, 37, 41, 43, 47, 53, 59, 61, 67}

We have seen that every counting number greater than 1 is either a prime number or a composite number. Now we shall find that, except for the order of the factors, every counting number greater than 1 can be written as a product of powers of prime numbers in one and only one way.

Consider the various ways of factoring 24:

$$24 = 1 \times 24$$
$$24 = 2 \times 12$$
$$24 = 3 \times 8$$
$$24 = 4 \times 6$$
$$24 = 2 \times 2 \times 6$$
$$24 = 2 \times 3 \times 4$$
$$24 = 2 \times 2 \times 2 \times 3 = 2^3 \times 3$$

The last factorization in terms of the prime numbers 2 and 3 could be written as $2 \times 3 \times 2^2$ and in other ways. However, these ways are equivalent, since the order of the factors does not affect the product. Thus 24 can be expressed as the product of a unique set of powers of its prime factors.

One of the easiest ways to find the prime factors of a number is to consider the prime numbers

2, 3, 5, 7, 11, 13, 17, 19, 23, 29, 31, . . .

in order and use each one as a factor as many times as possible. Then for 24 we would have the following:

$$\begin{aligned} 24 &= 2 \times 12 \\ &= 2 \times 2 \times 6 \\ &= 2 \times 2 \times 2 \times 3 \end{aligned}$$

Some people prefer to write these steps using division:

$$\begin{array}{r} 2\ \underline{)\ 24} \\ 2\ \underline{)\ 12} \\ 2\ \underline{)\ 6} \\ 3 \end{array}$$

Since 3 is a prime number, no further steps are needed and $24 = 2^3 \times 3$.

Example 2: Express 3850 in terms of its prime factors.

Solution:

$$\begin{array}{r} 2\ \underline{)\ 3850} \\ 5\ \underline{)\ 1925} \\ 5\ \underline{)\ 385} \\ 7\ \underline{)\ 77} \\ 11 \end{array} \qquad 3850 - 2 \times 5^2 \times 7 \times 11$$

In general, if a counting number n is greater than 1, then n has a prime number p_1 as a factor. Suppose

$$n = p_1 n_1$$

Then if n is a prime number, $n = p_1$ and $n_1 = 1$. If n is not a prime number, then n_1 is a counting number greater than 1. In this case n_1 is either a prime number or a composite number. Suppose

$$n_1 = p_2 n_2 \quad \text{and thus} \quad n = p_1 p_2 n_2$$

where p_2 is a prime number. As before, if $n_2 \neq 1$, then

$$n_2 = p_3 n_3 \quad \text{and thus} \quad n = p_1 p_2 p_3 n_3$$

where p_3 is a prime number, and so forth. We may continue this process until some $n_k = 1$, since there are only a finite number of counting numbers less than n and

$$n > n_1 > n_2 > n_3 > \cdots > n_k = 1$$

Then we have an expression for n as a product of prime numbers:

$$n = p_1 p_2 p_3 \cdots p_k$$

We call this the **prime factorization** of n, that is, the factorization of n into its prime factors. Except for the order of the factors, the prime factorization of any counting number greater than 1 is unique; that is *any counting number greater than* 1 *may be expressed as a product of a unique set of powers of its prime factors.* As in the examples, we usually write the prime factorization as a product of powers of prime numbers.

Example 3: Find the prime factorization of 5280.

Solution:

$$
\begin{array}{r|l}
2 & 5280 \\
2 & 2640 \\
2 & 1320 \\
2 & 660 \\
2 & 330 \\
3 & 165 \\
5 & 55 \\
 & 11
\end{array}
\qquad 5280 = 2^5 \times 3 \times 5 \times 11
$$

EXERCISES

Let A = the set of numbers divisible [illegible] 2, *B = the set of numbers divisible by* 3, *D = the set of numbers divisible b*[illegible] *F = the set of numbers divisible by* 12, *and H = the set of numbers divisib*[illegible]*y* 15. *Express as an English sentence in terms of divisibility, each of the statements in Exercises* 1 *through* 6.

1. $H \subset B$ *2.* $H \subset D$ *3.* $B \cap D = H$

4. $F \subset A$ *5.* $F \subset B$ *6.* $F \subset (A \cap B)$

7. List the composite numbers between 20 and 50.

8. Adapt the Sieve of Eratosthenes to find the prime numbers less than or equal to 200.

9. Is every odd number a prime number? Is every prime number an odd number?

10. Exhibit a pair of prime numbers that differ by 1 and show that there is only one such pair possible.

11. Here is a famous theorem that has not yet been proved: Every even number greater than 2 is expressible as the sum of two prime numbers. (This theorem is often called **Goldbach's conjecture.**) Express each even number from 4 to 40 inclusive as a sum of two prime numbers.

12. Here is another famous theorem. Two prime numbers that differ by 2, such as 17 and 19, are called *twin primes*. It is believed, but has not yet been proved, that there are infinitely many twin primes. Find a pair of twin primes that are between **(a)** 35 and 45; **(b)** 65 and 75; **(c)** 95 and 105.

13. A set of three prime numbers that differ by 2 is called a *prime triplet*. Exhibit a prime triplet and explain why it is the only possible triplet of primes.

14. It has been conjectured but not proved that every odd number greater than 5 is expressible as the sum of three prime numbers. Verify this conjecture for the numbers 7, 9, 11, 13, and 15.

***15.** What is the largest prime that you need to consider to be sure that you have excluded all composite numbers less than or equal to **(a)** 200; **(b)** 500; **(c)** 1000?

Express each number as the product of two counting numbers in as many different ways as possible.

16. 12	***17.*** 15	***18.*** 18
19. 24	***20.*** 17	***21.*** 31

Find the prime factorization of each number.

22. 96	***23.*** 64	***24.*** 415
25. 213	***26.*** 938	***27.*** 2425
28. 257	***29.*** 618	***30.*** 3000

PEDAGOGICAL EXPLORATIONS

1. Many elementary school texts use the idea of a branching tree to help students think about factors. If the branches terminate with prime numbers, then we have a **prime-factor tree**. Two such trees are shown below. Draw prime-factor trees for 30, 60, and 96.

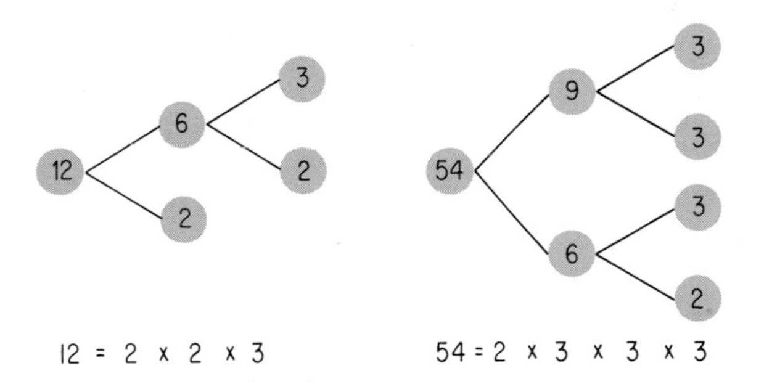

2. No one has ever been able to find a formula that will produce only prime numbers. At one time it was thought that the expression $n^2 - n + 41$ would give only prime numbers for the set of counting numbers as replacements for n. Show that prime numbers are obtained when n is replaced by 1, 2, 3, 4, and 5. Show that the formula fails for $n = 41$.

3. Euclid proved that the set of prime numbers is infinite. Refer to a textbook or a book on the history of mathematics and study this simple, yet elegant, proof.
4. At the time of the writing of this book the largest known prime was $2^{19,937} - 1$, a number with 6002 digits in decimal notation. Look for and read a reference to man's search for large prime numbers through the use of computers.
5. A theorem about prime numbers first stated by Pierre de Fermat in 1640 states that if p is a prime number, then for every integer a, $a^p - a$ is divisible by p. For example, if $p = 2$, then $a^2 - a$ is divisible by 2 for all integers a. Test that this is so for at least five different values of a. Then let $p = 3$ and test again. Then let $p = 4$, not a prime, and show by a single counterexample that the theorem does not hold for composite numbers.
6. Numbers of the form $2^{(2^n)} + 1$ for $n = 0, 1, 2, 3, \ldots$ are known as Fermat numbers. Fermat conjectured, incorrectly, that all such numbers would be prime. For $n = 0$, we have $2^{(2^0)} + 1 = 2^1 + 1 = 3$, a prime number. Show that prime numbers are obtained for $n = 1, n = 2$, and $n = 3$.
7. It can be proved that every counting number greater than 11 is the sum of two composite numbers. Show, by example, that this is true for the counting numbers 12 through 25.

4-3 applications of prime factorizations

We can make effective use of prime factorizations in many arithmetic situations. Let us first consider the various factors of counting numbers. Recall that a counting number t is said to be a factor of another counting number n if and only if there is a counting number k such that $n = t \times k$. For example, the number 3 is a factor of 18 because there exists a number, namely 6, such that $18 = 3 \times 6$.

The set of counting numbers that are factors of a given counting number may be found by listing pairs of numbers whose product is the given counting number. This is illustrated below for 12 and 18:

	12	18
	1×12	1×18
	2×6	2×9
	3×4	3×6
Factors:	12: {1, 2, 3, 4, 6, 12}	18: {1, 2, 3, 6, 9, 18}

The set of **common factors** of 12 and 18 consists of those numbers that are factors of both 12 and 18, that is,

{1, 2, 3, 6}

The largest member of this set, 6, is called the *greatest common factor* (G.C.F.) of the two numbers. In general, the **greatest common factor** of two or more counting numbers is the largest counting number that is a factor of each of the given numbers.

We may use the prime factorization of two counting numbers to find their greatest common factor. First express each number by its prime factorization. Then consider the prime numbers that are factors of both of the given numbers, and take the product of those prime numbers with each raised to the highest power that is a factor of *both* of the given numbers. For $12 = 2^2 \times 3$ and $18 = 2 \times 3^2$ we have G.C.F. $= 2 \times 3$, that is, 6.

Example 1: Find the greatest common factor of 60 and 5280.

Solution:

$$\begin{array}{r|l} 2 & 60 \\ \hline 2 & 30 \\ \hline 3 & 15 \\ \hline & 5 \end{array} \qquad 60 = 2^2 \times 3 \times 5$$

As in Example 3 of §4-2 we have $5280 = 2^5 \times 3 \times 5 \times 11$. Then the highest power of 2 that is a common factor of 60 and 5280 is 2^2; 3 is a common factor; 5 is a common factor; 11 is not a common factor. The greatest common factor of 60 and 5280 is $2^2 \times 3 \times 5$, that is, 60.

Example 2: Find the greatest common factor of 3850 and 5280.

Solution:

$$3850 = 2 \times 5^2 \times 7 \times 11$$
$$5280 = 2^2 \times 3 \times 5 \times 11$$

The greatest common factor of 3850 and 5280 is $2 \times 5 \times 11$, that is, 110.

Example 3: Find the G.C.F. of 12, 36, and 60.

Solution: First write the prime factorization of each number:

$$12 = 2^2 \times 3$$
$$36 = 2^2 \times 3^2$$
$$60 = 2^2 \times 3 \times 5$$

The G.C.F. is $2^2 \times 3$, that is, 12.

We may use the greatest common factor when we reduce (simplify) a fraction. For example,

$$\frac{60}{4880} = \frac{(2^2 \times 5) \times 3}{(2^2 \times 5) \times (2^2 \times 61)} = \frac{3}{2^2 \times 61} = \frac{3}{244}$$

Since 3 is the only prime factor of the numerator and 3 is not a factor of the denominator, the fraction $\frac{3}{244}$ is in **lowest terms.** The numerator and the denominator do not have any common prime factors and are said to be **relatively prime.**

Example 4: Reduce the fraction $\frac{60}{168}$ to lowest terms.

Solution:

$$60 = 2^2 \times 3 \times 5$$
$$168 = 2^3 \times 3 \times 7$$

The greatest common factor is $2^2 \times 3$.

$$\frac{60}{168} = \frac{(2^2 \times 3) \times 5}{(2^2 \times 3) \times 2 \times 7} = \frac{5}{14}$$

Let us now turn our attention to the concept of a multiple of a number. Recall that if a counting number t is a factor of a counting number n, then n is said to be a multiple of t. Consider the set of multiples of 12 and the set of multiples of 18:

12: {12, 24, 36, 48, 60, 72, 84, 96, 108, 120, . . .}
18: {18, 36, 54, 72, 90, 108, 126, . . .}

The set of **common multiples** of 12 and 18 consists of those numbers that are multiples of both 12 and 18, that is,

{36, 72, 108, . . .}

The smallest member of this set, 36, is called the *least common multiple* (L.C.M.) of the two numbers. In general, the **least common multiple** of two or more counting numbers is the smallest number that is a multiple of each of the given numbers.

We may use the prime factorization of two numbers to find their least common multiple. First express each number by its prime factorization. Then consider the prime factors that are factors of either of the given numbers, and take the product of these prime numbers with each raised to the highest power that occurs in *either* of the prime factorizations. Thus, for $12 = 2^2 \times 3$ and $18 = 2 \times 3^2$, we have L.C.M. $= 2^2 \times 3^2$, that is, 36.

Example 5: Find the least common multiple of 3850 and 5280.

Solution: As in Examples 1 and 2,

$$3850 = 2 \times 5^2 \times 7 \times 11$$
$$5280 = 2^5 \times 3 \times 5 \times 11$$

The least common multiple of 3850 and 5280 is

$$2^5 \times 3 \times 5^2 \times 7 \times 11$$

that is, 184,800.

Example 6: Find the L.C.M. of 12, 18, and 20.

Solution: First write the prime factorization of each number:

$$12 = 2^2 \times 3$$

$$18 = 2 \times 3^2$$
$$20 = 2^2 \times 5$$

The L.C.M. is $2^2 \times 3^2 \times 5$, that is, 180.

We use the least common multiple of the denominators of two fractions when we add or subtract fractions. For example, the least common multiple of 12 and 18 is 36:

$$\frac{7}{12} + \frac{5}{18} = \frac{21}{36} + \frac{10}{36} = \frac{31}{36}$$

The answer is in reduced form, since 31 and 36 are relatively prime.

Example 7: Simplify: $\frac{37}{5280} - \frac{19}{3850}$.

Solution: We use the least common multiple as found in Example 5:

$$\frac{37}{5280} - \frac{19}{3850} = \frac{37}{2^5 \times 3 \times 5 \times 11} - \frac{19}{2 \times 5^2 \times 7 \times 11}$$

$$= \frac{37 \times 5 \times 7}{2^5 \times 3 \times 5^2 \times 7 \times 11} - \frac{19 \times 2^4 \times 3}{2^5 \times 3 \times 5^2 \times 7 \times 11}$$

$$= \frac{1295 - 912}{2^5 \times 3 \times 5^2 \times 7 \times 11} = \frac{383}{184{,}800}$$

The instruction "simplify" is used as in Example 7 to mean "perform the indicated operations and express the answer in simplest form." In the case of fractions, "express in simplest form" means "reduce to lowest terms."

EXERCISES

Find the set of factors of:

1. 20 ***2.*** 24 ***3.*** 25 ***4.*** 36

5. 28 ***6.*** 64 ***7.*** 60 ***8.*** 72

Write the prime factorizations and find the G.C.F. of:

9. 68 and 96 ***10.*** 123 and 215

11. 96 and 1425 ***12.*** 123 and 1425

13. 215 and 1425 ***14.*** 68 and 112

15. 12, 15, and 20 ***16.*** 18, 24, and 45

17. 12, 18, and 36 ***18.*** 15, 45, and 60

List the first five elements in the set of multiples of:

19. 7

20. 8

21. 15

22. 20

Write the prime factorizations and find the L.C.M. of:

23. 68 and 96

24. 123 and 215

25. 96 and 1425

26. 123 and 1425

27. 215 and 1425

28. 68 and 112

29. 12, 15, and 20

30. 18, 24, and 45

31. 12, 18, and 36

32. 15, 45, and 60

Simplify, using the concepts of this section.

33. $\frac{369}{645}$

34. $\frac{96}{1425}$

35. $\frac{5}{8} + \frac{7}{12}$

36. $\frac{7}{8} - \frac{5}{6}$

37. $\frac{9}{10} - \frac{1}{15}$

38. $\frac{5}{12} + \frac{9}{16}$

39. $\frac{11}{12} - \frac{9}{18}$

40. $\frac{5}{68} + \frac{11}{76}$

41. $\frac{7}{123} - \frac{2}{215}$

42. $\frac{41}{215} + \frac{19}{1425}$

43. What is the *least* common factor of any two counting numbers?

44. What is the *greatest* common multiple of any two counting numbers?

PEDAGOGICAL EXPLORATIONS

Extend the following table for counting numbers from 1 through 17.

Counting numbers	Factors	Number of factors	Sum of factors
1	1	1	1
2	1, 2	2	3
•	•	•	•
•	•	•	•
•	•	•	•

1. Can any counting number greater than 1 have only one factor?

2. Can any prime number be identified in terms of the number n of its factors?
3. Can any composite number be identified in terms of the number n of its factors?
4. Give a rule for determining whether or not a counting number is a prime number if the number of its factors is known.
5. Find a way to distinguish in terms of the number of factors the counting numbers that are squares of counting numbers.

4-4 whole numbers

The counting numbers serve man well and are actually all that he needs for many purposes. However, with just this collection of numbers at his disposal he is unable to ask for *half* of a piece of pie, or to say that he has *no* money in his pocket because $\frac{1}{2}$ and 0 are not counting numbers. We need to extend our number system and shall do so by first introducing a numeral 0 to represent the number, zero, of elements in the empty set. Zero is the cardinal number of any empty set.

We also need the number zero to solve sentences such as $5 + n = 5$. If $5 + n = 5$, then $n = 0$ and $5 + 0 = 5$.

Identity, +: $a + 0 = 0 + a = a$. The number zero is the **additive identity** and is also called the **identity element for addition.** This property is often referred to as the **addition property of zero.**

When we include 0 with the set of counting numbers, we form a new set W of **whole numbers:**

$$W = \{0, 1, 2, 3, 4, 5, \ldots\}$$

Note that the only difference between the set of counting numbers and the set of whole numbers is the element 0. Often we say that the set of whole numbers consists of the *union* of the set of counting numbers and zero.

As in §1-5 we can graph sets of whole numbers on a *number line* since every whole number is the coordinate of some point on the number line. For example, let us draw the graph of the set of whole numbers *between* 1 and 5. The word "between" indicates that we are not to include the points for 1 and 5. Thus we draw a number line and place solid dots at the points that correspond to 2, 3, and 4 as in the figure.

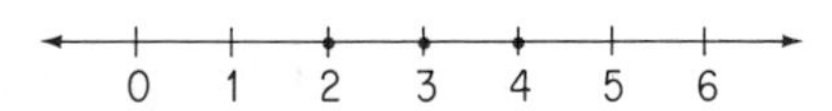

The set of whole numbers has all of the properties that were described for the set of counting numbers, as well as having an identity element for addition. These properties are listed at the top of the next page.

Closure, +	Closure, ×
Commutative, +	Commutative, ×
Associative, +	Associative, ×
Identity, +	Identity, ×
Distributive, × with respect to +	

The distinguishing feature between the set of counting numbers and the set of whole numbers is the presence of 0 as a whole number. The number 0 not only serves as the additive identity, but also has the property:

Zero, ×: $a \times 0 = 0 \times a = 0$. The product of any whole number a and 0 is 0. This property is frequently referred to as the **multiplication property of zero.**

EXERCISES

In Exercises 1 through 10 graph each set of numbers on a number line.

1. {1, 2, 3, 4, 5}
2. {3, 7, 8}
3. {0, 2, 4, 6, 8}
4. {0, 1}
5. The set of counting numbers less than 6.
6. The set of whole numbers less than 6.
7. The set of whole numbers between 0 and 5.
8. The set of counting numbers between 1 and 3.
9. The set of counting numbers between 0 and 1.
10. The set of whole numbers greater than or equal to 5 and less than 9.

11. Show that there is a one-to-one correspondence between the elements of the set of whole numbers less than 5 and the elements of the set of counting numbers less than 6.
12. If a and b represent whole numbers and $a \times b = 0$, what can you conclude about a and b?
13. Repeat Exercise 12 if $a \times b = 1$; if $a \times b = 2$; if $a \times b = 3$; and if $a \times b = 4$.
14. Does every point on the number line have a whole number as its coordinate? Is every whole number the coordinate of some point on the number line?
15. Let n represent any whole number. How many whole numbers are there between n and $n + 1$?
16. If n represents any whole number, then $n + 1$ is called the **successor** of n. Does every whole number have a successor? Is every whole number the successor of some other whole number?

17. Show that there is a one-to-one correspondence between the elements of the set of whole numbers and the elements of the set of counting numbers.

sec. 4-4 whole numbers

18. Consider $A \times \varnothing$ and $\varnothing \times A$ for any set A. Identify these sets and the associated property of whole numbers.

PEDAGOGICAL EXPLORATIONS

1. Many interesting games and puzzles that involve counting numbers only can be used in the classroom. One such puzzle involves using four fours, and any operation with which the student is familiar. The objective is to represent as many counting numbers as possible in this way. Here are several examples:

$$1 = \frac{44}{44} \qquad 2 = (4 \div 4) + (4 \div 4)$$

$$3 = (4 + 4 + 4) \div 4 \qquad 4 = 4 + [(4 - 4) \div 4]$$

See how far you can extend this list.

2. Another interesting puzzle is to represent the set of counting numbers using the digits of a year *in the order in which they appear.* Here are several examples, using the bicentennial year 1976:

$$1 = 1^{976} \qquad 2 = 1^9 + 7 - 6 \qquad 3 = 1 + \sqrt{9} - 7 + 6$$

How many others can you find?

Cross-number puzzles are very popular with elementary school students and serve as a novel and interesting way to provide practice with fundamental skills. The next figure shows such a puzzle for addition; the student is to find the sum of each row, column, and diagonal. Adding across we have the sums 8 + 9 = 17 and 7 + 6 = 13. Adding down we have 8 + 7 = 15 and 9 + 6 = 15. These partial sums, 15 + 15 and 17 + 13, are then added across and down to give the same sum in each direction, 30.

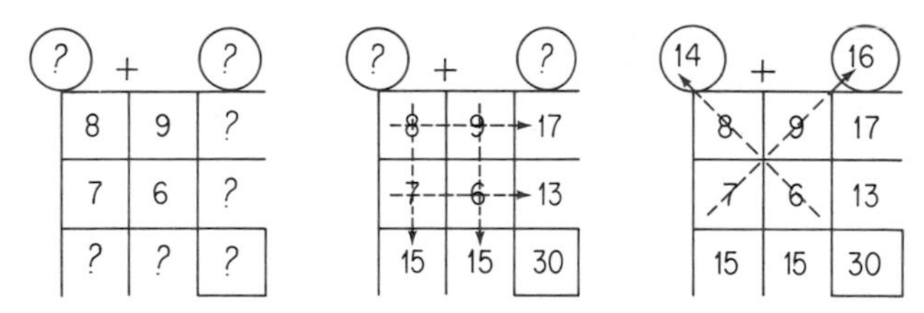

This is shown in the box in the lower right-hand corner. As a final check of this work, the student is asked to add along the diagonals, 7 + 9 and 8 + 6. These diagonal sums are placed in the circles, and their sum, 14 + 16, should give the same number found in the box, 30.

3. Prepare a similar cross-number puzzle for multiplication.

4. Prepare cross-number puzzles for subtraction and division. In these cases the diagonal check does not seem to work. See if you can suggest a modification of the procedures used for addition and multiplication so that the diagonals may be used as checks.

5. Elementary school students have fun with puzzles that have missing entries. Complete the following and make up similar ones of your own for use in class.

(a)
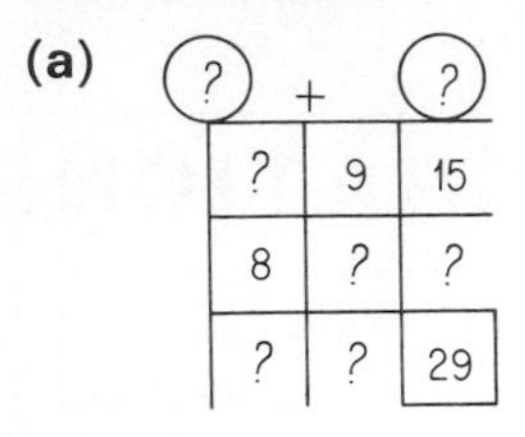

(b)
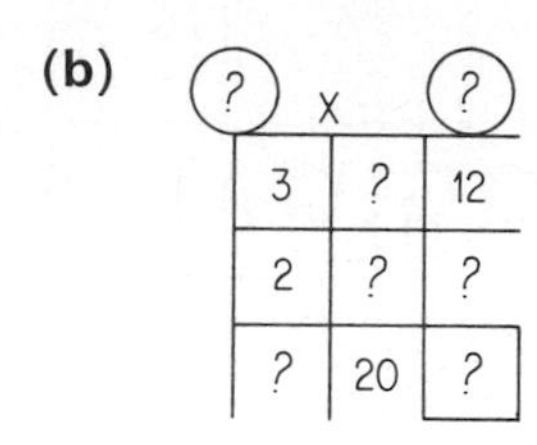

4-5 addition and subtraction

We are now ready to discuss in some detail the *algorithms* (procedures) for addition and subtraction of whole numbers. To do so we shall first explore several different approaches that may be used to justify the manner in which we add. One approach is to use expanded notation and the properties of the set of whole numbers. Supply a reason for each step shown in the following illustrative example where the numbers 35 and 49 are called the **addends** in the sum 35 + 49.

$$\begin{aligned}
35 + 49 &= [(3 \times 10) + (5 \times 1)] + [(4 \times 10) + (9 \times 1)] \\
&= [(3 \times 10) + (4 \times 10)] + [(5 \times 1) + (9 \times 1)] \\
&= (7 \times 10) + (14 \times 1) \\
&= (7 \times 10) + [(10 \times 1) + (4 \times 1)] \\
&= (7 \times 10) + [(1 \times 10) + (4 \times 1)] \\
&= (8 \times 10) + (4 \times 1) \\
&= 84
\end{aligned}$$

Of course we normally do not solve problems of addition in this cumbersome manner. Instead, our usual procedure looks something like this:

$$\begin{array}{r} 1 \\ 35 \\ +49 \\ \hline 84 \end{array}$$

We think: "5 + 9 is 14; put down the 4 and carry 1. Then 1 + 3 + 4 = 8."

In reality, as in the detailed explanation above, we first add the numbers in the ones column. We then "exchange" 14 ones for 1 ten and 4 ones. Then, in the tens column, we add 1 ten, 3 tens, and 4 tens to obtain 8 tens. The result is then 8 tens and 4 ones, or 84. This process may be shown in detail as follows:

$$\begin{array}{rl} 35 = & (3 \times 10) + (5 \times 1) \\ +49 = & (4 \times 10) + (9 \times 1) \\ \hline & (7 \times 10) + (14 \times 1) \end{array}$$

Exchange (14×1) for $(1 \times 10) + (4 \times 1)$:

$$\begin{array}{rl} & (1 \times 10) \\ 35 = & (3 \times 10) + (5 \times 1) \\ +49 = & (4 \times 10) + (9 \times 1) \\ \hline & (8 \times 10) + (4 \times 1) = 84 \end{array}$$

The following sequence of steps is yet another way to justify the usual addition algorithm and frequently helps clarify the procedures used.

$$\begin{array}{rl} 35 = & 30 + 5 \\ +49 = & 40 + 9 \\ \hline & 70 + 14 = 70 + (10 + 4) \\ & \qquad = (70 + 10) + 4 \\ & \qquad = 80 + 4 \\ & \qquad = 84 \end{array}$$

Note that the procedure shown here for addition in our decimal system of whole numbers is the same as that used for §3-7 for addition in base five. Indeed, one of the major justifications for a unit on other number bases in the curriculum is precisely that of showing this analogy between the algorithms used for computation with different place value systems of numeration in order to strengthen the student's understanding of our usual base ten system. For convenience 10^1 is usually written as 10 and 10^0 as 1. Example 1 provides a further illustration of this approach.

Example 1: Use expanded notation to find the sum 387 + 259.

Solution: First add the column of ones.

$$\begin{array}{rl} 387 = & (3 \times 10^2) + (8 \times 10) + (7 \times 1) \\ +259 = & (2 \times 10^2) + (5 \times 10) + (9 \times 1) \\ \hline & \qquad\qquad\qquad\qquad (16 \times 1) \end{array}$$

Next write (16×1) as $(1 \times 10) + (6 \times 1)$. "Carry" the (1×10) to the tens column, and add the column of tens.

$$\begin{array}{rrr} & (1 \times 10) & \\ (3 \times 10^2) + & (8 \times 10) + & (7 \times 1) \\ (2 \times 10^2) + & (5 \times 10) + & (9 \times 1) \\ \hline & (14 \times 10) + & (6 \times 1) \end{array}$$

Now write (14×10) as $(10 \times 10) + (4 \times 10)$; then as $(1 \times 10^2) + (4 \times 10)$. "Carry" the (1×10^2) to the 10^2 column (the hundreds column) and add again.

$$\begin{array}{l} (1 \times 10^2) \\ (3 \times 10^2) + (8 \times 10) + (7 \times 1) \\ (2 \times 10^2) + (5 \times 10) + (9 \times 1) \\ \hline (6 \times 10^2) + (4 \times 10) + (6 \times 1) = 646 \end{array}$$

Note, in Example 1, that the usual method of adding columns and "carrying" is possible because of the place value of our decimal system of notation. In abbreviated form, this example would be completed in this way:

(a) Add the ones.

$$\begin{array}{r} 387 \\ +259 \\ \hline 16 \end{array}$$

(b) Carry 1 ten to the tens' column and add.

$$\begin{array}{r} 1 \\ 387 \\ +259 \\ \hline 6 \\ 14 \end{array}$$

(c) Carry 1 hundred to the hundreds' column and add.

$$\begin{array}{r} 1 \\ 387 \\ +259 \\ \hline 6 \\ 4 \\ 6 \\ \hline 646 \end{array}$$

In reality, we are really regrouping at each step and exchanging 10 ones for 1 ten, and then 10 tens for 1 hundred. The following display should serve to further clarify this process.

$$\begin{array}{rl} 387 = & 300 + 80 + 7 \\ +259 = & 200 + 50 + 9 \\ \hline & 500 + 130 + 16 = 500 + 130 + (10 + 6) \\ & = 500 + (130 + 10) + 6 \\ & = 500 + 140 + 6 \\ & = 500 + (100 + 40) + 6 \\ & = (500 + 100) + 40 + 6 \\ & = 600 + 40 + 6 \\ & = 646 \end{array}$$

A final interesting approach is to add the numbers of ones, tens, and hundreds separately, taking advantage of place value, and then add the totals.

$$\begin{array}{rl} 387 & \\ +259 & \\ \hline 16 & (7 \text{ ones} + 9 \text{ ones} = 16 \text{ ones} = 16) \\ 130 & (8 \text{ tens} + 5 \text{ tens} = 13 \text{ tens} = 130) \\ 500 & (3 \text{ hundreds} + 2 \text{ hundreds} = 5 \text{ hundreds} = 500) \\ \hline 646 & \end{array}$$

Let us now turn our attention to the process used in subtraction of whole numbers such as 73 — 28. The usual algorithm for subtraction can be explained by first writing each term in this expanded form:

$$\begin{array}{rl} 73 = & (7 \times 10) + (3 \times 1) \\ -28 = & (2 \times 10) + (8 \times 1) \\ \hline \end{array}$$

We need to subtract 8 ones from 3 ones. Since this is not possible with whole numbers, we "borrow" from the tens' column. Rewrite 73 in this way:

$$\begin{aligned} 73 = (7 \times 10) + (3 \times 1) &= (6 \times 10) + (1 \times 10) + (3 \times 1) \\ &= (6 \times 10) + (10 \times 1) + (3 \times 1) \\ &= (6 \times 10) + (13 \times 1) \end{aligned}$$

Our work now looks like this:

$$\begin{array}{rl} 73 = & (6 \times 10) + (13 \times 1) \\ \underline{-28} = & \underline{(2 \times 10) + (\ 8 \times 1)} \\ & (4 \times 10) + (\ 5 \times 1) = 45 \end{array}$$

In actual practice, of course, we do not write all of the steps shown. Instead we usually cross out the 7, and replace it by 6. This shows that we are borrowing 1 ten from the 7 tens, leaving 6 tens. We then show a small numeral 1 alongside the ones' column to denote the borrowed 10 as 10 ones.

$$\begin{array}{r} 73 \\ \underline{-28} \end{array} \qquad \begin{array}{r} \overset{6}{\not 7}\,{}^{1}3 \\ \underline{2\ 8} \\ 4\ 5 \end{array}$$

Example 2: Express 237 and 805 in terms of hundreds, tens, and ones. Then find the difference 805 — 237.

Solution:

(a)
$$\begin{array}{rcl} 805 & = & 800 + \ \ 0 + 5 \\ \underline{-237} & = & \underline{200 + 30 + 7} \end{array}$$

(b) Next, rewrite 800 as 700 + 10 tens, or 100.

$$\begin{array}{r} 700 + 100 + 5 \\ \underline{200 + \ \ 30 + 7} \end{array}$$

(c) Now write 100 as 90 + 10. Add 10 + 5. Then subtract.

$$\begin{array}{r} 700 + 90 + 15 \\ \underline{200 + 30 + \ \ 7} \\ 500 + 60 + \ \ 8 = 568 \end{array}$$

In actual practice, the work demonstrated in Example 2 would normally be completed by the following corresponding steps:

(a) $\begin{array}{r} 805 \\ \underline{-237} \end{array}$ (b) $\begin{array}{r} \overset{7}{\not 8}\,{}^{1}05 \\ \underline{2\ 37} \end{array}$ (c) $\begin{array}{r} \overset{7}{\not 8}\,\overset{9}{\not 1}\!\!\not 0\,{}^{1}5 \\ \underline{2\ 3\ 7} \\ 5\ 6\ 8 \end{array}$

EXERCISES

Use expanded notation to complete each addition problem.

1. $\begin{array}{r} 45 \\ +38 \\ \hline \end{array}$ **2.** $\begin{array}{r} 56 \\ +29 \\ \hline \end{array}$ **3.** $\begin{array}{r} 375 \\ +287 \\ \hline \end{array}$

4. $\begin{array}{r} 509 \\ +238 \\ \hline \end{array}$ **5.** $\begin{array}{r} 1309 \\ +2578 \\ \hline \end{array}$ **6.** $\begin{array}{r} 4793 \\ +8147 \\ \hline \end{array}$

Use expanded notation to complete each subtraction problem.

7. $\begin{array}{r} 95 \\ -32 \\ \hline \end{array}$ **8.** $\begin{array}{r} 85 \\ -37 \\ \hline \end{array}$ **9.** $\begin{array}{r} 304 \\ -128 \\ \hline \end{array}$

10. $\begin{array}{r} 350 \\ -179 \\ \hline \end{array}$ **11.** $\begin{array}{r} 5023 \\ -2709 \\ \hline \end{array}$ **12.** $\begin{array}{r} 8301 \\ -2076 \\ \hline \end{array}$

PEDAGOGICAL EXPLORATIONS

1. Many elementary school students have special difficulties with subtraction whenever 0 appears. Explore an elementary mathematics textbook for grades 3 or 4 and note what attention is given to instruction and practice with addition and subtraction problems that involve zeros.
2. Prepare a set of worksheets that might be used for individualized study by an elementary school student who is having difficulty with subtraction problems that involve zeros in the larger of the two numbers.
3. In a subtraction problem such as $85 - 37$, 85 is sometimes called the *minuend*, and 37 is called the *subtrahend*. Explore an elementary textbook series to determine what attention, if any, is given to these words.
4. Subtraction is often introduced as the inverse operation of addition. For example, since $5 + 7 = 12$ and $7 + 5 = 12$, it then follows that $12 - 5 = 7$ and $12 - 7 = 5$. Prepare a worksheet of problems for a specified elementary grade level that emphasizes these related number facts.

More and more course guides and curriculum outlines are now being introduced through the use of **behavioral objectives**. That is, objectives of instruction are stated in terms of desired student behavior. Such objectives are easy to assess in that the statement should indicate the type of performance required by the student to demonstrate attainment of the objective.

There are certain key action words that are useful in constructing behavioral objectives. These include the following:

Identify	Describe
Distinguish	State a principle or rule
Construct	Apply the rule
Name	Demonstrate
Order	Interpret

Objectives as traditionally stated have been vague and difficult to evaluate. For example, the objective "the student should understand the relationship between rational numbers and decimals" does not clearly spell out the expected behavior on the part of the student after studying such a unit of work. As a contrast, consider the objective "the student should be able to demonstrate the expanded notation for a given decimal numeral." Here we know what is expected of the student, and can clearly construct a test item to evaluate his knowledge of this particular item. For a further reference on this topic see Readings and Projects 13 in the listing at the end of this chapter.

5. Write five behavioral objectives for the material of this section as it would be presented to an elementary school class. Use a different action word for each objective.

6. Prepare a 30-minute test that can be used to evaluate an elementary student's knowledge of addition and subtraction with whole numbers.

4-6 multiplication and division

Our task in this section will be to carefully examine the algorithms that we use for multiplication and division of whole numbers. Let us begin with a product such as 7×12. A computer handles this problem by treating the operation of multiplication as repeated addition. The product 7×12 may be thought of as the sum obtained by using 7 as an addend 12 times, or by using 12 as an addend 7 times:

$$7 \times 12 = 7 + 7 + 7 + 7 + 7 + 7 + 7 + 7 + 7 + 7 + 7 + 7$$

$$12 \times 7 = 12 + 12 + 12 + 12 + 12 + 12 + 12$$

In §3-2 we found such products through the use of Napier's rods, in this manner.

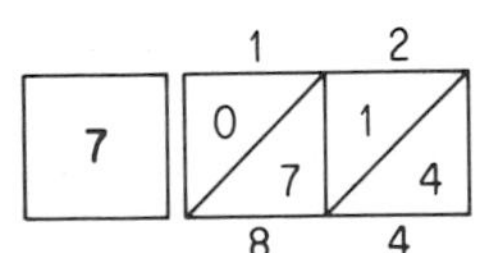

Actually we are making use of the distributive property in disguised form when we multiply with Napier's rods. Note how this is shown in the next display. Note also the use of grouping by tens and ones.

$$
\begin{aligned}
7 \times 12 &= 7 \times (10 + 2) \\
&= (7 \times 10) + (7 \times 2) \\
&= 70 + 14 \\
&= 70 + 10 + 4 \\
&= 80 + 4 \\
&= (8 \times 10) + (4 \times 1) \\
&= 84
\end{aligned}
$$

Example 1: Use the distributive property and show groupings by multiples of powers of 10 for the product 7×235.

Solution:

$$
\begin{aligned}
7 \times 235 &= 7 \times (200 + 30 + 5) \\
&= (7 \times 200) + (7 \times 30) + (7 \times 5) \\
&= 1400 + 210 + 35 \\
&= 1400 + 200 + 10 + 30 + 5 \\
&= 1600 + 40 + 5 \\
&= (16 \times 100) + (4 \times 10) + (5 \times 1) \\
&= 1645
\end{aligned}
$$

Let us return once again to the product 7×12 and see how this can be shown using the traditional vertical arrangement.

$$
\begin{array}{rl}
12 & \\
\times\ 7 & \\
\hline
14 & (7 \times 2) \\
70 & (7 \times 10) \\
\hline
84 & (7 \times 12)
\end{array}
$$

In actual practice, we normally think of 7×2 as 14. We then write the 4 and "carry" the 1. Actually we are carrying 10, and indicate this by writing the numeral 1 in the tens' column as in these two steps:

Step 1:

$$
\begin{array}{r}
{}^{1}12 \\
\times\ 7 \\
\hline
4
\end{array}
$$

7×2 (ones) = 14 (ones); write 4 (ones) and carry 1 (ten).

Step 2:

$$
\begin{array}{r}
{}^{1}12 \\
\times\ 7 \\
\hline
84
\end{array}
$$

7×1 (ten) = 7 (tens); plus the 1 (ten) carried equals 8 (tens).

Let us explore a slightly more complex example, 23×34. One method to explain the usual algorithm for finding such products is to use an expanded form of the distributive property as follows:

$$
\begin{aligned}
23 \times 34 &= (20 + 3) \times (30 + 4) \\
&= (20 \times 30) + (20 \times 4) + (3 \times 30) + (3 \times 4) \\
&= 600 + 80 + 90 + 12 \\
&= 600 + 170 + 12 \qquad \text{(adding numbers of tens)} \\
&= 600 + (100 + 70) + (10 + 2) \\
&= (600 + 100) + (70 + 10) + 2 \\
&= 700 + 80 + 2 \\
&= 782
\end{aligned}
$$

Note where each of the terms appears in this vertical arrangement:

$$
\begin{array}{rl}
34 & \\
\times\ 23 & \\
\hline
12 & (3 \times 4) \\
90 & (3 \times 30) \\
80 & (20 \times 4) \\
600 & (20 \times 30) \\
\hline
782 &
\end{array}
$$

In actual practice, these steps are condensed. Also, because of the place-value nature of our decimal system we need not insert all the 0's as shown in the preceding example. Instead, our work appears as follows:

$$
\begin{array}{rl}
34 & \\
\times\ 23 & \\
\hline
102 & (3 \times 34) \\
68 & \leftarrow \\
\hline
782 &
\end{array}
$$

This is really $20 \times 34 = 680$. The place-value position allows us to think of $2 \times 34 = 68$ instead of $20 \times 34 = 680$.

Just as a computer does a multiplication problem by repeated addition, so it will complete a division problem by repeated subtraction. For example, the quotient $35 \div 5$ can be found by subtracting 5 repeatedly as shown below.

$$
\begin{array}{r}
35 \\
-\ 5 \\
\hline
30 \\
-\ 5 \\
\hline
25 \\
-\ 5 \\
\hline
20 \\
-\ 5 \\
\hline
15 \\
-\ 5 \\
\hline
10 \\
-\ 5 \\
\hline
5 \\
-\ 5 \\
\hline
0
\end{array}
$$

By counting we note that 5 has been subtracted 7 times. Therefore, $35 \div 5 = 7$.

In the example shown, the remainder was 0. If the same process were to be used for the quotient $37 \div 5$, we would continue to subtract 5's until the final remainder was less than 5. In this case we would find the quotient to be 7 and the remainder, 2. We call 37 the **dividend,** 5 the **divisor,** 7 the **quotient,** and 2 the **remainder.**

To arrive at the usual algorithm used for division, we begin by considering the meaning of the division problem $235 \div 5$. As one concrete representation of this problem, we may imagine that we have 235 sticks of gum and wish to package these in bundles of 5. We wish to determine how many such bundles of 5 we will have. One way to arrive at the answer is to take successive guesses as follows:

Since $10 \times 5 = 50$, and $50 < 235$, we certainly have at least 10 bundles of 5. By subtraction, we would then have 185 remaining sticks of gum to package.

$$\begin{array}{r} 10 \\ 5\overline{)235} \\ 50 \\ \hline 185 \end{array}$$

Repeat this procedure as often as possible, subtracting successive quantities of 50 each, until a known multiple of 5 is reached.

$$\begin{array}{r} 10 \\ 5\overline{)185} \\ 50 \\ \hline 135 \end{array} \qquad \begin{array}{r} 10 \\ 5\overline{)135} \\ 50 \\ \hline 85 \end{array} \qquad \begin{array}{r} 10 \\ 5\overline{)85} \\ 50 \\ \hline 35 \end{array} \qquad \begin{array}{r} 7 \\ 5\overline{)35} \\ 35 \\ \hline 0 \end{array}$$

Next, add all of the quotients used to determine the final quotient.

$$10 + 10 + 10 + 10 + 7 = 47; \qquad 235 \div 5 = 47$$

These individual divisions and subtractions can be shown in a single veritcal arrangement with each partial quotient placed above the line.

$$\begin{array}{r} \left.\begin{array}{r} 7 \\ 10 \\ 10 \\ 10 \\ 10 \end{array}\right\} 47 \\ 5\overline{)235} \\ 50 \\ \hline 185 \\ 50 \\ \hline 135 \\ 50 \\ \hline 85 \\ 50 \\ \hline 35 \\ 35 \\ \hline \end{array}$$

With practice one can use this process more efficiently by determining the largest multiple of some power of 10 to use at each step. Thus, in the example under discussion, we note that $40 \times 5 = 200$, whereas $50 \times 5 = 250$. Thus our first partial quotient is 40 as shown next:

$$\begin{array}{r} \left.\begin{array}{r} 7 \\ 40 \end{array}\right\} 47 \\ 5\overline{)235} \\ 200 \\ \hline 35 \\ 35 \\ \hline \end{array}$$

The only advantage to the display just shown over the preceding one is that of convenience and brevity; both provide the same final result. Indeed, it would be cumbersome but possible to go to the extreme of subtracting 5 from 235 repeatedly, which is the actual manner that such a problem is performed by some computers.

$$\begin{array}{r} 1 \\ 5\overline{)235} \\ 5 \\ \hline 230 \end{array} \qquad \begin{array}{r} 1 \\ 5\overline{)230} \\ 5 \\ \hline 225 \end{array} \qquad \begin{array}{r} 1 \\ 5\overline{)225} \\ 5 \\ \hline 220 \end{array} \quad \cdots \quad \text{and so on, for 47 times!}$$

Our traditional algorithm is actually an abbreviation of the procedure explained thus far. Thus, consider the example previously done and note the zeros that are crossed out as being unnecessary due to our place value system.

$$\begin{array}{r} 7 \\ 4\not{0} \\ 5\overline{)235} \\ 20\not{0} \\ \hline 35 \\ 35 \\ \hline \end{array} \qquad \text{In condensed form:} \quad \begin{array}{r} 47 \\ 5\overline{)235} \\ 20 \\ \hline 35 \\ 35 \\ \hline \end{array}$$

Our final result indicates that we should have 47 bundles of 5, and this indicates a procedure for checking a division problem. Assuming there is no remainder, the product of the quotient 47 and the divisor 5 should be equal to the dividend 235.

Let us examine one more division problem in detail, $4961 \div 23$. The explanation will be given in individual steps as follows:

Step 1: Consider the largest multiple of 100 that can be used as a quotient. Since $200 \times 23 = 4600$, and $300 \times 23 = 6900$, we see that the first partial quotient is 200. (This is what is really meant when we divide 23 into 49, as shown in the abbreviated process on the right.)

$$\begin{array}{r} 200 \\ 23\overline{)4961} \\ 4600 \\ \hline 361 \end{array} \qquad \qquad \begin{array}{r} 2 \\ 23\overline{)4961} \\ 46 \\ \hline 3 \end{array}$$

Step 2: Next consider multiples of 10 that can be divided into 361. Since $10 \times 23 = 230$ and $20 \times 23 = 460$, we find the next partial quotient to be 10. (In the abbreviated process we "bring down" 6, and divide 23 into 36.)

```
     10
    200             21
23)4961         23)4961
   4600            46
    361             36
    230             23
    131             13
```

Step 3: We now consider the largest multiple of 23 that we can divide into 131, and find this to be 5; $5 \times 23 = 115$. (In the abbreviated form we "bring down" 1, and divide 23 into 131.) Then add the partial quotients to obtain a quotient of 215, and a remainder of 16.

```
      5 ⎫
     10 ⎬ 215  (quotient)
    200 ⎭                        215
23)4961                      23)4961
   4600                         46
    361                          36
    230                          23
    131                         131
    115                         115
     16  (remainder)             16
```

We can interpret the answer to this problem as meaning that 4961 can be broken up into 215 bundles of 23, with a remainder of 16. This provides a clue for checking when there is a remainder:

```
         quotient
divisor)dividend          dividend = (quotient × divisor) + remainder
           .
           .
           .
         remainder
```

```
    215
23)4961      Check:     215
   46                   ×23
    36                  645
    23                 430
   131                 4945
   115                  +16
    16                 4961   4961 = (215 × 23) + 16
```

If we let $n =$ the dividend, $d =$ the divisor, $q =$ the quotient, and $r =$ the remainder, then the rule for checking may be stated in this form:

$$n = (q \times d) + r, \qquad \text{where } 0 \leq r < d$$

Example 2: Check this division: $387 \div 15 = 25$, remainder 12.

Solution: Let $n = 387$, $q = 25$, $d = 15$, and $r = 12$. Then the problem checks because of the following equality:

$$\begin{aligned} 387 &= (25 \times 15) + 12 \\ &= 375 + 12 \\ &= 387 \end{aligned}$$

EXERCISES

Use repeated addition to find each product.

1. 5×20	***2.*** 6×15
3. 4×7	***4.*** 7×4

Use repeated subtraction to find each quotient.

5. $40 \div 5$	***6.*** $40 \div 8$
7. $120 \div 20$	***8.*** $105 \div 15$

Use the distributive property and multiples of powers of 10 to find each product.

9. 8×15	***10.*** 7×25
11. 9×36	***12.*** 5×435
13. 12×15	***14.*** 16×23
15. 35×45	***16.*** 42×57

Find each quotient by using multiples of 100 and 10 as on page 187.

17. $7712 \div 32$	***18.*** $13{,}803 \div 43$
19. $23{,}328 \div 54$	***20.*** $8970 \div 26$
21. $5683 \div 27$	***22.*** $7943 \div 18$

Check each division. If incorrect, find the correct quotient and remainder.

23. $1107 \div 23 = 48$, remainder 3	***24.*** $1895 \div 29 = 68$, remainder 10
25. $3163 \div 37 = 83$, remainder 18	***26.*** $11{,}828 \div 83 = 142$, remainder 42

PEDAGOGICAL EXPLORATIONS

1. Examine a recently published elementary textbook series and determine each of the following:
   (a) The grade level at which various multiplication facts are introduced.
   (b) The use made of the distributive property in explaining multiplication algorithms.
   (c) The manner in which the division algorithm is first introduced.
   (d) The attention given to multiplication and division as inverse operations.
2. Division is often introduced as the inverse operation of multiplication. For example, since $3 \times 4 = 12$ and $4 \times 3 = 12$, it then follows that $12 \div 3 = 4$ and $12 \div 4 = 3$. Prepare a worksheet of problems for a specified elementary grade level that emphasizes these related number facts.
3. Numerical curiosities can be used in elementary classes as a means of motivating work on fundamental skills. An example of such a device begins by having the student write any three-digit number, such as 475. Then repeat the same set of digits to form a six-digit number, such as 475,475. Now have the student divide this number by 7. Then divide the resulting quotient by 11. Finally, divide the next quotient by 13. Regardless of the number started with, the result will always be the original three-digit number.
   (a) Test the procedure described for several other examples.
   (b) Try to figure out why this "trick" works.
4. An interesting device to motivate drill on multiplication consists of starting with a "magic" number. For example, consider the number 12,345,679. If this number is multiplied by 18, that is, 2×9, the product will consist of 2's only. If multiplied by 27, that is, 3×9, the resulting product will have 3's only.
   (a) Verify this description by multiplying 12,345,679 by 45, that is 5×9 and by 63, that is, 7×9.
   (b) Use the number 15,873 and multiply by successive multiples of 7 to discover another "magic" number.
5. At one time it was fashionable to call the numbers used in a product the *multiplier* and *multiplicand*. Examine several elementary mathematics textbooks to determine if such terminology is still used. If not, are other terms used instead for multiplication problems?
6. Prepare a 30-minute test that evaluates an elementary school student's knowledge of the basic multiplication facts, as well as the multiplication algorithm.
7. Repeat Exploration 6 for division.

4-7 equivalence and order relations

We conclude this chapter with a discussion of various relations and their properties. In mathematics we have, as examples of relations:

is equal to, $=$
is greater than, $>$
is less than or equal to, $\leq$
is parallel to, $||$
is perpendicular to, $\perp$

In everyday language we consider many relations as well, such as:

is a brother of
is a friend of
is more expensive than
is lighter than
is further than

In general, we consider a relation R and a set S, and then explore the properties of R for elements of S. The first property we consider is the **reflexive property:**

$a \text{ R } a$ for all elements a in S.

For example, the *equality relation* ("is equal to" or "is another name for") is reflexive for the set of whole numbers because every whole number is equal to itself; $a = a$ for all whole numbers a. On the other hand, the *order relation* "is less than" is *not* reflexive because a number cannot be less than itself. For example, for $a = 7$, it is not true that $7 < 7$.

In everyday language, the relation "is a brother of" is not a reflexive relation because a person is not considered to be his own brother. However, the relation "is in the same class as" is a reflexive relation because a person is certainly in the same class as himself.

The equality relation also satisfies the **symmetric property:**

If $a \text{ R } b$, then $b \text{ R } a$ for all a and b in S.

For example, if $2 + 3 = 5$, then $5 = 2 + 3$. On the other hand, the order relation "is less than" is not symmetric as can be shown by a single counter-example. For example, it is sufficient to note that $2 < 5$, but it is not true that $5 < 2$.

In everyday language, the relation "is a sister of" is not a symmetric relation. For example, if Sue is a sister of Don, it does not follow that Don is a sister of Sue. However, "is a roommate of" is a symmetric relation because if Ellen is a roommate of Michelle, then certainly Michelle is a roommate of Ellen.

Finally we consider the **transitive property:**

If $a \text{ R } b$ and $b \text{ R } c$, then $a \text{ R } c$ for all a, b, and c in S.

The equality relation has this property as can be illustrated by this example:

$$\text{If } 3 + 5 = 4 + 4, \text{ and } 4 + 4 = 6 + 2, \text{ then } 3 + 5 = 6 + 2.$$

It is interesting to note that the order relation "is less than" is also transitive. For example, $2 < 5$, $5 < 8$, and $2 < 8$.

In everyday language, the relation "is a neighbor of" (meaning, "lives next door to") is not a transitive relation. Thus the Adams can live next door to the Browns, and the Browns can live next door to the Jones, but it does not follow that the Adams live next door to the Jones. Show that the relation "is a descendant of" is a transitive relation.

Any relation that satisfies the three properties that we have just studied is said to be an **equivalence relation.** For example, the equality relation is reflexive ($a = a$), symmetric (if $a = b$, then $b = a$), and transitive (if $a = b$ and $b = c$, then $a = c$). Thus the equality relation is said to be an equivalence relation. For a general relation R the properties of an equivalence relation may be summarized as follows:

Reflexive: $a \text{ R } a$ for all elements a in S.
Symmetric: If $a \text{ R } b$, then $b \text{ R } a$ for all a and b in S.
Transitive: If $a \text{ R } b$ and $b \text{ R } c$, then $a \text{ R } c$ for all a, b, and c in S.

We illustrate these properties for various binary relations in the following examples. A **binary relation** is one that associates a single element with any two given elements of a set.

Example 1: Test the relation "is perpendicular to" for lines in a plane to see if this relation is (a) reflexive; (b) symmetric; (c) transitive.

Solution: (a) The relation is not reflexive since a line is not considered to be perpendicular to itself.
(b) If line m is perpendicular to line n, then line n is perpendicular to line m. The relation is symmetric.
(c) In the figure below, we have $m \perp n$, $n \perp s$, but m is not perpendicular to s. The relation is not transitive.

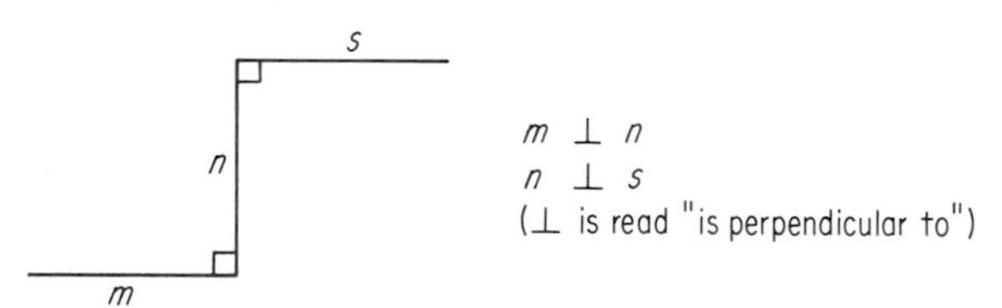

Example 2: Test the relation "weight within five grams of" to determine whether or not it is an equivalence relation.

Solution: The relation is reflexive, since an object weighs within five grams of its own weight. Furthermore if A weighs within five grams of B, then B weighs within five grams of A; the relation is symmetric. However, one counterexample is sufficient to show that the relation is not transitive. Thus assume the weights of A, B, and C are 145, 149, and 153 grams, respectively. Then A weighs within five grams of B, B weighs within five grams of C, but A does not weigh within five grams of C. The relation is therefore not an equivalence relation.

Relations are widely used for many types of elements. For example, "is the same age as" is an equivalence relation among people. Consider three students Don, John, and Bill. Note that even though you do not know their ages, you do know that Don is the same age that he is. If Don and John are the same age, then John and Don are the same age. Also, if Don and John are the same age and John and Bill are the same age, then Don and Bill are the same age.

The relation of "is younger than" is not an equivalence relation since, for example, Don is not younger than himself (not reflexive). Also if Don is younger than Sue, then Sue is not younger than Don (not symmetric). However, if Don is younger than Sue and Sue is younger than Debbie, then Don is younger than Debbie. The relation "is younger than" is transitive.

For numbers the order relations *is less than* and *is greater than* have special importance because of their numerous interpretations such as *is younger than* and *is older than*. You may think of one number such as 2 as less than 5 because the graph of 2 is on the left of the graph of 5 on a number line. From a different point of view, 2 is less than 5 because there is a counting number 3 than can be added to 2 to obtain 5, that is,

$$2 < 5 \quad \text{because} \quad 2 + 3 = 5$$

We may also write $5 > 2$, 5 is greater than 2, if and only if $2 < 5$.

Example 3: Tell whether the relation "is greater than" ($>$) is (a) reflexive, (b) symmetric, (c) transitive.

Solution: (a) It is not reflexive. For example, the statement $5 > 5$ is not true.
(b) It is not symmetric. For example, $5 > 2$, but it is not true that $2 > 5$.
(c) It is transitive. For example, $8 > 5$, $5 > 2$, and $8 > 2$.

EXERCISES

1. Let $A = \{1\}$, $B = \{1, 2\}$, and $C = \{1, 2, 3\}$. Consider the statements:
(a) $A \text{ R } A$ **(b)** $A \text{ R } B$ **(c)** $A \text{ R } C$ **(d)** $B \text{ R } A$ **(e)** $B \text{ R } B$
(f) $B \text{ R } C$ **(g)** $C \text{ R } A$ **(h)** $C \text{ R } B$ **(i)** $C \text{ R } C$
Which of these statements are true for the relation $\subseteq$, is a subset of?

2. As in Exercise 1, which of the statements are true for the relation $\subset$, is a proper subset of?

3. As in Exercise 1, which of the statements are true for $\not\subset$, is not a proper subset of?

For Exercises 4 *through* 12 *tell whether each relation is* **(a)** *reflexive,* **(b)** *symmetric,* **(c)** *transitive,* **(d)** *an equivalence relation:*

4. For people, is a daughter of.
5. For students, is a classmate of.
6. For numbers, is not equal to, $\neq$.
7. For numbers, is less than or equal to, $\leq$.
8. For books, is heavier than.
9. For meals, is less expensive than.
10. For lines in a plane, is parallel to, $||$, if a line is defined to be parallel to itself.
11. For lines in a plane, is parallel to, $||$, if a line is not parallel to itself.
12. For numbers, is divisible by.
13. Consider the three relations $<, =, >$ and tell which relations can be used in place of R to make each statement a true statement.
    (a) $5 \text{ R } 4$
    (b) $25 \text{ R } 42$
    (c) $(10 + 3) \text{ R } (3 + 10)$
    (d) $[2 + (3 \times 5)] \text{ R } [8 + (6 + 2)]$
    (e) $[(5 \times 2) + 1] \text{ R } [(5 \times 2) + (5 \times 1)]$
14. In Exercise 13 exactly one relation could be used in each case. This is a general property of counting numbers called the **trichotomy law;** for any counting numbers a, b exactly one of these relations must hold: $a < b$, $a = b$, $a > b$. State which of the three relations hold(s) for the following in the order given:
    (a) $5 + (7 \times 6)$ and $6 + (7 \times 5)$
    (b) b and $b + 1$
    (c) $b + d$ and $a + d$ if $b < a$
    (d) $b \times d$ and $a \times d$ if $b > a$

PEDAGOGICAL EXPLORATIONS

The three properties of an equivalence relation must be satisfied in all cases in which the symbol = is used. Other symbols may be used for equivalence relations. In many classes it is worthwhile to use a variety of examples (age, height, nationality, shape, size, cardinal number, and so forth) to illustrate that being alike in any respect is an equivalence relation. Then, since there are so many ways in which elements can be alike, it often seems worthwhile to use different symbols to show the manner in which the elements are alike. For example, for geometric figures we use:

$=$ for two names for the same figure.
$=$ for the measures of two figures of the same size
$\sim$ for two figures of the same shape.
$\cong$ for two figures of the same shape and size.

Order relations are used with young children in many ways. For example, they may be asked to arrange a set of blocks in order from smallest to largest, to line up in order from shortest to tallest, and so forth.

1. Give three examples of common uses of equivalence relations in first grade classes.
2. Repeat Exercise 1 for order relations.

The Secondary School Mathematics Curriculum Improvement Study has prepared experimental materials for gifted mathematics students in grades 7 through 12. Their headquarters is located at Teachers College, Columbia University, New York City and they operate under the direction of Dr. Howard F. Fehr. In their seventh grade materials they use **arrow diagrams** to show relations in the following way:

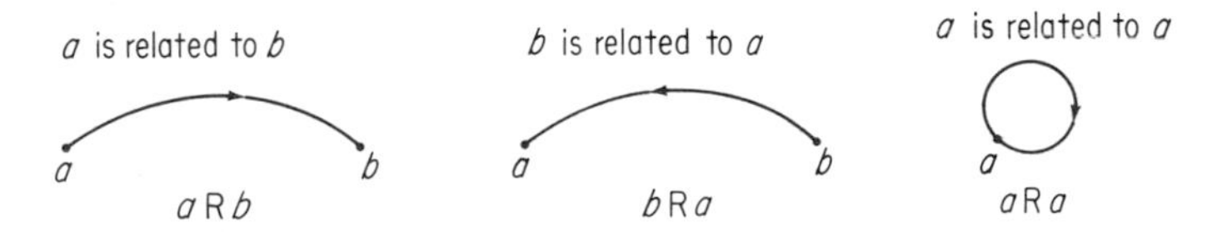

If a relation is reflexive, there is a loop at every element of the given set under discussion. If a relation is symmetric, then each arrow that goes from an element *a* to an element *b* must also have an arrow that goes from *b* to *a*. A relation is transitive if whenever there is an arrow from *a* to *b*, and one from *b* to *c*, then there is also an arrow from *a* to *c*.

3. Here is a graph for the relation "divides" for the set $M = \{2, 3, 6, 12\}$. From the graph, show that the relation is reflexive and transitive but not symmetric.

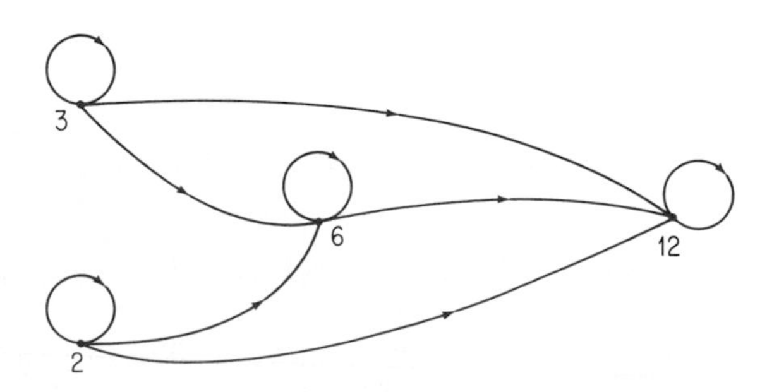

Draw arrow graphs for each of the following relations on the sets given:

4. Is greater than for $A = \{2, 5, 6, 9, 10\}$.
5. Is greater than or equal to for $B = \{3, 8, 9, 12\}$.
6. Is within three kilograms of for the set of weights $C = \{130, 132, 134\}$.

CHAPTER TEST

In Exercises 1 through 3 classify each statement as true or false.

1. (a) Every counting number is a whole number.
   (b) Every whole number is a counting number.
2. (a) Every prime number is odd.
   (b) The product of two prime numbers is a composite number.
3. (a) The set of counting numbers contains an identity element for addition.
   (b) The set of whole numbers is closed with respect to subtraction.
4. List the set of prime numbers that are less than 19.
5. List the set of composite numbers between 10 and 20.
6. Write the set of factors of 27.
7. Write the prime factorization of 600.
8. Find the greatest common factor of 120 and 140.
9. Find the greatest common factor of 24, 30, and 42.
10. Find the least common multiple of 90 and 1500.

Name the property of whole numbers that is illustrated by each statement.

11. $7 \times (5 \times 8) = (7 \times 5) \times 8$
12. $9 \times (7 + 6) = 9 \times (6 + 7)$

Find the replacement set for the whole number n to make each sentence true.

13. $8(5 + 7) = (8)(5) + (8)(n)$
14. $9(3 + n) = (9)(3) + (9)(n)$

Graph each set of numbers on a number line.

15. The set of whole numbers less than 4.
16. The set of counting numbers between 5 and 8.

For the set of whole numbers tell whether each relation is (**a**) *reflexive,* (**b**) *symmetric,* (**c**) *transitive.*

17. Is greater than or is equal to, $\geq$. **18.** Is a factor of.

Use expanded notation to complete each problem.

19. $\begin{array}{r} 173 \\ +258 \\ \hline \end{array}$

20. $\begin{array}{r} 153 \\ -79 \\ \hline \end{array}$

21. Show how to find the product 8×13 by repeated addition.

22. Show how to find the quotient $150 \div 25$ by repeated subtraction.

23. Use the distributive property to find the product 8×537.

24. Usc multiples of 100 and 10 to find the quotient $13{,}398 \div 58$.

25. Complete the following statement that tells how one checks a division problem:

dividend = (______ × ______) + ______

READINGS AND PROJECTS

1. Review several recently published elementary school mathematics textbooks and determine:
   (a) The grade level at which the concept of prime number is first introduced.
   (b) The uses made, if any, of the concepts of G.C.F. and L.C.M.
2. Read Chapter 5 (Short Cuts and Bypaths) of the twenty-seventh yearbook of the National Council of Teachers of Mathematics and prepare a report on this topic.
3. Read and report on the article by Cheryl L. Bradford entitled "Keith's Discovery of the Sieve of Eratosthenes" on pages 239 through 241 in the March 1974 issue of *The Arithmetic Teacher.*
4. Read *The Overhead Projector in the Mathematics Classroom* by George Lenchner, a 1974 publication of the National Council of Teachers of Mathematics. Prepare a demonstration-report that shows at least five different uses of the overhead projector.
5. Prepare, for use with the overhead projector, a set of transparencies that can be used to display the Sieve of Eratosthenes. Use the counting numbers from 1 through 100 on your first (base) transparency, with 1 crossed out. Then prepare a sequence of overlays such that each successive one blocks out multiples of 2, 3, 5, and 7 that are greater than the numbers themselves. The numbers that remain should show the set of primes less than 100.

6. Prepare a report on the history of prime numbers. Among the sources you might consider are Capsules 14 (The Infinitude of Primes) and 15 (Prime and Composite Numbers) in the thirty-first yearbook of the National Council of Teachers of Mathematics.

7. See what you can find out about *figurate numbers*, and prepare a short report on this topic. Several sources to consider include *Teaching Mathematics* by Max A. Sobel and Evan M. Maletsky, Prentice-Hall, Inc., 1975, and the article by Cloman Weaver entitled "Figurate Numbers" on pages 661 through 666 of the November 1974 issue of *The Mathematics Teacher.*

8. The January 1974 issue of *The Arithmetic Teacher* focuses upon the problems of evaluation that all teachers of mathematics must face.
   (a) Read and prepare a review of two of the articles in this issue.
   (b) Compose a test that attempts to measure an elementary school student's knowledge of the four fundamental operations with whole numbers by the end of the sixth grade.

9. Review a set of recently published elementary mathematics textbooks and report on the attention given to word problems at various grade levels. Then prepare for a specific grade level a set of word problems whose solutions involve the four fundamental operations with whole numbers.

10. Prepare a set of pictures (drawings) that an elementary school student might be asked to use as a basis to compose his own problems. For each drawing that you make, list at least two different questions that might be thought of by a student and that can be solved using operations with whole numbers.

11. A great deal of attention has been given in recent years to the psychological bases for learning mathematics. Read Chapter 10 (Implications of the Psychology of Learning for the Teaching of Mathematics) of the twenty-fourth yearbook of the National Council of Teachers of Mathematics. Also try to find at least three articles in journals that deal with the psychology of learning mathematics. One good source to consider is the *Journal for Research in Mathematics Education,* a publication of the National Council of Teachers of Mathematics. Then prepare a report on modern theories of learning and include suggestions for effective teaching based upon psychological foundations.

12. Read Chapter 3 (Behavioral Objectives) of the thirty-fifth yearbook of the National Council of Teachers of Mathematics. Use this as the basis for a 10-minute class report on the meaning, uses, and limitations of behavioral objectives.

13. Read Chapter 3 (Research on Mathematics Learning) and Chapter 7 (Operations on Whole Numbers) of the thirty-seventh yearbook of the National Council of Teachers of Mathematics. Prepare a report on the teaching of arithmetic to primary school children.

14. Read "1976—A Nice Number For The Bicentennial" by Lynn A. Richbart on pages 162 and 163 of the February 1976 issue of *The Mathematics Teacher*.

5

Integers: Properties and Operations

Although many adults can correctly claim that they first worked with positive and negative integers in a high school algebra course, we now find that most schools introduce such concepts in the early elementary grades. This is as it should be inasmuch as we are able to find so many applications of these concepts in everyday life activities. For example, each of the following uses of negative integers should be recognizable to the reader:

> Your budget has a deficit of \$18; your balance is -18.
> You have enough antifreeze for a temperature of 20 degrees below zero, that is, $-20°$.
> California's elevation varies from 282 feet below sea level, -282 feet, at Badwater to 14,495 feet at the top of Mount Whitney.

Unfortunately, too many of us have learned techniques for working with integers without really understanding the rationale behind such processes. As an example, can the reader explain why the product of two negative numbers is a positive number? That is, why does -2 times -3 equal $+6$? This chapter endeavors to explain the reasoning behind this and other operations with integers. Thus we shall extend our number system from the set of whole numbers to the set of integers, and explore properties of, and operations with, this collection of numbers.

5-1 the integers as a commutative group

Once again we extend our number system. With just the set of whole numbers at our disposal, we are still unable to find replacements for n that make these sentences true:

$$5 + n = 2$$
$$n + 4 = 0$$
$$2 - 7 = n$$

That is, the solution set for each of these sentences is the empty set if only the set of whole numbers may be used as the set of possible replacements for the *variable n.*

We begin our extension of the number system by drawing a number line (as in §1-5) and then considering points to the left of the origin 0. For each point on a number line that is the coordinate of a whole number n, locate a point to the left of the origin that is the same distance from 0 as the point with coordinate n. We call the coordinate of this new point **negative n**, written $-n$, and refer to the number that $-n$ represents as the **opposite** of n.

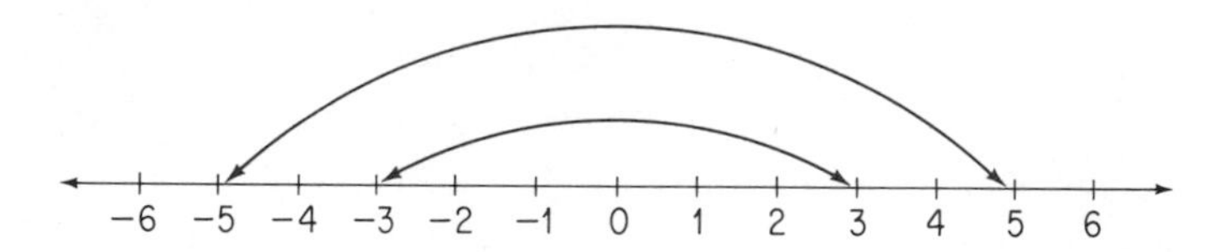

In the figure the point with coordinate negative 3 is located three units to the left of 0. We say that the opposite of 3 is -3; also, the opposite of -3 is 3. Similarly, the point with coordinate -5 is located five units to the left of 0, and we say that 5 and -5 are opposites of each other. Finally we agree that the opposite of 0 is 0; that is, $-0 = 0$.

We call the set of whole numbers together with the set of the opposites of the whole numbers, the set I of *integers.*

$$I = \{\ldots, -3, -2, -1, 0, 1, 2, 3, \ldots\}$$

Note that the graph of the set of integers continues indefinitely both in the positive direction (to the right of 0) and in the negative direction (to the left of 0).

The set of integers may be considered as the union of these sets of numbers:

Positive integers:	$\{1, 2, 3, 4, 5, \ldots\}$
Zero:	$\{0\}$
Negative integers:	$\{\ldots, -5, -4, -3, -2, -1\}$

At times, especially for emphasis, plus signs are used to show positive integers. Thus the set of positive integers may be written in this way:

$$\{+1, +2, +3, +4, +5, \ldots\}$$

We shall consider such expressions as $+4$ and 4 as merely two names for the same number, the number with its graph located at the point four units to the right of the origin on a number line. The positive and negative integers are at times referred to as *signed numbers*. Note that zero is neither a positive integer nor a negative integer.

Example 1: Graph the set of integers between -3 and 2.

Solution:

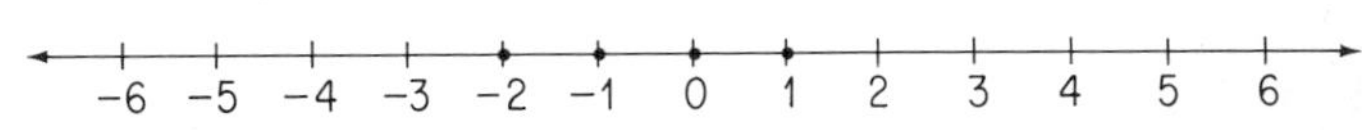

The set of integers has one important property that the set of whole numbers did not possess; every integer has an opposite with respect to addition. This is stated formally by the following property:

Inverse, +: $n + (-n) = 0$. Every integer n has an opposite $-n$ such that $n + (-n) = 0$. This property is sometimes referred to as the **addition property of opposites.**

Each of the integers n and $-n$ is said to be the **additive inverse** or opposite of the other. That is, the additive inverse of 3 is -3, the additive inverse of -3 is 3, and $3 + (-3) = 0$. Indeed, the distinguishing feature between the set of integers and the set of whole numbers is that every integer has an additive inverse *in the set.*

We are now able to solve the sentences such as those given at the beginning of this section.

Sentence	*Solution*
$5 + n = 2$	$n = -3$
$n + 4 = 0$	$n = -4$
$2 - 7 = n$	$n = -5$

In the next two sections we shall carefully review the operations of addition, subtraction, multiplication, and division with integers. However, based on past experience and intuition, the reader should see that we have the following properties of the set of integers under addition.

Closure, +: The sum of any two integers is an integer.

Associative, +: $a + (b + c) = (a + b) + c$ for all integers a, b, and c.

Identity, +: The set of integers contains an element 0 such that $a + 0 = 0 + a = a$ for every integer a.

Inverse, +: For each integer a, the set of integers contains an element $-a$ such that $a + (-a) = 0$.

Commutative, +: $a + b = b + a$ for all integers a and b.

We indicate that the set of integers has the first four properties by saying that the set of integers forms a **group under addition**. Furthermore, since addition of integers is also commutative, we say that the set of integers forms a *commutative group* with respect to addition. A **group under multiplication** must have four properties corresponding to those for a group under addition.

Example 2: Show that subtraction of integers is not commutative.

Solution: We can show this by a single counterexample such as $5 - 2 = 3$ and $2 - 5 = -3$; $5 - 2 \neq 2 - 5$.

In general, we have the following properties of the set of integers under multiplication:

Closure, ×: The product of any two integers is an integer.

Associative, ×: $a \times (b \times c) = (a \times b) \times c$ for all integers $a, b,$ and c.

Identity, ×: The set of integers contains an element 1 such that $a \times 1 = 1 \times a = a$ for every integer a.

Commutative, ×: $a \times b = b \times a$ for all integers a and b.

Example 3: Show that the set of integers does *not* form a group with respect to multiplication.

Solution: Multiplication of integers is closed and associative, and there exists an identity element 1 since $n \times 1 = 1 \times n = n$ for each integer n. However, the set of integers does not contain the multiplicative inverse for each of its elements. For example, there is no integer n such that $3 \times n = 1$.

We are now ready to explore a number of interesting proofs that illustrate the type of reasoning that mathematicians use. First we need several definitions; then we shall consider several proofs and suggest others in the exercises. Some of these proofs will be needed in a later chapter when we prove that $\sqrt{2}$ cannot be expressed as a quotient of integers.

An integer is an **even integer** if it is a multiple of 2, that is, if it may be expressed as $2k$, where k stands for an integer. Then the set of even integers is

$$\{\ldots, -6, -4, -2, 0, 2, 4, 6, \ldots\}$$

An integer that is not even is said to be an **odd integer**. Each odd integer may be expressed in the form $2k + 1$, where k stands for an integer. Then the set of odd integers is

$$\{\ldots, -7, -5, -3, -1, 1, 3, 5, 7, \ldots\}$$

Example 4: Prove that the sum of any two even integers is an even integer.

Proof: Any two even integers m and n may be expressed as $2k$ and $2r$, where k and r stand for integers. Then

$$m + n = 2k + 2r = 2(k + r)$$

where $k + r$ stands for an integer, since the sum of any two integers is an integer. Therefore, $m + n$ is an even integer.

Example 5: Prove that the square of any even integer is an even integer.

Proof: Any even integer may be expressed as $2k$, where k stands for an integer. Then the square of the integer may be expressed as $(2k)^2$, where

$$(2k)^2 = (2k)(2k) = 2(2k^2)$$

Since k, k^2, and $2k^2$ all stand for integers, $(2k)^2$ stands for an even integer.

EXERCISES

1. Show that there is a one-to-one correspondence between the elements of the set of positive integers and the elements of the set of negative integers.

2. Show that the set of whole numbers does not form a group under addition.

Graph each set of numbers on a number line.

3. The set of integers between -3 and 5.

4. The set of integers -3 through 5 inclusive.

5. The set of integers that are the opposites of the first six counting numbers.

6. The set of integers that are the opposites of the members of the set $M = \{-4, -3, -2\}$.

Classify each of the statements in Exercises 7 through 17 as true or false.

7. Every counting number is an integer.

8. Every whole number is an integer.

9. Every integer is a whole number.

10. Every integer is either positive or negative.

11. Every integer is the opposite of some integer.

12. The set of integers is the same set as the set of the opposites of the integers.

13. The set of negative integers is the same as the set of the opposites of the whole numbers.

14. The set of integers is closed under multiplication.
15. The set of integers is closed under division.
16. The set of even integers forms a group under addition.
17. The set of odd integers forms a group under addition.

18. What is the intersection of the set of positive integers and the set of negative integers?
19. Is the union of the set of positive integers and the set of negative integers equal to the set of integers? Explain your answer.
20. Show that there is a one-to-one correspondence between the elements of the set of whole numbers and the elements of the set of integers.
21. Show that there is a one-to-one correspondence between the set of integers and the set of counting numbers.

Prove each statement.

22. The sum of any two odd integers is an even integer.
23. The product of any two even integers is an even integer.
24. The square of any odd integer is an odd integer.
25. If the square of an integer is odd, the integer is odd; if the square of an integer is even, the integer is even.

PEDAGOGICAL EXPLORATIONS

1. Refer to a recently published elementary mathematics textbook series and determine the grade level at which negative integers are first introduced. Also report on the manner in which they are introduced.
2. List as many practical situations as you can think of that might involve the use of negative integers, and that would be meaningful to elementary school students.
3. Prepare a 20-minute lesson plan that introduces the concept of negative integers at a specific early elementary grade level. Include the use of at least one visual aid in your lesson.

5-2 addition and subtraction

It is assumed that the reader has had some past experience involving operations with integers. However, it is frequently enlightening to review such procedures in detail paying special attention to understanding the meaning behind such processes. We begin with addition of integers.

The number line can be used quite effectively to help us add any two integers. Represent the first addend by an arrow that starts at 0, the origin. From the end of the first arrow, draw a second arrow to show the second addend. Positive numbers are shown by arrows that go to the right, and negative numbers by arrows that go to the left. Finally, the sum is found as the coordinate of the point at the end of the second arrow. Examples 1 and 2 illustrate this number-line method for addition of integers. Each arrow represents a *directed line segment* and is also called a **vector**.

Example 1: Illustrate on a number line: $(+3) + (-7)$.

Solution: Start at 0 and draw an arrow that goes three units to the right. Then show an arrow that goes seven units to the left. The second arrow ends at -4. Thus, $(+3) + (-7) = -4$.

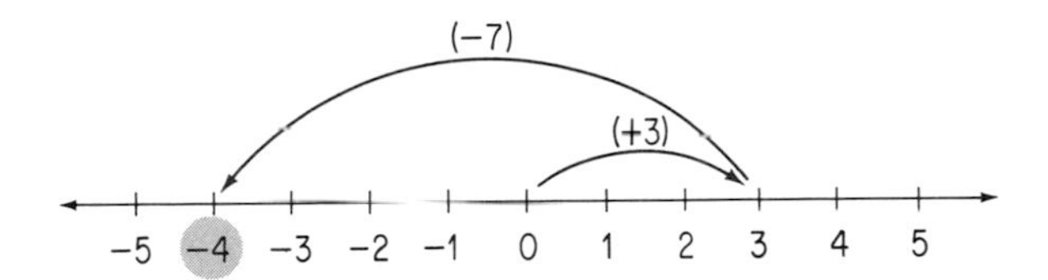

Example 2: Illustrate on a number line: $(-2) + (-3)$.

Solution:

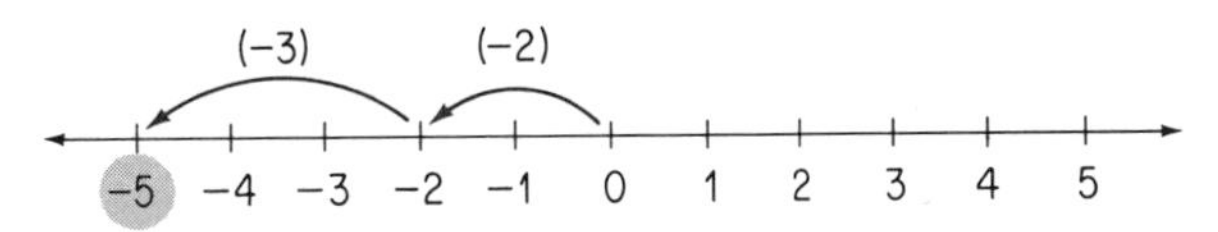

From the figure we see that $(-2) + (-3) = -5$.

Example 3: Find the sum: $(-3) + (+3)$.

Solution: Think of a number line. From the origin, move three units to the left to represent -3. From this point, move three units to the right, back to the origin. Thus, $(-3) + (+3) = 0$. One can also obtain the answer by applying the inverse property for addition. That is, $-(-3) = +3$, and the given sum is 0.

Example 4: Find the sum: $(+3) + (-5) + (-2)$.

Solution: Use the associative property and find the sum of two addends at a time. This grouping is essential because of the binary nature of addition; only two numbers can be added at a time.

$$\begin{aligned}(+3) + (-5) + (-2) &= [(+3) + (-5)] + (-2) \\ &= (-2) + (-2) \\ &= -4\end{aligned}$$

Verify this answer by using the grouping $(+3) + [(-5) + (-2)]$.

To explore the procedure for subtraction of integers, we first recall from ordinary arithmetic that addition and subtraction are inverse operations; each one will undo the other. For example:

$$17 - 9 = n \quad \text{if and only if} \quad n + 9 = 17$$

Similarly:

$$(-2) - (+3) = n \quad \text{if and only if} \quad n + (+3) = -2$$

Any subtraction problem may be solved by translating it into an equivalent addition problem.

To find n in the equation $n + (+3) = -2$ we need to find the number n that must be added to $+3$ to obtain -2. On a number line we need to determine the number of units and direction to move from $+3$ to -2.

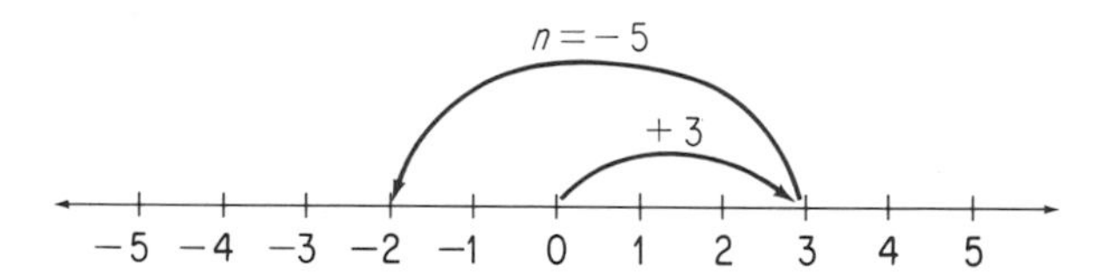

We see that the answer is five units in the negative direction, so that $n = -5$. Therefore, $(-2) - (+3) = -5$.

In this example just completed we made use of the following property for the inverse operations of addition and subtraction:

If $a - b = n$, then $n + b = a$.

We shall apply this idea for the solution of Example 5.

Example 5: Find the difference: $(-2) - (-8)$.

Solution: Let $(-2) - (-8) = n$. Then $n + (-8) = -2$. Thus we need to find a number that must be added to -8 to obtain -2. On a number line, you must move six units to the right to go from -8 to -2. Therefore $n = 6$, and $(-2) - (-8) = +6$.

Let us examine the subtraction problems that we have been studying and see if we can discover another procedure for solving such problems. In each case we list the subtraction problem and a corresponding addition problem. Note that the results obtained are the same in each case.

Subtraction	*Addition*
$(+17) - (+9) = +8$	$(+17) + (-9) = +8$
$(-2) - (+3) = -5$	$(-2) + (-3) = -5$
$(-2) - (-8) = +6$	$(-2) + (+8) = +6$

Comparing the set of examples, you should note that you can *subtract* a number b from a number a by *addition* of the opposite (additive inverse) of b to a.

In general we define the subtraction $a - b$ of two integers a and b as follows:

$$a - b = a + (-b)$$

Note that $-b$ can be thought of as $(-1) \times b$. Thus we may also define subtraction in this way:

$$a - b = a + (-1) \times b$$

Example 6: Subtract: $(+5) - (-3)$.

Solution: By the definition of subtraction, $a - b = a + (-b)$. Therefore, $(+5) - (-3) = (+5) + (+3) = +8$.

Example 7: Subtract:

$$\begin{array}{r} +5 \\ \underline{-3} \end{array}$$

Solution: This is precisely the same problem as in Example 6, but written in vertical form. Therefore, the difference is $+8$.

EXERCISES

Illustrate each sum on a number line.

1. $(+5) + (-7)$

2. $(-3) + (-4)$

3. $(+3) + (+4)$

4. $(-2) + (-5)$

Find each sum.

5. $(+8) + (-12)$

6. $(-3) + (-7)$

7. $(-8) + (+12)$

8. $(-23) + (+23)$

9. $(-15) + (-12)$

10. $(+13) + (-20)$

11. $(-5) + (-3) + (-7)$

12. $(+6) + (-8) + (-7)$

13. $(+12) + (-12) + (-9)$

14. $(-15) + (-7) + (+8)$

15. $(+5) + (-7) + (-6) + (+8)$

16. $(-7) + (-3) + (+12) + (-9)$

Add.

17. $\begin{array}{r} +12 \\ \underline{-15} \end{array}$

18. $\begin{array}{r} -15 \\ \underline{-10} \end{array}$

19. $\begin{array}{r} -13 \\ \underline{+25} \end{array}$

20. $\begin{array}{r} +16 \\ \underline{-24} \end{array}$

Find each difference.

21. $(+5) - (+2)$

22. $(-5) - (-2)$

23. $(+5) - (-2)$

24. $(-5) - (+2)$

25. $(-12) - (+15)$

26. $(+13) - (+7)$

27. $(+15) - (+25)$

28. $(-15) - (-25)$

29. $(+11) - (-5)$

30. $(+11) - (-20)$

Subtract.

31. $\begin{array}{r} -12 \\ \underline{-11} \end{array}$

32. $\begin{array}{r} +7 \\ \underline{-9} \end{array}$

33. $\begin{array}{r} -19 \\ \underline{+12} \end{array}$

34. $\begin{array}{r} +13 \\ \underline{+17} \end{array}$

Perform the indicated operations.

35. $[(-5) + (-3)] - (+3)$

36. $(-5) + [(-3) - (+3)]$

37. $[(-8) - (-2)] + (-5)$

38. $(-8) - [(-2) + (-5)]$

39. $[(+5) + (-8)] - [(-3) + (-7)]$

40. $[(-7) - (-3)] + [(+5) - (-8)]$

41. Use a counterexample to show that the set of integers is not commutative with respect to subtraction.

PEDAGOGICAL EXPLORATIONS

1. An interesting approach to addition of integers is to use the concept of positive and negative arrows. For example, consider the following: These two arrows together illustrate +2:

$\longrightarrow \qquad \longrightarrow$

These three negative unit arrows together illustrate −3:

$\longleftarrow \qquad \longleftarrow \qquad \longleftarrow$

Positive and negative arrows are then combined, as indicated by the circles, to give zero pairs as follows:

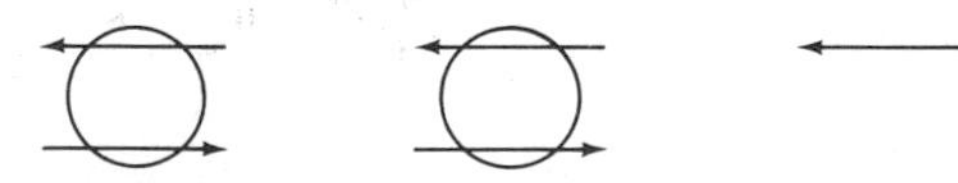

We have one negative arrow left. Thus $(-3) + (+2) = -1$.

Prepare a short unit introducing addition of integers in this way.

2. Here is a nomograph that can be used to find sums of integers. Just connect the point representing one addend on the *A* scale, with the corresponding point for the other addend on the *B* scale. The point where the line crosses the S scale will give the sum. The figure shows the sum $(+4) + (-6) = -2$. Make your own nomograph using graph paper and use it to find various sums of integers.

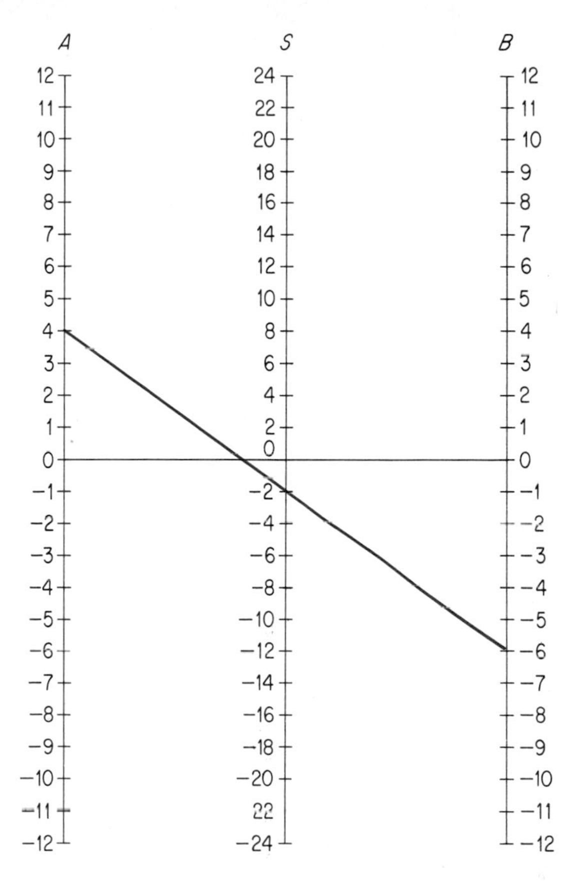

3. See if you can use your knowledge of elementary geometry to explain why the nomograph shown works as it does.
4. Try to determine how the nomograph shown can be used for subtraction of integers.
5. Prepare a nomograph that will enable you to find $a + 2b$ directly for any two integers a and b.

5-3 multiplication and division

There are a variety of ways in which we can consider the rules for multiplication of integers. Applications of numbers to physical situations can be helpful. Consider, for example, the situation of earning $5 a day for two days. This can be expressed as the product of two positive integers, $(+2) \times (+5)$. We may think of multiplication by a positive integer as repeated addition:

$$(+2) \times (+5) = (+5) + (+5) = +10$$

Thus the result is a gain of \$10, a positive number. In general, the product of two positive integers is a positive integer.

To evaluate the product of a positive and a negative integer, consider the situation of losing \$5 each day for two days. This can be expressed as the product $(+2) \times (-5)$. Using repeated addition:

$$(+2) \times (-5) = (-5) + (-5) = -10$$

Thus the result is a loss of \$10, a negative number. In general, the product of a positive integer and a negative integer is a negative integer. Then, since multiplication is commutative, the product of a negative integer and a positive integer is also a negative integer. For example:

$$(-5) \times (+2) = (+2) \times (-5) = -10$$

Justifying the product of two negative numbers is somewhat more difficult. Informally, we might say that if you lose \$5 a day for two days, then two days *ago* you had \$10 more than you have today. Therefore, $(-2) \times (-5) = +10$. However, such arguments are often unconvincing. Another approach is to consider patterns in a multiplication table such as the following. The entries in the lower right-hand corner represent the products of positive integers.

X	−3	−2	−1	0	+1	+2	+3
−3				0			
−2				0			
−1				0			
0	0	0	0	0	0	0	0
+1				0	+1	+2	+3
+2				0	+2	+4	+6
+3				0	+3	+6	+9

X	−3	−2	−1	0	+1	+2	+3
−3				0	−3	−6	−9
−2				0	−2	−4	−6
−1				0	−1	−2	−3
0	0	0	0	0	0	0	0
+1	−3	−2	−1	0	+1	+2	+3
+2	−6	−4	−2	0	+2	+4	+6
+3	−9	−6	−3	0	+3	+6	+9

Next, note the pattern of entries as one reads from right to left, or from the bottom up. In the +3 row and column the entries are decreasing by 3's; in the +2 row and column they are decreasing by 2's; and in the +1 row and column they are decreasing by 1's. We use this observation to complete two more portions of the table, and note that the second table shows the product of a positive integer and a negative integer to be a negative integer.

To complete the table, note the pattern of entries once again. For example, consider the first row and read the entries from right to left:

$$\_\_\ \_\_\ \_\_\ 0\ \ -3\ \ -6\ \ -9$$

From right to left, the entries are increasing by 3's. Also, in the first column, the entries are increasing by 3's as you read from the bottom up. Similar observations for the other rows and columns allow us to complete the multiplication table as shown next. Note that the pattern shows the product of two negative integers to be a positive integer.

×	−3	−2	−1	0	+1	+2	+3
−3	+9	+6	+3	0	−3	−6	−9
−2	+6	+4	+2	0	−2	−4	−6
−1	+3	+2	+1	0	−1	−2	−3
0	0	0	0	0	0	0	0
+1	−3	−2	−1	0	+1	+2	+3
+2	−6	−4	−2	0	+2	+4	+6
+3	−9	−6	−3	0	+3	+6	+9

Example 1: Find each product. (a) $(+4) \times (+7)$; (b) $(-3) \times (+6)$; (c) $(+5) \times (-8)$; (d) $(-7) \times (-8)$.

Solution: (a) The product of two positive integers is a positive integer; $(+4) \times (+7) = +28$.
(b) The product of a negative integer and a positive integer is a negative integer; $(-3) \times (+6) = -18$.
(c) The product of a positive integer and a negative integer is a negative integer; $(+5) \times (-8) = -40$.
(d) The product of two negative integers is a positive integer; $(-7) \times (-8) = +56$.

Example 2: Find the product: $(-5) \times (-7) \times (-2)$.

Solution: Use the associative property and then find the product of two factors at a time.

$$\begin{aligned}(-5) \times (-7) \times (-2) &= [(-5) \times (-7)] \times (-2)\\ &= (+35) \times (-2)\\ &= -70\end{aligned}$$

Verify this solution by using the grouping $(-5) \times [(-7) \times (-2)]$.

The rules for multiplication of integers can be confirmed by mathematical procedures rather than by the somewhat intuitive approach we have just developed. Let us consider, for example, a mathematical justification for the product of two negative integers. We begin by exploring a specific example:

$$(-2)[(+5) + (-5)]$$

Because we wish to preserve the distributive property for the set of integers, we agree that there are two ways to obtain an answer to this problem. If we add first, within the brackets, and then multiply, we have the following:

$$\begin{aligned}(-2)[(+5) + (-5)] &= (-2)(0)\\ &= 0\end{aligned}$$

Now we shall use the distributive property, multiply first, and then add. The final result must be the same, namely 0.

$$\begin{aligned}(-2)[(+5) + (-5)] &= [(-2)(+5)] + [(-2)(-5)]\\ &= (-10) + (?)\end{aligned}$$

We have been forced to use a question mark for the product $(-2) \times (-5)$ because this is precisely the product that we are seeking. We do know that the final sum must be 0 if the distributive property is to hold. Furthermore, we know that $(-10) + (+10) = 0$. Therefore we conclude that $(-2) \times (-5)$ must be equal to $+10$; the product of two negative integers is a positive integer.

We can generalize this approach by considering any positive integers c and b, and the expression $(-c) \times [(b) + (-b)]$. We then evaluate this product in two different ways:

1. $$\begin{aligned}(-c) \times [(b) + (-b)] &= (-c) \times 0\\ &= 0\end{aligned}$$

2. $$\begin{aligned}(-c) \times [(b) + (-b)] &= [(-c) \times (b)] + [(-c) \times (-b)]\\ &= -cb + (?)\end{aligned}$$

The final result must be 0, so that $-cb$ must be the opposite of $(-c) \times (-b)$. Therefore $(-c) \times (-b) = cb$, and the product of two negative integers is a positive integer.

Just as subtraction is the inverse operation of addition, so is division the inverse operation of multiplication. Therefore, every division problem can be solved by first translating it into an appropriate multiplication problem. In ordinary arithmetic of whole numbers, the quotient $12 \div 3$ can be thought of as the number n for which $3 \times n = 12$.

$$3\overline{)12}^{\,n} \quad \text{can be written as} \quad 3 \times n = 12$$

Indeed we can list the members of this family of number facts:

$$\begin{aligned}3 \times 4 &= 12\\ 4 \times 3 &= 12\\ 12 \div 4 &= 3\\ 12 \div 3 &= 4\end{aligned}$$

In a similar manner we can find the quotient $(-12) \div (+3)$ by considering the product $(+3) \times n = -12$. From our rules for multiplication we note that $n = -4$, and thus see that this quotient of a negative integer and a positive integer is a negative integer.

In general, $a \div b$ is an integer if and only if there exists some integer n such that $b \times n = a$.

Example 3: Find the quotient $(-12) \div (-3)$.

Solution: Let $(-12) \div (-3) = n$. Then $(-3) \times n = (-12)$, so that $n = +4$. Thus the quotient of these two negative integers is a positive integer.

Later we can define the quotient $a \div b$ of two integers a and b, $b \neq 0$, in this way:

$$a \div b = a \times \frac{1}{b}$$

Then a quotient such as $(-12) \div (+3)$ can be found as follows:

$$(-12) \div (+3) = (-12) \times (+\tfrac{1}{3}) = -4$$

Regardless of the approach used, we conclude that the quotient of two positive integers is positive, the quotient of a positive and a negative integer is negative, and the quotient of two negative integers is positive. The set of integers is *not* closed with respect to division, as noted in the following example. The desire to make division always possible will lead to further extensions of the number system in the next chapter.

Example 4: Show that the set of integers is not closed with respect to the operation of division.

Solution: We can show this by a single counterexample: $8 \div 3$ cannot be named by an integer. That is, there is no integer n such that $3 \times n = 8$.

EXERCISES

Find each product.

1. $(-5) \times (+9)$

2. $(-5) \times (-9)$

3. $(+5) \times (-9)$

4. $(+5) \times (+9)$

5. $(-8) \times (+12)$

6. $(+12) \times (-12)$

7. $(-25) \times (-25)$

8. $(-15) \times (+15)$

9. $(+10) \times (-17)$

10. $(-11) \times (-11)$

11. $(-3) \times (-7) \times (-5)$

12. $(-3) \times (+7) \times (-7)$

13. $(-8) \times (-17) \times 0$

14. $(-2) \times (-2) \times (-2)$

15. $(-1) \times (-2) \times (-3) \times (-4)$

16. $(+3) \times (-3) \times (-5) \times (-5)$

Find each quotient.

17. $(-24) \div (+3)$

18. $(+24) \div (+3)$

19. $(+24) \div (-3)$

20. $(-24) \div (-3)$

21. $(-36) \div (-18)$

22. $(+72) \div (-12)$

23. $(-144) \div (-12)$

24. $(+125) \div (-25)$

25. $(-100) \div (+100)$

26. $0 \div (-5)$

27. $[(-60) \div (-3)] \div (-4)$

28. $(-100) \div [(-50) \div (+10)]$

29. $[(-48) \div (+6)] \div (-1)$

30. $(+144) \div [(+24) \div (-3)]$

31. $[(+24) \div (-2)] \div [(-18) \div (-3)]$

32. $[(-150) \div (+5)] \div [(+225) \div (-15)]$

Perform the indicated operations.

33. $(-8) + [(-5) + (-7)]$

34. $[(-8) + (-5)] + (-7)$

35. $(-2) \times [4 \times (-5)]$

36. $[(-2) \times 4] \times (-5)$

37. $12 \div [6 \div (-2)]$

38. $(12 \div 6) \div (-2)$

39. Use the results of Exercises 37 and 38 to form a conjecture about the associativity of integers with respect to division. Then find another example to help confirm your conjecture.

40. Use a counterexample to show that the set of integers is not commutative with respect to division.

PEDAGOGICAL EXPLORATIONS

1. Many elementary textbooks introduce the set of negative integers by using raised negative signs. Thus the set N of negative integers would be written in this way:

$$\{\ldots, {}^{-}5, {}^{-}4, {}^{-}3, {}^{-}2, {}^{-}1\}$$

Discuss the pedagogical value of introducing the set in this manner, thus reserving the minus sign, −, for subtraction. Refer to a first year algebra text to see how the transition is made from raised negative signs to lowered ones.

2. Flow charts can be used effectively to motivate practice with integers, as in the illustrative example at the top of the next page. The student is to use the table for given inputs for which he is to find the corresponding outputs.

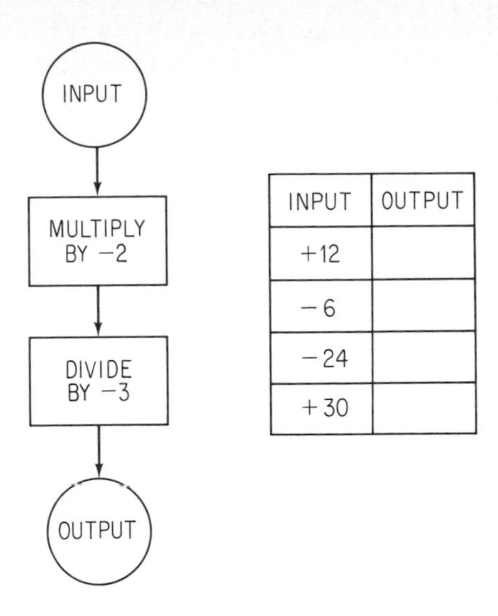

INPUT	OUTPUT
+12	
−6	
−24	
+30	

Make up several flow charts of your own that involve computation with integers.

3. The product of a positive integer and a negative integer can be shown on a number line as repeated addition. The figure that follows shows the product $5 \times (-2)$ as five "steps" of two units each, in the negative direction from the origin. That is, $5 \times (-2) = -10$.

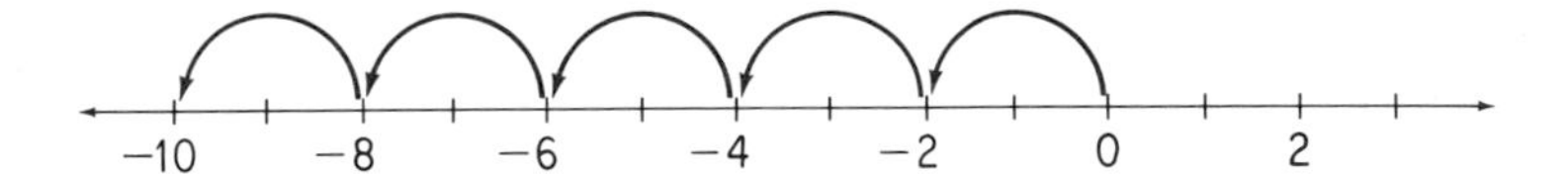

Prepare a plan for a 15–20-minute lesson that uses this number line approach to show such products for a class at a specified elementary grade level.

5-4 absolute value

There is one more concept that we need to introduce in order to be able to give a formal definition of addition and multiplication of integers. To do so, consider a number line and think of the points that are at a distance of three units from the origin. You should see that there are two such points, with coordinates 3 and −3.

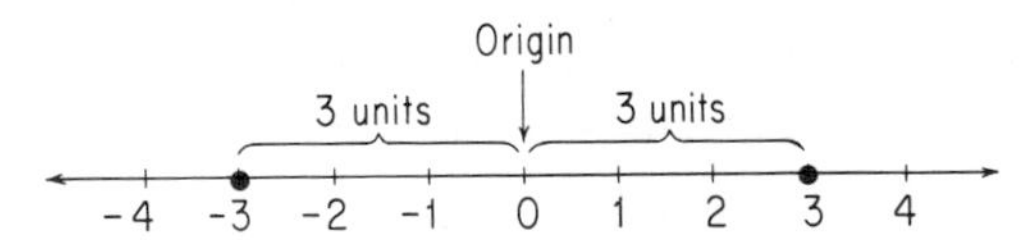

We use the symbol $|x|$, read "the *absolute value* of x," to represent the distance of the point with coordinate x from the origin on the number line, without concern for direction. Thus for the preceding diagram we may write

$$|3| = 3 \quad \text{and} \quad |-3| = 3$$

to indicate that both points are three units from the origin. Also, it should be clear that $|0| = 0$ since the origin is zero units from itself.

Example 1: Evaluate: $|-12|$.

Solution: The point on a number line with coordinate -12 is 12 units from the origin. Therefore, $|-12| = 12$. (Note that the absolute value of a number can never be negative.)

Example 2: Evaluate: $|-7| + |-2| + |+5|$.

Solution: First we note the following:

$$|-7| = 7, \qquad |-2| = 2, \qquad |+5| = 5$$

Thus

$$|-7| + |-2| + |+5| = 7 + 2 + 5 = 14$$

Example 3: Evaluate: $|3 - 7|$.

Solution: First we evaluate the expression $3 - 7$ as $3 + (-7) = -4$. Thus $|3 - 7| = |-4| = 4$.

In general, we define the **absolute value** of any integer k as follows:

$$\begin{aligned} |k| &= k, && \text{if } k \text{ is positive} \\ |k| &= -k, && \text{if } k \text{ is negative} \\ |k| &= 0, && \text{if } k = 0 \end{aligned}$$

Notice in the second case that if k represents a negative number, then $-k$ represents a positive number. For example, if $x = -3$, $|-3| = -(-3) = 3$.

We are now able to use absolute value notation to define the sum of two integers a and b under these four possible cases:

Case 1: $a + b = |a| + |b|$, when $a \geq 0$ and $b \geq 0$.
For example, let $a = +3$ and $b = +5$:

$$(+3) + (+5) = |+3| + |+5| = 3 + 5 = 8$$

Case 2: $a + b = -(|a| + |b|)$, when $a \leq 0$ and $b \leq 0$.
For example, let $a = -3$ and $b = -5$:

$$(-3) + (-5) = -(|-3| + |-5|) = -(3 + 5) = -8$$

Case 3: $a + b = |a| - |b|$, when $a > 0$, $b < 0$, and $|a| > |b|$.
For example, let $a = +5$ and $b = -3$:

$$(+5) + (-3) = |+5| - |-3| = 5 - 3 = 2$$

Case 4: $a + b = -(|b| - |a|)$, when $a > 0$, $b < 0$, and $|b| > |a|$.
For example, let $a = +3$ and $b = -5$:

$$(-5) + (+3) = -(|-5| - |+3|) = -(5 - 3) = -2$$

Under normal circumstances we find the sum of two integers by informal methods, as in §5-2. However it is important to note that we can offer formal definitions for finding such sums. In a similar manner we can use absolute value notation to define formally the product of two integers a and b in all possible cases:

Case 1: $a \times b = |a| \times |b|$, when $a > b$ and $b > 0$.
For example, let $a = +3$ and $b = +5$:

$$(+3) \times (+5) = |+3| \times |+5| = 3 \times 5 = 15$$

Case 2: $a \times b = |a| \times |b|$, when $a < 0$ and $b < 0$.
For example, let $a = -3$ and $b = -5$:

$$(-3) \times (-5) = |-3| \times |-5| = 3 \times 5 = 15$$

Case 3: $a \times b = -(|a| \times |b|)$, when $a > 0$ and $b < 0$, or when $a < 0$ and $b > 0$.
For example, let $a = +3$ and $b = -5$:

$$(+3) \times (-5) = -(|+3| \times |-5|) = -(3 \times 5) = -15$$

For another example, let $a = -3$ and $b = +5$:

$$(-3) \times (+5) = -(|-3| \times |+5|) = -(3 \times 5) = -15$$

Note that the definition does not include any mention of multiplication by zero. This case is covered by the multiplication property of zero, which states that for any integer n, $n \times 0 = 0$.

The various cases for sums and products of positive and negative integers may be summarized as follows:

The sum of two positive integers is a positive integer.
The sum of two negative integers is a negative integer.
The sum of a positive integer and a negative integer may be either a positive integer or a negative integer.
The product of two positive integers is a positive integer.
The product of two negative integers is a positive integer.
The product of a positive integer and a negative integer is a negative integer.

EXERCISES

Evaluate each of the following.

1. $|-9|$ **2.** $|+25|$ **3.** $-|+8|$ **4.** $-|-6|$

5. $|+8| + |+5|$ **6.** $|(+8) + (+5)|$

7. $|-8| + |-5|$ **8.** $|(-8) + (-5)|$

9. $|(-8) + (+5)|$ **10.** $|-8| + |+5|$

11. $-(|+8| + |-5|)$ **12.** $-|(+8) + (-5)|$

13. $-(|-8| + |-5|)$ **14.** $-|(-8) + (-5)|$

Use the definition of addition to find each sum.

15. $(+7) + (+5)$ **16.** $(-7) + (-5)$

17. $(+7) + (-5)$ **18.** $(+7) + (-9)$

Use the definition of multiplication to find each product.

19. $(+9) \times (+6)$ **20.** $(-9) \times (-6)$

21. $(+9) \times (-6)$ **22.** $(-9) \times (+6)$

23. The absolute value of a number n is sometimes defined as being the greater of the number n or its opposite. Show with several examples that this definition is equivalent to the one given in this section.

24. Does $|a| + |b| = |a + b|$? Answer this question for each of these three cases: **(a)** $a > 0, b > 0$; **(b)** $a < 0, b < 0$; **(c)** $a > 0, b < 0$.

Find two replacements for n to make each of these sentences true.

25. $|n + 3| = 9$ **26.** $|n + 9| = 3$

27. $|n - 3| = 9$ **28.** $|n - 9| = 3$

PEDAGOGICAL EXPLORATIONS

1. In many modern programs flow charts are used extensively to help explain steps in algorithms. Use various replacements for *a* and *b* and verify that the following flow chart gives the correct sum in each case.

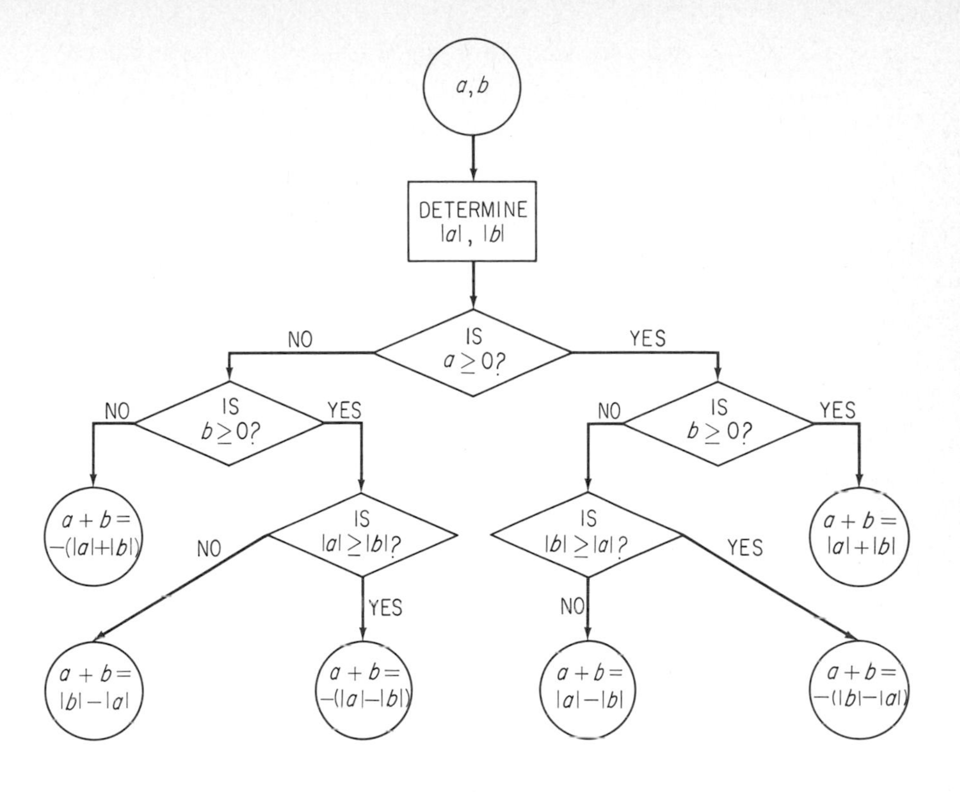

2. Make a flow chart for finding the difference $a - b$ of two integers a and b.
3. Make a flow chart for finding the product $a \times b$ of two integers a and b.
4. Make a flow chart for finding the quotient $a \div b$ of two integers a and b, where $b \neq 0$.

5-5 integers as exponents

Now that we have extended our number system to include the set of integers, let us see how these numbers may be used as exponents. We are already familiar with the use of the counting numbers (positive integers) as exponents from §3-5. Recall that an exponent is a number that tells how many times another number, called the base, is used as a factor in a product. In general,

$$a^m = \overbrace{a \times a \times a \times a \times a \times a \times \cdots \times a}^{m \text{ factors}}$$

where m is a counting number. This is read as "a to the mth power" where a is the *base*, m is the *exponent*, and a^m is the *power*. For example, a^5 is read as "a to the fifth power" and means $a \times a \times a \times a \times a$.

Through the use of specific examples, we can readily develop rules for multiplying and dividing. Thus, consider the product $a^3 \times a^2$:

$$a^3 = a \times a \times a, \qquad a^2 = a \times a$$
$$a^3 \times a^2 = (a \times a \times a) \times (a \times a) = a \times a \times a \times a \times a = a^5$$

In general, we may state the following rule for multiplication:

For all a, and for counting numbers m and n, $a^m \times a^n = a^{m+n}$.

Note that in multiplication the bases must be the same. Thus we may simplify $a^3 \times a^2$ by adding exponents to obtain a^5. However, an expression such as $a^3 \times b^2$ cannot be further simplified and would normally be written as a^3b^2.

Example 1: Simplify: $x^3 \times x^4$.

Solution:

$$x^3 \times x^4 = x^{3+4} = x^7$$

Although an intermediate step is shown here for illustrative purposes, it is usually omitted in actual practice.

Example 2: Simplify: $(2a^2b^3)(3a^4b^5)$.

Solution: As in Example 1, it is possible to proceed directly to the solution but an intermediate step is shown to indicate groupings of like terms.

$$(2a^2b^3)(3a^4b^5) = (2)(3)(a^2)(a^4)(b^3)(b^5) = 6a^6b^8$$

Next, let us consider the quotient $a^6 \div a^2$, $a \neq 0$. By writing the factors of each term we have the following:

$$\begin{aligned}\frac{a^6}{a^2} &= \frac{a \times a \times a \times a \times a \times a}{a \times a}\\ &= \frac{a}{a} \times \frac{a}{a} \times a \times a \times a \times a\\ &= 1 \times 1 \times a \times a \times a \times a\\ &= 1 \times a^4\\ &= a^4\end{aligned}$$

From this example it seems clear that to divide powers of the same base we merely subtract the exponents, as in Example 3.

Example 3: Simplify: (a) $\frac{x^{12}}{x^3}$; (b) $\frac{12a^6b^5}{3a^2b^3}$.

Solution: (a) $\frac{x^{12}}{x^3} = x^{12-3} = x^9$

(b) $\frac{12a^6b^5}{3a^2b^3} = \left(\frac{12}{3}\right)\left(\frac{a^6}{a^2}\right)\left(\frac{b^5}{b^3}\right) = 4a^4b^2$

Thus far we have restricted our use of exponents to the set of counting numbers. Let us now extend the use of exponents to the set of all integers. First we give meaning to the use of 0 as an exponent. (Note that in §3-5 we *defined* $10^0 = 1$.)

Consider the quotient $a^4 \div a^4$, $a \neq 0$. We know that if a number different from zero is divided by itself, the result is 1. However, if we wish to find such quotients by subtracting exponents, then we have the following:

$$\frac{a^4}{a^4} = a^{4-4} = a^0$$

Since we want to have a unique result, regardless of the procedure used, we are forced to adopt the following definition.

For all a, $a \neq 0$, $a^0 = 1$.

Note the restriction $a \neq 0$; the expression 0^0 is said to be undefined.

Next we shall attempt to give meaning to the use of negative integers as exponents. For example, what do we mean by a^{-4}? It hardly makes sense to think of using the base a as a factor -4 times. Instead we approach the problem by consideration of a specific example, $a^2 \div a^6$:

$$\begin{aligned}
\frac{a^2}{a^6} &= \frac{a \times a}{a \times a \times a \times a \times a \times a} \\
&= \frac{a}{a} \times \frac{a}{a} \times \frac{1}{a \times a \times a \times a} \\
&= 1 \times 1 \times \frac{1}{a \times a \times a \times a} \\
&= 1 \times \frac{1}{a^4} \\
&= \frac{1}{a^4}
\end{aligned}$$

Suppose we were to solve this same problem by subtracting exponents. We would then have the following result:

$$\frac{a^2}{a^6} = a^{2-6} = a^{-4}$$

Again, it is essential that we obtain a unique result. We therefore define a^{-4} to mean

$$\frac{1}{a^4}$$

In general, we now define negative exponents in this way:

For all integers m, and for $a \neq 0$,

$$a^{-m} = \frac{1}{a^m}$$

Example 4: Simplify: (a) 5^{-2}; (b) 10^{-3}.

Solution: (a) $5^{-2} = \frac{1}{5^2} = \frac{1}{25}$; (b) $10^{-3} = \frac{1}{10^3} = \frac{1}{1000}$.

We may now summarize this discussion by considering a rule with three possible cases:

For all a, $a \neq 0$, and for counting numbers m and n,

$$\begin{aligned} \frac{a^m}{a^n} &= a^{m-n} && \text{if } m > n \\ &= 1 && \text{if } m = n \\ &= \frac{1}{a^{n-m}} && \text{if } m < n \end{aligned}$$

Having defined the meaning of zero and of negative integers as exponents, we may now remove the restrictions originally stated and assert that these rules hold for all integers used as exponents. To summarize:

For all a, and for all integers m and n:

$$\begin{aligned} a^m \times a^n &= a^{m+n} && \\ a^m \div a^n &= a^{m-n} && (a \neq 0) \\ a^0 &= 1 && (0^0 \text{ undefined}) \\ a^{-m} &= \frac{1}{a^m} && (a \neq 0) \end{aligned}$$

The following example illustrates further applications of these rules. The word "simplify" is used here to imply that all possible operations are to be performed, and that the final result should be free of negative exponents.

Example 5: Simplify: (a) $(xy)^{-3}$; (b) $(x^2)^3$.

Solution: (a) By the definition of a negative exponent $(xy)^{-3} = \frac{1}{(xy)^3}$.

$$\begin{aligned} (xy)^{-3} &= \frac{1}{(xy)^3} \\ &= \frac{1}{(xy)(xy)(xy)} \\ &= \frac{1}{x^3y^3} \end{aligned}$$

(b) By the definition of an exponent, we may think of this example as meaning $x^2 \times x^2 \times x^2 = x^{2+2+2} = x^6$.

The results found in Example 5 lead to the following two additional rules for integral exponents:

For all $a \neq 0$, $b \neq 0$, m, and n:

$$(a^m)^n = a^{mn}$$

$$(ab)^m = a^m b^m$$

Perform the indicated operations and simplify where possible.

1. $a^5 \times a^3$ **2.** $a^4 \times a^5$ **3.** $3^2 \times 3^3$

4. $2^2 \times 2^3$ **5.** $10^2 \times 10^1$ **6.** $10^3 \times 10^4$

7. $(3x^5)(2x^4)$ **8.** $(2x^7)(5x^3)$ **9.** $(5a^3b^5)(2a^4b^2)$

10. $(3m^4n^8)(4m^3n^2)$ **11.** $\frac{x^{10}}{x^5}$ **12.** $\frac{x^{12}}{x^2}$

13. $\frac{10^6}{10^2}$ **14.** $\frac{10^5}{10^3}$ **15.** $\frac{5^8}{5^5}$

16. $\frac{3^6}{3^3}$ **17.** $\frac{20a^4b^5}{4a^2b^2}$ **18.** $\frac{12m^2n^8}{2m^2n^4}$

19. $\frac{8ab^3c^{10}}{2ab^2c^{10}}$ **20.** $\frac{9x^{12}y^{15}}{3x^{10}y^5}$ **21.** $n^2 \times n^3 \times n^4$

22. $(x^2y)(x^3y^2)(xy^3)$ **23.** $(a^3)^4$ **24.** $(3x^3)^2$

25. $(x^2)^5 + (y^5)^2$ **26.** $(2m)^3 + (3n)^2$ **27.** 8^{-2}

28. 10^{-4} **29.** $(-5)^{-3}$ **30.** 12^0

31. $(mn)^{-2}$ **32.** $(3x)^{-3}$ **33.** $x^{-2}y^{-3}$

34. Does $2^4 = 4^2$? Does this show that raising to a power is a commutative operation? Explain your answer.

PEDAGOGICAL EXPLORATIONS

Scientific notation is used to represent very large and very small numbers in concise form. To express a number in **scientific notation** we write it as the product of a number between 1 and 10 and some power of 10. These illustrations should clarify this procedure:

$$7{,}000{,}000 = 7 \times 10^6$$
$$23{,}000{,}000 = 2.3 \times 10^7$$
$$0.000007 = 7 \times 10^{-6}$$
$$0.00000023 = 2.3 \times 10^{-7}$$

This notation is especially helpful in computations with very large or very small numbers. Note, for example, its use in this product:

$$\begin{aligned} 5{,}000{,}000 \times 8{,}000{,}000 &= (5 \times 10^6)(8 \times 10^6) \\ &= 40 \times 10^{12} \\ &= 4 \times 10^{13} \quad \text{in scientific notation} \end{aligned}$$

Unless otherwise instructed, we assume that in ordinary decimal notation zeros that are not otherwise needed are used only to locate the decimal point.

1. Write each number in scientific notation:
   (a) 900,000 (b) 8,000,000,000 (c) 45,000
   (d) 0.00003 (e) 0.00000000005 (f) 0.0000065
2. Use scientific notation to compute each result. Express the answers in scientific notation.
   (a) $300{,}000 \times 30{,}000$ (b) $12{,}000{,}000 \times 5{,}000{,}000$
   (c) $45{,}000{,}000 \div 90{,}000$ (d) $720{,}000 \div 8{,}000$
   (e) 0.000002×0.000004 (f) $0.000000045 \div 0.000005$
3. Make a collection of very large numbers and very small numbers that are used in science, expressing each in scientific notation.
4. Explore a recently published elementary mathematics textbook series to determine what use, if any, is made of negative exponents.

CHAPTER TEST

Classify each statement as true or false.

1. (a) The set of whole numbers is a subset of the set of integers.
   (b) The absolute value of each integer is a whole number.
2. (a) The opposite of each whole number is a negative integer.
   (b) The opposite of each integer is an integer.

Graph each set of numbers on a number line.

3. The set of integers between -4 and 2.
4. The set of integers that are opposites of the first three counting numbers.

Illustrate each sum on a number line.

5. $(-3) + (+7)$
6. $(-2) + (-4)$

Perform the indicated operations.

7. $(-20) + (+3)$
8. $(+7) + (-18)$
9. $(-3) - (-5)$
10. $(-2) - (+8)$
11. $(-5) \times (+8)$
12. $(-9) \times (-8)$
13. $(+36) \div (-4)$
14. $(-42) \div (-7)$
15. $(-3) + [(+2) + (-7)]$
16. $[(-9) - (-2)] - (+3)$
17. $(-5) \times [(-2) \times (-1)]$
18. $[(-40) \div (+5)] \div (-2)$

19. Evaluate: (a) $|-7| + |-3|$ (b) $-|(-7) + (-3)|$

Perform the indicated operations and simplify.

20. $(3x^2)(5x^4)$

21. $(2a^3b^3)(3a^0b^2)(4ab)$

22. $\dfrac{6x^4y^{12}}{2xy^3}$

23. $3^0x^{-3}y^5$

For what real values of x is each statement true?

24. $|-3x| = 3x$

25. $|x - 3| = x - 3$

READINGS AND PROJECTS

1. Explore a recently published elementary textbook series to determine the grade at which the set of integers is first introduced. Report on how negative integers are first introduced to elementary school students.
2. Read Booklet No. 2 (The Whole Numbers) of the twenty-ninth yearbook of the National Council of Teachers of Mathematics. Complete at least three of the exercise sets given there.
3. Read Booklet No. 9 (The System of Integers) of the thirtieth yearbook of the National Council of Teachers of Mathematics. Complete at least three of the exercise sets given there.
4. Construct a chart for multiplying integers (as shown on page 210) on a clear sheet of acetate. Use this as a base, and construct three overlays, one for each of the blank regions of the chart to show the product of two negative numbers; a negative number by a positive number; and a positive number by a negative number. Then demonstrate the use of an overhead projector in introducing products of signed numbers.
5. Read Chapter 12 (Tricks and Why They Work) of the twenty-seventh yearbook of the National Council of Teachers of Mathematics. Then invent a trick of your own that involves integers.
6. A Babylonian tablet known as *Plimpton 322* gives evidence that the Babylonians were aware of Pythagorean triples over one thousand years before the time of Pythagoras. Read and report on this interesting archeological find that was uncovered and described in 1945. It is discussed in most history of mathematics books. In particular, see *Episodes from the Early History of Mathematics* by Asger Aaboe, Random House, Inc., 1964.
7. Read about the life of Evariste Galois, the mathematician given credit for his original work on group theory. A fascinating account of his life, and death at age 20, is given in the book by Leopold Infeld, *Whom the Gods Love: The Story of Evariste Galois*, Whittlesey, 1948.
8. Read Chapter 8 (Language and Symbolism in Mathematics) of the twenty-fourth yearbook of the National Council of Teachers of Mathematics. Report on the authors' discussion of *number* versus *numeral.*

6 Rational Numbers: Properties and Operations

There are many practical problems that cannot be solved when only the integers are available:

1. A cook cannot use a recipe that calls for three cups of flour if only one-half the quantity is desired. The problem $3 \div 2$ has no solution in the system of integers.
2. A child cannot find an integer to represent one-half of a candy bar; $1 \div 2$ is not an integer.
3. A driver does not have an integer to represent the average number of miles per gallon of gas for a car that travels 200 miles using 11 gallons; $200 \div 11$ has no solution in the system of integers.

Formally we say that the set of integers is *not* closed with respect to division. Thus we need a further extension of the number system to make division always possible, except by zero. In this chapter we make such an extension and call our new set of numbers the *set of rational numbers*.

For many adults operations with fractions and with decimals remain a mystery throughout their lifetimes, possibly because they learned rules without understanding the meaning behind various procedures. Thus we find most people able to say that "invert and multiply" is one way to divide by a fraction, but unable to explain why this is so. One objective of this chapter is to make clear the underlying reasons for such operations.

In today's movement toward a metric system of measurement it becomes increasingly important to be able to work effectively with decimals. The metric system is considered in detail in Chapter 8. In the present chapter we set the stage with a detailed exploration of operations with decimals.

6-1 the system of rational numbers

A rational number is a number that can be expressed in the form

$$\frac{a}{b}$$

(often written a/b) where a is an integer and b is a counting number. Then

$$\frac{2}{3}, \quad \frac{1}{2}, \quad \frac{-7}{3}, \quad \frac{0}{1}, \quad \text{and} \quad \frac{215}{524}$$

are examples of rational numbers. Many elementary school textbooks refer to positive numbers represented by such fractions as whole numbers and **fractional numbers.** We use rational numbers to include whole numbers, fractional numbers, and their opposites. Later, in §6-5, we shall explore the relationship between rational numbers and decimals.

Example 1: Show that every integer is a rational number.

Solution: Every integer n can be written in the form $n/1$, the quotient of the integer and 1, and thus is a rational number. For example,

$$7 = \frac{7}{1} \quad \text{and} \quad -5 = \frac{-5}{1}$$

Statements of equality such as

$$n = \frac{n}{1}$$

are used when two symbols represent the same number. Each rational number may be represented in many ways. For example,

$$\frac{6}{2}, \quad \frac{15}{5}, \quad \frac{30}{10}, \quad \text{and} \quad \frac{150}{50}$$

each represent the number $\frac{3}{1}$. In general, the equality of rational numbers a/b is *defined* as follows:

$$\frac{a}{b} = \frac{c}{d} \quad \text{if and only if} \quad ad = bc$$

Then

$$\frac{6}{2} = \frac{15}{5} \quad \text{since} \quad 6 \times 5 = 2 \times 15.$$

Order relations ($<$, $>$) among rational numbers are defined in a similar way. (Remember that b and d are counting numbers.)

$$\frac{a}{b} < \frac{c}{d} \quad \text{if and only if} \quad ad < bc$$

$$\frac{a}{b} > \frac{c}{d} \quad \text{if and only if} \quad ad > bc$$

Then

$$\frac{5}{8} < \frac{2}{3} \quad \text{since} \quad 5 \times 3 < 8 \times 2$$

$$\frac{5}{7} > \frac{2}{3} \quad \text{since} \quad 5 \times 3 > 7 \times 2$$

Note that the definition of a rational number prohibits the denominator from being 0, since 0 is not a counting number. Division by 0 is not permitted and it is important that we understand why this is so. To do so we should first see the relationship between the operations of division and multiplication.

$$a \div b \quad \text{is the number } n \text{ such that} \quad b \times n = a.$$

For example, $12 \div 3$ is the number n such that $3 \times n = 12$. Indeed, this is the way that elementary school children check their division problems.

$$3\overline{)12}^{\;4} \qquad \textit{Check:} \quad 3 \times 4 = 12$$

Now consider the problem $7 \div 0$.

$$0\overline{)7}^{\;n} \quad 7 \div 0 \quad \text{is the number } n \text{ such that} \quad 0 \times n = 7.$$

But $0 \times n = 0$ for all n, and can never be equal to 7.

In the preceding discussion we could have used any number other than 7 with similar results. Therefore we say that division by 0 is not possible. With this one exception, division is now always possible within the set of rational numbers, which was *not* the case for integers. Thus the problem $1 \div 2$ now has a solution, namely $\frac{1}{2}$.

Example 2: Show that $0 \div 0$ does not represent a unique rational number.

Solution: First we note that $0 \div 0$ is the number, n such that $0 \times n = 0$. But $0 \times n = 0$ for *all n*. Therefore $0 \div 0$ has no unique solution and is often said to be *indeterminate*.

The set of rational numbers includes the set of integers as one of its subsets. That is, every integer is a rational number as in Example 1. However, not every rational number is an integer. Thus the set of rational numbers includes such elements as $\frac{2}{3}$ and $\frac{-2}{3}$ which are not integers.

By expanding our number system from the set of integers to the set of rational numbers, we have included a *multiplicative inverse* or *reciprocal* for each integer, and indeed each rational number, different from zero. The multiplicative inverse of 2 is $\frac{1}{2}$, of $\frac{2}{3}$ is $\frac{3}{2}$, and of $\frac{-3}{4}$ is $\frac{4}{-3}$. In each case the product of the number and its multiplicative inverse is 1, the *identity element for multiplication:*

$$2 \times \frac{1}{2} = 1 \qquad \frac{2}{3} \times \frac{3}{2} = 1 \qquad \left(\frac{-3}{4}\right) \times \left(\frac{4}{-3}\right) = 1$$

Inverse, ×: $\frac{a}{b} \times \frac{b}{a} = 1$. Every rational number $\frac{a}{b}$, $\frac{a}{b} \neq 0$, has a multiplicative inverse $\frac{b}{a}$ such that $\frac{a}{b} \times \frac{b}{a} = 1$.

(We need to make the restriction that $\frac{a}{b}$ is not equal to 0 because $0 \times n = 0$ for any rational number n, and thus 0 does not have a reciprocal in the set of rational numbers.)

We may summarize the properties of the set of rational numbers as follows:

Closure, +: The sum of any two rational numbers is a rational number.

Closure, ×: The product of any two rational numbers is a rational number.

Commutative, +: $a + b = b + a$ for all rational numbers a and b.

Commutative, ×: $a \times b = b \times a$ for all rational numbers a and b.

Associative, +: $a + (b + c) = (a + b) + c$ for all rational numbers, a, b, and c.

Associative, ×: $a \times (b \times c) = (a \times b) \times c$ for all rational numbers a, b, and c.

Identity, +: The set of rational numbers contains an element 0 such that $a + 0 = 0 + a = a$ for every rational number a.

Identity, ×: The set of rational numbers contains an element 1 such that $a \times 1 = 1 \times a = a$ for every rational number a.

Inverse, +: For every rational number a, there is another rational number, $-a$, such that $a + (-a) = 0$.

Inverse, (≠0), ×: For every rational number a, $a \neq 0$, there is another rational number $\frac{1}{a}$ such that $a \times \frac{1}{a} = 1$.

Distributive: For all rational numbers a, b, and c we have $a \times (b + c) = (a \times b) + (a \times c)$.

By studying these properties we note that the set of rational numbers forms a group under addition. However this set does not form a group with respect to multiplication since zero does not have a multiplicative inverse in the set.

We describe most of the properties of the set of rational numbers by saying that the set of rational numbers forms a field. In general, any set of numbers forms a **field** if:

The set is closed under addition and multiplication.
The associative law holds for addition and multiplication.
The commutative law holds for addition and multiplication.
The identity elements for addition (0) and multiplication (1) are members of the set.
Each element of the set has its inverse under addition in the set, and each element except 0 has its inverse under multiplication in the set.
The elements of the set satisfy the distributive law for multiplication with respect to addition.

EXERCISES

Select a relation R (=, <, >) to make each statement true.

1. $\frac{1}{2}$ R $\frac{2}{3}$ *2.* $\frac{2}{3}$ R $\frac{5}{7}$ *3.* $\frac{1}{3}$ R $\frac{2}{7}$
4. $\frac{3}{4}$ R $\frac{7}{10}$ *5.* $\frac{5}{11}$ R $\frac{6}{13}$ *6.* $\frac{1}{8}$ R $\frac{2}{17}$

Each symbol a/b has a **numerator** *a and a* **denominator** *b.*

7. Identify the numerator of **(a)** $\frac{2}{3}$; **(b)** $\frac{1}{5}$; **(c)** $\frac{2}{7}$; **(d)** $\frac{0}{7}$.

8. Identify the denominator of each symbol (numeral) in Exercise 7.

Copy the following table. Use "√" to show that the set of elements named at the top of the column has the property listed at the side. Use "×" if the set does not have the property.

	Property	Counting numbers	Whole numbers	Integers	Positive rationals	Rational numbers
9.	Closure, +					
10.	Associative, +					
11.	Identity, +					
12.	Inverse, +					
13.	Commutative, +					
14.	Commutative group, +					
15.	Closure, x					
16.	Associative, x					
17.	Identity, x					
18.	Inverse, (≠ o), x					
19.	Commutative, x					
20.	After excluding o, commutative group, x					
21.	Distributive					

PEDAGOGICAL EXPLORATIONS

1. Some elementary textbooks define a rational number in this way:

> A rational number is one that can be written in the form a/b, where a and b are integers and $b \neq 0$.

Show that this definition has the same meaning as the one given in this section.

2. Explore a recently published elementary textbook series and determine the grade level at which different fractional concepts are introduced, as well as the manner in which any visual aids are used to introduce fractions.

3. An interesting way to introduce the need for fractional numbers is to present a class with a set of questions and ask whether a fractional number or a whole number is needed to answer the question. For example, consider such questions as these:

> How many students are in this class?
> A pound of butter is usually packed as four bars. Each bar is what part of a pound?

Prepare a set of five similar questions that illustrate this difference between counting numbers and fractional numbers.

4. Show, with several specific examples, that if a, b, c, and d are counting numbers, and if $a/b < c/d$, then $a/b < (a + c)/(b + d) < c/d$. Then try to prove that this inequality holds true for all counting numbers. (*Hint:* Recall that $a/b < c/d$ if and only if $ad < bc$.)

6-2 rational numbers and the number line

Consider a number line and locate a point midway between the points with coordinates 0 and 1. The coordinate of the midpoint is called $\frac{1}{2}$. Then we may locate all points that correspond to "halves," such as $\frac{3}{2} = 1\frac{1}{2}$, $\frac{5}{2} = 2\frac{1}{2}$, $\frac{7}{2} = 3\frac{1}{2}$, and so forth. Also we may locate points with coordinates $-\frac{1}{2}$, $-1\frac{1}{2}$, $-2\frac{1}{2}$, $-3\frac{1}{2}$ and so forth. The graph of any rational number $n/2$, where n is an integer, may be found in this way.

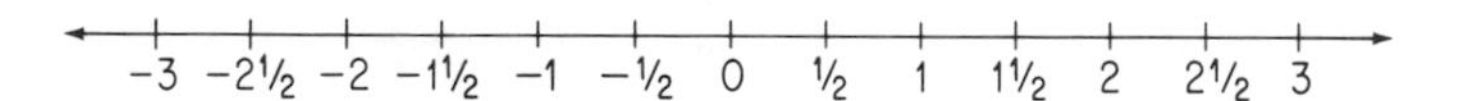

Rational numbers $n/3$ may be graphed on a number line. The graphs of $\frac{1}{3}$ and $\frac{2}{3}$ may be found by dividing the unit interval from 0 to 1 into three congruent parts. Then the points corresponding to $\frac{4}{3}, \frac{5}{3}, \frac{6}{3}, \frac{7}{3}, \ldots$ and also the points corresponding to $-\frac{1}{3}, -\frac{2}{3}, -\frac{3}{3}, -\frac{4}{3}, \ldots$ may be located.

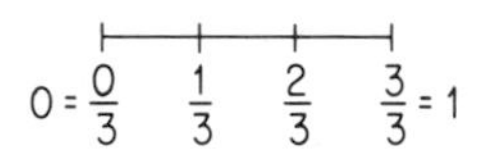

Similarly, the points corresponding to multiples of $\frac{1}{4}, \frac{1}{5}, \frac{1}{6}$, and so forth may be located on a number line. Each rational number has a graph on the number line. Note that the set of rational numbers includes the set of integers as a subset.

A number that is expressed as the sum of an integer and a fraction ($2\frac{1}{2} = 2 + \frac{1}{2}$) is often called a **mixed number,** or is said to be expressed in **mixed form.** As in §5-2 for integers, we may illustrate the addition of mixed numbers on a number line. Consider, for example, the sum $(+1\frac{3}{4}) + (-2\frac{1}{4})$. Start at the origin, 0, and draw an arrow that goes $1\frac{3}{4}$ units to the right. Then draw an arrow that goes $2\frac{1}{4}$ units to the left. The second arrow ends at $-\frac{1}{2}$. Thus, $(+1\frac{3}{4}) + (-2\frac{1}{4}) = -\frac{1}{2}$.

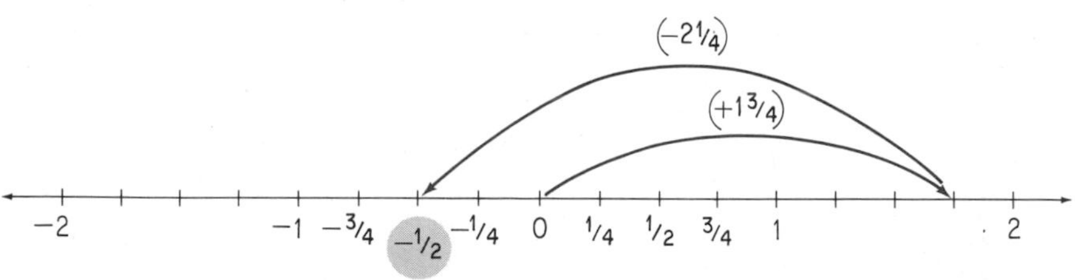

Example 1: Illustrate on a number line: $(-1\frac{1}{2}) + (-1\frac{1}{4})$.

Solution:

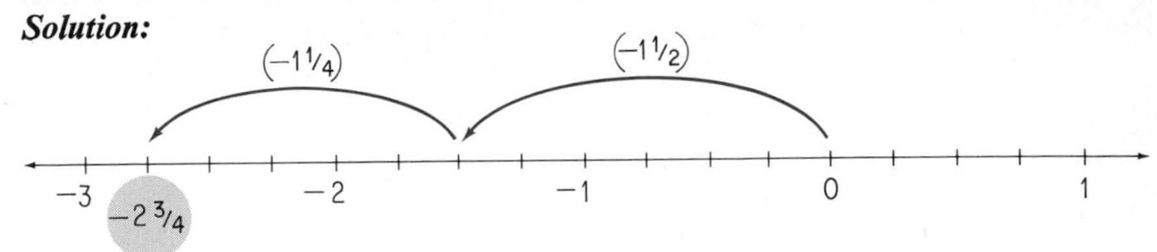

From the figure we see that $(-1\frac{1}{2}) + (-1\frac{1}{4}) = -2\frac{3}{4}$. Note that the scale on the number line used was divided into fourths because of the units used in the given problem.

The set of rational numbers can be classified as being positive (greater than 0), negative (less than 0), or 0. The number 0 is a rational number, but it is neither positive nor negative. As in the case of integers, positive rational numbers are the coordinates of points on the right of the origin and negative rational numbers are coordinates of points on the left of the origin.

Our number line has now become dense with points and "resembles" a complete line. The word "resembles" is used inasmuch as we shall find later that there are still points on the number line that do not have rational numbers as their coordinates. However, the set of rational numbers is said to be **dense,** since between any two elements of the set, there is always another element of the set.

Example 2: Locate a rational number between $\frac{17}{19}$ and $\frac{18}{19}$.

Solution: Change each rational number to one with a denominator of 38.

$$\frac{17}{19} \times \frac{2}{2} = \frac{34}{38}, \qquad \frac{18}{19} \times \frac{2}{2} = \frac{36}{38}$$

Clearly, $\frac{35}{38}$ lies between the two given rational numbers. Note also that

$$\frac{35}{38} = \frac{1}{2}(\frac{17}{19} + \frac{18}{19})$$

Example 3: Locate a rational number between $\frac{17}{19}$ and $\frac{35}{38}$.

Solution: Write each rational number as one with a denominator of 76.

$$\frac{17}{19} \times \frac{4}{4} = \frac{68}{76}, \qquad \frac{35}{38} \times \frac{2}{2} = \frac{70}{76}$$

Note that $\frac{69}{76}$ lies between the two given rational numbers. Also,

$$\frac{69}{76} = \frac{1}{2}(\frac{17}{19} + \frac{35}{38})$$

We can summarize the results of the two preceding examples in this manner:

$$\frac{17}{19} < \frac{35}{38} < \frac{18}{19}$$

$$\frac{17}{19} < \frac{69}{76} < \frac{35}{38}$$

This process can be extended indefinitely. For example, to locate a rational number between $\frac{17}{19}$ and $\frac{69}{76}$, write each with a denominator of 152. Then find one-half the sum of the two numbers. This ability to locate a rational number between any two given rational numbers is what is meant by the *density property*.

In this section we have found that every rational number is the coordinate of some point on the number line, but *not* every point on the number line can be named by a rational number. In the next chapter we shall extend our number system so as to establish a one-to-one correspondence between the points of the number line and a new set of numbers, thereby completing the number line.

EXERCISES

Classify each statement as true or false.

1. Every integer is a rational number.
2. Every rational number is an integer.
3. The reciprocal of every rational number except 0 is a rational number.
4. The number 0 is not a rational number.
5. The multiplicative inverse of a positive rational number is negative.

Copy the following table. Use "$\surd$" to show that the number listed at the top of a column is a member of the set listed at the side. Use "$\times$" if the number is not a member of the set.

	6.	7.	8.	9.	10.
Set	-3	8	0	$\frac{2}{3}$	$-\frac{3}{4}$
Counting numbers					
Whole numbers					
Integers					
Rational numbers					

11. Name two properties of the set of rational numbers that are not properties of the set of integers.
12. Find three rational numbers between $\frac{7}{9}$ and $\frac{8}{9}$.
13. Find three rational numbers between 0 and $\frac{1}{100}$.

sec. 6-2
rational numbers
and the number line

14. Name a rational number between the rational numbers a/b and c/d.

15. Is there a next rational number after 0? Use a property of the set of rational numbers to explain your answer.

Find a replacement for n that will make each of the sentences below true where n must be a member of the set named at the top of the column. If there is no such replacement, write "none."

	Sentence	Counting numbers	Whole numbers	Integers	Rational numbers
16.	$n - 4 = 1$				
17.	$n + 5 = 5$				
18.	$n + 6 = 1$				
19.	$5n = 3$				
20.	$n \times n = 2$				

Illustrate each sum on a number line.

21. $(+1\frac{1}{4}) + (+1\frac{3}{4})$

22. $(-1\frac{3}{4}) + (-\frac{3}{4})$

23. $(+1\frac{1}{3}) + (-3\frac{2}{3})$

24. $(-1\frac{2}{3}) + (+3\frac{1}{3})$

PEDAGOGICAL EXPLORATIONS

1. Make a collection of different ways that rational numbers are used in practical situations. For example, stock market reports are quoted in fractional terms and generally end with a final column that indicates the daily rise or fall of particular stocks in terms of positive or negative numbers.
2. Prepare a worksheet of ten examples based on a stock market report from a daily newspaper, and suitable for upper elementary or junior high school students.
3. The Greek philosopher Zeno, in the fifth century B.C., proposed several paradoxes that were based on the density property. Prepare a report on these paradoxes, using a history of mathematics book as your reference. This report should be appropriate for upper elementary and junior high school students.
4. Take a small strip of paper about 12 inches long and label it 0 at the left end and 1 at the right end. Then fold the paper once to obtain the midpoint of the strip. Label this point as $\frac{1}{2}$. Continue with this paper-

folding experiment to construct points representing the fourths and points representing the eighths.

5. A number line such as the following can be very helpful in giving young students an intuitive concept of order:

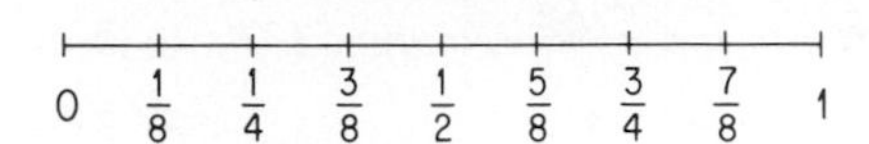

$\frac{1}{2}$ is to the left of $\frac{7}{8}$; therefore $\frac{1}{2} < \frac{7}{8}$.
$\frac{1}{2}$ is to the right of $\frac{1}{8}$; therefore $\frac{1}{2} > \frac{1}{8}$.

Prepare a short 10-minute lesson that illustrates order of fractional numbers through the use of a number-line approach.

6. Explain how the density property enables us to use rational numbers as measures of line segments.

6-3 fractions: multiplication and division

Any rational number may, by definition, be expressed as a **fraction** in the form a/b, where the **numerator** a is an integer and the **denominator** b is a counting number. When expressed in terms of "parts of a whole," the *denominator* names the fractional parts (halves, thirds, quarters, etc.) under consideration, and the *numerator* tells the number of those parts. For example, $\frac{3}{4}$ may be thought of as three quarters or three parts that are each one quarter of a complete unit; that is:

$$\frac{3}{4} = 3 \times \frac{1}{4}$$

In general, we may think of a/b in this way:

$$\frac{a}{b} = a \times \frac{1}{b}$$

In the figure that follows, the rectangular region has been divided into fourths. Three of these parts have been shaded to denote $\frac{3}{4}$ of the original rectangular region.

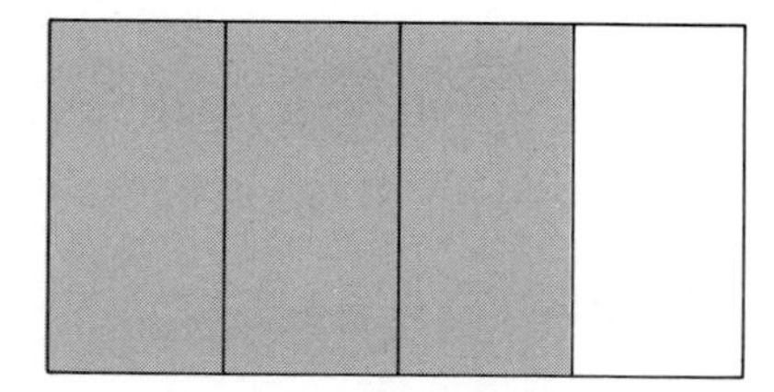

If we divide the original rectangular region into halves so that each of the fourths is also divided into halves, we have:

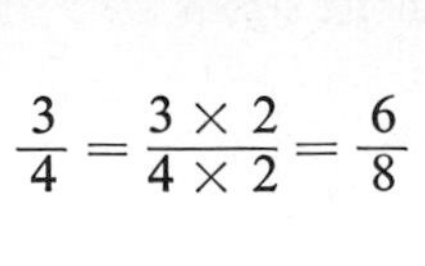

$$\frac{3}{4} = \frac{3 \times 2}{4 \times 2} = \frac{6}{8}$$

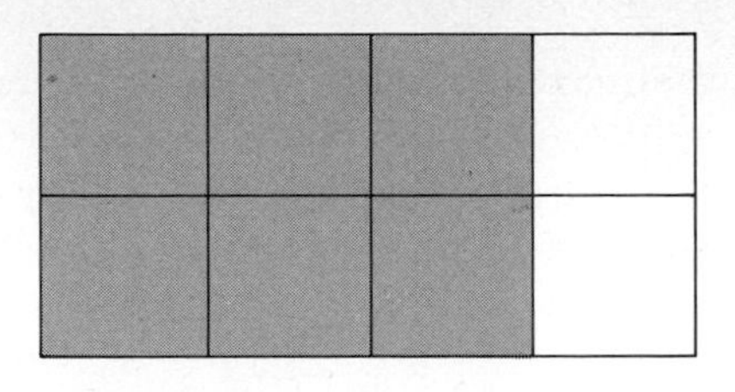

If we divide the original rectangular region into thirds so that each of the fourths is also divided into thirds, we have:

$$\frac{3}{4} = \frac{3 \times 3}{4 \times 3} = \frac{9}{12}$$

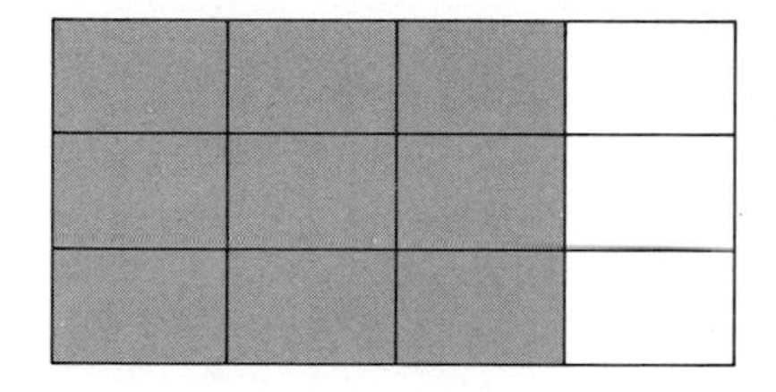

In general, we have the familiar rule

$$\frac{a}{b} \times \frac{k}{k} = \frac{ak}{bk} = \frac{a}{b} \qquad \text{for any integer } k \neq 0.$$

This rule indicates that multiplication by 1 in the form k/k for $k \neq 0$ leaves the number unchanged. We use the rule in the form

$$\frac{a \times k}{b \times k} = \frac{a}{b}$$

to "reduce" fractions by dividing numerator and denominator by k. For example,

$$\frac{12}{30} = \frac{2 \times 6}{5 \times 6} = \frac{2}{5} \times \frac{6}{6} = \frac{2}{5} \times 1 = \frac{2}{5}$$

$$\frac{72}{18} = \frac{4 \times 18}{1 \times 18} = \frac{4}{1} \times \frac{18}{18} = \frac{4}{1} \times 1 = 4$$

$$\frac{16}{80} = \frac{1 \times 16}{5 \times 16} = \frac{1}{5} \times \frac{16}{16} = \frac{1}{5} \times 1 = \frac{1}{5}$$

A fraction a/b, where b is a counting number, is said to be **reduced to lowest terms** if it is not possible to find an integer $k > 1$ and integers c and d such that

$$\frac{a}{b} = \frac{ck}{dk}$$

Each of the fractions that we have just considered was reduced to lowest terms. This rule also allows us to find as many ways of writing a rational number as we like. For example,

$$\frac{2}{3} = \frac{4}{6} = \frac{6}{9} = \frac{8}{12} = \frac{10}{15} = \cdots$$

This procedure for writing rational numbers in many ways enables us to express any two rational numbers a/b and c/d as fractions with like denominators:

$$\frac{a}{b} = \frac{a \times d}{b \times d} = \frac{ad}{bd}$$

$$\frac{c}{d} = \frac{c \times b}{d \times b} = \frac{b \times c}{b \times d} = \frac{bc}{bd}$$

Example 1: Write $\frac{2}{3}$ and $\frac{4}{5}$ as fractions with like denominators.

Solution:

$$\frac{2}{3} = \frac{2 \times 5}{3 \times 5} = \frac{10}{15} \qquad \frac{4}{5} = \frac{4 \times 3}{5 \times 3} = \frac{12}{15}$$

Multiplication of two fractional numbers such as $\frac{2}{3} \times \frac{3}{4}$ is sometimes introduced by means of diagrams such as the following. First draw a rectangular region and shade $\frac{3}{4}$ of the figure as shown.

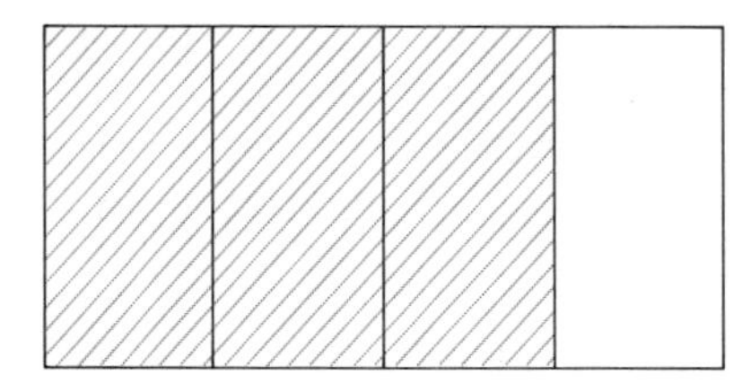

Next shade $\frac{2}{3}$ of the figure as in the following diagram.

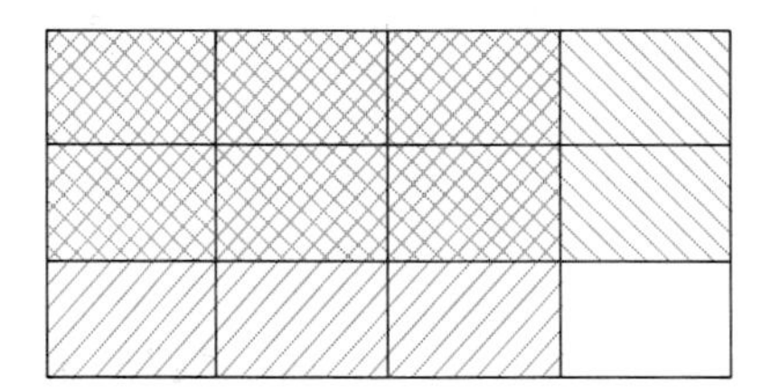

The final picture shows that the rectangular region has been divided into 12 congruent parts, and that six of these parts are shaded in both directions. Thus $\frac{2}{3} \times \frac{3}{4} = \frac{6}{12}$.

Such diagrams are used to justify the usual rule for multiplication of rational numbers written in fractional form:

$$\frac{a}{b} \times \frac{c}{d} = \frac{ac}{bd}$$

Division of rational numbers often causes difficulty. Many people "invert and multiply" without really knowing why they do so. We consider this rule in terms of two specific examples. Suppose that we start with two integers such as $6 \div 2$. Then

$$6 \div 2 = \frac{6}{2} = \frac{6 \times 1}{1 \times 2} = \frac{6}{1} \times \frac{1}{2} = 6 \times \frac{1}{2}$$

Thus $6 \div 2 = 6 \times \frac{1}{2}$ where $\frac{1}{2}$ is the reciprocal of 2. In general, for any integers a and $b \neq 0$,

$$a \div b = a \times \frac{1}{b}$$

Division of rational numbers may also be expressed as a product. For example, consider

$$\frac{2}{3} \div \frac{5}{7}$$

As for integers we may write this quotient in the form

$$\frac{\frac{2}{3}}{\frac{5}{7}}$$

and seek methods for rewriting this new fraction so that its denominator becomes 1. Then we multiply numerator and denominator by the reciprocal of $\frac{5}{7}$, that is, $\frac{7}{5}$. These steps may be displayed as follows:

$$\frac{2}{3} \div \frac{5}{7} = \frac{\frac{2}{3}}{\frac{5}{7}}$$ Since these are just two different ways of expressing a quotient.

$$= \frac{\frac{2}{3}}{\frac{5}{7}} \times 1$$ By the multiplication property of 1.

$$= \frac{\frac{2}{3}}{\frac{5}{7}} \times \frac{\frac{7}{5}}{\frac{7}{5}}$$ Since $\frac{7}{5} \div \frac{7}{5} = 1$.

$$= \frac{\frac{2}{3} \times \frac{7}{5}}{\frac{5}{7} \times \frac{7}{5}}$$ By the rule for multiplication of fractional numbers.

$$= \frac{\frac{2}{3} \times \frac{7}{5}}{1}$$ Since the product of $\frac{5}{7}$ and $\frac{7}{5}$ is $\frac{35}{35} = 1$.

$$= \frac{2}{3} \times \frac{7}{5}$$ Since any number divided by 1 is equal to that number.

Note that we began with a division problem, and showed that this could be written as an equivalent multiplication problem. In general we have the following rule for division of rational numbers:

$$\frac{a}{b} \div \frac{c}{d} = \frac{a}{b} \times \frac{d}{c}$$

In other words, to divide by a number different from zero we multiply by its reciprocal. This is a formal statement of the common procedure described by "invert and multiply." This procedure may be justified in general as follows:

$$\frac{\frac{a}{b}}{\frac{c}{d}} = \frac{\frac{a}{b} \times \frac{d}{c}}{\frac{c}{d} \times \frac{d}{c}} = \frac{\frac{a}{b} \times \frac{d}{c}}{1} = \frac{a}{b} \times \frac{d}{c} = \frac{a \times d}{b \times c}$$

A division problem expressed as

$$\frac{\frac{2}{3}}{\frac{5}{7}}$$

may also be reduced to lowest terms by multiplying by 1 in a form such that the numerator and the denominator become integers. For example, notice that the least common multiple (L.C.M., §4-3) of the denominators is 21:

L.C.M. (3,7) = 21

$$\frac{\frac{2}{3}}{\frac{5}{7}} \times \frac{21}{21} = \frac{14}{15}$$

This example illustrates the following general procedure:

$$\frac{\frac{a}{b}}{\frac{c}{d}} = \frac{\frac{a}{b}}{\frac{c}{d}} \times \frac{b \times d}{b \times d} = \frac{a \times d}{b \times c}$$

Note that the form obtained is the same as that obtained by the "invert and multiply" procedure.

Example 2: Simplify:

$$\frac{\frac{2}{5}}{\frac{3}{4}}$$

Solution: Note that the word "simplify" is used here and elsewhere to indicate that all operations are to be performed and the resulting expression reduced to lowest terms, if possible. We shall consider two methods of solution for this example.

$$\frac{\frac{2}{5}}{\frac{3}{4}} = \frac{2}{5} \div \frac{3}{4} = \frac{2}{5} \times \frac{4}{3} = \frac{8}{15}$$

$$\frac{\frac{2}{5}}{\frac{3}{4}} = \frac{\frac{2}{5}}{\frac{3}{4}} \times \frac{20}{20} = \frac{8}{15}$$ Note that the L.C.M. (5, 4) = 20.

Finally, we need to note that negative rational numbers, such as $\frac{-5}{4}$ may also be expressed in two other ways simply by changing the location of the minus sign. For example:

$$\frac{-5}{4} = \frac{-5}{4} \times 1 = \frac{-5}{4} \times \frac{-1}{-1} = \frac{(-5)(-1)}{(4)(-1)} = \frac{5}{-4}$$

Therefore,

$$\frac{-5}{4} = \frac{5}{-4}$$

We can also show that yet another form for this rational number is $-\frac{5}{4}$. To do so first show that $\frac{-5}{4}$ and $\frac{5}{4}$ are opposites of one another; that is, their sum is 0:

$$\frac{5}{4} + \frac{-5}{4} = 5\left(\frac{1}{4}\right) + (-5)\left(\frac{1}{4}\right) = [(5) + (-5)]\left(\frac{1}{4}\right) = (0)\left(\frac{1}{4}\right) = 0$$

Therefore, $\frac{-5}{4}$ is the opposite of $\frac{5}{4}$; that is, $\frac{-5}{4} = -\frac{5}{4}$.

In general, there are three signs associated with any fraction c/d:

1. The sign of the fraction, as in $-\frac{c}{d}$.

2. The sign of the numerator, as in $\frac{-c}{d}$.

3. The sign of the denominator, as in $\frac{c}{-d}$.

From the preceding discussion and multiplication by 1 in the form $\frac{-1}{-1}$,

$$\frac{-c}{d} \times \frac{-1}{-1} = \frac{(-c) \times (-1)}{d \times (-1)} = \frac{c}{-d}$$

and

$$-\frac{c}{d} = \frac{-c}{d} = \frac{c}{-d}$$

In other words, any two of the three signs associated with a fraction may be changed without changing the rational number represented by the fraction.

Example 3: Reduce $\frac{6}{-8}$.

Solution: The denominator should be a counting number.

$$\frac{6}{-8} = \frac{-6}{8} = \frac{-3}{4}$$

This result is usually written as $-\frac{3}{4}$.

Example 4: Multiply: $\frac{2}{3} \times \frac{3}{4} \times \left(-\frac{1}{8}\right)$.

Solution: We use the associative property and multiply the rational numbers two at a time.

$$\begin{aligned}\frac{2}{3} \times \frac{3}{4} \times \left(-\frac{1}{8}\right) &= \left(\frac{2}{3} \times \frac{3}{4}\right) \times \left(-\frac{1}{8}\right)\\ &= \frac{2 \times 3}{3 \times 4} \times \left(-\frac{1}{8}\right)\\ &= \frac{6}{12} \times \left(-\frac{1}{8}\right)\\ &= \frac{1}{2} \times \left(-\frac{1}{8}\right)\\ &= -\frac{1}{16}\end{aligned}$$

Note that with practice some of the steps shown are often omitted.

EXERCISES

Reduce each fraction to lowest terms.

1. $\frac{18}{60}$ **2.** $\frac{45}{70}$ **3.** $\frac{48}{64}$ **4.** $-\frac{17}{34}$ **5.** $\frac{-16}{80}$

Write each pair of fractions as fractions with like denominators.

6. $\frac{7}{8}, \frac{3}{4}$ **7.** $\frac{2}{3}, \frac{3}{5}$ **8.** $\frac{5}{6}, \frac{1}{4}$ **9.** $-\frac{2}{3}, -\frac{3}{4}$

Perform the indicated operations, and simplify.

10. $\frac{11}{12} \times \frac{3}{7}$ **11.** $\frac{9}{10} \times \frac{2}{3}$ **12.** $\frac{12}{5} \times \frac{15}{8}$

13. $\frac{3}{5} \div \frac{3}{5}$ **14.** $\frac{3}{5} \div \frac{5}{3}$ **15.** $\frac{12}{7} \div \frac{9}{10}$

16. $\frac{8}{9} \div \frac{4}{3}$ **17.** $\frac{-3}{5} \times \frac{2}{5}$ **18.** $\frac{5}{-2} \times \frac{-3}{5}$

19. $\left(-\frac{3}{4}\right) \times \frac{2}{5}$ **20.** $\left(-\frac{7}{8}\right) \times \left(-\frac{4}{9}\right)$ **21.** $\frac{8}{9} \div \frac{-2}{3}$

22. $\frac{-5}{12} \div \frac{10}{-3}$ **23.** $\left(-\frac{5}{6}\right) \div \frac{1}{3}$ **24.** $\left(-\frac{4}{5}\right) \div \left(-\frac{2}{5}\right)$

25. $\frac{4}{5} \times \frac{3}{4} \times \frac{2}{3}$ **26.** $\frac{7}{8} \times \frac{4}{9} \times \left(-\frac{3}{2}\right)$ **27.** $\frac{5}{9} \div \left(\frac{2}{3} \div \frac{1}{2}\right)$

28. $\left(\frac{5}{9} \div \frac{2}{3}\right) \div \frac{1}{2}$ **29.** $\left(\frac{7}{9} \times \frac{3}{4}\right) \div \frac{1}{2}$ **30.** $\frac{7}{9} \times \left(\frac{3}{4} \div \frac{1}{2}\right)$

31. The results obtained for Exercises 27 and 28 provide a counterexample to show that the set of rational numbers is *not* associative for division. Provide another counterexample of your own to show this fact.

***32.** Study the results for Exercises 29 and 30. Then show, in general, that

$$\left(\frac{a}{b} \times \frac{c}{d}\right) \div \frac{e}{f} = \frac{a}{b} \times \left(\frac{c}{d} \div \frac{e}{f}\right)$$

***33.** First use your own specific example, and then in general determine whether or not the following relationship is true:

$$\frac{a}{b} \div \left(\frac{c}{d} \times \frac{e}{f}\right) = \left(\frac{a}{b} \div \frac{c}{d}\right) \times \frac{e}{f}$$

Simplify by multiplying numerator and denominator by the least common denominator of each fraction involved.

34. $\dfrac{\frac{1}{2}}{\frac{3}{8}}$ **35.** $\dfrac{\frac{4}{5}}{\frac{2}{3}}$ **36.** $\dfrac{\frac{7}{8}}{\frac{7}{8}}$ **37.** $\dfrac{\frac{4}{7}}{\frac{5}{6}}$ **38.** $\dfrac{\left(-\frac{3}{5}\right)}{\left(-\frac{7}{10}\right)}$

PEDAGOGICAL EXPLORATIONS

1. Draw a rectangular region 6 centimeters long and 3 centimeters wide for each part of this exercise. Let the entire rectangular region represent 1 unit and shade portions to represent:

(a) $\frac{1}{2}$ (b) $\frac{1}{3}$ (c) $\frac{2}{3}$ (d) $\frac{1}{6}$
(e) $\frac{1}{9}$ (f) $\frac{5}{9}$ (g) $\frac{1}{18}$ (h) $\frac{7}{18}$

2. Draw a rectangular region as in Exploration 1 above. Let a part of the region that is 5 centimeters long and 2 centimeters wide represent 1 unit and shade portions to represent:

(a) 1 (b) $\frac{1}{5}$ (c) $\frac{2}{5}$ (d) $\frac{6}{5}$

(e) $\frac{1}{2}$ (f) $\frac{3}{2}$ (g) $\frac{1}{10}$ (h) $\frac{13}{10}$

3. A diagram was shown in this section to illustrate the product $\frac{2}{3} \times \frac{3}{4}$. Prepare a lesson suitable for a specific elementary grade level that demonstrates the product $\frac{3}{4} \times \frac{5}{7}$ through the use of a similar diagram.
4. State the objectives of this section in behavioral language. (See pages 182–183.) Then prepare a 20-minute test that evaluates attainment of these objectives.
5. Prepare a collection of visual aids that can be used to present basic concepts of fractional numbers to elementary school children.
6. Here is a flow chart with directions for reducing a fraction.

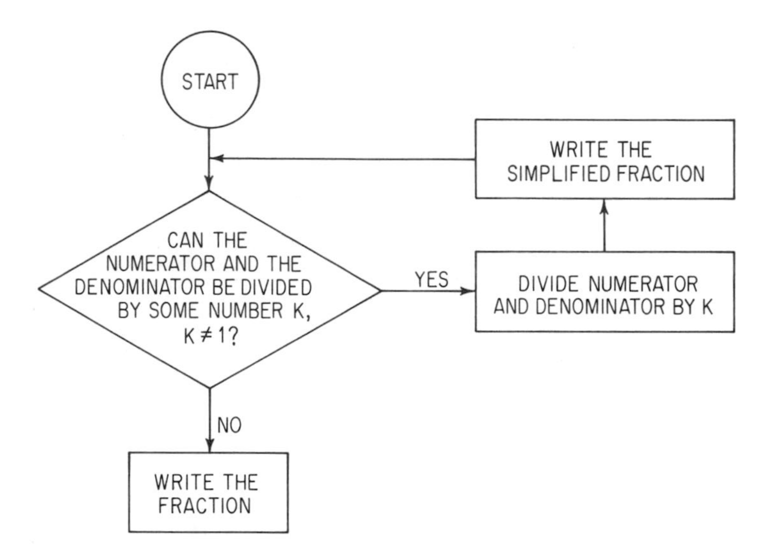

For example, if you start with $\frac{18}{24}$, you could divide numerator and denominator by 2 at first. Then you could divide by 3. Thus,

$$\frac{18}{24} \longrightarrow \frac{9}{12} \longrightarrow \frac{3}{4}$$

Use the flow chart to reduce each fraction.

(a) $\frac{102}{186}$ (b) $\frac{84}{120}$ (c) $\frac{96}{144}$ (d) $\frac{120}{156}$

6-4 fractions: addition and subtraction

The addition of two rational numbers that are represented by fractions with like denominators is accomplished by adding the numerators and using the common denominator. For example:

$$\frac{3}{8} + \frac{2}{8} = \frac{3+2}{8} = \frac{5}{8}$$

This problem may be illustrated by using a rectangular region divided into eighths as in the given figure. Three of the parts are shaded with horizontal lines to represent $\frac{3}{8}$. Two of the parts are shaded with vertical lines to represent $\frac{2}{8}$. The sum $\frac{3}{8} + \frac{2}{8}$ is represented by the five shaded parts; $\frac{3}{8} + \frac{2}{8} = \frac{5}{8}$.

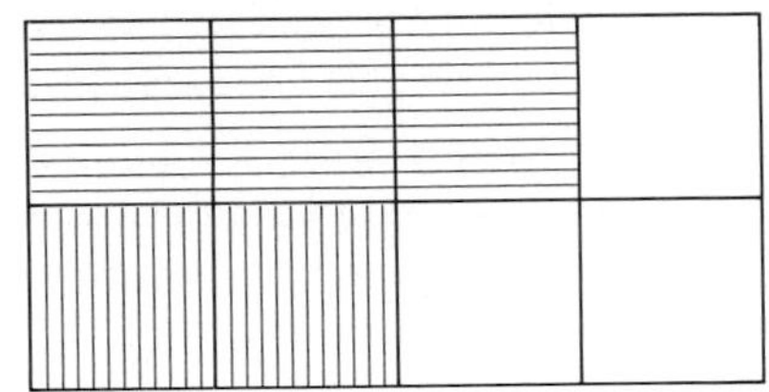

Also, as in §6-2, we may illustrate this sum on a number line that is divided into eighths, as in the figure.

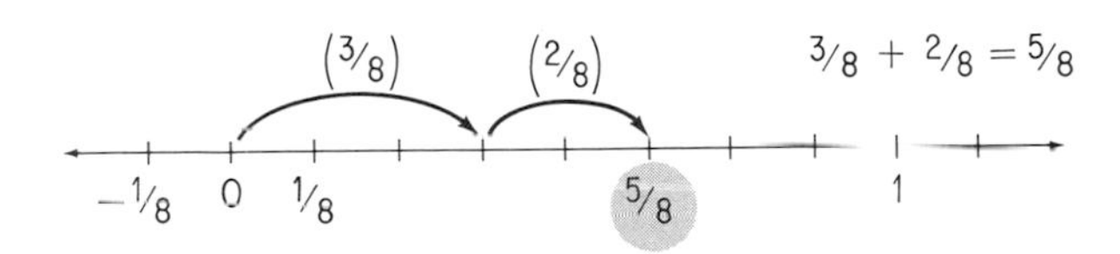

In summary, to add rational numbers expressed by fractions with like denominators we add the numerators and use the common denominator:

$$\frac{a}{c} + \frac{b}{c} = \frac{a+b}{c}$$

To subtract rational numbers expressed by fractions with like denominators we subtract the numerators and use the common denominator:

$$\frac{a}{c} - \frac{b}{c} = \frac{a-b}{c}$$

Example 1: Find the sum $\frac{-3}{5} + \frac{-1}{5}$.

Solution:

$$\frac{-3}{5} + \frac{-1}{5} = \frac{(-3)+(-1)}{5} = \frac{-4}{5} = -\frac{4}{5}$$

Example 2: Find the difference $\frac{2}{8} - \frac{7}{8}$.

Solution:

$$\frac{2}{8} - \frac{7}{8} = \frac{2-7}{8} = \frac{-5}{8} = -\frac{5}{8}$$

Both of the above rules can be justified through the use of the distributive property. Think of a/c as $a \times (1/c)$ and b/c as $b \times (1/c)$:

$$\frac{a}{c} + \frac{b}{c} = \left(a \times \frac{1}{c}\right) + \left(b \times \frac{1}{c}\right)$$
$$= (a + b) \times \frac{1}{c}$$
$$= \frac{a + b}{c}$$

The rule for subtraction follows in exactly the same manner.

Let us next consider addition and subtraction with fractions that have different denominators, such as $\frac{2}{3}$ and $\frac{3}{4}$. To add these fractions we first rewrite them so that they have like denominators.

$$\frac{2}{3} + \frac{3}{4} = \left(\frac{2}{3} \times \frac{4}{4}\right) + \left(\frac{3}{4} \times \frac{3}{3}\right)$$
$$= \frac{2 \times 4}{3 \times 4} + \frac{3 \times 3}{4 \times 3}$$
$$= \frac{8}{12} + \frac{9}{12}$$
$$= \frac{17}{12} \qquad \left(\frac{a}{c} + \frac{b}{c} = \frac{a + b}{c}\right)$$

In general, let us find the sum of any two rational numbers represented by fractions a/b and c/d.

$$\frac{a}{b} + \frac{c}{d} = \left(\frac{a}{b} \times \frac{d}{d}\right) + \left(\frac{c}{d} \times \frac{b}{b}\right)$$
$$= \frac{ad}{bd} + \frac{bc}{bd}$$
$$= \frac{ad + bc}{bd}$$

This procedure may also be justified by using properties of fractions, the distributive law, and the interpretation of each fraction a/b as the product $a \times (1/b)$:

$$\frac{a}{b} + \frac{c}{d} = \frac{ad}{bd} + \frac{bc}{bd}$$
$$= \left(ad \times \frac{1}{bd}\right) + \left(bc \times \frac{1}{bd}\right)$$
$$= (ad + bc) \times \frac{1}{bd}$$
$$= \frac{ad + bc}{bd}$$

In a similar manner, the difference of two rational numbers expressed in fractional form may be found as follows:

$$\frac{a}{b} - \frac{c}{d} = \frac{ad}{bd} - \frac{bc}{bd} = \frac{ad - bc}{bd}$$

Example 3: Use the rule for addition of rational numbers to find $\frac{2}{3} + \frac{3}{4}$.

Solution:

$$\frac{2}{3} + \frac{3}{4} = \frac{a}{b} + \frac{c}{d} = \frac{ad + bc}{bd}$$

$$a = 2, \qquad b = 3, \qquad c = 3, \qquad d = 4$$

$$ad + bc = (2 \times 4) + (3 \times 3), \qquad bd = 3 \times 4$$

$$\frac{2}{3} + \frac{3}{4} = \frac{(2 \times 4) + (3 \times 3)}{3 \times 4} = \frac{8 + 9}{12} = \frac{17}{12}$$

Problems that involve addition and subtraction with fractions are often written in vertical form, such as in the following display for the difference $\frac{3}{4} - \frac{1}{5}$.

$$\begin{array}{rccc} & \frac{3}{4} & \frac{3}{4} \times \frac{5}{5} = & \frac{15}{20} \\ - & \underline{\frac{1}{5}} & \frac{1}{5} \times \frac{4}{4} = & \underline{\frac{4}{20}} \\ & & & \frac{11}{20} \quad \left(\frac{3}{4} - \frac{1}{5} = \frac{11}{20}\right) \end{array}$$

A *mixed number* can be written in fractional form through the use of addition rules, as in the next example.

Example 4: Express $4\frac{2}{3}$ as a fraction.

Solution:

$$4\frac{2}{3} = 4 + \frac{2}{3} = \left(\frac{4}{1} \times \frac{3}{3}\right) + \frac{2}{3} = \frac{12}{3} + \frac{2}{3} = \frac{14}{3}$$

We may also express a fraction in *mixed form* if the numerator of the fraction is greater than the denominator, as in Example 5.

Example 5. Express $\frac{17}{4}$ in mixed form.

Solution:

$$\frac{17}{4} = \frac{16}{4} + \frac{1}{4} = 4 + \frac{1}{4} = 4\frac{1}{4}$$

Note that $\frac{17}{4}$ is an indicated division of 17 by 4. Therefore the result can also be obtained by division of these whole numbers.

Example 6: Add: $(-2\frac{2}{3}) + (-3\frac{1}{2})$.

Solution: There are several possible ways to complete this addition.
(a) Write each addend in fractional form.

$$-2\frac{2}{3} = -\frac{8}{3}, \qquad -3\frac{1}{2} = -\frac{7}{2}$$

$$\left(-2\frac{2}{3}\right) + \left(-3\frac{1}{2}\right) = \left(-\frac{8}{3}\right) + \left(-\frac{7}{2}\right)$$
$$= \left(-\frac{16}{6}\right) + \left(-\frac{21}{6}\right)$$
$$= -\frac{37}{6}$$
$$= -6\frac{1}{6}$$

(b) In vertical form, this addition appears as follows.

$$\begin{array}{rcr} -2\frac{2}{3} & = & -2\frac{4}{6} \\ -3\frac{1}{2} & = & -3\frac{3}{6} \\ \hline & & -5\frac{7}{6} = -6\frac{1}{6} \end{array}$$

Note that solution (b) is generally the preferred approach when adding or subtracting with mixed numbers.

At times it may become necessary to make appropriate exchanges during a subtraction problem, as we shall illustrate for the example $8\frac{1}{2} - 4\frac{2}{3}$.

$$\begin{array}{rcr} 8\frac{1}{2} & = & 8\frac{3}{6} \\ 4\frac{2}{3} & = & 4\frac{4}{6} \\ \hline \end{array}$$

After the fractional parts have been rewritten so that they have a common denominator, we find that we must subtract $\frac{4}{6}$ from $\frac{3}{6}$. To do this we think of $8\frac{3}{6}$ as follows:

$$8\frac{3}{6} = 7 + 1 + \frac{3}{6} = 7 + \frac{6}{6} + \frac{3}{6} = 7 + \frac{9}{6} = 7\frac{9}{6}$$

Of course, most of this work is done mentally but it is important to understand the precedure. In this example we exchange the number 1 for $\frac{6}{6}$, and write our work in this way:

$$\begin{array}{rrl} 8\frac{1}{2} & 8\frac{3}{6} & = 7\frac{9}{6} \\ -\ 4\frac{2}{3} & 4\frac{4}{6} & = 4\frac{4}{6} \\ \hline & & \quad 3\frac{5}{6} \end{array} \qquad \left(8\frac{1}{2} - 4\frac{2}{3} = 3\frac{5}{6}\right)$$

EXERCISES

Perform the indicated operation and simplify.

1. $\frac{2}{3} + \frac{1}{4}$ **2.** $\frac{1}{6} + \frac{3}{4}$ **3.** $\frac{8}{9} + \frac{2}{3}$

4. $\frac{5}{12} + \frac{7}{8}$ **5.** $\frac{5}{6} - \frac{2}{3}$ **6.** $\frac{3}{5} - \frac{1}{4}$

7. $\frac{13}{8} - \frac{3}{4}$ **8.** $\frac{7}{9} - \frac{3}{5}$ **9.** $\left(-\frac{5}{8}\right) + \left(-\frac{7}{8}\right)$

10. $\left(-\frac{2}{3}\right) + \left(-\frac{1}{5}\right)$ **11.** $\frac{3}{10} - \frac{7}{10}$ **12.** $\left(-\frac{1}{4}\right) - \left(+\frac{1}{3}\right)$

Express as a fraction $\frac{a}{b}$.

13. $3\frac{1}{2}$ **14.** $7\frac{1}{8}$ **15.** $2\frac{7}{8}$ **16.** $-4\frac{2}{3}$ **17.** $-1\frac{9}{10}$

Express in mixed form.

18. $\frac{12}{5}$ **19.** $\frac{13}{4}$ **20.** $\frac{7}{2}$ **21.** $-\frac{15}{7}$ **22.** $-\frac{19}{3}$

Add and simplify.

23. $\begin{array}{r} 4\frac{1}{8} \\ +3\frac{3}{4} \\ \hline \end{array}$ **24.** $\begin{array}{r} 5\frac{1}{6} \\ +4\frac{2}{3} \\ \hline \end{array}$ **25.** $\begin{array}{r} 12\frac{1}{2} \\ -\ 5\frac{3}{4} \\ \hline \end{array}$ **26.** $\begin{array}{r} 15 \\ -\ 7\frac{3}{5} \\ \hline \end{array}$ **27.** $\begin{array}{r} 9\frac{2}{3} \\ -5\frac{7}{8} \\ \hline \end{array}$

Perform the indicated operations and simplify.

28. $\frac{2}{3} + \left(\frac{3}{4} + \frac{7}{8}\right)$

29. $\left(\frac{2}{3} + \frac{3}{4}\right) + \frac{7}{8}$

30. $\frac{7}{8} - \left(\frac{3}{4} - \frac{1}{2}\right)$

31. $\left(\frac{7}{8} - \frac{3}{4}\right) - \frac{1}{2}$

32. $\left(2\frac{1}{2} + 3\frac{3}{4}\right) - \left(1\frac{1}{8} + 2\frac{1}{4}\right)$

33. $\left(5\frac{1}{3} - 1\frac{1}{2}\right) + \left(7\frac{1}{4} - 2\frac{2}{3}\right)$

34. $\dfrac{\frac{1}{2} + \frac{2}{3}}{\frac{3}{4}}$

35. $\dfrac{\frac{3}{4}}{\frac{1}{2} + \frac{2}{3}}$

36. $\dfrac{\frac{1}{3} + \frac{3}{4}}{\frac{1}{2} + \frac{2}{3}}$

37. $\dfrac{\frac{3}{4} - \frac{1}{3}}{\frac{2}{3} - \frac{1}{2}}$

Express as a single quotient.

***38.** $\frac{a}{b} + \left(\frac{c}{d} + \frac{e}{f}\right)$

***39.** $\left(\frac{a}{b} + \frac{c}{d}\right) + \frac{e}{f}$

***40.** $\frac{a}{b}\left(\frac{c}{d} + \frac{e}{f}\right)$

***41.** $\left(\frac{a}{b} \times \frac{c}{d}\right) + \left(\frac{a}{b} \times \frac{e}{f}\right)$

***42.** $\dfrac{\frac{a}{b} + \frac{c}{d}}{\frac{e}{f}}$

***43.** $\dfrac{\frac{a}{b}}{\frac{c}{d} + \frac{e}{f}}$

PEDAGOGICAL EXPLORATIONS

1. State the objectives of this section in behavioral language. (See pages 182–183.) Then prepare a 20-minute test that evaluates attainment of these objectives.
2. Examine a recently published elementary mathematics textbook series and determine the grade level at which addition and subtraction with fractions is introduced and developed.
3. Note that multiplication and division with fractions was developed in this text prior to addition and subtraction. Can you explain the reason for this order?
4. Prepare a collection of ten word problems suitable for a specified elementary grade level, and whose solutions involve addition and subtraction with fractions.

5. Assume that you find some students in an elementary mathematics class committing errors such as these:

(a) $\frac{1}{2} + \frac{1}{3} = \frac{1}{5}$ (b) $\frac{3}{10} - \frac{1}{5} = \frac{2}{5}$

(c) $\frac{1\not{6}}{\not{6}4} = \frac{1}{4}$ (d) $\frac{1\not{9}}{\not{9}5} = \frac{1}{5}$

Prepare a 10-minute lesson that might be used to help correct and avoid such errors.

6-5 operations with decimals

We have seen that any rational number can be represented by a fraction. Any rational number can also be represented by a decimal. In particular, fractions that have denominators that are powers of 10 can be represented as decimals by our knowledge of place value. For example, consider the following:

$$\tfrac{3}{10} = 0.3, \qquad \tfrac{3}{100} = 0.03, \qquad \tfrac{3}{1000} = 0.003$$

The following chart summarizes the most commonly used units of decimal values.

Tenths	$\frac{1}{10}$	$\frac{1}{10}$	0.1
Hundredths	$\frac{1}{100}$	$\frac{1}{10^2}$	0.01
Thousandths	$\frac{1}{1000}$	$\frac{1}{10^3}$	0.001
Ten-thousandths	$\frac{1}{10,000}$	$\frac{1}{10^4}$	0.0001
Hundred-thousandths	$\frac{1}{100,000}$	$\frac{1}{10^5}$	0.00001
Millionths	$\frac{1}{1,000,000}$	$\frac{1}{10^6}$	0.000001

As in §5-5, we may use our knowledge of integers as exponents to write a decimal numeral in *expanded notation.* Recall that $10 = 10^1$, $1 = 10^0$, $\frac{1}{10} = 10^{-1}$, and $\frac{1}{10^2} = 10^{-2}$ and consider the decimal 3256.78 written with this notation:

$$\begin{aligned} 3256.78 &= (3 \times 1000) + (2 \times 100) + (5 \times 10) \\ &\qquad + (6 \times 1) + (7 \times \tfrac{1}{10}) + (8 \times \tfrac{1}{100}) \\ &= (3 \times 10^3) + (2 \times 10^2) + (5 \times 10^1) \\ &\qquad + (6 \times 10^0) + (7 \times 10^{-1}) + (8 \times 10^{-2}) \end{aligned}$$

This decimal is read "three thousand two hundred fifty six *and* seventy-eight hundredths." We use the word "and" to represent the decimal point.

Example 1: Write in decimal notation:

$$(5 \times 10^0) + (3 \times 10^{-1}) + (2 \times 10^{-2}) + (7 \times 10^{-3})$$

Solution: Replace the powers of 10 by their fractional equivalents. Then the decimal may be written in this manner:

$$(5 \times 1) + (3 \times \tfrac{1}{10}) + (2 \times \tfrac{1}{100}) + (7 \times \tfrac{1}{1000}) = 5 + 0.3 + 0.02 + 0.007$$
$$= 5.327$$

This answer is read "five and three hundred twenty seven thousandths."

We next need to consider procedures for computing with decimals. Addition with decimals becomes a relatively easy task once an algorithm for the addition of whole numbers has been established. For example, consider the addition problem 0.35 + 0.49. Arrange the addends in this form:

$$\begin{array}{r} 0.35 \\ +\ 0.49 \\ \hline \end{array}$$

Because of the place value of our decimal system we merely add in the hundredths column, carry as needed, and add in the tenths column.

$$\begin{array}{r} 1 \\ 0.35 \\ +\ 0.49 \\ \hline 0.84 \end{array}$$

Notice that we actually line up place values by "lining up the decimal points" and then adding as with whole numbers.

The addition problem 0.35 + 0.49 may also be done through application of the distributive property

$$ac + bc = (a + b)c$$

where

$$\begin{aligned} 0.35 + 0.49 &= 35(0.01) + 49(0.01) \\ &= (35 + 49)(0.01) = 84(0.01) = 0.84 \end{aligned}$$

Notice that this use of the distributive property is equivalent to doing the problem in hundredths; that is, 0.35 + 0.49 is (35 + 49) hundredths. Finally, note that this same problem may be stated and solved in terms of fractions:

$$0.35 + 0.49 = \frac{35}{100} + \frac{49}{100} = \frac{84}{100} = 0.84$$

The subtraction of decimals is also based upon the procedures that are used for whole numbers. We may line up the decimal points and then proceed to subtract as with whole numbers. Example 2 illustrates this method of subtraction as well as other procedures for subtraction with decimals.

Example 2: Subtract: 0.73 − 0.28.

Solution: Three methods of solution are given. In fractions we have

$$\frac{73}{100} - \frac{28}{100} = \frac{73 - 28}{100} = \frac{45}{100} = 0.45$$

Lining up the decimal points, we have

$$\begin{array}{r} 0.73 \\ -\ 0.28 \\ \hline 0.45 \end{array}$$

Using the distributive property, we have

$$\begin{aligned} 0.73 - 0.28 &= 73(0.01) - 28(0.01) \\ &= 73(0.01) + (-28)(0.01) \\ &= [73 + (-28)](0.01) = 45(0.01) = 0.45 \end{aligned}$$

To explain the usual algorithm for multiplication of decimals, we examine several examples done as fractions and attempt to discover a pattern.

Example 3: Multiply: (a) 0.7 × 0.8; (b) 0.7 × 0.08; (c) 0.07 × 0.08.

Solution: (a) $0.7 \times 0.8 = \frac{7}{10} \times \frac{8}{10} = \frac{56}{100} = 0.56$.

Note that a number of *tenths* multiplied by a number of *tenths* gives a number of *hundredths.*

(b) $0.7 \times 0.08 = \frac{7}{10} \times \frac{8}{100} = \frac{56}{1000} = 0.056$.

Note that a number of *tenths* multiplied by a number of *hundredths* gives a number of *thousandths.*

(c) $0.07 \times 0.08 = \frac{7}{100} \times \frac{8}{100} = \frac{56}{10000} = 0.0056$.

Note that a number of *hundredths* multiplied by a number of *hundredths* gives a number of *ten-thousandths.*

The above example illustrates the rule for multiplying two rational numbers that are written as decimals. First, ignore the decimal points and multiply as with whole numbers. Then the number of decimal places in the final product will be equal to the sum of the numbers of decimal places in the two factors that are multiplied.

Example 4: Place the decimal point in the correct position in the product: 0.23 × 5.7 = 1311.

Solution: There are two decimal places in 0.23 and one decimal place in 5.7. The product should then have three decimal places; 0.23 × 5.7 = 1.311.

Division with decimals is probably one of the most difficult arithmetic skills for many people. Once again we can justify the usual algorithm by using fractional equivalents. Consider the division problem 0.46 ÷ 2. This problem can be expressed and solved by means of fractions:

$$\frac{46}{100} \div 2 = \frac{46}{100} \times \frac{1}{2} = \frac{46 \times 1}{100 \times 2} = \frac{23}{100} = 0.23$$

We may think of any division problem in the form

$$\text{divisor}\,\overline{)\,\text{dividend}}^{\;\text{quotient}}$$

Thus the problem under consideration may be written as

$$2\,\overline{)\,0.46}^{\;0.23}$$

Notice that the divisor is a whole number; the division may be done as with whole numbers with the decimal point in the quotient placed directly over the decimal point in the dividend. Whenever a decimal is divided by a whole number, we divide as with whole numbers and place the decimal point in the quotient directly over the decimal point in the dividend.

Any problem of division of ordinary decimals may be solved by finding an equivalent problem in which the divisor is a whole number and following the above procedure. Consider the problem

$$1.75 \div 2.5$$

We first write the problem in the form

$$\frac{1.75}{2.5}$$

and multiply the fraction by $\frac{10}{10}$ to obtain a fraction with a whole number as denominator:

$$\frac{1.75}{2.5} = \frac{1.75}{2.5} \times \frac{10}{10} = \frac{17.5}{25}$$

This rewriting of the fraction

$$\frac{1.75}{2.5} \quad \text{in the form} \quad \frac{17.5}{25}$$

corresponds to rewriting the division problem

$$2.5\,\overline{)\,1.75} \quad \text{in the form} \quad 25\,\overline{)\,17.5}$$

In other words, multiplying both the dividend and the divisor by 10 is equivalent to multiplying the numerator and the denominator of the corresponding fraction by 10. In each case the problem is unchanged.

Most people think of the above process as "moving" the decimal point one unit to the right in both the dividend and the divisor. In general, since the numerator and the denominator of the corresponding fraction may be multiplied by any power of 10, the decimal point may be moved as many places as necessary to make the divisor a whole number and moved the same number of places in the dividend.

Example 5: Divide: $3.726 \div 0.23$.

Solution: We move the decimal point two places to the right in the divisor to make the divisor a whole number. Therefore we move the decimal point two places to the right in the dividend. Actually we have multiplied both the dividend and the divisor by 100. Then we divide as for whole numbers and place the decimal point in the quotient directly over the new location of the decimal point in the dividend.

```
          16.2
0.23 ) 3.72 6
       23
       142
       138
         46
         46
```

Thus $3.726 \div 0.23 = 372.6 \div 23 = 16.2$.

EXERCISES

Write in expanded notation and include all zero digits.

1. 9.34 **2.** 49.6

3. 235.78 **4.** 3402.07

5. 0.009 **6.** 5.0103

7. 0.0002 **8.** 3000.03

9. 0.404 **10.** 2.0202

Write in decimal notation.

11. $(6 \times 10^{-1}) + (8 \times 10^{-2}) + (5 \times 10^{-3})$

12. $(7 \times 10^{1}) + (3 \times 10^{0}) + (2 \times 10^{-1}) + (0 \times 10^{-2}) + (7 \times 10^{-3})$

13. $(9 \times 10^{1}) + (0 \times 10^{0}) + (0 \times 10^{-1}) + (5 \times 10^{-2}) + (3 \times 10^{-3})$

14. $(4 \times 10^{-2}) + (8 \times 10^{-3}) + (2 \times 10^{-4})$

15. $(6 \times 10^{3}) + (9 \times 10^{2}) + (0 \times 10^{1}) + (1 \times 10^{0}) + (9 \times 10^{-1})$

16. $(8 \times 10^{0}) + (0 \times 10^{-1}) + (5 \times 10^{-2}) + (3 \times 10^{-3})$

Find each sum or difference by using decimals and check by using fractions.

17. $0.6 + 0.8$ **18.** $0.13 + 0.17$

19. $0.05 + 0.08$ **20.** $0.132 + 0.546$

21. $0.8 - 0.3$

22. $0.18 - 0.09$

23. $0.17 - 0.12$

24. $0.324 - 0.187$

Find each product or quotient by using decimals and check by using fractions.

25. 0.3×0.5

26. 0.07×0.05

27. 0.5×0.13

28. 0.03×0.004

29. $0.9 \div 0.3$

30. $0.25 \div 0.05$

31. $0.36 \div 0.9$

32. $0.36 \div 0.09$

Place the decimal point in the correct position in each product or quotient. (You may need to prefix or annex zeros.)

33. $3.45 \times 2.87 = 99015$

34. $27.3 \times 0.367 = 100191$

35. $0.257 \times 3.65 = 93805$

36. $72.09 \times 308 = 2220372$

37. $244.72 \div 5.6 = 437$

38. $3.393 \div 8.7 = 39$

39. $0.7905 \div 0.93 = 85$

40. $1469.63 \div 0.281 = 523$

Perform the indicated operations.

41. $3.75 + (2.87 + 9.56)$

42. $12.76 - (9.05 - 7.87)$

43. $5.01 \times (2.3 \times 0.68)$

44. $(22.14 \div 0.27) \div 0.2$

45. $3.75 \times (2.74 + 8.35)$

46. $5.95 \times (5.73 - 1.28)$

47. $(8.56 + 7.93) - (2.47 + 3.09)$

48. $(3.78 - 1.99) + (5.03 - 2.78)$

Select the best answer.

49. 0.89×0.9 is about **(a)** 0.8; **(b)** 8; **(c)** 0.08; **(d)** 0.9.

50. 79.6×9.8 is about **(a)** 8; **(b)** 80; **(c)** 800; **(d)** 8000.

51. 9.01×6.99 is about **(a)** 0.63; **(b)** 6.3; **(c)** 63; **(d)** 630.

52. 0.09×10.01 is about **(a)** 0.009; **(b)** 0.09; **(c)** 0.9; **(d)** 9.

53. $19.79 \div 3.98$ is about **(a)** 5; **(b)** 0.5; **(c)** 0.05; **(d)** 50.

54. $0.079 \div 0.02$ is about **(a)** 4; **(b)** 40; **(c)** 0.4; **(d)** 0.04.

55. $7.962 \div 1.9$ is about **(a)** 4; **(b)** 40; **(c)** 0.4; **(d)** 0.04.

56. $123.9 \div 0.2$ is about **(a)** 62; **(b)** 620; **(c)** 6.2; **(d)** 0.62.

PEDAGOGICAL EXPLORATIONS

1. Review a recently published elementary mathematics textbook series. Determine the manner in which decimals are first introduced in the

program, as well as the grade placement for the various operations with decimals.

2. Prepare a 30-minute test that evaluates an elementary student's knowledge of the fundamental skills of addition, subtraction, multiplication, and division with decimals.
3. Prepare a bulletin board display that shows the use of decimals in daily life activities.
4. Flow charts can be used with elementary school students to motivate practice with decimals. In the following example the student is to use the flow chart to find the output for each given input.

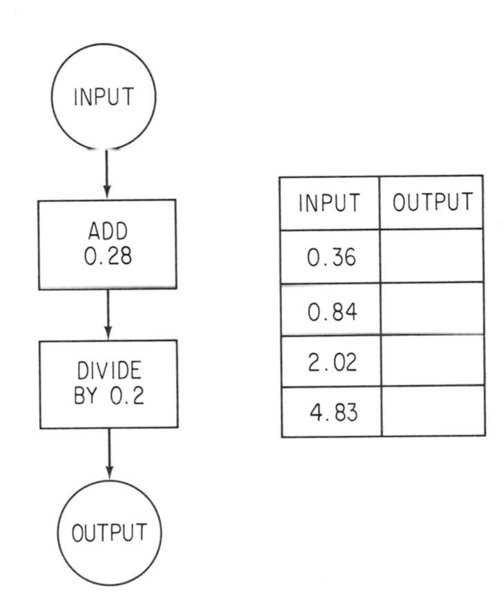

INPUT	OUTPUT
0.36	
0.84	
2.02	
4.83	

Make up several flow charts of your own for a specified elementary grade level, and covering all four of the fundamental operations with decimals.

6-6 rational numbers and repeating decimals

There is an interesting relationship between rational numbers and their decimal representations that we shall explore in this section. First we note that every rational number in the form a/b can be written as a decimal. If the denominator b is a power of 10, then we may write the decimal directly, as in these cases:

$$\tfrac{3}{10} = 0.3, \qquad \tfrac{13}{100} = 0.13, \qquad \tfrac{123}{1000} = 0.123$$

The denominator may also be a factor of some power of 10. If so, we can multiply numerator and denominator by a power of 2 or a power of 5 in order to write an equivalent fraction with denominator that is a power of 10.

Example 1: Write as a decimal: (a) $\frac{3}{5}$, (b) $\frac{5}{8}$.

Solution: (a) $\dfrac{3}{5} = \dfrac{3}{5} \times \dfrac{2}{2} = \dfrac{6}{10} = 0.6$

(b) $\dfrac{5}{8} = \dfrac{5}{8} \times \dfrac{125}{125} = \dfrac{625}{1000} = 0.625.$

In Example 1(b) we could also have found the decimal equivalent by division. Thus $\frac{5}{8}$ is the indicated division $5 \div 8$.

$$\begin{array}{r}
0.625 \\
8\,\overline{)\,5.000} \\
\underline{48} \\
20 \\
\underline{16} \\
40 \\
\underline{40}
\end{array}
\qquad \frac{5}{8} = 0.625$$

Next let us consider a fraction such as $\frac{1}{3}$. For this fraction there is no whole number by which we can multiply numerator and denominator and obtain a fraction with a power of 10 as its denominator. Thus we resort to long division.

$$\begin{array}{r}
0.3333\ldots \\
3\,\overline{)\,1.0000\ldots} \\
\underline{9} \\
10 \\
\underline{9} \\
10 \\
\underline{9} \\
10 \\
\underline{9} \\
1
\end{array}$$

We soon note that this division process gives us an endless supply of 3's. In fact, we find that the quotient is an **infinite repeating decimal,** 0.3333. . . . We may write this as follows:

$$\tfrac{1}{3} = 0.333\ldots = 0.\bar{3}$$

In an infinite repeating decimal, we use a bar to indicate that a sequence of digits repeats endlessly. The bar is customarily used over the first repeating digits, or set of digits, on the right of the decimal point.

Example 2: Write $\frac{3}{11}$ as a repeating decimal.

Solution: By division we find that the quotient repeats the digits 2 and 7 endlessly. We write this quotient as follows:

$$\tfrac{3}{11} = 0.272727\ldots = 0.\overline{27}$$

We may also think of any *terminating decimal* as having repeated zeros and thus as being a repeating decimal. For example:

$$\tfrac{1}{4} = 0.25\bar{0}, \qquad \tfrac{3}{8} = 0.375\bar{0}$$

It is easy to show that every rational number (this includes integers), that is expressed in fractional form can be represented as a repeating (or terminating) decimal. Consider, for example, any rational number such as $\frac{12}{7}$. When we divide 12 by 7, the possible remainders are 0, 1, 2, 3, 4, 5, 6. If the remainder is 0, the division is exact; if any remainder occurs a second time, the terms after it will repeat also. Since there are only seven possible remainders when you divide by 7, the remainders must repeat or be exact by the seventh decimal place. Consider the determination of the decimal value of $\frac{12}{7}$ by long division:

$$\begin{array}{r}
1.\overline{714285} \\
7\,\overline{)\,12.000000} \\
\underline{7} \\
\textcircled{5}0 \\
\underline{49} \\
10 \\
\underline{7} \\
30 \\
\underline{28} \\
20 \\
\underline{14} \\
60 \\
\underline{56} \\
40 \\
\underline{35} \\
\textcircled{5}
\end{array}$$

The fact that the remainder 5 occurred again implies that the same steps will be used again in the long division process and the digits 714285 will be repeated over and over; that is, $\frac{12}{7} = 1.\overline{714285}$. Similarly, any rational number p/q can be expressed as a terminating or repeating decimal, and at most q decimal places will be needed to identify it.

We can also show that any terminating or repeating decimal can be written as a rational number in the form a/b.

If a decimal is terminating, you can write it as a fraction with a power of 10 as the denominator. For example, if $n = 0.75\bar{0}$, then $100n = 75$, and $n = \frac{75}{100}$, which reduces to $\frac{3}{4}$. If a fraction can be expressed as a terminating decimal, its denominator must be a factor of a power of 10.

If a decimal is repeating, it can be written as a rational number. For example, if a decimal n repeats one digit, we can find $10n - n$. Suppose $n = 3.2\bar{4}$; then $10n = 32.\bar{4}$ and we have:

$$\begin{array}{r@{\;=\;}l} 10n & 32.4\bar{4} \\ n & 3.2\bar{4} \\ \hline 9n & 29.2\bar{0} \end{array}$$

$$n = \frac{29.2}{9} = \frac{292}{90} = \frac{146}{45}$$

We can avoid the use of decimals in this way:

$$\begin{array}{r@{\;=\;}l} 100n & 324.\bar{4} \\ 10n & 32.\bar{4} \\ \hline 90n & 292.\bar{0} \end{array}$$

$$n = \frac{292}{90} = \frac{146}{45}$$

If a decimal n repeats two digits, we find $10^2n - n$; if it repeats three digits, we find $10^3n - n$; and so forth.

Example 3: Express $0.\overline{36}$ as a fraction.

Solution: Let $n = 0.\overline{36}$; then $100n = 36.\overline{36}$ and we have:

$$\begin{array}{r@{\;=\;}l} 100n & 36.\overline{36} \\ n & 0.\overline{36} \\ \hline 99n & 36.\bar{0} \end{array}$$

$$n = \frac{36}{99} = \frac{4}{11}$$

We may summarize the discussion of this section in this way:

Every rational number can be named by a repeating decimal.
Every repeating decimal is the name of a rational number.

Note that for this summary we classify a terminating decimal, such as 0.25, as a repeating decimal. That is, $0.25 = 0.25\bar{0}$.

EXERCISES

Express each fraction as a decimal.

1. $\frac{3}{5}$	**2.** $\frac{13}{20}$	**3.** $\frac{7}{8}$	**4.** $\frac{17}{200}$	**5.** $\frac{27}{15}$
6. $\frac{1}{6}$	**7.** $\frac{5}{7}$	**8.** $\frac{1}{13}$	**9.** $\frac{9}{13}$	**10.** $\frac{13}{7}$

Express each decimal as a fraction.

11. $0.\overline{72}$ **12.** $0.\overline{81}$ **13.** $0.\overline{423}$ **14.** $0.\overline{414}$

15. $0.\overline{531}$ **16.** $0.\overline{9}$ **17.** $6.\overline{1}$ **18.** $8.\overline{2}$

19. $3.\overline{14}$ **20.** $3.\overline{25}$ **21.** $0.\overline{321}$ **22.** $65.\overline{268}$

23. Write the first ten digits to the right of the decimal point in the decimal **(a)** $0.\overline{45}$; **(b)** $0.3\overline{572}$; **(c)** $0.8\overline{0}$; **(d)** $0.42\overline{89}$.

24. Write the digit in the fifteenth position to the right of the decimal point in the decimal **(a)** $0.52\overline{0}$; **(b)** $0.\overline{78}$; **(c)** $0.\overline{342}$; **(d)** $0.71\overline{68}$.

25. Arrange in order, from smallest to largest.

$2.59 $2.89, $2.37, $2.65, $2.08

26. Write each decimal to two decimal places and arrange in order, from smallest to largest.

4.37, 4.42, 4.4, 4.39, 4.51, 4.3

27. Write each decimal to at least three decimal places and arrange in order, from smallest to largest.

1.78, $1.\overline{7}$, $1.\overline{8}$, $1.7\overline{8}$, $1.\overline{78}$

28. Write each decimal to at least six decimal places and arrange in order, from smallest to largest.

0.234, $0.\overline{234}$, $0.\overline{23}$, $0.\overline{24}$, 0.24

Name a rational number that lies between each of these pairs of rational numbers.

***29.** 0.234 and 0.235

***30.** $0.\overline{234}$ and $0.\overline{235}$

***31.** $0.23\overline{0}$ and $0.24\overline{0}$

***32.** 0.234 and $0.23\overline{4}$

***33.** Consider the decimal 0.252252225 . . . . This is an infinite decimal that has a pattern but that does not repeat a fixed set of digits, and thus does not represent a rational number. **(a)** How many 2's will there be immediately preceding the fifteenth 5? **(b)** How many 2's will there be altogether preceding the fifteenth 5?

***34.** Name an infinite decimal for a number between 0.1 and 0.2 that is not a rational number.

PEDAGOGICAL EXPLORATIONS

1. Write the decimal representation for each of the rational numbers $\frac{1}{7}$, $\frac{2}{7}$, $\frac{3}{7}$, $\frac{4}{7}$, $\frac{5}{7}$, and $\frac{6}{7}$. See if you can find a pattern that describes the manner in which the digits in each representation are related.

2. Repeat Exploration 1 for the multiples of $\frac{1}{13}$ from $\frac{1}{13}$ to $\frac{12}{13}$.

3. Show that $0.\overline{9} = 1$. Then try to prepare an explanation of this fact to satisfy a seventh or eighth grader. Consider these possibilities:

(a)
$$\begin{array}{r} \frac{1}{3} = 0.\overline{3} \\ + \frac{2}{3} = 0.\overline{6} \\ \hline 1 = 0.\overline{9} \end{array}$$

(b)
$$\begin{array}{r} \frac{1}{3} = 0.\overline{3} \\ 3 \times \frac{1}{3} = 3 \times 0.\overline{3} \\ 1 = 0.\overline{9} \end{array}$$

4. If the denominator of a fraction can be expressed as the product of powers of 2 and 5 only, then it can be written as a terminating decimal.
   (a) Show that this is so for several examples, such as $\frac{3}{50}$ and $\frac{33}{60}$.
   (b) Explain why

$$\frac{n}{2^q 5^p}$$

for any whole numbers n, p, and q can be represented by a terminating decimal.

6-7 rational numbers as per cents

Any rational number can be represented as a decimal and therefore as a number of hundredths:

$$\frac{3}{4} = 0.75 = \frac{75}{100}$$

$$\frac{3}{2} = 1.5 = 1.50 = \frac{150}{100}$$

$$\frac{1}{3} = 0.333\overline{3} = \frac{33.3\overline{3}}{100} = \frac{33\frac{1}{3}}{100}$$

Any rational number that is represented by a fraction may be considered as a **ratio** of its numerator to its denominator. For example, $\frac{1}{2}$ is the ratio of 1 to 2; $\frac{5}{8}$ is the ratio of 5 to 8. Then any number of hundredths is the ratio of that number to 100 and is called a **per cent** (from the Latin *per centum* for per hundred). The symbol for per cent is %. Thus:

$$\frac{3}{4} = \frac{75}{100} = 75\%$$

$$\frac{3}{2} = \frac{150}{100} = 150\%$$

$$\frac{1}{3} = \frac{33\frac{1}{3}}{100} = 33\tfrac{1}{3}\%$$

To express a rational number as a per cent, write the number as a decimal or as a number of hundredths, and then as a per cent. To express a per cent as a fraction, write the per cent as a number of hundredths and reduce the fraction, if possible:

$$50\% = \frac{50}{100} = \frac{1}{2}$$

$$100\% = \frac{100}{100} = 1$$

$$700\% = \frac{700}{100} = 7$$

$$125\% = \frac{125}{100} = \frac{5}{4}$$

$$0.3\% = \frac{0.3}{100} = \frac{3}{1000}$$

A statement such as

$$50\% = \frac{1}{2}$$

may be expressed as

$$\frac{50}{100} = \frac{1}{2}$$

and read as "50 is to 100 as 1 is to 2." Such an equality of ratios (fractions) is often called a **proportion.** By the definition of the equality of fractions we have the **proportion property**

$$\frac{a}{b} = \frac{c}{d} \quad \text{if and only if} \quad ad = bc$$

The problem of expressing a fraction as a per cent may be considered as *solving* (finding a missing part of) a proportion.

Example 1: Write $\frac{5}{8}$ as a per cent.

Solution:

$$\frac{5}{8} = \frac{n}{100}$$

$$5 \times 100 = 8 \times n$$

$$8n = 500$$

$$n = 62\tfrac{1}{2}$$

$$\frac{5}{8} = 62\tfrac{1}{2}\%$$

Another way to express $\frac{5}{8}$ as a per cent is to recognize the fraction as the indicated quotient $5 \div 8$. Then divide, and recall that per cent means hundredths.

$$\begin{array}{r} 0.625 \\ 8\overline{)5.000} \\ \underline{4\,8} \\ 20 \\ \underline{16} \\ 40 \\ \underline{40} \end{array}$$

Thus $\frac{5}{8} = 0.625 = 62.5\%$, or $62\frac{1}{2}\%$.

We are now ready to solve problems that involve per cents. Most such problems can be solved by remembering that a per cent is the ratio of some number to 100. In the following three examples we will solve three different "types" of problems by setting up proportions and using the proportion property.

Example 2: Find 25% of 80.

Solution: We wish to find a number n that compares to 80 in the same way that 25% compares to 100%. The following line segment is used to represent 80, which we consider to be 100% or all of the number.

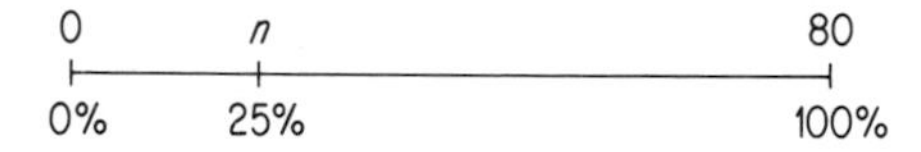

Thus:

$$\frac{n}{80} = \frac{25}{100} \qquad \text{(n compares to 80 in the same way that 25\% compares to 100\%.)}$$
$$100 \times n = 80 \times 25$$
$$100n = 2000$$
$$n = 20 \qquad (2000 \div 100 = 20)$$

Therefore 25% of 80 is 20.

Example 3: 20 is what per cent of 80?

Solution: Consider this diagram, with the known parts shown.

0 20 80

0% n% 100%

Thus:

$$\frac{20}{80} = \frac{n}{100} \qquad \text{(20 compares to 80 in the same way that n\% compares with 100\%.)}$$
$$20 \times 100 = 80 \times n$$
$$80n = 2000$$
$$n = 25 \qquad (2000 \div 80 = 25)$$

Therefore 20 is 25% of 80.

Example 4: 20 is 25% of what number?

Solution: Again, consider the diagram comparing known parts.

Thus:

$$\frac{20}{n} = \frac{25}{100} \qquad \text{(20 compares to n in the same way that 25\% compares with 100\%.)}$$

$$20 \times 100 = n \times 25$$
$$25n = 2000$$
$$n = 80 \qquad (2000 \div 25 = 80)$$

Therefore 20 is 25% of 80.

In each of the examples shown, other means of solution are possible. Because the numbers selected in these examples caused no problem, the other methods may be easier. For example, consider again the problem of finding 25% of 80. We can also obtain the answer by either of the following methods:

Using fractions: $25\% = \frac{25}{100} = \frac{1}{4}$

$\frac{1}{4}$ of $80 = \frac{1}{4} \times 80 = 20$

Using decimals: $25\% = 0.25$

25% of $80 = 0.25 \times 80 = 20$

Although these approaches may seem somewhat easier here, the procedure described in Examples 2, 3, and 4 are applicable to all problems, regardless of their complexity. Consider, for example, one additional problem solved by the proportion method and where the answer is not readily seen in advance.

Example 5: There are 22 students that arrive by school bus. These students represent 55% of the students in the class. How many students are there in the class?

Solution: Here we consider the ratio of students bussed to the total number in the class as $22/n$. This is equal to the ratio of 55% to the total, which is always 100%. Thus we have the following:

$$\frac{22}{n} = \frac{55}{100}$$
$$22 \times 100 = 55 \times n$$
$$55n = 2200$$
$$n = 40 \qquad (2200 \div 55 = 40)$$

There are 40 students in the class.

EXERCISES

Write each per cent as a decimal.

1. 57% ***2.*** 89% ***3.*** 3% ***4.*** 225% ***5.*** 0.95%

6. 1% ***7.*** 250% ***8.*** 125% ***9.*** 100% ***10.*** 0.5%

Write each per cent as a fraction in simplest form.

11. 50% **12.** 40% **13.** 95% **14.** 65% **15.** 99%

16. 130% **17.** 150% **18.** 125% **19.** 8% **20.** 0.8%

Write each decimal as a per cent.

21. 0.35 **22.** 0.01 **23.** 0.45 **24.** 1.35 **25.** 0.9

26. 0.08 **27.** 1.45 **28.** 0.006 **29.** 0.001 **30.** 1.00

Write each fraction as a per cent.

31. $\frac{9}{100}$ **32.** $\frac{36}{100}$ **33.** $\frac{92}{100}$ **34.** $\frac{125}{100}$ **35.** $\frac{150}{100}$

36. $\frac{16}{20}$ **37.** $\frac{38}{50}$ **38.** $\frac{9}{10}$ **39.** $\frac{17}{10}$ **40.** $\frac{42}{20}$

In Exercises 41 through 52 write and solve a proportion.

41. Find 40% of 60.
42. Find 80% of 140.
43. Find 35% of 80.
44. Find 120% of 75.
45. 20 is what per cent of 160?
46. 15 is what per cent of 120?
47. 120 is what per cent of 160?
48. 160 is what per cent of 120?
49. 25 is 20% of what number?
50. 40 is 75% of what number?
51. 80 is 125% of what number?
52. 80 is 20% of what number?

53. Bill spent 40% of his allowance for a pen. If his allowance is $5.00, how much did he spend for the pen?

54. Doris' savings account pays 6% interest per year. Last year she left her account untouched and received $2.70 interest. How much money did she have in the account before receiving the interest?

55. Karen spent $2.40 for a book, accounting for 30% of the money that she had with her. How much money did she have with her?

56. Wendy bought a coat on sale for $35.55. The regular price of the coat was $45.00.
  (a) What per cent discount did she receive? That is, the amount she saved on this sale is what per cent of the regular price of the coat?
  (b) The sale price is what per cent of the regular price of the coat?

57. The school population in a certain elementary school rose from 700 to 800 in a recent year. What was the per cent of increase? That is, the increase was what per cent of the original enrollment?

58. Jan went on a diet and reduced her weight from 150 pounds to 135 pounds. What was the per cent of decrease in her weight? That is, her loss of weight was what per cent of her original weight?

***59.** A television set that regularly costs $250 is advertised on sale at a 15% discount. A week later it is further reduced by 10% of the sale price. Find a single discount that is equivalent to these two successive discounts.

***60.** A storekeeper pays $80 for a coat and sells it at a markup of 15%. Later he marks it up an additional 10% of the selling price. Find a single markup equivalent to these two successive markups.

sec. 6-7 rational numbers as per cents

PEDAGOGICAL EXPLORATIONS

1. Per cent problems are often solved by using the formula $p = r \times b$, where p is the percentage, r is the rate, and b is the base. For example, consider the statement 25% of 80 is 20. Here 25% is the rate, 80 is the base, and 20 is the percentage. Now use this formula to solve each of the examples given in this section.

2. Prepare a bulletin board display showing various uses of per cent found in local newspapers and magazines.

3. The following figure illustrates a **per cent chart**. Here is how this chart may be used to find 50% of 60. First a line is drawn from 0 on the per cent scale to 60 on the base scale. Next 50 is located on the per cent scale. Then the answer is found by reading across to the line and down to 30 on the base scale. Thus, 50% of 60 is 30.

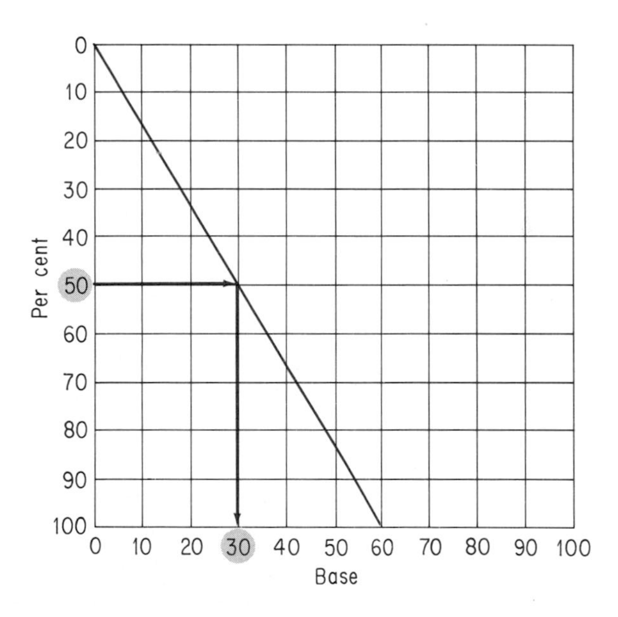

Construct a per cent chart to illustrate each of the following problems:
(a) Find 25% of 80.
(b) What per cent of 50 is 40?
(c) 20 is 40% of what number?

4. Many traditional junior high school texts solve per cent problems by considering three different "cases." Find out what these three types of problems are, and then identify the examples of this section according to this classification.

CHAPTER TEST

Classify each statement as true or false.

1. **(a)** Between any two rational numbers there is always another rational number.

(b) There is a one-to-one correspondence between the set of rational numbers and the points of the number line.

2. **(a)** The number 0 does not have a multiplicative inverse.

(b) Every rational number has an additive inverse.

Name the property illustrated by each statement for rational numbers a, b, and c.

3. $(a + b) + c = a + (b + c)$ **4.** $a(b + c) = ab + ac$

Copy the following table. Use "√" to show that the number at the top of the column is a member of the set named at the side. Use "×" if the number is not a member of the set.

	5.	**6.**	**7.**	**8.**
Set	$-\frac{2}{3}$	0	5	-4
Counting numbers				
Whole numbers				
Integers				
Rational numbers				

9. Illustrate this sum on a number line: $(+2\frac{1}{4}) + (-5\frac{1}{2})$.

Find all possible replacements for n that will make each of the sentences below true, where n must be a member of the set named at the top of each column. When there is no such replacement, write "none."

		(a)	(b)	(c)
	Sentence	*Counting numbers*	*Integers*	*Rational numbers*
10.	$n - 6 = 1$			
11.	$n + 7 = 7$			
12.	$2n = 9$			

Perform the indicated operations and simplify.

13. $\left(-\frac{3}{4}\right) \times \left(-\frac{8}{9}\right)$ **14.** $\left(\frac{-5}{12}\right) \div \left(\frac{3}{-4}\right)$

15. $\left(-\frac{3}{4}\right) + \left(-\frac{2}{3}\right)$ **16.** $\frac{3}{8} - \frac{1}{2}$

17. Express as a single quotient.

(a) $\frac{a}{b} + \frac{c}{d}$ (b) $\frac{a}{b} \div \frac{c}{d}$

18. Write in expanded notation: 32.157.

19. Write in decimal notation: $(2 \times 10^{-1}) + (3 \times 10^{-2}) + (4 \times 10^{-3})$.

Perform the indicated operations.

20. $7.25 + (3.68 - 1.79)$ **21.** $(4.3 \times 2.08) \div 0.02$

22. Express each decimal as a fraction.

(a) $0.\overline{63}$ (b) $0.\overline{612}$

23. Express each fraction as a repeating decimal.

(a) $\frac{7}{12}$ (b) $\frac{7}{13}$

24. Write each per cent as a decimal and then as a fraction in simplest form.

(a) 45% (b) 100% (c) 8% (d) 120% (e) 0.1%

25. Write a proportion and solve each problem.

(a) What is 30% of 80? (b) 12 is 15% of what number?

READINGS AND PROJECTS

1. Read Booklet No. 4 (Algorithms for Operations with Whole Numbers) of the twenty-ninth yearbook of the National Council of Teachers of Mathematics. Complete at least three of the exercise sets given there.
2. Read "Magic Squares: Extensions into Mathematics", by Daiyo Sawada on pages 183 through 188 of the March 1974 issue of *The Arithmetic Teacher.* Then construct a 3 by 3 magic square with each entry a fraction.
3. A fraction is one type of numeral (symbol) for a rational number. We add numbers; we do not add symbols. Therefore, when we wish to be particularly precise, we say that we operate *with* fractions. Similarly, we say that we operate *with* decimals. Select an elementary mathematics textbook for grades 5 or 6 and explore the manner in which they overcome this language problem. In particular, note the manner in which they use the words "fraction," "fractional number," and "rational number."

4. Find out what you can about the use of **Cuisenaire rods** in the teaching of computation with fractions. If possible, demonstrate their use in class. Information can be obtained from Cuisenaire Company of America, Inc., 12 Church Street, New Rochelle, New York, 10805.

5. A great deal has been written about the need to provide students with an opportunity to make discoveries. Prepare a 20-minute lesson, suitable for a specified elementary grade level, featuring a laboratory approach that leads to some arithmetic discovery. For a discussion of possible experiments see Chapter 4 (Laboratory Experiments) of *Teaching Mathematics* by Max A. Sobel and Evan Maletsky, Prentice-Hall, Inc., 1975.

6. Read Chapter 9 (Manipulative Devices in Elementary School Mathematics) of the thirty-fourth yearbook of the National Council of Teachers of Mathematics. Then prepare several aids that could be used in an elementary mathematics classroom to help teach one of the skills or concepts presented in this chapter.

7. Show that there is a one-to-one correspondence between the set of counting numbers and the set of rational numbers. For an excellent discussion of this topic see Chapter 2 (Beyond the Googol) of *Mathematics and the Imagination* by Edward Kasner and James Newman, Simon and Schuster, Inc., 1940.

8. Read Chapter 13 (An Example from Arithmetic) of the thirty-third yearbook of the National Council of Teachers of Mathematics. Then discuss, in class, the various teaching methods employed in the lessons described.

9. Read one of the following chapters from the twenty-ninth yearbook of the National Council of Teachers of Mathematics:
Booklet No. 4 (Algorithms For Operations With Whole Numbers)
Booklet No. 6 (The Rational Numbers)
Booklet No. 7 (Numeration Systems For Rational Numbers)
Complete at least three of the exercise sets for the chapter you chose to read.

10. Read Booklet No. 10 (The System Of Rational Numbers) of the thirtieth yearbook of the National Council of Teachers of Mathematics. Complete at least three of the exercise sets given there.

11. Read "Repeating Decimals" by Marvin R. Wizenread on pages 678 through 682 of the December 1973 issue of *The Arithmetic Teacher.* Then prepare a lesson on operations with repeating decimals suitable for an upper elementary or junior high school grade.

12. Read "Notes from National Assessment: Addition and Multiplication with Fractions" by Thomas P. Carpenter, Terrence G. Coburn, Robert E. Reys, and James W. Wilson on pages 137 through 142 of the February 1976 issue of *The Arithmetic Teacher*. Summarize the related research and recommendations for instruction that are presented in the article.

13. Read Chapter 8 (Fractional Numbers) of the thirty-seventh yearbook of the National Council of Teachers of Mathematics. Add at least two additional learning activities to the ones suggested for each part of the developmental sequence of the fraction concept.

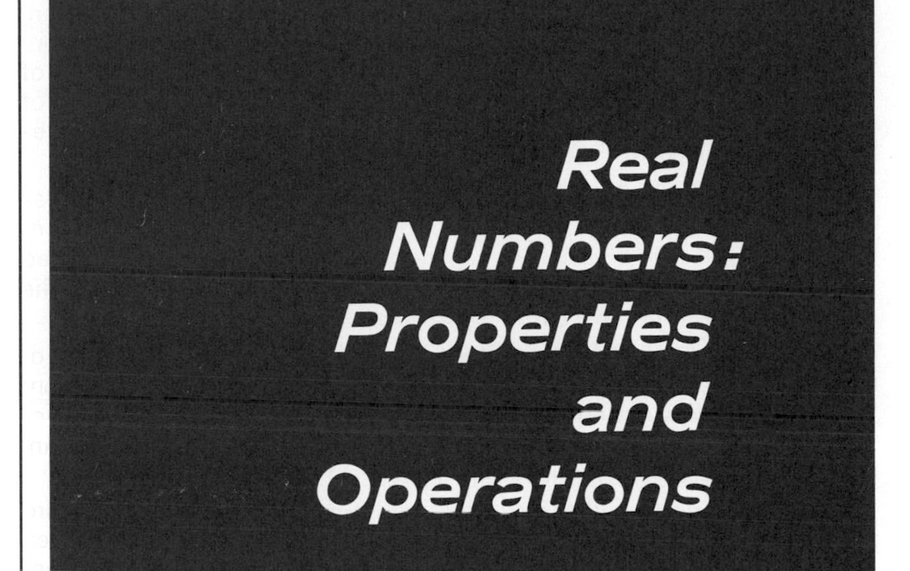

7

Any measurement can be represented by a rational number of units. Even massive electronic digital computers use only rational numbers. However, we still need to extend our concept of number beyond the set of rational numbers. There are many mathematical problems that we are unable to solve using only rational numbers. For example, there is no rational number n for which either of the following equations is true:

$$n^2 - 3 = 0 \qquad (n^2 = 3)$$
$$n^2 - 5 = 0 \qquad (n^2 = 5)$$

That is, there are no rational numbers a/b such that $(a/b)^2$ is equal to 3 or is equal to 5.

In this chapter we shall extend our number system to the *real number system* so that all *positive* numbers have square roots, and indeed nth roots, for any counting number n. *Real numbers* may be considered as:

Coordinates of points on a line in our ordinary geometry.
Lengths of line segments (for positive real numbers).
Numbers represented by infinite decimals.

Each of these equivalent representations of real numbers and some of the properties of the real number system are considered.

7-1 the real number line

Number scales and number lines have been used throughout our introduction of number concepts. A line of points with integral coordinates does not fit our concept of a line in the physical universe since there are obvious holes in a line of integral points. A line of points with rational numbers as coordinates is dense and three thousand years ago appeared to fit the concept of a line in the physical universe, the real world.

There is a legend that the discovery of the need for numbers that are not rational numbers was made by a Pythagorean, who was brutally punished for his unorthodox discovery. The Pythagoreans were a secret society of scholars about twenty-five hundred years ago. Their concept of a number was as a number of units—a counting number. Numbers such as $\frac{5}{2}$ were considered as five halves; that is, as five units where each unit was $\frac{1}{2}$ instead of 1. Today we still use this concept when we add two rational numbers that are expressed as fractions. We find a lowest common denominator (a common unit for the two addends) and express each addend as an integral number of this new unit.

As a mystical sect the Pythagoreans believed that all events and all elements of the universe depended upon numbers (whole numbers). Thus, in particular, the lengths of any two line segments should be expressible as integral multiples of some common unit of length, just as any two fractions can be expressed as integral multiples of the same unit (their lowest common denominator). This belief was shattered when it was discovered that there could not exist a common unit for a diagonal and a side of a square. In other words, for any given square there cannot exist a unit line segment such that the lengths of both diagonal and a side of the square are integral multiples of that unit of length.

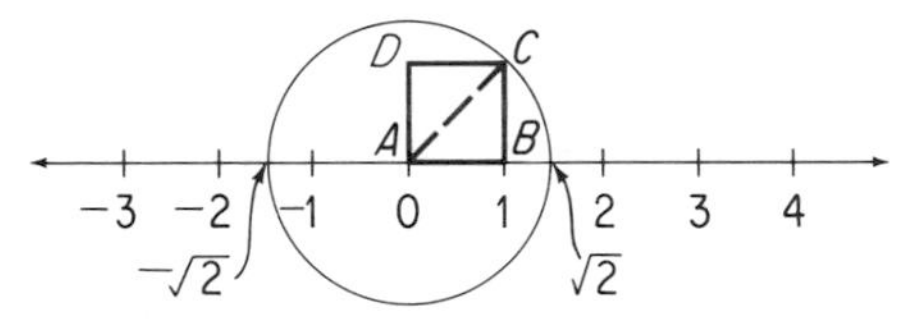

We use this discovery of the Pythagoreans in our development of a real number line by constructing a unit square as in the figure and drawing a circle with its center at the origin and a diagonal of the square as radius. Then in our ordinary geometry the circle appears to intersect the line in two points. We shall later find that these points have coordinates $\sqrt{2}$ and $-\sqrt{2}$, which are not rational numbers.

One method of introducing the real numbers is to call the lines of our ordinary geometry **real lines**, the lines that appear to fit the real world in which we live. If we select an origin and a unit point on any real line, then we have a **real number line**. The numbers that are coordinates of the points of a real number line are **real numbers**. In other words, we assume that:

> Any line in our ordinary geometry can be considered as a real number line.

Each point of a real number line has a unique real number as its *coordinate*.

Each real number has a unique point of a real number line as its *graph*.

The significance of these assumptions may be recognized as we compare the graphs of the statement

$$x \leq 3$$

using replacements for x from the major sets of numbers that we have studied.

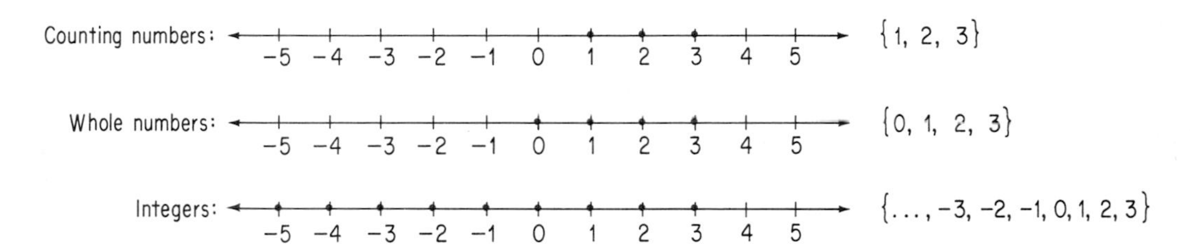

In the last figure the heavily shaded arrowhead at the left indicates that the set includes all negative integers.

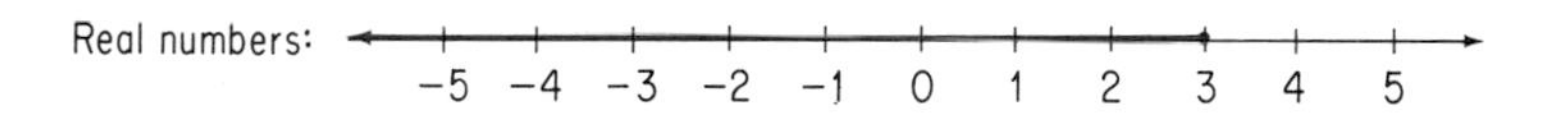

The heavily shaded arrow drawn to the left indicates that the set includes 3 and all real numbers that are less than 3.

In each of the previous figures the small arrowheads have been used to indicate that the line extends indefinitely in both senses (left and right). When, in some other books, arrowheads are used only on the right, they are used to indicate the positive sense, or orientation, of the line instead of its indefinite extent. The following examples show other graphs on a number line.

Example 1: Graph the set of real numbers 2 through 5.

Solution: The graph is a line segment with endpoints indicated by the heavy dots at the points with coordinates 2 and 5.

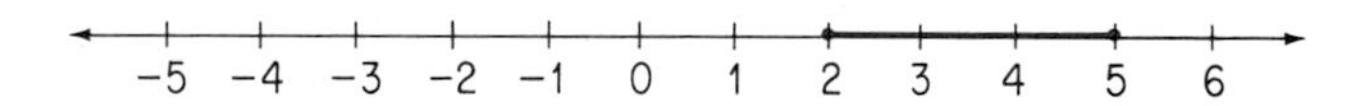

Note in Example 1 that the language "2 through 5" is used to indicate that the endpoints are to be included in the graph. If they are to be excluded, the set would be described as in Example 2.

Example 2: Graph the set of real numbers between 2 and 5.

Solution: This graph may be obtained from the graph in Example 1 by excluding the points with coordinates 2 and 5. Use hollow dots to show that the points with coordinates 2 and 5 are not included in the graph.

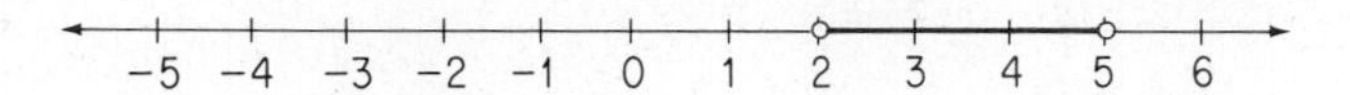

Example 3: Graph the set of real numbers greater than or equal to 2.

Solution: A solid dot is placed at the point with coordinate 2 to indicate that this point is included in the graph of the set. Then a heavily shaded arrow is drawn to indicate that all points with coordinates greater than 2 are to be included.

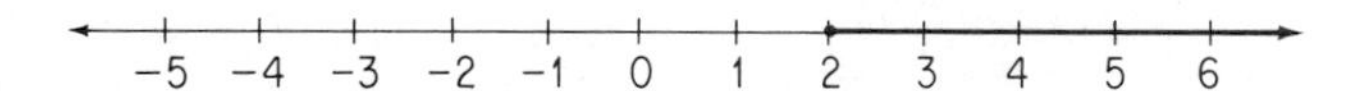

Example 4: Graph the set of real numbers less than 1.

Solution:

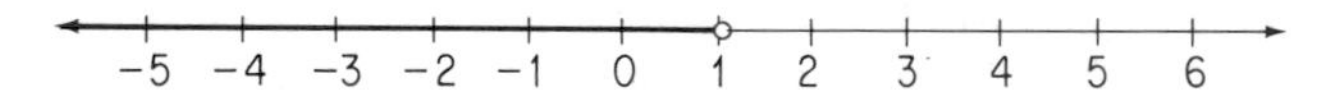

All of the preceding graphs were on a number line. We can also use real numbers to locate points on a plane. To do so we represent the points of a plane by ordered pairs of real numbers in the following way.

First we draw two perpendicular number lines, called the **x-axis** and the **y-axis**, as in the figure. These **coordinate axes** divide the entire plane into four parts, called **quadrants,** that are numbered as shown. The point where the two axes intersect is called the **origin** and is said to have coordinates (0, 0).

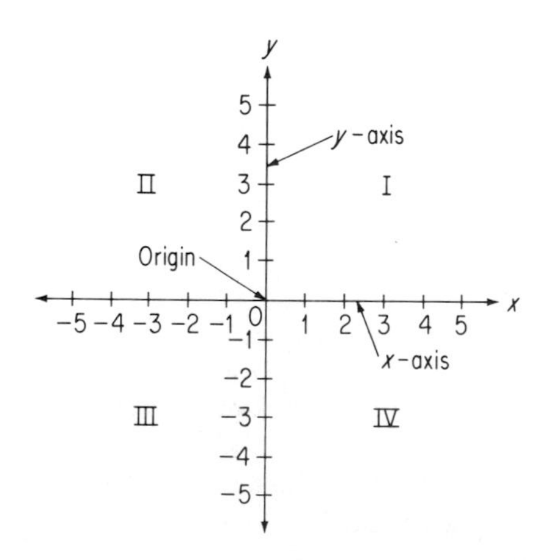

We locate points on this plane by means of ordered pairs of numbers (x, y), where we agree that the order is first x, the directed distance along the x-axis, and then y, the directed distance along the y-axis. Each element of the ordered pair tells us how many units to move from the origin, and in which direction.

The points A: (3, 2), B: (−3, 2), C: (−3, −2), and D: (3, −2) are located in the next four graphs. Notice that one point is located in each of the four quadrants.

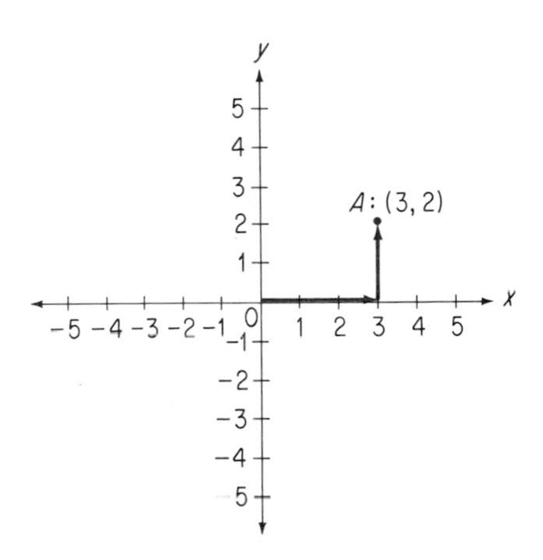

A: (3, 2) To locate the point A, start at the origin, then move three units to the right and two units up.

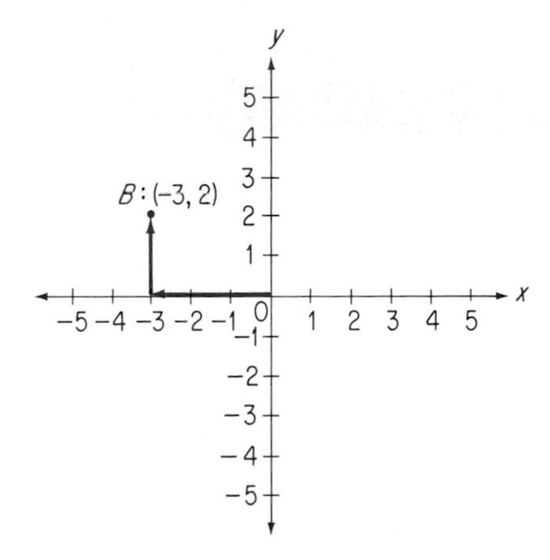

B: (−3, 2) To locate the point B, start at the origin, then move three units to the left and two units up.

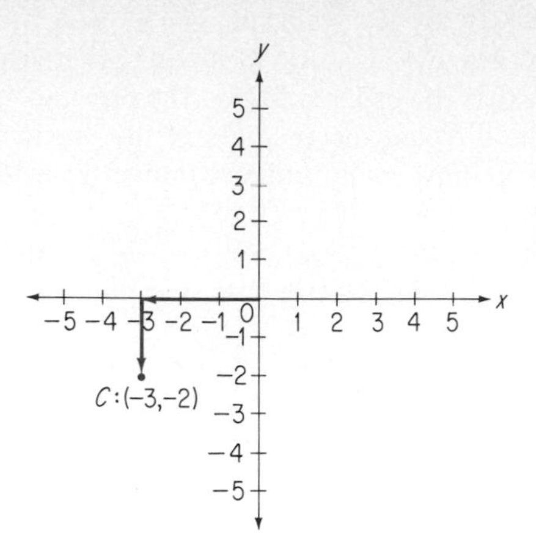

C: (−3, −2) To locate the point *C*, start at the origin, then move three units to the left and two units down.

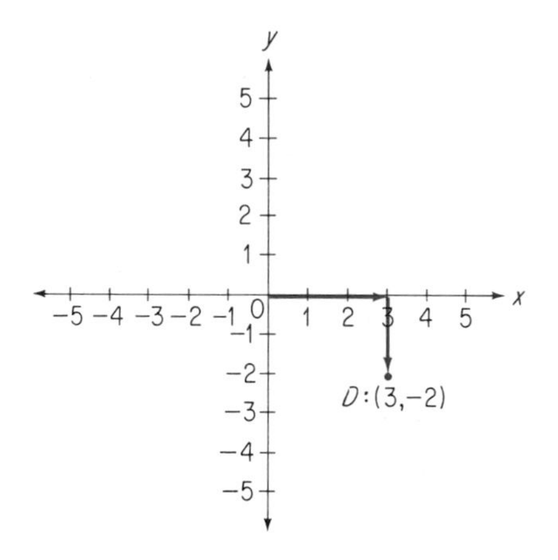

D: (3, −2) To locate the point *D*, start at the origin, then move three units to the right and two units down.

There is a one-to-one correspondence between the real numbers and the points on a number line. There is also a one-to-one correspondence between the ordered pairs of real numbers and the points on a **coordinate plane.**

EXERCISES

Graph each of the following sets of numbers on a number line.

1. The set of counting numbers less than 6.
2. The set of counting numbers less than or equal to 4.
3. The set of counting numbers 2 through 6.

4. The set of counting numbers between 2 and 6.

5. The set of whole numbers less than 4.

6. The set of whole numbers less than or equal to 5.

7. The set of whole numbers between 0 and 5.

8. The set of whole numbers 0 through 5.

9. The set of integers between −2 and 4.

10. The set of integers −3 through 3.

11. The set of integers −7 through −1.

12. The set of integers less than −2.

Graph the real numbers:

13. Between −5 and 4.
14. Less than 3.
15. Greater than −1.
16. Less than or equal to 0.
17. Greater than or equal to −3.
18. −3 through 4.

Draw a pair of coordinate axes and locate each of the following points.

19. P: (2, −1)
20. Q: (−3, −4)
21. R: (4, −3)
22. S: (5, 2)

Use the figure below and give approximate coordinates for each point shown.

23. *A* **24.** *B* **25.** *C* **26.** *D* **27.** *E* **28.** *F*

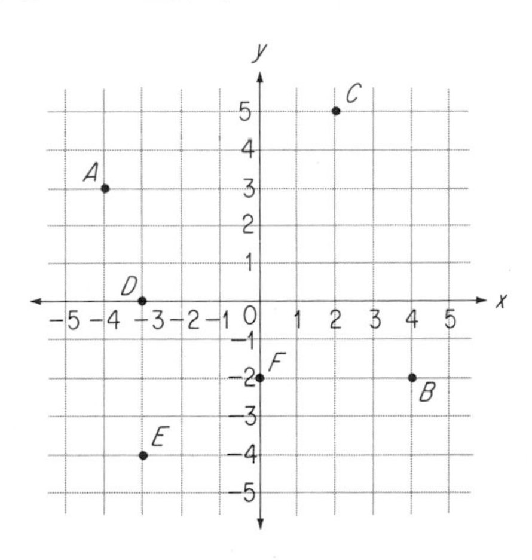

PEDAGOGICAL EXPLORATIONS

1. Think of a city street (call it School Street) and use city blocks as units of distance even though the blocks may not be of equal lengths. Suppose that School Street extends east and west with your school at

one of the major street corners. Describe how you might use coordinates on School Street as a number line to direct a stranger how to drive from your school to any street that crosses School Street. Assume that only one street crosses School Street at each intersection and that each street actually crosses.

2. As in Exploration 1 describe how you might direct the stranger to any desired point on School Street.
3. Think of a city with Main Street and School Street intersecting at right angles at the center of the city. Suppose that the streets east of Main Street are named in order First Street, Second Street, Third Street, and so forth. Also suppose that the streets north of School Street are named in order First Avenue, Second Avenue, Third Avenue, and so forth. Draw a map of the northeast part of the city. Describe how you might use coordinates to direct a stranger from the intersection of School Street and Main Street to any desired street intersection in the northeast part of the city. (*Note:* This procedure may be extended for all parts of the city using a street plan such as the one in Salt Lake City, Utah.)
4. As in Exploration 3 describe how you might direct the stranger to any desired point on any street in the northeast part of the city.
5. Many interesting figures can be sketched on a coordinate plane by drawing line segments joining points in a specified order. Select a suitable figure for a pre-holiday class activity. Then make a worksheet identifying the points with their coordinates in order (by number or letter) so that the students can draw the figure.

Assume that a line segment one unit long is given. Then, with reference to this unit of length, explain why you accept or do not accept each of the following statements as a true statement.

6. Every line segment has a real number as its length.
7. Every positive real number may be represented by a line segment.

7-2 lengths of line segments

Any line segment AB has a length d that we may think of either as the distance from A to B or as the distance from B to A. The assumption that each line segment has a number as its length is equivalent to the assumption that every point on a number line has a number as its coordinate. For example, every line segment AB of length d has the same length as the line segment OD where O is the origin and D is the point with coordinate d on a number line.

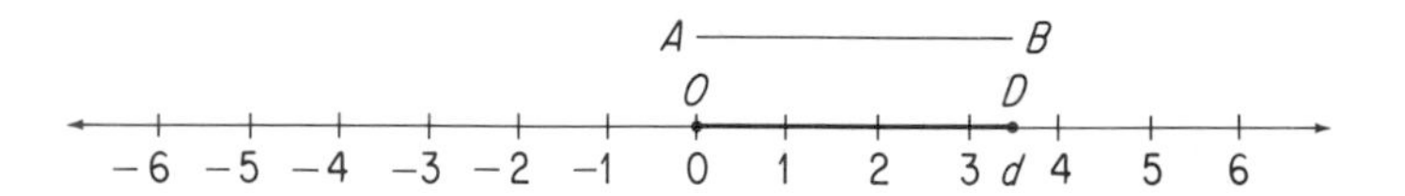

The Pythagoreans used the assumption that each line segment has a number as its length when they discovered the need for numbers that were not rational numbers. Much earlier the Egyptians and Babylonians had used triples of counting numbers a, b, c such that

$$a^2 + b^2 = c^2$$

Such triples are now called **Pythagorean triples**. The numbers 3, 4, and 5 form a Pythagorean triple since $3^2 + 4^2 = 5^2$. There are legends that the Egyptians used such triples to obtain right angles for remarking the locations of fields after the annual floods of the Nile River had subsided.

Today we state the **Pythagorean theorem** as:

> The sum of the squares of the lengths of the two short sides (**legs**) of any right triangle is equal to the square of the length of the longest side (**hypotenuse,** side opposite the right angle).

Since the Pythagoreans were active over a thousand years before the development of algebraic notation, they thought of the theorem in terms of its geometric representation. Thus they stated the theorem as:

> The sum of the areas of the squares on the legs of any right triangle is equal to the area of the square on the hypotenuse.

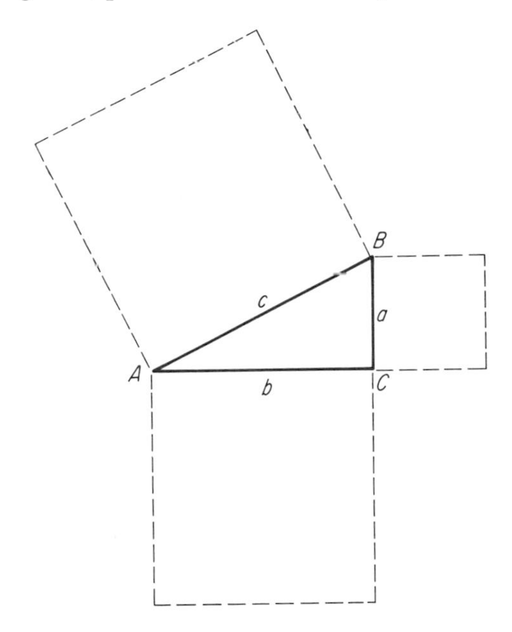

We may use the Pythagorean theorem to construct line segments of length $\sqrt{n}$ for any given counting number n and any given unit of length. Several examples are shown in the following figures. A general approach is considered in Pedagogical Exploration 1.

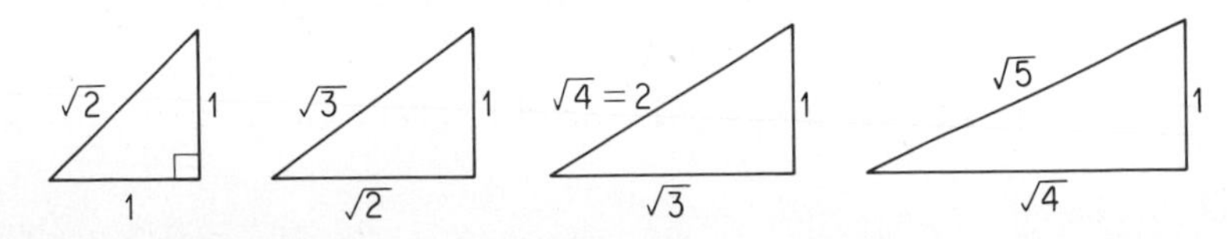

In the labels for the figures we have used the **radical symbol** $\sqrt{\ }$ to indicate the positive real number that is the **square root** of the given number. In other words, for any counting number n

$$\sqrt{n} > 0 \quad \text{and} \quad (\sqrt{n})^2 = n$$

This definition is extended to all positive real numbers. Zero may be included with $\sqrt{0} = 0$ since $0^2 = 0$. However, the square root of a negative number cannot be a real number since the square of any real number is positive or zero.

The need for extending the set of rational numbers to some more extensive set of numbers was discovered by recognizing that a number was needed whose square was 2. Thus the proof that the real number $\sqrt{2}$ is not a rational number is an *existence proof* that there must exist at least one real number that is not a rational number. Let us now prove that $\sqrt{2}$ is not a rational number. To do so we make use of the indirect proof used by Euclid. The proof is not difficult, but is fairly subtle and will require some careful thought on the reader's part. We assume the opposite of what we are trying to prove. That is, we assume that $\sqrt{2}$ is a rational number and show that this leads to a contradiction. Thus suppose that $\sqrt{2} = p/q$, a rational number. We first reduce p/q so that $p/q = a/b$ where a and b are not both even integers. Then $\sqrt{2} = a/b$.

If $\sqrt{2} = a/b$, then $2 = a^2/b^2$ by squaring both sides of the equation. (Remember that $\sqrt{2} \times \sqrt{2} = 2$.)

Then $2b^2 = a^2$, and $2b^2$ is even since it has a factor 2.

Now a^2 is even since it is equal to $2b^2$. However, if the square of an integer is even, then the integer is even. (See §5-1, Exercise 25.) Thus a is even.

Since a is even, we may write $a = 2k$, where k is some integer. Thus

$2b^2 = (2k)^2$ by this substitution
$2b^2 = 4k^2$ by squaring $2k$
$b^2 = 2k^2$ by division of each member of the equation by 2

Now b^2 is even since it is equal to a number with a factor of 2. Again, if the square of an integer is even, then the integer is even; thus b is even. This is contrary to the assumption that a and b were not both even. Therefore our assumption that there exists a rational number whose square is 2 has led to a contradiction. Thus the real number whose square is 2 can not be a rational number.

The set of all numbers that are coordinates of points on the number line is the set of *real numbers*. Some real numbers such as

$$3, \frac{2}{3}, 0, \text{ and } -5$$

are rational numbers; all other real numbers are **irrational numbers**. For example,

$$\sqrt{2}, \quad 3\sqrt{2}, \quad \tfrac{1}{2}\sqrt{2}, \quad \sqrt{3}, \quad \sqrt{5}, \quad \text{and} \quad \pi$$

are irrational numbers. The set of real numbers is the union of the set of rational numbers and the set of irrational numbers as indicated by this array:

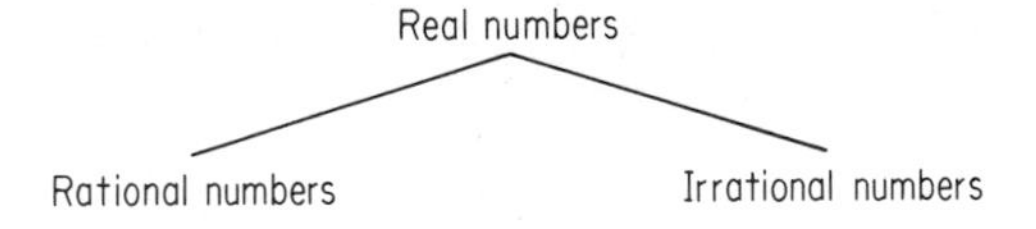

Example 1: Prove that $1 + \sqrt{2}$ is an irrational number.

Solution: We use an indirect proof and assume that $1 + \sqrt{2}$ is rational. Thus we assume it can be written in the form a/b, as the quotient of two integers, with $b \neq 0$. Assume

$$1 + \sqrt{2} = \frac{a}{b}$$

Then

$$\sqrt{2} = \frac{a}{b} - 1$$

But $(a/b) - 1$ names a rational number because the set of rational numbers is closed with respect to subtraction. Therefore $\sqrt{2}$ must be a rational number. But this contradicts the fact that $\sqrt{2}$ is an irrational number. Therefore our original assumption was incorrect, and $1 + \sqrt{2}$ must be an irrational number.

Note that we can locate (graph) on a number line a point with coordinate $1 + \sqrt{2}$. We repeat the construction used in §7-1, but placed one unit to the right.

Example 2: Graph $1 + \sqrt{2}$ on the number line.

Solution:

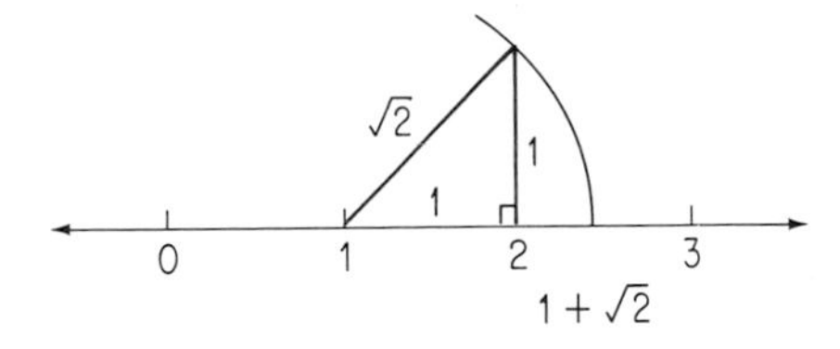

As a general "rule of thumb" we assume that:

> If a positive integer n is not the square of a positive integer, then $\sqrt{n}$ is an irrational number.

Then, for example, $\sqrt{3}$, $\sqrt{5}$, $\sqrt{6}$, and $\sqrt{15}$ are irrational numbers because 3, 5, 6, and 15 are not squares of positive integers. Similarly, $b \times b \times b = b^3$

(read as "b cubed"), $(\sqrt[3]{b})^3 = b$, and we assume that if a positive integer n is not the cube of a positive integer, then $\sqrt[3]{n}$ is an irrational number. Then, for example $\sqrt[3]{2}$, $\sqrt[3]{5}$, and $\sqrt[3]{75}$ are irrational numbers. There are infinitely many irrational numbers that may be expressed using the radical symbol; for example, $n\sqrt{103}$ where n is a counting number. There are also infinitely many irrational numbers that cannot be expressed using the radical symbol; for example, $n\pi$ where n is a counting number and π is the ratio of the circumference to the diameter of a circle.

EXERCISES

Classify each statement as true or false.

1. Every rational number is a real number.
2. Every real number is a rational number.
3. Not every integer is a real number.
4. Every real number is either a rational number or an irrational number.
5. Every point on the number line has a real number as its coordinate.
6. Every real number has a point on a number line as its graph.
7. For a given unit of length every positive real number may be represented by the length of a line segment.

Copy the following table. Use "$\surd$" to show that the number at the top is a member of the set listed at the side. Use "$\times$" if it is not a member of the set.

	8.	**9.**	**10.**	**11.**	**12.**
Set	5	$-\frac{2}{3}$	$\sqrt{9}$	0	$\sqrt{5}$
Counting numbers					
Whole numbers					
Integers					
Rational numbers					
Real numbers					

Prove that each number is an irrational number.

13. $\sqrt{2} + 2$ **14.** $\sqrt{2} - 1$ **15.** $\sqrt{2} - \frac{1}{2}$
16. $5\sqrt{2}$ **17.** $5 - 2\sqrt{2}$ **18.** $\frac{3}{2} + \sqrt{2}$

Construct a number line and graph the number given in the indicated exercise.

sec. 7-2 lengths of line segments

19. Exercise 13. ***20.*** Exercise 14. ***21.*** Exercise 15.

22. Exercise 16. ***23.*** Exercise 17. ***24.*** Exercise 18.

PEDAGOGICAL EXPLORATIONS

1. The construction for $\sqrt{2}$ can be extended to find segments whose measures are $\sqrt{3}$, $\sqrt{4}$, $\sqrt{5}$, etc. Merely continue to construct right triangles, using the hypotenuse of the preceding triangle as one leg and a segment of one unit as the other leg. The following figure shows the construction for $\sqrt{3}$.

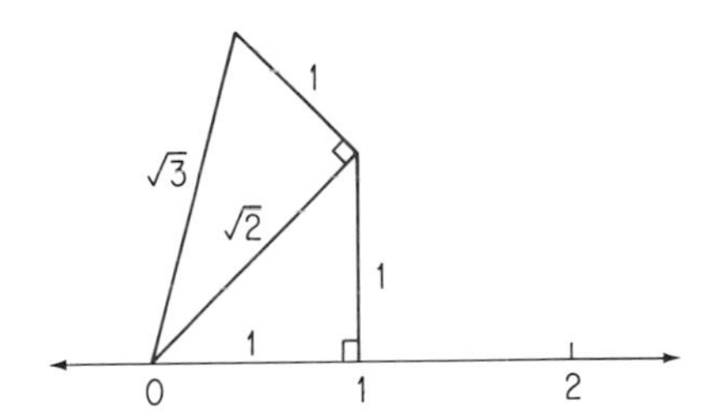

Continue this method of construction to find lengths of $\sqrt{5}$ and $\sqrt{7}$.

2. Look through recently published mathematics textbooks for grades 4 through 8 and write a summary of the ways in which irrational numbers (not necessarily by name) are introduced and used.

3. Explain why, for any rational number $r \neq 0$, each of the following must be an irrational number:

(a) $r\sqrt{2}$ (b) $r + \sqrt{2}$

(c) $\sqrt{2} - r$ (d) $\sqrt{2} \div r$

4. Just above a number line use dots to describe the graphs of the points $r\sqrt{2}$ for

(a) Any integer $r \neq 0$ (b) Any rational number $r \neq 0$

5. Repeat Exploration 4 for $r\sqrt{3}$.

6. Repeat Exploration 4 for $r\sqrt{5}$.

7. Note that there is an irrational number as the coordinate of each point graphed in Explorations 4, 5, and 6 and indeed for each point $r\sqrt{n}$ where the rational number $r \neq 0$ and the integer n is not the square of an integer. How does the number of irrational numbers appear to compare with the number of rational numbers?

7-3 decimal representations of real numbers

The relationship between the set of rational numbers and the set of repeating decimals was considered in §6-6. Indeed, any rational number can be named by a repeating decimal, and conversely. Recall that we agreed that a terminating decimal may be thought of as a decimal that repeats the digit zero. Thus, for example, we have the following decimal representations of several rational numbers:

$$\tfrac{3}{4} = 0.75\overline{0}, \qquad \tfrac{1}{3} = 0.3\overline{3}, \qquad \tfrac{1}{7} = 0.142857\overline{142857}$$

We also found a way to express every repeating decimal as the quotient of two integers, and thus as the name of a rational number. For example, consider the repeating decimal $0.72\overline{72}$ and find $100n - n$.

$$\begin{aligned} 100n &= 72.72\overline{72} \\ n &= 0.72\overline{72} \\ \hline 99n &= 72 \\ n &= \frac{72}{99} = \frac{8}{11} \end{aligned} \qquad \text{Thus } 0.72\overline{72} = \frac{8}{11}.$$

Now we wish to consider the decimal representation for irrational numbers. To do so, we return again to the number $\sqrt{2}$ and try to find its decimal equivalent through arithmetic procedures. First we know that $\sqrt{2}$ must fall somewhere between 1 and 2 since $1^2 = 1$, $2^2 = 4$, and $(\sqrt{2})^2 = 2$. Thus we may write this first inequality:

$$1 < \sqrt{2} < 2$$

Next, by arithmetic, we may verify that $(1.4)^2 = 1.96$ and $(1.5)^2 = 2.25$. This gives rise to the next inequality:

$$1.4 < \sqrt{2} < 1.5$$

Continuing in this way, we can verify each of the following by multiplication:

$$\left.\begin{aligned}(1.41)^2 &= 1.9881 \\ (1.42)^2 &= 2.0164\end{aligned}\right\} \quad 1.41 < \sqrt{2} < 1.42$$

$$\left.\begin{aligned}(1.414)^2 &= 1.999396 \\ (1.415)^2 &= 2.002225\end{aligned}\right\} \quad 1.414 < \sqrt{2} < 1.415$$

$$\left.\begin{aligned}(1.4142)^2 &= 1.99961644 \\ (1.4143)^2 &= 2.00034449\end{aligned}\right\} \quad 1.4142 < \sqrt{2} < 1.4143$$

$$\left.\begin{aligned}(1.41421)^2 &= 1.9999899241 \\ (1.41422)^2 &= 2.0000182084\end{aligned}\right\} \quad 1.41421 < \sqrt{2} < 1.41422$$

We are now caught up in a never-ending process. The decimal representation for $\sqrt{2}$ does not terminate, and it does not repeat. Given enough time and patience we may express $\sqrt{2}$ correct to as many decimal places as we wish, but at each step we merely have a rational number approximation to this irrational number. For example, correct to seven decimal places we have the following:

$$\sqrt{2} = 1.4142135\ldots$$

We use the three dots to show that the decimal is an infinite one. The absence of a bar over any of the digits shows that it is a nonrepeating decimal.

We first defined an irrational number as a real number that is *not* a rational number. We may now define an irrational number directly by saying that an irrational number is a real number that is named by a nonrepeating, nonterminating decimal.

In summary, every real number can be represented by a decimal. If the real number is a rational number, then it may be represented by a terminating or by a repeating decimal. If the real number is an irrational number, then it may be represented by a nonterminating, nonrepeating decimal.

It is often convenient to represent irrational numbers by a sequence of digits that has a pattern but does not repeat any particular sequence of digits. For example, each of the following names an irrational number:

$$0.20220222022220222220\ldots$$

$$0.305300530005300005\ldots$$

$$0.404004000400004000004\ldots$$

Example 1: How many zeros are there altogether between the decimal point and the one hundredth 5 in this sequence?

$$0.05005000500005\ldots$$

Solution: There is one zero preceding the first 5, two zeros preceding the second 5, etc. The total number of zeros is the sum

$$1 + 2 + 3 + \cdots + 100$$

By the method used in §1 of the introductory chapter, the sum is 5050.

We may use decimal representations to show the density of the set of real numbers. That is, the set of real numbers is said to be dense; between any two real numbers, there is always another real number. We have already shown that this is true for two rational numbers (see §6-2). In Example 2 we find an irrational number between two given rational numbers.

Example 2: Find an irrational number between 0.47 and 0.48.

Solution: It is easier to see the solution if the given numbers are written as decimals that repeat the digit 0.

$$0.47 = 0.470000\bar{0}$$

$$0.48 = 0.480000\bar{0}$$

We need to write a decimal for an irrational number that is greater than 0.47 and less than 0.48. This decimal must be a nonrepeating, nonterminating decimal. There are many possible answers. Here are two:

$$0.47247224722247222247\ldots$$

$$0.47505005000500005\ldots$$

The pattern shown allows you to write as many digits of each decimal as you wish. Each pattern insures that the decimal will never terminate nor repeat any particular sequence of digits.

EXERCISES

Classify each decimal as the name of a rational number or an irrational number.

1. 0.745
2. $0.74\overline{545}$
3. 0.7454554555. . .
4. $0.745\overline{745}$
5. 0.7454554555
6. 0.7454454445. . .

Tell whether each number can be represented by (a) a terminating decimal; (b) a repeating decimal; (c) a nonterminating, nonrepeating decimal. Give the decimal representation of each number that can be expressed as a terminating or a repeating decimal.

7. $\frac{3}{8}$
8. $\frac{5}{12}$
9. $\sqrt{8}$
10. $\sqrt{3}$
11. $\sqrt{100}$
12. $\frac{13}{16}$

Write the first fifteen digits to the right of the decimal point in each decimal.

13. 0.131331333. . .
14. 0.2052005200052. . .

Write the digit in the sixteenth position to the right of the decimal point for each decimal.

15. 0.969969996. . .
16. 0.6146114611146. . .

Perform the necessary multiplications to verify each inequality.

17. $1.7 < \sqrt{3} < 1.8$
18. $1.73 < \sqrt{3} < 1.74$
19. $2.2 < \sqrt{5} < 2.3$
20. $2.23 < \sqrt{5} < 2.24$

Use the method shown in this section to find each square root, correct to the nearest tenth.

21. $\sqrt{7}$
22. $\sqrt{10}$
23. $\sqrt{11}$
24. $\sqrt{17}$

List the numbers of each set in order, from smallest to largest.

25. 0.45, $0.454554555\ldots$, 0.45455, $0.\overline{45}$, $0.4\overline{5}$

26. 2.525, 2.5252, $2.\overline{52}$, $2.5252252225\ldots$, 2.5

27. 0.067, $0.06\overline{7}$, $0.067677677\ldots$, 0.06, $0.0\overline{6}$

28. Which of the following names a rational number between 0.37 and 0.38? **(a)** 0.375; **(b)** $0.\overline{37}$; **(c)** $0.373773777\ldots$; **(d)** $0.37\overline{8}$.

29. Which of the following names an irrational number between 0.234 and 0.235? **(a)** 0.2345; **(b)** $0.\overline{234}$; **(c)** $0.234040040004\ldots$; **(d)** $0.23454554555\ldots$.

30. Name two rational numbers between $0.52424424442\ldots$ and $0.52525525552\ldots$.

31. Name two irrational numbers between 0.48 and 0.49.

32. Name two irrational numbers between $0.\overline{78}$ and $0.\overline{79}$.

PEDAGOGICAL EXPLORATIONS

1. (a) Write an infinite nonrepeating decimal for a number between 5 and 6.
   (b) Explain why your answer for part (a) is considered to be infinite.
   (c) Explain why your answer for part (a) is considered to be nonrepeating?
2. Describe a procedure for writing an infinite nonrepeating decimal for a number between any two given whole numbers.
3. Repeat Exploration 2 for any two given rational numbers.
4. Repeat Exploration 2 for any two irrational numbers that are given as decimals.
5. Explain why you think that the set of irrational numbers is dense or is not dense.

7-4 properties of the real numbers

We started with the set of counting numbers in §4-1. In the set of counting numbers addition and multiplication were defined so that we had the eight properties:

Closure, +: $a + b$ is a unique element of the set.

Closure, ×: $a \times b$ is a unique element of the set.

Associative, +: $(a + b) + c = a + (b + c)$.

Associative, ×: $(a \times b) \times c = a \times (b \times c)$.

Commutative, +: $a + b = b + a$.

Commutative, ×: $a \times b = b \times a$.

Identity, ×: There is an element 1 in the set such that $1 \times a = a \times 1 = a$.

Distributive: $a \times (b + c) = (a \times b) + (a \times c)$.

The first extension of our concept of a number was to the set of whole numbers in §4-4. This extension enabled us to keep the eight properties of counting numbers and to gain one additional property:

Identity, +: There is an element 0 in the set such that $0 + a = a + 0 = a$.

Then in Chapter 5 we extended our concept of number to the set of integers. The nine properties of the whole numbers were retained and one additional property gained:

Inverse, +: For each element a there is an element $(-a)$ in the set such that $a + (-a) = 0$.

In Chapter 6 we again extended our concept of number to include all rational numbers, retained all previously mentioned properties, and gained one more:

Inverse (≠ 0), ×: For each element a different from zero, the additive identity, there is an element $\frac{1}{a}$ in the set such that $a \times \frac{1}{a} = 1$.

We summarized these eleven properties by saying that the set of rational numbers formed a *field*. We also noted that the rational numbers were *dense*, that is, between any two rational numbers there is at least one other rational number.

In the present chapter we have discovered that even with the density property, the rational numbers are not sufficient to serve as coordinates of all points on a line in our ordinary geometry. Accordingly, we introduced the real numbers in three equivalent ways:

As coordinates of points on a line (§7-1).
As lengths of line segments (§7-2).
As infinite decimals (§7-3).

The set of coordinates for points on a line is now complete and the set of real numbers is said to be **complete**. There is a real number as the coordinate of each point of the line and each point of the line has a real number as its coordinate. The set of real numbers has the properties that we have considered for the set of rational numbers. Furthermore, a one-to-one correspondence can be established between the set of points on an ordinary line and the set of real numbers.

We may think of the original set of numbers as a subset of the new set at each step of the development of our concept of number. These relationships are shown in the following array:

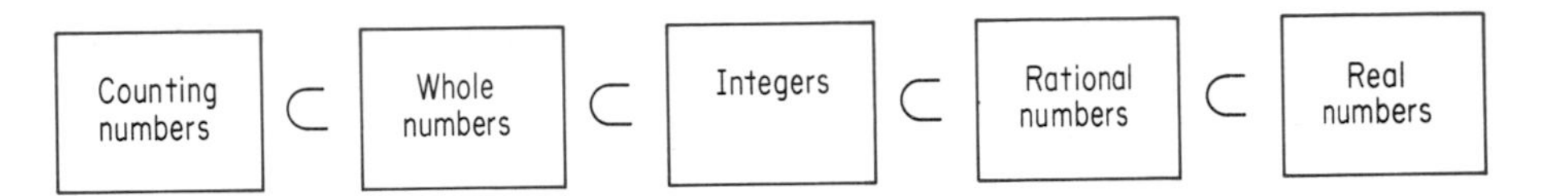

The numbers of each set are *ordered* in the same manner as their graphs on a number line, that is, for the number line in its usual position, $a < b$ and $b > a$ if and only if the point with coordinate a is on the left of the point with coordinate b. The set of real numbers with the operations $+$ and $\times$ forms a **complete linearly-ordered field**.

The real numbers may be classified in several different ways. For example, any real number is:

Positive, zero, or negative.
A rational number or an irrational number.
Expressible as a terminating, a repeating, or a nonterminating, nonrepeating decimal.

Properties of numbers seem a bit remote without a consideration of the uses of the sets of numbers. Frequently these uses are related to the types of sentences that can be solved. Several comparisons are shown in the next array.

Sentence	*Counting numbers*	*Whole numbers*	*Integers*	*Rational numbers*	*Real numbers*
$n + 3 = 5$	2	2	2	2	2
$n + 5 = 5$	None	0	0	0	0
$n + 7 = 5$	None	None	-2	-2	-2
$2n = 5$	None	None	None	2.5	2.5
$n^2 = 5$	None	None	None	None	$\sqrt{5}, -\sqrt{5}$
$n^2 = -5$	None	None	None	None	None

Note the existence of relatively simple equations such as $n^2 = -5$, or $n^2 + 5 = 0$, that do not have any real numbers as solutions. To obtain solutions for all such equations, the set of real numbers is extended to the set of complex numbers, usually in secondary school algebra classes.

In order to give examples of, and use, the properties of rational numbers and irrational numbers, we need to be able to identify several irrational numbers "on sight." This is possible using two properties that were mentioned briefly in §7-2:

> The square root of an integer is either an integer or an irrational number.
> If t is an irrational number, then $t + n$, $t - n$, $t \times n$, and $t \div n$ are also irrational numbers for any integer $n \neq 0$ and indeed for any rational number $n \neq 0$.

There are other types of irrational numbers but these will suffice for our present purposes.

Example. Identify each number as a rational number or an irrational number: (a) $\sqrt{19}$; (b) $2 + \sqrt{7}$; (c) $3 + \sqrt{9}$; (d) $5\sqrt{11}$; (e) $\sqrt{13} - 2$; (f) $5 - \sqrt{15}$.

Solution: (a) Irrational number, since 19 is not the square of an integer;
(b) irrational number, since 7 is not the square of an integer;
(c) rational number, since 9 is the square of an integer;
(d), (e), (f) irrational numbers, since 11, 13, and 15 are not squares of integers.

EXERCISES

Find all possible replacements for n that will make each of the sentences below true when n must be a member of the set named at the top of the column. If there is no such replacement, write "none."

	Sentence	Counting numbers	Whole numbers	Integers	Rational numbers	Real numbers
1.	$n + 3 = 1$					
2.	$3n = 8$					
3.	$n \times n = 2$					
4.	$n \times n = -1$					

Identify each number as a rational number or an irrational number.

5. (a) $\sqrt{11}$ (b) $5\sqrt{11}$

6. (a) $-4 + \sqrt{16}$ (b) $11 - \sqrt{63}$

7. (a) $7\sqrt{36}$ (b) $5\sqrt{37}$

8. (a) $\sqrt{20}$ (b) $\sqrt{63} + 5$

Use numbers such as those in Exercises 5 through 8 and give at least three examples of each of the following statements.

9. A product of irrational numbers may be:
(a) A rational number. (b) An irrational number.

10. A sum of two irrational numbers may be:
(a) A rational number. (b) An irrational number.

11. A difference of two irrational numbers may be:
(a) A rational number. (b) An irrational number.

12. A quotient of two irrational numbers may be:
(a) A rational number. (b) An irrational number.

Classify each statement as true or false and give at least one example or counterexample.

13. Every real number is either an irrational number or an integer.

14. Every real number is either positive or negative.

15. Every irrational number is either positive or negative.

16. The set of real numbers is a subset of the set of rational numbers.

17. The additive inverse of an irrational number is an irrational number.

18. The multiplicative inverse of an irrational number is also an irrational number.

19. The set of irrational numbers is dense.

20. The set of irrational numbers forms a field.

Complete the chart at the top of the next page to summarize the various properties of the number systems that we have studied thus far. In each case use "$\surd$" to show that the set of elements named at the top of the column has the property listed at the side. Use "$\times$" if the set does not have the stated property. Note that the first 11 properties listed are the field properties.

		Counting numbers	Whole numbers	Integers	Rational numbers	Real numbers
21.	Closure, +					
22.	Closure, x					
23.	Commutative, +					
24.	Commutative, x					
25.	Associative, +					
26.	Associative, x					
27.	Identity, + (0)					
28.	Identity, x (1)					
29.	Inverse, +					
30.	Inverse, ($\neq$0), x					
31.	Distributive					
32.	Field					
33.	Density					
34.	Completeness					

PEDAGOGICAL EXPLORATIONS

Irrational numbers may be represented by special symbols such as $\sqrt{2}$, $\sqrt[3]{5}$, and π or by infinite nonrepeating decimals. Usually we approximate irrational numbers by rational numbers and obtain the approximations by "rounding off" the decimal representations of the irrational numbers. For example, if c is the circumference (distance around) of a circle and d is the diameter (greatest distance across) of the circle, then $c = \pi d$ where π is an irrational number;

$\pi = 3.141\ 592\ 653\ 589\ 793\ 238\ 462\ 643\ 383\ 279\ 56\ldots$

You are probably accustomed to rounding off π to the nearest hundredth

$$\pi \approx 3.14$$

or possibly to the nearest ten thousandth

$$\pi \approx 3.1416$$

The process of rounding off frequently causes difficulty in elementary school mathematics. A number line picture is helpful in explaining the process to students. Suppose that in the next figure we wish to round 22 and 27 to the nearest 10; that is, we wish to find the nearest multiple of 10. Since 22 is closer to 20 than it is to 30, we round this number down to 20.

On the other hand, 27 is closer to 30 than it is to 20 so that we round this up to 30.

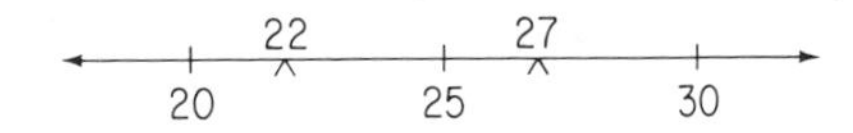

The natural question that arises is: How should we round off a number that is exactly halfway between two multiples of ten? It is a common practice in such cases to automatically round up. Then we would round 25 to 30 and 35 to 40. Such a practice is followed in nearly all financial transactions. In later work in mathematics, especially in statistics, a bias in the data is introduced by this procedure. The usual practice in such situations is to round so that the preceding numeral is even whenever a 5, alone or followed only by zeros, is the first digit to the eliminated. In other words, rounded to the nearest 10, we would round 25 down to 20, but would round 35 up to 40. Then about half of the time we round down and about half of the time we round up. Use this procedure to round in the exercises that follow.

1. Round to the nearest 100: (a) 348; (b) 257; (c) 175; (d) 265.
2. Round to the nearest tenth: (a) 0.37; (b) 0.23; (c) 0.55; (d) 0.85.
3. Round to the nearest hundredth: (a) 0.376; (b) 0.284; (c) 0.025; (d) 0.015.

Consider the previously indicated decimal expression for π and round to the indicated number of places to the right or the decimal point.

4. 3
5. 5
6. 7
7. 9
8. The ancient Chinese used the fraction $\frac{355}{113}$ as an approximation for π. Find this value to ten decimal places and compare it with the correct value.

7-5 powers and roots

The product

$$(x)(x) = x^2$$

is called the *second power* of x, or x squared, for any real number x. Similarly, as in §5-5, for any counting number n the product of n factors, each equal to x, is expressed as x^n. Then x is the *base*, n is the *exponent*, and x^n is the *nth power* of x. Positive integral powers of 10 were considered when the decimal notation for counting numbers was considered in expanded form (§3-5). Positive, zero, and negative integral powers of integers were considered in §5-5. We now consider fractional powers, that is, fractions as exponents.

The rules for multiplication and division of integral powers were considered in §5-5. These rules are often called the *laws of exponents*. Fractional and other powers are defined so that these laws hold for any positive real number as base and any real numbers as exponents. Proofs of these properties are left for the mathematical specialists. Special problems may arise when roots of negative numbers are concerned, as for $\sqrt{-2}$ and $\sqrt{-3}$. However, there are some powers and roots of negative numbers that can be readily obtained from the corresponding powers and roots of positive numbers. For example, an integral power of a negative number is a positive number if the exponent is an even number, a negative number if the exponent is an odd number:

$$(-2)^6 = 64, \qquad (-2)^5 = -32$$
$$(-2)^{-5} = -\frac{1}{32}, \qquad (-3)^{-4} = \frac{1}{81}$$

Also for any odd number n the nth root of a negative number is the negative of the nth root of the absolute value of the number:

$$\sqrt[5]{-32} = -2, \qquad \sqrt[3]{-8} = -2$$

We assume that such operations with negative numbers can be easily understood as soon as operations with positive numbers are understood. Thus we assume that for any *positive* real numbers a and b and any real numbers m and n:

$$a^m \times a^n = a^{m+n}$$
$$a^m \div a^n = a^{m-n}$$
$$a^0 = 1$$
$$a^{-m} = \frac{1}{a^m}$$
$$(a^m)^n = a^{mn}$$
$$(ab)^m = a^m b^m$$

Then in addition to the laws stated in Chapter 5 we assume that

$$a^{1/n} = \sqrt[n]{a}$$
$$\left(\frac{a}{b}\right)^n = \frac{a^n}{b^n}$$

The first law states the relationship between exponents and radicals. The law that a power of a quotient is equal to the quotient of the powers can be derived from the law for a power of a product:

$$\left(\frac{a}{b}\right)^n = (a \times b^{-1})^n = a^n \times (b^{-1})^n = a^n b^{-n} = \frac{a^n}{b^n}$$

We have assumed that fractional exponents may be associated with roots of the base. For example,

$$\sqrt{5} \times \sqrt{5} = 5, \qquad 5^{1/2} \times 5^{1/2} = 5^1 = 5$$
$$\sqrt{4} \times \sqrt{4} = 4, \qquad 4^{1/2} \times 4^{1/2} = 4^1 = 4$$

You know that the solution set $\{2, -2\}$ of the equation $x^2 = 4$ has two members. The symbols $\sqrt{4}$ and $4^{1/2}$ are both defined to represent exactly one number, the positive square root of 4; that is,

$$\sqrt{4} = 4^{1/2} = 2$$

Whenever the radical symbol $\sqrt{}$ is used for the square root of a positive number, the positive square root is meant. This is consistent with the introduction of square roots as lengths of line segments, since lengths are always nonnegative. The fractional exponent $\frac{1}{2}$ may be used in place of the radical symbol for square root at any time, and conversely.

Powers and roots, expressed either in terms of radicals or in terms of fractional exponents, are treated the same as other numbers in computations. For example,

$$(2 \times 11^3) + (3 \times 11^3) = 5 \times 11^3, \qquad 6\sqrt{17} - 2\sqrt{17} = 4\sqrt{17}$$

$$\frac{15\sqrt{6}}{3\sqrt{2}} = \frac{5\sqrt{2}\sqrt{3}}{\sqrt{2}} = 5\sqrt{3}, \qquad 12^{1/2} \times 3^{1/2} = 36^{1/2} = 6$$

Cube roots occur when volumes of cubes are considered. A solid cube with e unit cubes along each edge contains e^3 unit cubes. If we use V for the volume of the cube in cubic units and e for the length of the edge in linear units, then

$$V = e^3$$

e
e
e

A cube having each edge 2 centimeters long has volume 8 cubic centimeters where $8 = 2^3$. Next suppose that we wanted the volume of a cube to be 2 cubic centimeters. Then

$$2 = e^3, \qquad e = \sqrt[3]{2}, \qquad e = 2^{1/3}$$

Cube roots are treated the same as other numbers in computations. There is one major difference between our use of cube roots and our use of square roots—negative numbers have real numbers as cube roots but do not have real numbers as square roots. For example,

$$(-8)^{1/3} = -2 \quad \text{but} \quad (-9)^{1/2} \quad \text{does not represent a real number.}$$

When we simplify an expression involving powers and roots we seek a form that can be used or approximated as easily as possible. This conventionally means that in the simplified form:

Operations have been performed whenever possible.
There are no negative exponents.
Roots have been taken wherever possible.
There are no radicals or fractional exponents in any denominator.

Example 1: Simplify $\frac{6}{5\sqrt{3}}$.

First Solution: $\sqrt{3} \approx 1.732050808\ldots \qquad 5\sqrt{3} \approx 8.660254040\ldots$

$$\frac{6}{5\sqrt{3}} \approx 0.692820323\ldots$$

This is certainly a very tedious procedure for anyone who does not have a calculator. Accordingly, we look for a procedure that does not involve division by an irrational number. In other words, we **rationalize the denominator**, that is, clear the denominator of radicals. This can be done in two ways.

Second Solution: We may multiply numerator and denominator by $\sqrt{3}$ so that the denominator will be an integer and thus *cleared of radicals.*

$$\frac{6}{5\sqrt{3}} = \frac{6}{5\sqrt{3}} \times \frac{\sqrt{3}}{\sqrt{3}} = \frac{6\sqrt{3}}{5 \times 3} = \frac{2}{5}\sqrt{3}$$

Alternatively, we may think of $6 = 2 \times 3 = 2 \times \sqrt{3} \times \sqrt{3}$. Then:

$$\frac{6}{5\sqrt{3}} = \frac{2\sqrt{3} \times \sqrt{3}}{5\sqrt{3}} = \frac{2}{5}\sqrt{3}$$

Example 2: Simplify: $3 \times 5^{-1/2}$.

Solution:

$$3 \times 5^{-1/2} = 3 \times \frac{1}{5^{1/2}} \qquad \text{or} \quad 3 \times \frac{1}{\sqrt{5}}$$

$$= 3 \times \frac{1}{5^{1/2}} \times \frac{5^{1/2}}{5^{1/2}} \quad \text{or} \quad 3 \times \frac{1}{\sqrt{5}} \times \frac{\sqrt{5}}{\sqrt{5}}$$

$$= 3 \times \frac{\sqrt{5}}{5} \qquad \text{or} \quad \frac{3\sqrt{5}}{5} \quad \text{or} \quad \frac{3}{5}\sqrt{5}$$

Example 3: Simplify: $\frac{16^{1/3}}{2^{5/3}}$.

Solution:

$$\frac{16^{1/3}}{2^{5/3}} = \left(\frac{16}{2^5}\right)^{1/3} = \left(\frac{1}{2}\right)^{1/3} = \frac{1}{2^{1/3}} = \frac{1}{2^{1/3}} \times \frac{2^{2/3}}{2^{2/3}} = \frac{2^{2/3}}{2} = \frac{1}{2}\sqrt[3]{4}$$

In Example 3 the solution could have been expressed entirely in terms of radicals or entirely in terms of fractional exponents. Our mixture of the two notations was unorthodox but purposeful for pedagogical reasons. The recognition of the equivalence of fractional exponents and radicals is one of our primary objectives. This equivalence enables us to use the laws of exponents in simplifying expressions that involve radicals.

Example 4: Simplify $\sqrt{\frac{6}{5}}$.

Solution:

$$\sqrt{\frac{6}{5}} = \left(\frac{6}{5}\right)^{1/2} = \frac{6^{1/2}}{5^{1/2}} = \frac{6^{1/2}}{5^{1/2}} \times \frac{5^{1/2}}{5^{1/2}} = \frac{(6 \times 5)^{1/2}}{5} = \frac{30^{1/2}}{5} = \frac{1}{5}\sqrt{30}$$

Example 5: Simplify: $\sqrt[3]{\frac{3}{4}}$.

First solution, in terms of fractional exponents:

$$\sqrt[3]{\frac{3}{4}} = \frac{3^{1/3}}{4^{1/3}} \frac{3^{1/3}}{(2^2)^{1/3}} = \frac{3^{1/3}}{2^{2/3}} \times \frac{2^{1/3}}{2^{1/3}} = \frac{3^{1/3} \times 2^{1/3}}{2^{3/3}} = \frac{6^{1/3}}{2} = \frac{1}{2}\sqrt[3]{6}$$

Second solution, in terms of radicals:

$$\sqrt[3]{\frac{3}{4}} = \frac{\sqrt[3]{3}}{\sqrt[3]{4}} = \frac{\sqrt[3]{3}}{\sqrt[3]{4}} \times \frac{\sqrt[3]{2}}{\sqrt[3]{2}} = \frac{\sqrt[3]{6}}{\sqrt[3]{8}} = \frac{\sqrt[3]{6}}{2} = \frac{1}{2}\sqrt[3]{6}$$

Third solution:

$$\sqrt[3]{\frac{3}{4}} = \frac{\sqrt[3]{3}}{\sqrt[3]{4}} \times \frac{\sqrt[3]{4^2}}{\sqrt[3]{4^2}} = \frac{\sqrt[3]{3 \times 16}}{4} = \frac{\sqrt[3]{6 \times 8}}{4} = \frac{2\sqrt[3]{6}}{4} = \frac{1}{2}\sqrt[3]{6}$$

The third solution of Example 5 illustrates the desirability of keeping the numbers as small as possible when working with roots, as was also the case in using the lowest common denominator in adding or subtracting with fractions.

Example 6: Assume that $\sqrt{2} \approx 1.41$ and find to the nearest hundredth $3 + \sqrt{\frac{25}{8}}$

Solution:

$$3 + \sqrt{\frac{25}{8}} = 3 + \frac{\sqrt{25}}{\sqrt{8}} = 3 + \frac{5}{\sqrt{8}} \times \frac{\sqrt{2}}{\sqrt{2}}$$

$$= 3 + \frac{5}{4}\sqrt{2} \approx 3 + 1.76 = 4.76$$

Note in Examples 7 and 8 that whenever the expression under the radical (the *radicand*) or the expression that is raised to a power can be simplified, this simplification should be done before there is any consideration of radicals or exponents.

Example 7: Simplify: $\sqrt{9 + 16}$.

Solution:

$$\sqrt{9 + 16} = \sqrt{25} = 5$$

Note that $\sqrt{9 + 16} \neq \sqrt{9} + \sqrt{16} = 3 + 4 = 7$.

Example 8: Simplify: $\left(\frac{88}{11}\right)^{2/3}$.

Solution:

$$\left(\frac{88}{11}\right)^{2/3} = (8^{1/3})^2 = 2^2 = 4$$

$$\text{or} \quad = (8^2)^{1/3} = 64^{1/3} = 4$$

EXERCISES

Express in terms of fractional exponents:

1. $\sqrt{11}$ **2.** $\sqrt[3]{6}$ **3.** $\sqrt[3]{5^2}$

4. $\sqrt{3 \times 5}$ **5.** $2 + \sqrt[3]{7}$ **6.** $\sqrt{1 + \sqrt{2}}$

Express in terms of radicals:

7. $19^{1/2}$ **8.** $21^{1/3}$ **9.** $7^{2/3}$

10. $1 + 5^{1/2}$ **11.** $6 - 3^{2/3}$ **12.** $(2 + 6^{1/2})^{1/3}$

Simplify:

13. $\sqrt{\frac{5}{9}}$ **14.** $\sqrt[3]{\frac{7}{8}}$ **15.** $\sqrt[3]{\frac{5}{9}}$

16. $\sqrt{\frac{21}{24}}$ **17.** $\sqrt[3]{\frac{-112}{14}}$ **18.** $\sqrt{\frac{7}{18}}$

19. $\sqrt[3]{\frac{-5}{7}}$ **20.** $\sqrt[3]{\frac{-24}{5}}$ **21.** $\sqrt{10 + 6}$

22. $\sqrt{81 + 144}$ **23.** $(25 + 144)^{1/2}$ **24.** $(100 - 36)^{2/3}$

Use $\sqrt{2} \approx 1.414$ *and* $\sqrt{3} \approx 1.732$ *and evaluate to the nearest hundredth:*

25. $5 - \sqrt{3}$ **26.** $6 - \sqrt{2}$ **27.** $(50)^{1/2}$

28. $(75)^{1/2}$ **29.** $2 + (8)^{1/2}$ **30.** $4 + \sqrt{12}$

31. $3 + \sqrt{\frac{1}{8}}$ **32.** $7 - \sqrt{\frac{1}{9}}$ **33.** $9 - \left(\frac{11}{22}\right)^{1/2}$

34. $6 + \left(\frac{7}{21}\right)^{1/2}$ ***35.** $2 + \sqrt{\frac{3}{8}}$ ***36.** $5 - \sqrt{6}$

PEDAGOGICAL EXPLORATIONS

1. Approximate values for square roots of positive numbers were found in §7-3 using the assumption that $a^2 < n < b^2$ for positive numbers a and b if and only if $a < \sqrt{n} < b$. Either from this assumption or from an examination of elementary school textbooks list the steps of a procedure (algorithm) for approximating $\sqrt{3}$ to as many decimal places as desired.
2. State an assumption that could be used as the basis for finding approximate values for cube roots of any given positive number.

3. Use your assumption from Exploration 2 and list the steps of a procedure for approximating values of cube roots of positive numbers to as many decimal places as desired.
4. Use your procedure from Exploration 3 to find to the nearest tenth:
   (a) $\sqrt[3]{13}$ (b) $\sqrt[3]{100}$
5. Find to the nearest tenth:
   (a) $\sqrt[4]{2}$ (b) $\sqrt[4]{90}$

CHAPTER TEST

1. Graph each set on a number line.
   (a) The set of integers between −2 and 3.
   (b) The set of real numbers greater than or equal to −1.

2. Use the given figure and give the coordinates of each of these points:
   (a) A (b) B (c) C (d) D (e) E

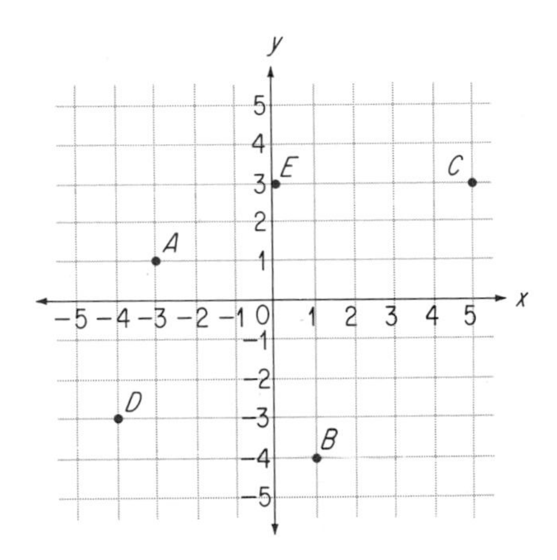

Classify each statement as true or false.

3. (a) Every irrational number is not an integer.
   (b) There is a one-to-one correspondence between the set of rational numbers and the set of points on a number line.

4. (a) The set of irrational numbers is a subset of the set of real numbers.
   (b) Every irrational number is the coordinate of some point on a number line.

Copy the following table. Use "$\surd$" to show that the number at the top of the column is a member of the set named at the side. Use "$\times$" if the number is not a member of the set.

	5.	6.	7.	8.	9.	10.
Set	$-\frac{2}{3}$	0	$\sqrt{25}$	$\sqrt{75}$	$\sqrt{3/4}$	$\sqrt{-9}$
Integers						
Rational numbers						
Irrational numbers						
Real numbers						

11. Construct a number line and locate on it the point with coordinate $3 + \sqrt{2}$.

12. Prove that $3 + \sqrt{2}$ is an irrational number.

13. Write the first fifteen digits to the right of the decimal point in the numeral 0.232232223. . . .

14. In the numeral in Exercise 13, what is:
(**a**) The 20th digit to the right of the decimal point?
(**b**) The 30th digit to the right of the decimal point?

15. List the given numbers in order from smallest to largest:

$$2.\overline{56}, \quad 2.5\overline{6}, \quad 2.56, \quad 2.566, \quad 2.565.$$

16. Name two irrational numbers between $0.\overline{56}$ and $0.5\overline{6}$.

Find for n all possible replacements that will make each of the following sentences true, where n must be a member of the set named at the top of each column. When there is no such replacement, write "none."

	Sentence	(a) Integers	(b) Rational numbers	(c) Irrational numbers	(d) Real numbers
17.	$n - 1 = 0$				
18.	$n + 3 = 3$				
19.	$5n = 8$				
20.	$n^2 = 3$				
21.	$n^2 + 9 = 0$				

Simplify:

22. $\sqrt{\frac{3}{4}}$

23. $\sqrt[3]{\frac{3}{4}}$

24. $(36 + 64)^{1/2}$

25. $(1 + 4^{3/2})^{1/2}$

READINGS AND PROJECTS

1. Read "A Chronology of π" on pages 89 through 95 of Howard Eves' *An Introduction to the History of Mathematics,* third edition, Holt, Rinehart & Winston, 1969. More recent evaluations of π are mentioned on page 698 of the November 1973 issue of *School Science and Mathematics* and include 16,167 places in 1959; 100,265 places in 1961; 250,000 places in 1965; and 500,000 places in 1966. Prepare a report on the irrational number π for use with an elementary school class. Include an explanation of why π still is sometimes referred to in Germany as "the Ludolphine number."
2. Read and try some of the suggested projects in "Making $\sqrt{2}$ Seem 'Real'" by Judith E. Jacobs and Elizabeth B. Herbert, pages 133 through 136 of the February 1974 issue of *The Arithmetic Teacher.*
3. For a proof that for any positive integer N the number $\sqrt[n]{N}$ is either a positive integer or an irrational number see Toshio Shibata's letter on page 119 of the February 1974 issue of *The Mathematics Teacher.*
4. Read "Pythagorean Puzzles" by Raymond Spaulding in the activities section, pages 143 through 146, of the February 1974 issue of *The Mathematics Teacher.*
5. Read "Teaching the Square Root Algorithm by the Discovery Approach" by Edward C. Tarte on pages 317 through 319 of the April 1974 issue of *The Mathematics Teacher.*
6. Read "Periodic Decimals" by John T. Anderson on pages 504 through 509 of the October 1974 issue of *The Mathematics Teacher.*
7. Read Booklet No. 11 (The System of Real Numbers) of the thirtieth yearbook of the National Council of Teachers of Mathematics. Complete at least three of the exercise sets given there.
8. Read Chapter 3 of *Mathematics and the Imagination* by Edward Kasner and James Newman, Simon and Schuster, Inc., 1940.
9. For use with an elementary class, prepare a short presentation on irrational numbers and on the beliefs and mathematical work of the Pythagoreans.
10. Read a major portion of and do at least twenty of the exercises in *Taxicab Geometry* by Eugene F. Krause, Addison-Wesley, 1975.

8

The Metric System and Measurement

Most people expect mathematics to serve two basic needs: to compute and to measure. In this chapter we pay special attention to this second need as we explore measurement in detail.

Many of the units of measurement that we find in use today are very inconvenient to use. Often some of these inconvenient units of measure are nearly obsolete. For example, in one state the right of way for town roads must be at least three rods wide. This requirement makes use of a unit of measure that is seldom used, consequently seldom understood by the average citizen.

Today most of the civilized world makes use of the metric system of measurement, and there is a great deal of pressure to have the United States officially adopt this system. We already make use of a number of its units of measure in our daily lives. For example:

Many people own or at least have heard of a 35 *millimeter* camera.
In some states road signs show distances in both miles and *kilometers*.
The contents of vitamins are usually given in terms of multiples of a *gram*.

Almost all of the scientific laboratories in this country use the metric system and have done so for many years. It seems virtually certain that congressional action will force the entire country to adopt the metric system at some future date. In the interim, a public information campaign is under way to acquaint the general public with the various units of measure involved. More and more elementary schools are incorporating units of study on the metric system into their curricula, and future teachers of elementary mathematics need to be fairly well acquainted with the details of the metric system if they are to be able to teach this to their students with competence. One of the major objectives of this chapter is to have the reader gain facility in working with various metric units of measure for length, weight, and capacity.

8-1 a short history of measurement

Throughout the ages man has had a need to measure. As the Nile River overflowed its banks each year, the ancient Egyptians needed to survey the flooded area to determine boundary marks, and they are believed to have used ropes stretched to form right angles in doing so. Because of the process used, these surveyors were actually referred to as "rope-stretchers." From the Pythagorean theorem we know that if a rope is knotted so as to form 3, 4, and 5 units respectively, a right triangle and therefore a right angle can be formed, as shown in the figure.

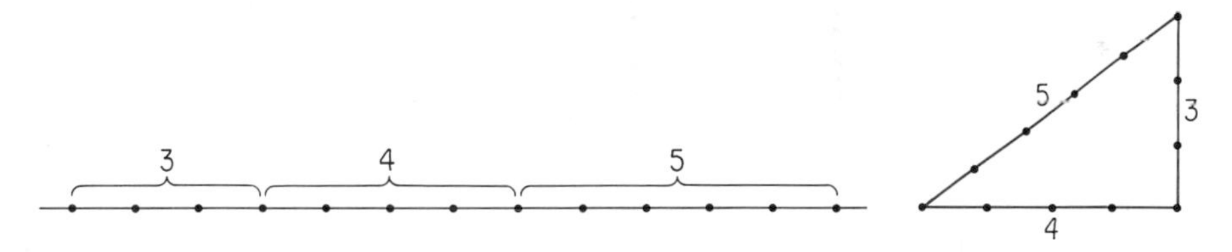

Early man used various parts of his own body as units of measure. Records of early Egyptian and Babylonian civilizations indicate that lengths were first measured by such units as a **palm** and a **span**, the distance covered by an outstretched hand.

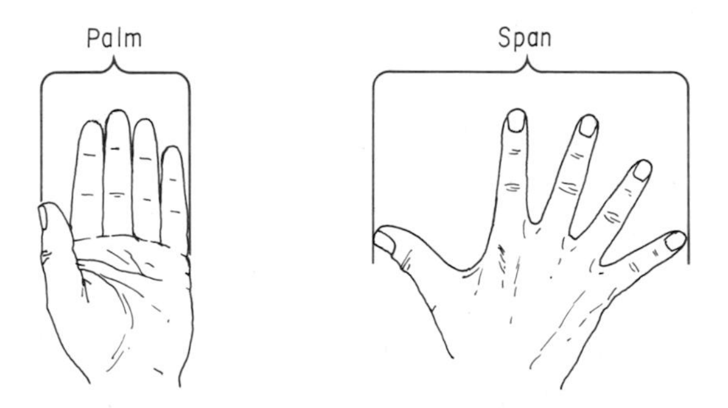

Another early unit of measurement widely used was a **cubit**, the distance from one's elbow to the tip of the middle finger when the hand is held straight. This is shown in the figure at the top of the next page. For most people, this is a distance of approximately 18 inches.

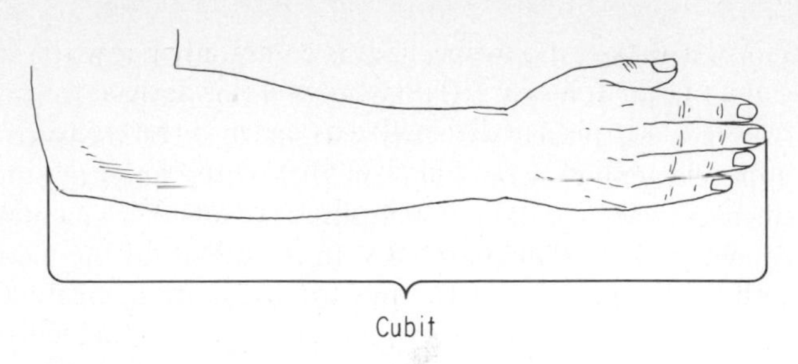

Cubit

Mention of the cubit has been found in excavations that date back to 3000 B.C., and it is a unit of measure frequently mentioned in the Bible. Thus in the sixth chapter of Genesis we find a description of Noah's ark that states "the length of the ark shall be 300 cubits, the breadth of it 50 cubits, and the height of it 30 cubits."

Interesting stories about units of measure have grown through the ages. Thus it is said that King Henry I of England decreed that the distance from the tip of his nose to the end of his thumb should be the official equivalent of one yard. One hundred years later, in the thirteenth century, King Edward I declared that one-third of a yard would be called the foot. Much later, in the sixteenth century, Queen Elizabeth I declared that the traditional *Roman mile* of 5000 feet would be replaced by one of 5280 feet. However it was the Romans who used 12 as a base for measurement, and our use of 12 inches to a foot is derived from them.

Although the use of a unit of measure such as the cubit was more meaningful than such units as "a bundle" or "a day's journey," it certainly was not a standardized unit and obviously varied from person to person. There was a great need throughout the world for the establishment of a uniform standard for all weights and measures.

About 300 years ago Gabriel Mouton, Vicar of St. Paul in Lyons, France, recognized the need for a standardized system of measurement that would be accepted throughout the world. In 1670 he proposed a decimal system of measurement that was based on a standard unit equal to the length of one minute of arc of a great circle of the earth. In 1790, in the midst of the French Revolution, the National Assembly of France asked the French Academy of Sciences to create a uniform standard for all weights and measures.

In establishing a uniform standard there are a number of fundamental principles that need to be kept in mind:

1. The standard unit used should be based on some invariant factor in the physical universe.
2. Basic units of length, capacity (volume), and weight (mass), should be interrelated.
3. The multiples and subdivisions of the basic unit should be in terms of the decimal system, that is, in terms of powers of 10.

With this in mind, the Academy proposed that the basic unit of length be one ten-millionth of an arc drawn from the North Pole to the equator.

They called this basic unit the **metre**, as it is commonly known in most parts of the world and in scientific work. The word metre is derived from the Greek *metron*, "a measure," and is usually spelled as **meter** in the United States. One meter is a little longer than a yard in size. One-tenth of a meter is one **decimeter**; one-tenth of a decimeter is one **centimeter (cm)**. We can also think of these units of measure in terms of groups of 10 in this way:

$$1 \text{ meter} = 10 \text{ decimeters}$$

$$1 \text{ decimeter} = 10 \text{ centimeters}$$

We conclude that 1 meter = 100 centimeters, or that 1 centimeter is one-hundredth of a meter.

The name **gram** was assigned to the basic metric unit of mass (weight). The gram was defined as the weight of one **cubic centimeter** of water. A cubic centimeter is a cube with each edge of length one centimeter.

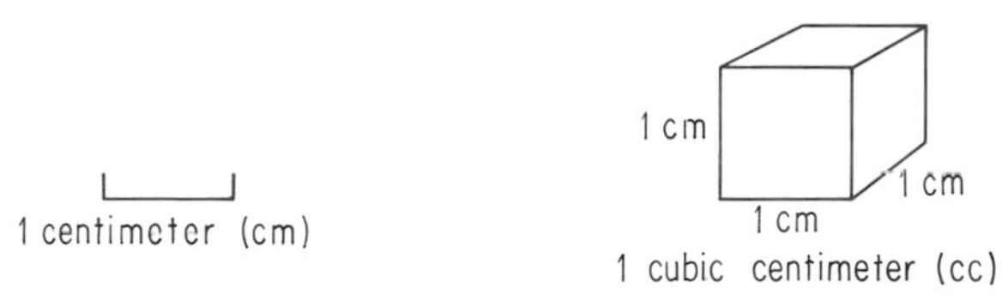

As a basic measure of capacity, the Academy chose the **liter** and defined this as a **cubic decimeter**, that is, the volume of a cube with each edge of length one decimeter.

In 1840 France made it compulsory to use the metric system, and soon thereafter many other nations followed suit. In 1875 the Metric Convention met and 17 nations (including the United States) signed the "Treaty of the Meter" which established permanent metric standards for length and weight. By 1900 most of the nations of Europe and South America, a total of 35 in all, had officially adopted the metric system of measurement. There is now an international General Conference on Weights and Measures to which the United States belongs.

About 1890 the International Bureau of Weights and Measures defined the meter as the distance between two marks on a platinum bar that was kept at a constant temperature of 32° Fahrenheit. This distance was equal to one ten-millionth of the distance from the North Pole to the equator. However, in 1960 the Eleventh General Conference on Weights and Measures abandoned the meter bar as the international standard of length, and redefined the meter in terms of 1 650 763.73 wavelengths of the orange-red line in the spectrum of krypton-86. With this definition lengths of one meter could be reproduced in scientific laboratories anywhere in the world. The 1960 Conference adopted a revision and simplification of the metric system, *Le Système International d'Unités* (International System of Units), now known as SI, which will be described in detail in the following section.

The SI system is really quite simple. First a basic unit, such as the meter, gram, or liter is adopted. Thereafter, multiples and subdivisions of this unit are always given in terms of powers of 10. Greek prefixes are used to denote multiples, and Latin prefixes to denote subdivisions. Some of the basic prefixes used are shown in this table:

tera-	one trillion	1 000 000 000 000	10^{12}
giga-	one billion	1 000 000 000	10^{9}
mega-	one-million	1 000 000	10^{6}
kilo-	one thousand	1 000	10^{3}
hecto-	one hundred	100	10^{2}
deka-	ten	10	10^{1}
BASE UNIT	one	1	10^{0}
deci-	one-tenth	0.1	10^{-1}
centi-	one-hundredth	0.01	10^{-2}
milli-	one-thousandth	0.001	10^{-3}
micro-	one-millionth	0.000 001	10^{-6}
nano-	one-billionth	0.000 000 001	10^{-9}
pico-	one-trillionth	0.000 000 000 001	10^{-12}

There are several observations that should be made concerning the table. Note the increasing (or decreasing) powers of 10 as one reads up (or down) the table. In the SI system spaces are used instead of commas to separate groups of three digits starting from the decimal point in either or both directions. The reason for this is the use, in certain countries, of commas to represent decimal points.

Although these units will be discussed in detail in the following section, let us briefly explore the use of these prefixes with the basic unit of one meter. Using the table, you should be able to see such relationships as the following:

1 *kilo*meter = 1000 meters (1 meter = 0.001 kilometer)

1 *centi*meter = 0.01 meter (1 meter = 100 centimeters)

In the metric system any measurement in terms of one unit may be expressed in terms of a larger unit by moving the decimal point an appropriate number of places to the left, as in Example 1.

Example 1: (a) Change 358 centimeters to meters.
(b) Change 7495 meters to kilometers.

Solution: (a) 100 centimeters = 1 meter
358 centimeters = 3.58 meters
(b) 1000 meters = 1 kilometer
7495 meters = 7.495 kilometers

Any measurement in terms of one unit may be expressed in terms of a smaller unit by moving the decimal point an appropriate number of places to the right, as in Example 2.

Example 2: (a) Change 8.35 meters to centimeters.
(b) Change 15.755 kilometers to meters.

Solution: (a) 1 meter = 100 centimeters
8.35 meters = 835 centimeters
(b) 1 kilometer = 1000 meters
15.755 kilometers = 15 755 meters

EXERCISES

Classify each statement as true or false.

1. 100 meters = 1 kilometer
2. 1 meter = 100 centimeters
3. 1 centimeter = 100 millimeters
4. 10 millimeters = 1 centimeter
5. 1 decimeter = 10 centimeters
6. 1 meter = 1000 millimeters
7. 1 millimeter = 0.01 centimeter
8. 10 meters = 1 decimeter
9. 1 centimeter = 0.01 meter
10. 1 centimeter = 0.01 decimeter

Complete each statement.

11. 5 meters = _____ centimeters
12. 3 kilometers = _____ meters
13. 4 centimeters = _____ millimeters
14. 475 centimeters = _____ meters
15. 8350 meters = _____ kilometers
16. 75 millimeters = _____ centimeters

Use decimal notation to change as indicated.

17. 358 centimeters to meters
18. 482 millimeters to centimeters
19. 3785 meters to kilometers
20. 7500 millimeters to meters
21. 85 kilometers to meters
22. 9.82 meters to centimeters
23. 5.8 centimeters to millimeters
24. 3.5 meters to millimeters

PEDAGOGICAL EXPLORATIONS

1. Collect as many illustrations of popular uses of measurements that are not standardized as you can find, such as "a pinch of salt."
2. Estimate the length of your car in terms of cubits. Then use your arm to find its length using this measure.
3. Repeat Exploration 2 for the distance across the front of your classroom.
4. Inasmuch as it takes 100 pennies to equal one dollar, we may write such equivalents as 249¢ = \$2.49 and \$3.75 = 375¢. Discuss how this idea may be used with an elementary class to help illustrate translations from a number of centimeters to a number of meters, as well as the reverse.
5. Explore a recently published set of elementary school mathematics textbooks, or a series of workbooks devoted to metric measures, and report on the manner in which metric measures are introduced:
   (a) At the primary grade level.
   (b) At the intermediate grade level.
   Discuss the use of, or absence of, decimal notation at each of these levels.

8-2 the metric system

The metric system is based upon powers of 10. Since our system of numeration is also based upon powers of 10, computations using metric units are relatively simple. For practical everyday use, there are surprisingly few units of measure to learn. The basic unit for measuring lengths is the meter (m). A meter is approximately 39 inches in length. The following *scale drawing* shows the comparison between one meter and one yard.

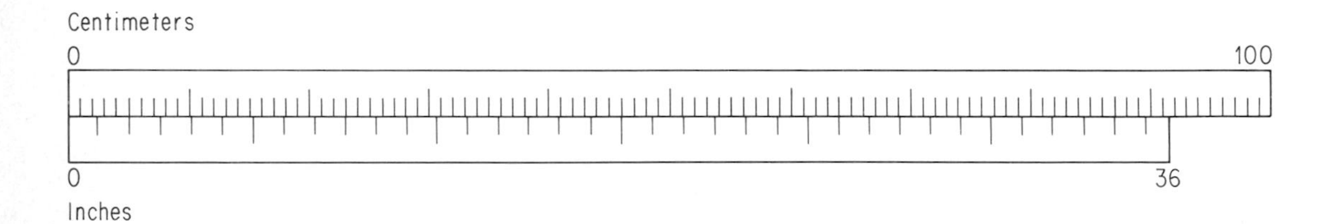

Normally the yard is divided into 36 parts, each one inch in length. The meter, as we have seen, is divided into 100 parts, each one centimeter (cm), in length. It is well to have an intuitive feeling for the size of one centimeter. The centimeter is about 0.4 of an inch. A nickel is about 2 cm in diameter (wide). By a relatively recent definition of one inch:

$$1 \text{ inch} = 2.54 \text{ centimeters}$$

Then

$$1 \text{ yard} = 0.9144 \text{ meters}$$

We often use the approximation

$$1 \text{ inch} \approx 2\tfrac{1}{2} \text{ centimeters}$$

The figures that follow show the *actual size* of one centimeter, and compare this to one inch.

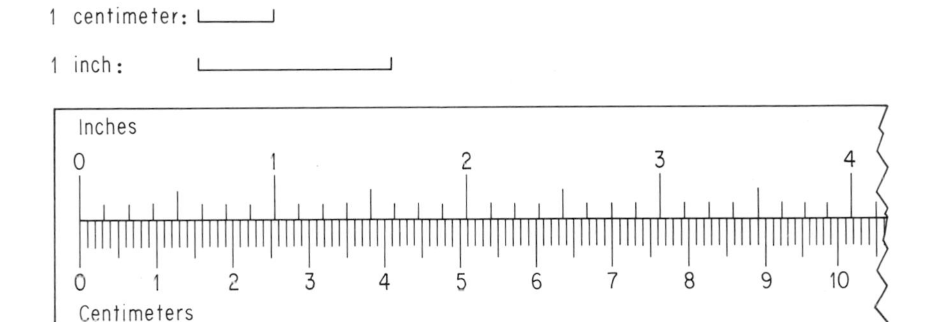

As in the previous figures each centimeter is divided into ten parts, each one **millimeter (mm)** in length.

$$1 \text{ centimeter} = 10 \text{ millimeters} \qquad (1 \text{ cm} = 10 \text{ mm})$$

$$1 \text{ millimeter} = 0.1 \text{ centimeter} \qquad (1 \text{ mm} = 0.1 \text{ cm})$$

Also as in the previous figures 10 cm $\approx$ 4 inches. To help you visualize the size of a millimeter note that a nickel is about 2 mm thick; 35 mm film is 35 mm in width. Note that the symbol cm has no period. The same convention (no periods) is followed for all symbols of metric units.

For measuring greater distances we use the kilometer (km) which is equal to 1000 meters. A kilometer is approximately equal to 0.6 of a mile, so that 10 kilometers is equal to about 6 miles. The relationship between the kilometer and the mile is shown by this scale drawing.

1 kilometer

1 mile

For normal, everyday usage in a country that has adopted the metric system, the average citizen uses the centimeter for small measurements, the meter for somewhat larger ones, and the kilometer for greater distances. Thus, in countries that have adopted the metric system, distances between cities are given in terms of kilometers. However, the meter is the basic unit of length. Several common multiples of this unit are shown in the following table. Note that the prefixes used for multiples of a meter are those for the basic units in the metric (SI) system as shown in §8-1.

1 *kilo*meter	=	1000 meters
1 *hecto*meter	=	100 meters
1 *deka*meter	=	10 meters
1 meter	=	1 meter
1 *deci*meter	=	0.1 meter
1 *centi*meter	=	0.01 meter
1 *milli*meter	=	0.001 meter

Example 1: The distance between New York City and San Francisco is 4120 kilometers. About how many miles is this?

Solution: Recall that 1 kilometer is approximately 0.6 of a mile. Since $0.6 \times 4120 = 2472$, the distance is about 2472 miles. Actually, a motorist would probably want a rough approximation only and would therefore round off the distance to 4000 kilometers. In miles, this would be approximately equal to $0.6 \times 4000 = 2400$ miles.

Example 2: Normally, metric rulers are marked in centimeters and subdivided into millimeters. Read each indicated point on this metric ruler in millimeters and in centimeters.

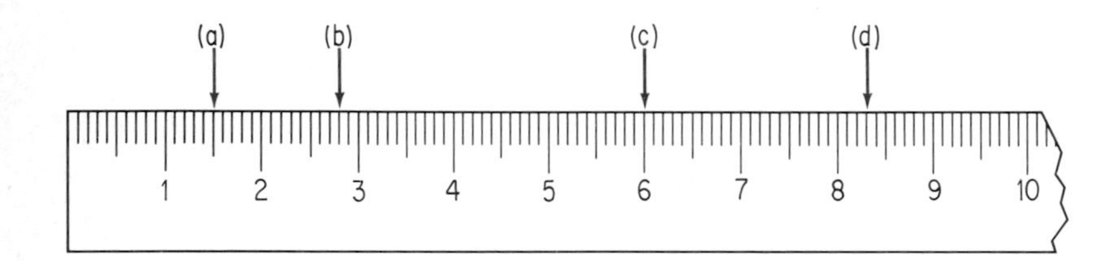

Solution: Recall that 1 centimeter = 10 millimeters.
(a) 15 mm = 1.5 cm; (b) 28 mm = 2.8 cm;
(c) 60 mm = 6.0 cm; (d) 83 mm = 8.3 cm.

Example 3: Complete: (a) 8 cm = ____ mm; (b) 35 mm = ____ cm; (c) 3 km = ____ m; (d) 2500 m = ____ km.

Solution: (a) Each centimeter is equal to 10 millimeters; 8 cm = 80 mm.
(b) 10 millimeters = 1 centimeter; 35 mm = 3.5 cm.
(c) Each kilometer is equal to 1000 meters; 3 km = 3000 m.
(d) 1000 meters = 1 kilometer; 2500 m = 2.5 km.

To measure volume, the basic unit used is the *liter*. One liter is just a little greater than a quart, and is the metric measure of capacity that one is most likely to encounter in daily activities. However, various multiples of one liter are obtained by using the set of prefixes previously listed. Again, note the use of decimal notation, with each unit being 10 times as great as the one listed beneath it in this table.

1 *kilo*liter	=	1000 liters
1 *hecto*liter	=	100 liters
1 *deka*liter	=	10 liters
1 liter	=	1 liter
1 *deci*liter	=	0.1 liter
1 *centi*liter	=	0.01 liter
1 *milli*liter	=	0.001 liter

One might encounter the **milliliter (ml)** in small measurements. For example, in cooking and baking, it would be important to know that 5 milliliters is approximately equivalent to 1 teaspoonful.

The basic unit of metric measure for *mass* is the *gram* (g). The *weight* of an object is determined by its mass and the force of gravity to which it is subjected. For example, an astronaut's body has the same mass on earth as when he is in a weightless condition orbiting the earth. However, most of us stay relatively close to the surface of the earth. Therefore we are concerned with objects that are subjected to an approximately constant force of gravity. Thus we shall use the term *weight* instead of using the technical term and concept of *mass*. Then the basic unit for weight is the *gram*, a nickel weighs about 5 grams. As before, other multiples of a gram are obtained as in the following table.

1 *kilo*gram	=	1000 grams
1 *hecto*gram	=	100 grams
1 *deka*gram	=	10 grams
1 gram	=	1 gram
1 *deci*gram	=	0.1 gram
1 *centi*gram	=	0.01 gram
1 *milli*gram	=	0.001 grams

Of these various multiples of the gram, the one that is most likely to be encountered when the United States finally "goes metric" is the **kilogram (kg)**. A kilogram is a little over 2 pounds, about 2.2 pounds, and is generally referred to as a **kilo**. Thus a shopper ordering 1 kilo of apples can expect to receive a little more than 2 pounds of apples; by ordering one-half of a kilo, one would receive about 1 pound.

The **milligram (mg)** is a frequently used unit of measure for various medicines and vitamins. A typical multipurpose vitamin pill might contain

250 mg of vitamin C. Inasmuch as 1 milligram is equal to one-thousandth of a gram, 250 mg is equal to one-fourth of a gram in weight.

Example 4: An athlete weighs 100 kilograms. About how many pounds is this?

Solution: One kilogram is approximately equal to 2.2 pounds. Therefore, 100 kilograms $\approx 2.2 \times 100 = 220$ pounds.

Example 5: Complete: (a) 3kg = ____ g; (b) 2500 g = ____ kg; (c) 5 g = ____ mg; (d) 7000 mg = ____ g.

Solution: (a) Each kilogram is equal to 1000 grams; 3 kg = 3000 g.
(b) 1000 grams = 1 kilogram; 2500 g = 2.5 kg.
(c) Each gram is equal to 1000 milligrams; 5 g = 5000 mg.
(d) 1000 milligrams = 1 gram; 7000 mg = 7g.

In the metric system temperatures are measured on a scale that runs from 0 degrees, where water freezes, to 100 degrees, where water boils. Measurement is in terms of degrees *Celsius* (centigrade). As of January 1, 1973 the American Society for Testing and Materials has adopted the notation °**C** for degrees **Celsius** and °**F** for degrees **Fahrenheit** as in SI notation. Thus a normal body temperature of 98.6 °F is about 37 °C. The scale that follows shows the relationship between our commonly used Fahrenheit scale, and the Celsius scale.

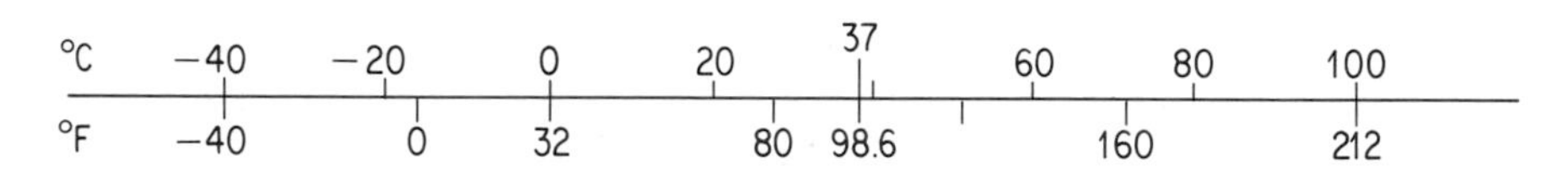

There is a formula that relates temperatures in degrees Fahrenheit (F) to degrees Celsius (C):

$$F = \tfrac{9}{5}C + 32$$

Example 6: Use the formula and convert 30 °C to the Fahrenheit scale.

Solution:

$$\begin{aligned} F &= \tfrac{9}{5}C + 32 \\ &= \tfrac{9}{5}(30) + 32 \\ &= 54 + 32 \\ &= 86 \end{aligned}$$

Therefore, 30 °C = 86 °F.

EXERCISES

Each of the following sentences uses a metric measure. State whether each seems likely, or unlikely.

1. A college football player weighs 200 kilograms.
2. A college basketball player is 2 meters tall.
3. An empty tank in a compact automobile will hold 40 liters of gasoline.
4. A quarter weighs 25 grams.
5. The average student drinks 5 liters of water daily.
6. The diameter of a nickel is about 5 centimeters.
7. A new pencil is about 20 centimeters long.
8. The distance between New York City and Boston is about 300 kilometers.
9. A typical man's wristwatch has a diameter of about 25 millimeters.
10. It is likely to snow at a temperature of 20 degrees Celsius.

Select the most likely answer for each situation.

11. The weight of a quarter is approximately:
    (a) 1 gram **(b)** 5 grams **(c)** 15 grams **(d)** 30 grams
12. The weight of this book is approximately:
    (a) 10 grams **(b)** 50 grams **(c)** 1 kilogram **(d)** 10 kilograms
13. Ten gallons of gasoline is approximately equal to:
    (a) 20 liters **(b)** 30 liters **(c)** 40 liters **(d)** 80 liters
14. New York City is in the midst of a heat wave in July. The temperature is likely to be:
    (a) 20 °C **(b)** 35 °C **(c)** 60 °C **(d)** 80 °C
15. Some doctors recommend that a person should spend an hour per day walking. In this time, one would probably walk:
    (a) 1 kilometer **(b)** 5 kilometers
    (c) 10 kilometers **(d)** 15 kilometers

Change each of the following as indicated.

16. 5 meters to centimeters
17. 8 kilometers to meters
18. 3 centimeters to millimeters
19. 80 millimeters to centimeters
20. 350 centimeters to meters
21. 15 000 meters to kilometers
22. 3000 grams to kilograms
23. 7 kilograms to grams
24. 7500 milligrams to grams
25. 2500 milliliters to liters

Read each indicated point on this metric scale in (a) millimeters and also in (b) centimeters.

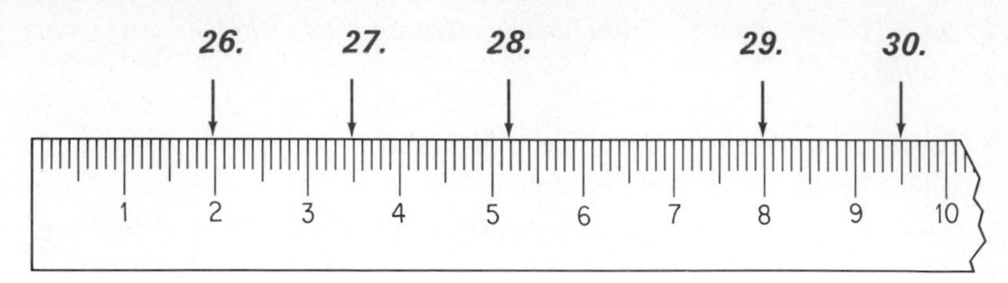

In Exercises 31 through 35 change each temperature in degrees Celsius to its equivalent in degrees Fahrenheit.

31. 0 °C ***32.*** 20 °C ***33.*** 25 °C ***34.*** 40 °C ***35.*** 100 °C

36. A weight of 150 kilograms is about how many pounds? (Use 1 kilogram as 2.2 pounds.)

37. A length of 5 inches is approximately how many centimeters? (Use 1 inch as 2.5 centimeters.)

38. Note that one inch is approximately equal to $2\frac{1}{2}$ centimeters, and complete these statements:
(a) 1 cm ≈ ____ inches **(b)** 40 cm ≈ ____ inches

39. A distance of 400 kilometers is about how many miles? (Use 1 kilometer ≈ 0.6 miles.)

40. Note that one kilometer is approximately equal to $\frac{3}{5}$ of a mile, and complete these statements:
(a) 1 mile ≈ ____ km **(b)** 10 miles ≈ ____ km

41. The prefix **mega** represents 10^6, that is, 1 000 000. Thus 1 megameter = 1 000 000 meters. Complete each statement:
(a) 1 megaliter = ____ liters **(b)** 1 megagram = ____ grams
(c) 1 megameter = ____ kilometers
(d) 5000 kilometers = ____ megameters

42. The prefix **micro** represents 10^{-6}, that is, 0.000 001. Complete each statement:
(a) 1 micrometer = ____ meter **(b)** 1 meter = ____ micrometers
(c) 5 grams = ____ micrograms
(d) 3 000 000 micrograms = ____ grams

PEDAGOGICAL EXPLORATIONS

1. Prepare a scale drawing showing the approximate relationship between one pound and one kilogram.

2. Collect as many popular sayings as you can find that involve units of measure. For example, here are two such sayings:
   (a) "An ounce of prevention is worth a pound of cure."
   (b) "Give him an inch and he'll take a foot."
   Then translate each into metric units of measure.
3. Begin a collection of objects that show metric units of measure. For example, some small packets of sugar show weight in grams and some bottles used for soft drinks show capacity in terms of liters.
4. There has been some controversy as to whether our schools should use "metre" or "meter" in introducing metric units of measure. See if you can find any information on this controversy.
5. Prepare a multiple choice test of ten items that endeavor to evaluate familiarity with metric measures. Use Exercises 11 through 15 of this section as a guide.
6. Prepare a bulletin board display that serves to generate interest in the metric system.
7. Use a metric ruler and find the dimensions of a one-dollar bill to the nearest millimeter.
8. Use a metric ruler and find the diameter of a quarter to the nearest millimeter.
9. Mark off, as carefully as possible, a 10 centimeter scale on a strip of paper about 2 cm wide. Then make nine more such scales and fasten them together to form a strip one meter long and subdivided into centimeters.
10. Bring three objects to class that are approximately 10 centimeters each in length.
11. Construct an open box with each edge 10 centimeters long. Then the volume of the box will be one cubic decimeter, which is, by definition, one liter.
12. Obtain a metric ruler and find the measure of the width of your palm, your hand span, and the distance from the tip of your elbow to the end of your outstretched middle finger, all to the nearest centimeter. Compare these measurements with those of other members of your class.

8-3 linear measures

Distances and lengths of objects are often called **linear measurements** since the shortest distance from one point to another is the length of the line segment joining them. Whenever distances or lengths are stated, some unit of measure needs to be given or assumed. For example, the length of a certain walking stick might be described by various people as

a little more than a yard	45 inches
about 4 feet	$1\frac{1}{4}$ yards
about 1 meter	5 spans
1.15 meters	

Any line may be considered as a number line relative to any given unit of measure. Then each point P_x on the line has a unique real number x as its *coordinate*; each real number x has a unique point P_x as its *graph*. Since each of the points of a number line may be identified by *one* real number, a line is often called a **one-space.**

Sets of points on a line form **line figures.** For example, any two points P_r and P_s on a number line are endpoints of a unique line segment $\overline{P_rP_s}$ with the absolute value of $s - r$ as its **linear measure.** We denote the linear measure of $\overline{P_rP_s}$ by $m(\overline{P_rP_s})$ (read as "measure of $\overline{P_rP_s}$") and write

$$m(\overline{P_rP_s}) = |s - r|$$

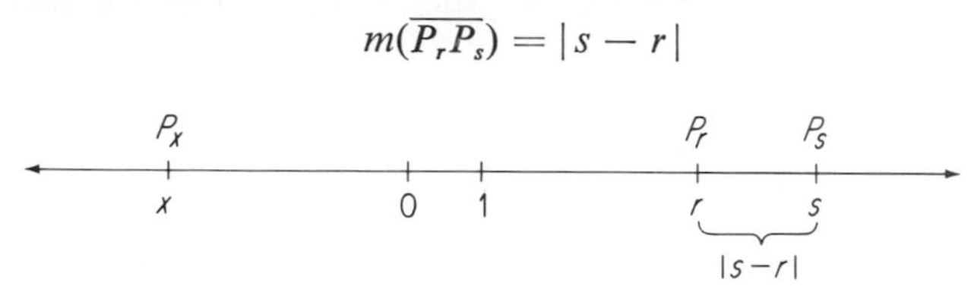

Example 1: Find the linear measure of (a) $\overline{P_4P_1}$; (b) $\overline{P_{-2}P_3}$.

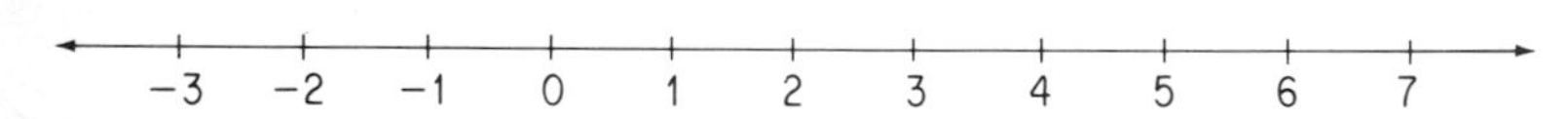

Solution: (a) $m(\overline{P_4P_1}) = |1 - 4| = |-3| = 3$
(b) $m(\overline{P_{-2}P_3}) = |3 - (-2)| = |5| = 5$

Any two line segments with equal measures are called **congruent line segments.** Since any line segment has a unique linear measure, any line segment is congruent to itself; for example, $\overline{AB} \cong \overline{BA}$ (read as "line segment AB is congruent to line segment BA").

Measures of line segments on a number line have been used in the discussion of addition of real numbers. In general, any line segment AB has a nonnegative real number $m(\overline{AB})$ as its linear measure. The **length** of the line segment is $m(\overline{AB})$ units for some unit of linear measure such as foot, inch, meter, mile, or kilometer. For example, a pencil with a linear measure of 8 in centimeters has length 8 centimeters.

A point has linear measure 0. In the figure $m(\overline{AB}) = 3$, $m(\overline{CD}) = 4$, $\overline{AB} \cap \overline{CD} = \overline{CB}$, $m(\overline{CB}) = 1$, $\overline{AB} \cup \overline{CD} = \overline{AD}$, $m(\overline{AD}) = 6$. Notice that $m(\overline{AB}) + m(\overline{CD}) \neq m(\overline{AD})$, since $m(\overline{AB} \cap \overline{CD}) \neq 0$.

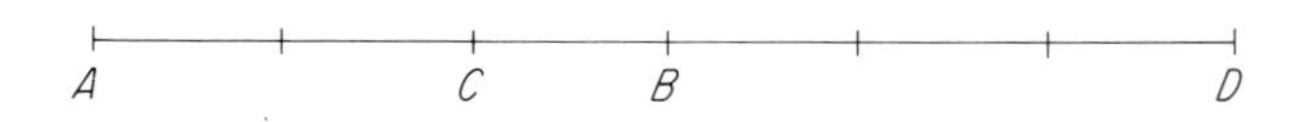

$$m(\overline{AD}) = m(\overline{AB} \cup \overline{CD}) = m(\overline{AB}) + m(\overline{CD}) - m(\overline{AB} \cap \overline{CD})$$

In general, if the measure of the intersection of two line segments is 0, then the linear measure of the union of the two line segments is equal to the sum of their linear measures.

Example 2: Show the relationship among the linear measures of $\overline{P_1P_2}$, $\overline{P_2P_5}$, and $\overline{P_1P_5}$.

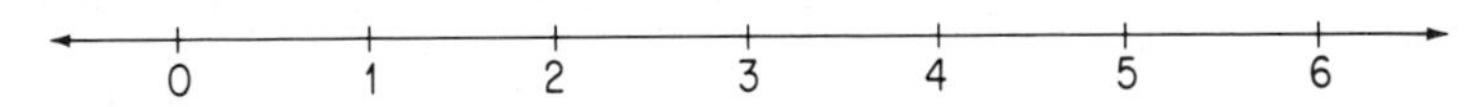

Solution: $m(\overline{P_1P_2}) = 1$, $m(\overline{P_2P_5}) = 3$
$m(\overline{P_1P_2} \cap \overline{P_2P_5}) = m(\overline{P_2P_2}) = 0$
$m(\overline{P_1P_5}) = 4 = m(\overline{P_1P_2}) + m(\overline{P_2P_5})$

Example 3: Show the relationship among the linear measures of $\overline{P_{-2}P_3}$, $\overline{P_2P_5}$, and $\overline{P_{-2}P_5}$.

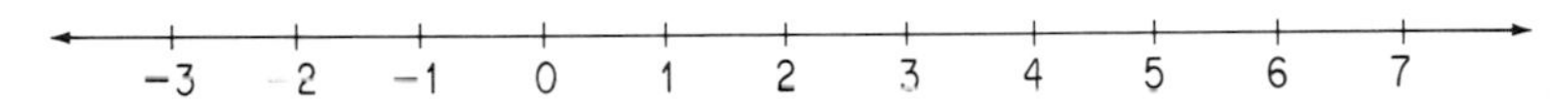

Solution: $m(\overline{P_{-2}P_3}) = 5$, $m(\overline{P_2P_5}) = 3$, $m(\overline{P_{-2}P_5}) = 7$
$m(\overline{P_{-2}P_3} \cap \overline{P_2P_5}) = m(\overline{P_2P_3}) = 1$
$m(\overline{P_{-2}P_5}) - m(\overline{P_{-2}P_3}) + m(\overline{P_2P_5}) - m(\overline{P_{-2}P_3} \cap \overline{P_2P_5})$

The *linear measure* of a figure is a number n; the *length* of the figure is n units. The length of a simple closed curve, such as a polygon, is its **perimeter.** Since any two sides of a polygon have at most one point in common, and a point has length 0, the perimeter of the polygon is the sum of the lengths of its sides.

Example 4: Find the perimeter of the given polygon.

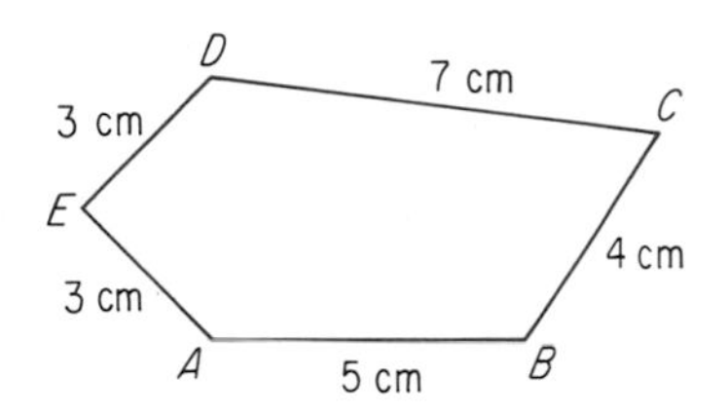

Solution: The perimeter in centimeters of polygon *ABCDE* is
$m(\overline{AB}) + m(\overline{BC}) + m(\overline{CD}) + m(\overline{DE}) + m(\overline{AE}) = 5 + 4 + 7 + 3 + 3 = 22$
Thus the perimeter is 22 centimeters.

The measuring of a line segment to determine its length is done with some standard scale such as a ruler marked off in centimeters. The reading of the linear measure from the ruler requires an estimation to the nearest unit used by the person doing the measuring. For example, in the following figure, find the measure of $\overline{PQ}$ relative to the scale below it.

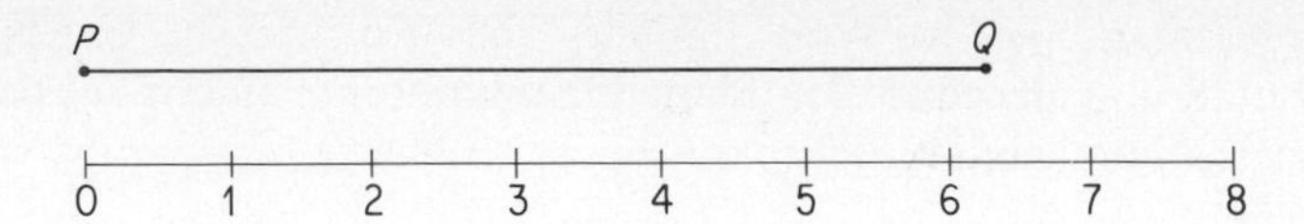

To the nearest unit $m(\overline{PQ}) = 6$. We write $m(\overline{PQ}) \approx 6$ (read as "the measure of the line segment PQ is approximately 6") to indicate that the length of $\overline{PQ}$ is about 6 units. Some people mentally subdivide the unit of length on a scale and *estimate* measures to smaller units. For example, different people might say:

$m(\overline{PQ}) \approx 6$ to the nearest half-unit.
$m(\overline{PQ}) \approx 6\frac{1}{4}$, that is, 6 and $\frac{1}{4}$ to the nearest quarter unit.
$m(\overline{PQ}) \approx 6.2$, that is, 6 and $\frac{2}{10}$ to the nearest tenth of a unit.

Whatever subdivisions are used, the measure is based upon someone's "reading the scale" and thus is approximate. All linear measurements are approximate since they are based upon estimation from reading a scale. Indeed, all measurements are approximate. Even though a measurement is made with a very precise instrument, there is always a final estimation and thus an approximation.

If the length of a segment is 6 centimeters to the nearest centimeter, then the *greatest possible error* is 0.5 centimeter (or 5 millimeters). Thus each of the segments shown below has a measure of 6 centimeters to the nearest centimeter. That is, each segment has a length that is closer to 6 than it is to 5 or 7 centimeters; each has a length between 55 and 65 millimeters.

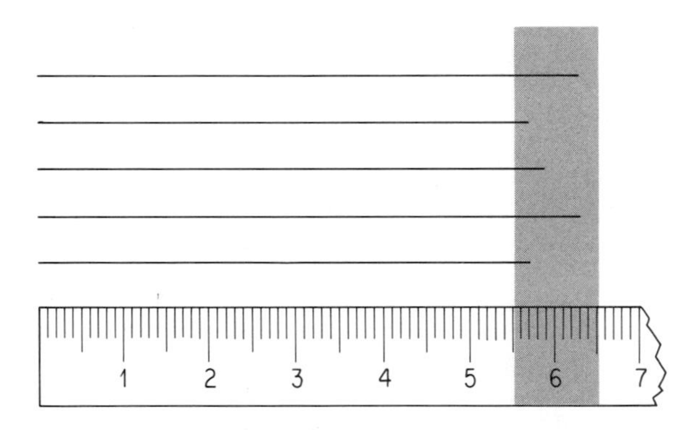

If a length is measured as 45 meters to the nearest meter, then the greatest possible error is 0.5 meter (or 5 decimeters). That is, the true length is somewhere between 44.5 and 45.5 meters. For any measurement, the **greatest possible error** is one-half of the smallest unit used. In other words, the absolute value of the difference between the actual length and the stated measurement is at most one-half of the smallest unit used.

We have defined linear measures to be approximate numbers. There exist special techniques for computations with approximate numbers, (§14-4).

In the present chapter we treat numbers obtained from measurements the same as other numbers and simply use the approximation symbol $\approx$ to indicate that the numbers are approximate rather than exact numbers.

EXERCISES

For points P_n on a number line give the linear measure of:

1. $\overline{P_1P_5}$ **2.** $\overline{P_{-1}P_4}$ **3.** $\overline{P_{-1}P_{-4}}$

4. $\overline{P_2P_{-3}}$ **5.** $\overline{P_{-7}P_{-1}}$ **6.** $\overline{P_3P_{-2}}$

Show the relationship among the linear measures of:

7. $\overline{P_2P_5}$, $\overline{P_5P_7}$, and $\overline{P_2P_7}$.

8. $\overline{P_{-3}P_2}$, $\overline{P_2P_5}$, and $\overline{P_{-3}P_5}$.

9. $\overline{P_1P_4}$, $\overline{P_2P_7}$, and $\overline{P_1P_7}$.

Find the perimeter of each given polygon.

10.

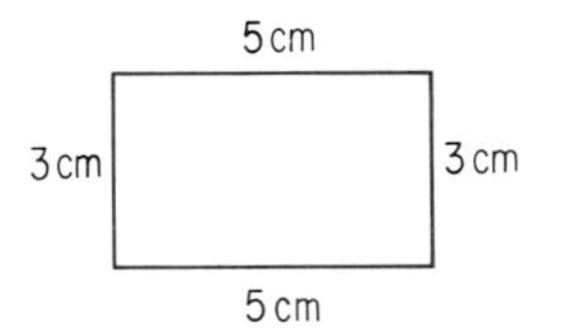

11.

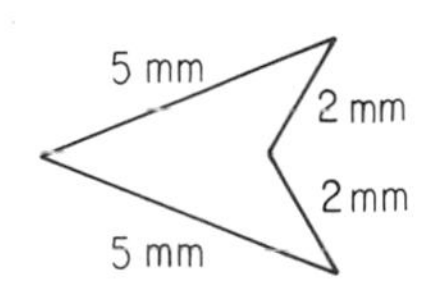

12.

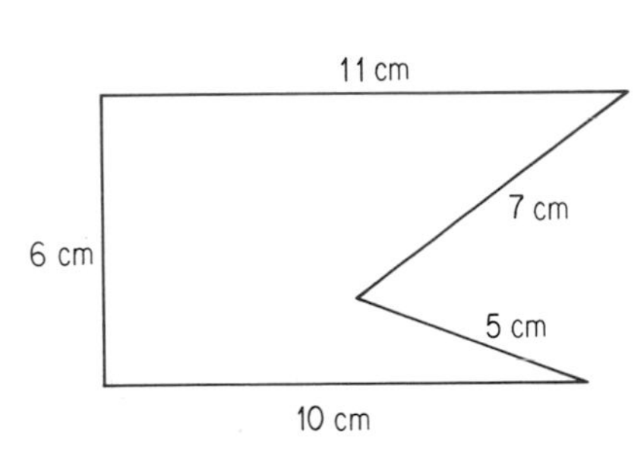

13.

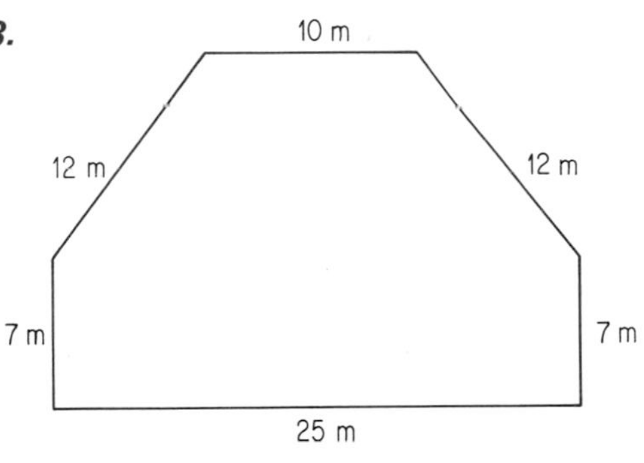

Express the measure of $\overline{PQ}$:

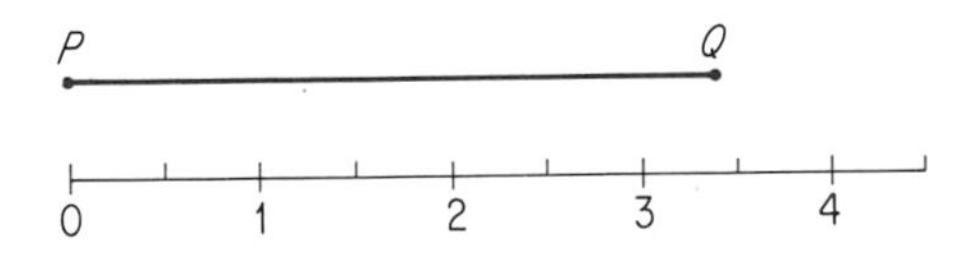

14. To the nearest unit.

15. To the nearest half-unit.

16. To the nearest quarter unit.

17. Find the greatest possible error of the measurements in Exercises 14 through 16.

Which of these measurements are probably approximate?

18. I have 27 students in my class.

19. My breakfast has 650 calories.

20. Portland has a population of 67,500.

21. There are 527 pages in my history book.

22. Bobby is 18 months old.

PEDAGOGICAL EXPLORATIONS

1. Select an elementary school grade level from 3 to 6. Then list as many uses of linear measures as you can think of for students at that grade level. Consult textbooks as necessary.
2. Look at mathematics and science books for the grade selected in Exploration 1 and try to extend your list. Be sure to include the word problems as you look through books.
3. Prepare a bulletin board display on the history of measurement.
4. Although a 12-inch ruler has each inch marked, not all of these divisions are necessary. The 8-inch ruler below can be used to measure any number of inches from 1 through 8. For example, the distance between the markings for 1 inch and 5 inches can be used to measure 4 inches.

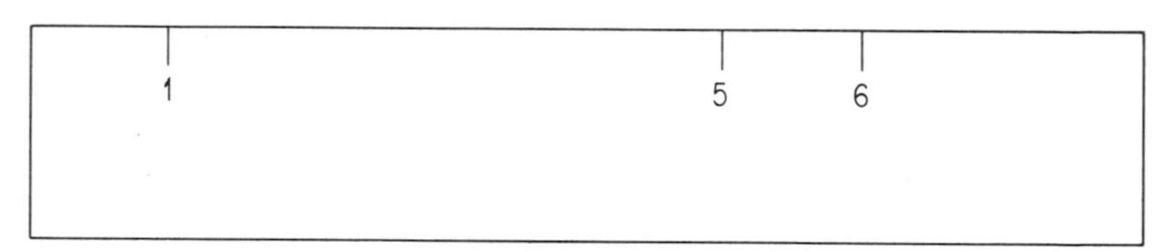

Construct a drawing of a 12-inch ruler with the fewest possible markings that will allow you to measure directly any length from 1 inch to 12 inches.

8-4 angular measures

The union of any two rays $\overrightarrow{PQ}$ and $\overrightarrow{PR}$ with a common endpoint P is a *plane angle* with the rays as *sides* and the point P as *vertex*. We write $\angle QPR = \angle RPQ$ to indicate that "$\angle QPR$" and "$\angle RPQ$" are two names for the same angle. We use "$m\angle QPR$" to denote the measure of the angle in degrees or some other specified unit of angular measure.

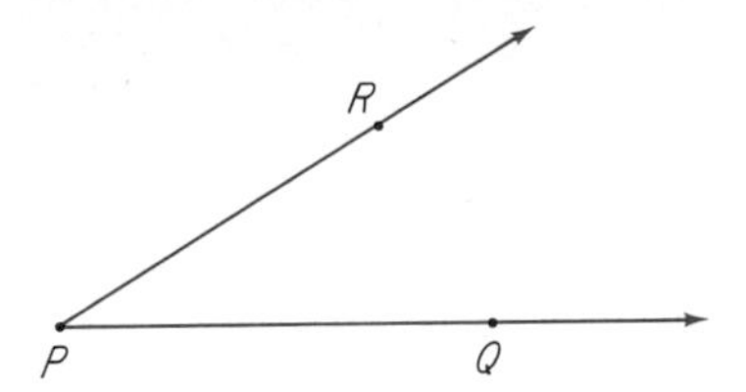

The measure of an angle may be considered as an amount of rotation from one side to the other. The first side is called the **initial side;** the other side is called the **terminal side.** By convention the positive direction of rotation is taken counterclockwise. Unless otherwise indicated, the amount of rotation is assumed to be less than a complete rotation, that is, one revolution. The common units of angular measure are related as follows:

$$1 \text{ revolution} = 360 \text{ degrees} = 4 \text{ right angles}$$

The SI unit of angular measure is the *radian.* However, we shall begin our introduction of angular measures with the more familiar unit, the degree.

When an angle is represented but an initial side is not indicated, the measure is usually taken as a positive number of degrees and as small as possible. Thus in the figure $m\angle ABC = 60$ unless curved arrows are added to indicate the initial side and a direction of rotation that requires a different angular measure.

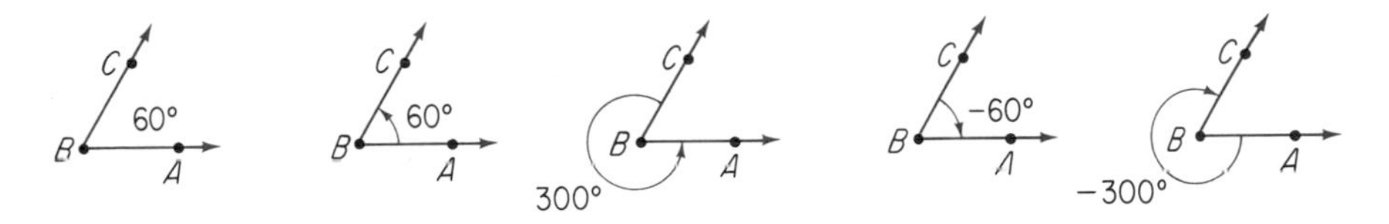

Protractors are used to measure angles just as rulers are used to measure line segments. One common type of protractor is shown in the figure.

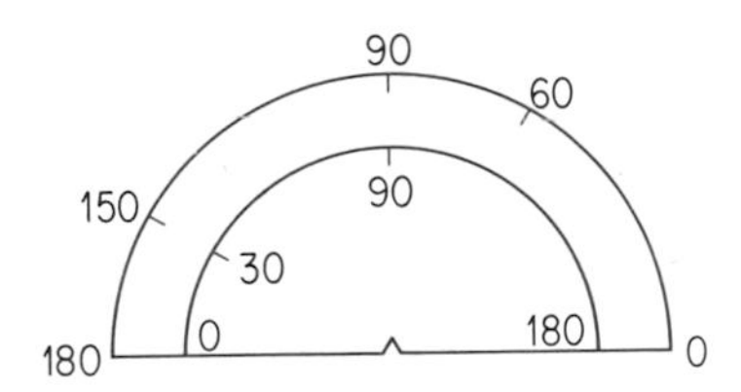

Several common types of angles are defined in terms of their angular measures in degrees. An angle ABC is

An **acute angle** if $0 < m\angle ABC < 90$.

A **right angle** if $m\angle ABC = 90$.

An **obtuse angle** if $90 < m\angle ABC < 180$.

A **straight angle** if $m\angle ABC = 180$.

Some textbooks postpone the use of the term straight angle and refer to the rays $\overrightarrow{BA}$ and $\overrightarrow{BC}$ as **opposite rays** when their union is a straight line.

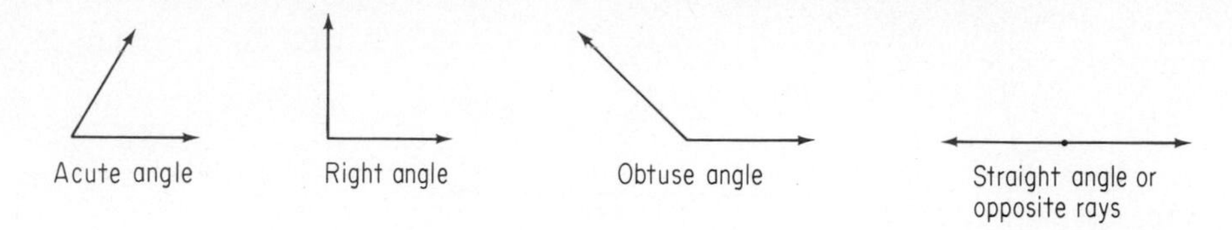

Any two angles with equal angular measures are called **congruent angles.** Since any angle has a unique angular measure, any angle is congruent to itself; for example $\angle QPR \cong \angle RPQ$ (read as "angle *QPR* is congruent to angle *RPQ*").

Congruent angles are used extensively in the study of geometric figures. For example, if two straight lines intersect, then four angles are formed. In the figure there are $\angle TRV$, $\angle VRS$, $\angle SRU$, and $\angle URT$. The sides of $\angle TRV$ are opposite rays of the sides of $\angle SRU$ and the angles are **vertical angles.**

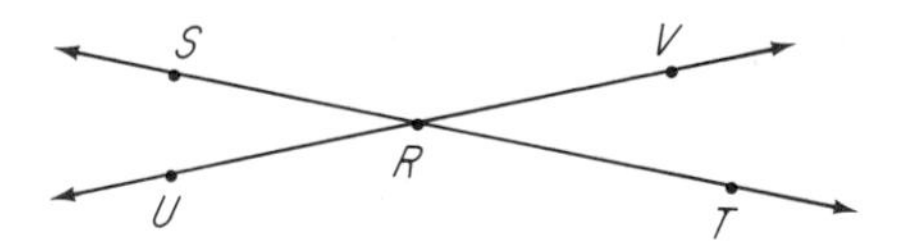

In a formal course in geometry it can be proved that:

Any two vertical angles are congruent.

Example 1: Draw straight angles *ABC* and *DBE*. Then identify two pairs of vertical angles and an angle that is congruent to $\angle ABD$.

Solution: $\angle ABD$ and $\angle CBE$ are vertical angles; $\angle ABE$ and $\angle CBD$ are vertical angles; $\angle ABD \cong \angle CBE$.

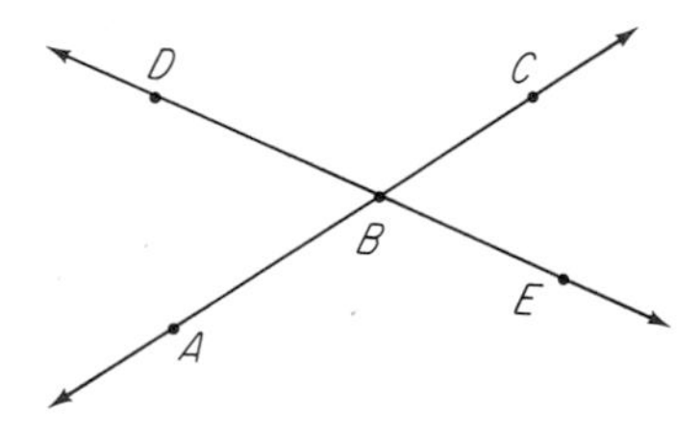

If the measure in degrees is known for any one of the angles formed when two straight lines intersect, then the measure of each of the four angles can be identified. Suppose in the figure for Example 1 that $m\angle EBC = 55$. Then $m\angle ABD = 55$ since vertical angles are congruent. Notice that $\angle EBD$ is a straight angle and a counterclockwise rotation about *B* from the ray $\overrightarrow{BE}$

to the ray $\overrightarrow{BD}$ could be considered as a rotation from $\overrightarrow{BE}$ to $\overrightarrow{BC}$ followed by a rotation from $\overrightarrow{BC}$ to $\overrightarrow{BD}$. Therefore

$$m\angle EBC + m\angle CBD = m\angle EBD = 180$$

Since $m\angle EBC = 55$,

$$55 + m\angle CBD = 180, \qquad m\angle CBD = 125$$

Any two angles with measures that add up to 180 are called **supplementary angles;** $\angle EBC$ and $\angle CBD$ are supplementary angles. However, these two angles also have a special relationship since they have a common side $\overrightarrow{BC}$ and the other two sides form a straight angle. Any two such angles are **adjacent supplementary angles.** Since $\angle EBC$ and $\angle EBA$ are also adjacent supplementary angles, $m\angle EBA = 125$. This checks with the fact that $\angle EBA$ and $\angle CBD$ are vertical angles. Thus each angle of the figure can be labeled with its measurement.

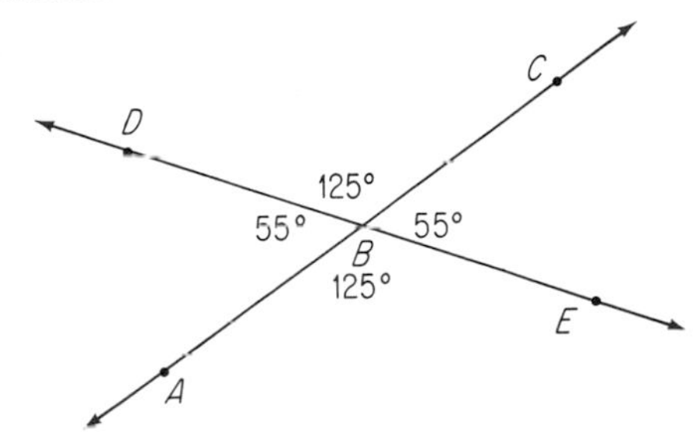

The measure of an angle can be expressed in units other than degrees. In most advanced work, and when working with metric measures the unit usually used for angular measures is a *radian*. An angle with measure of 1 **radian** is a **central angle** (one whose vertex is at the center of a circle) that intercepts on that circle an arc equal in length to the radius of the circle. In the figure, $r = m(\widehat{AB})$ (read as measure of arc AB) and $\angle AOB$ is an angle whose measurement is 1 radian.

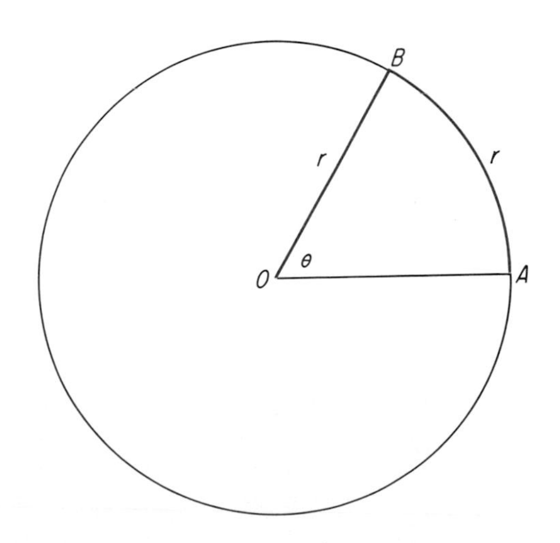

We can convert from degrees to radians and from radians to degrees by recalling a formula from elementary geometry for the *circumference* of a circle, that is, the distance around the circle. The formula states that the circumference C is equal to 2π times the radius r, where π is a constant irrational number. (We often approximate the value of π as $\frac{22}{7}$ or as 3.1416.) In symbols, this formula may be written as:

$$C = 2\pi r$$

It therefore follows that a complete rotation of 360° is equal to 2π radians. This provides us with the following conversion formulas:

$$360^\circ = 2\pi \text{ radians}$$

$$1^\circ = \frac{\pi}{180} \text{ radians}$$

$$1 \text{ radian} = \frac{180}{\pi} \text{ degrees}$$

Example 2: (a) Change 30° to radian measure.

(b) Change $\frac{\pi}{6}$ radians to degrees.

Solution: (a) $1^\circ = \frac{\pi}{180}$ radians; $30^\circ = 30\left(\frac{\pi}{180}\right) = \frac{\pi}{6}$ radians.

(b) $1 \text{ radian} = \frac{180}{\pi}$ degrees; $\frac{\pi}{6}$ radians $= \left(\frac{\pi}{6}\right)\left(\frac{180}{\pi}\right)$

$= 30$ degrees.

If we use 3.14 as π, and divide 180 by 3.14, we find that 1 radian is about 57 degrees. Similarly if we divide 3.14 by 180 we can show that 1 degree is approximately 0.0175 radians.

EXERCISES

If possible, sketch and label. If not possible, explain why it is not possible.

1. An acute angle ABC.
2. Two acute angles CDE and FGH that are not congruent.
3. A right angle IJK.
4. Two right angles LMN and OPQ that are not congruent.
5. An obtuse angle RST.
6. Two obtuse angles UVW and XYZ that are not congruent.
7. Two straight angles ABC and DEF that are not congruent.
8. Vertical angles FGH and IGJ.

Copy the given figure and label each angle with its measurement.

9. **10.**

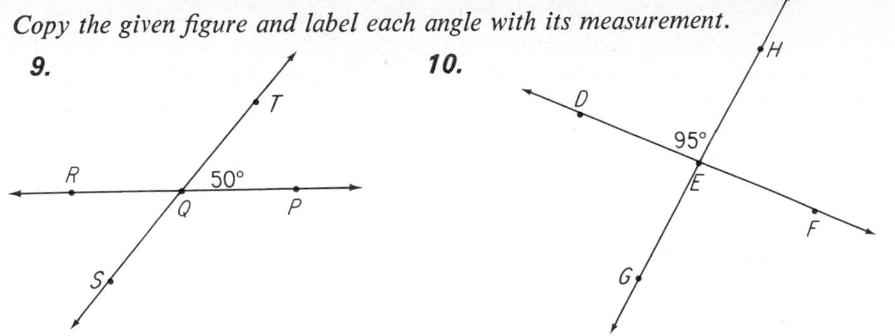

Find the radian measure for angles with each of the following measures in degrees.

11. 45° **12.** 90° **13.** 180° **14.** 270° **15.** 360°

16. 75° **17.** 225° **18.** −120° **19.** −315° **20.** −135°

Find the degree measure for angles with each of the following measures in radians.

21. $\frac{\pi}{2}$ **22.** $\frac{\pi}{4}$ **23.** $\frac{3\pi}{2}$ **24.** $\frac{3\pi}{4}$ **25.** π

26. $\frac{7\pi}{4}$ **27.** $\frac{2\pi}{3}$ **28.** -2π **29.** $-\frac{5\pi}{4}$ **30.** $-\frac{3\pi}{2}$

31. Use 3.14 as π and confirm that 1 radian is about 57 degrees.

32. Repeat Exercise 31 to show that 1 degree is about 0.0175 radians.

PEDAGOGICAL EXPLORATIONS

1. Many modern elementary textbooks make a careful distinction between *measure* and *measurement.* Explain this distinction, using a specific illustration. Then examine a recently published series of elementary textbooks to see what attention is given to this issue.
2. See if you can find a reference that explains the historical basis for our division of a circle into 360 congruent parts to obtain angle measurements in degrees.
3. Prepare a 25–30-minute lesson plan that can be used to teach an upper elementary school class how to use a protractor. Include both the use of a visual aid and student worksheets in your plan.
4. Locate a number of circular objects. Measure the circumference C and the diameter d of each object. Then compute the ratio C/d for each object and see how closely you approximate the value of π. (Note that since $C = 2\pi r$, then $C = \pi(2r) = \pi d$, and $\pi = C/d$.)

5. Mnemonic devices are often used by students as an aid to memorization. A very famous one that gives 13 digits for π is this: "See, I have a rhyme assisting my feeble brain, its task ofttimes resisting." Replace each word by the number of letters in that word. Thus "See" = 3, "I" = 1, "have" = 4, and so forth. This gives $\pi = 3.141592653489$ correct to 12 places to the right of the decimal point. See what other mnemonic devices you can find, or invent, that are helpful in memorizing important mathematical facts.

8-5 area measures

Area measure is a surface measure and is used for rectangular regions and other regions. The unit of area measure is the measure of a square region having an edge with length of 1 unit of linear measure. For example, 1 square centimeter, 1 square decimeter, and 1 square meter may each be used as units of area measure.

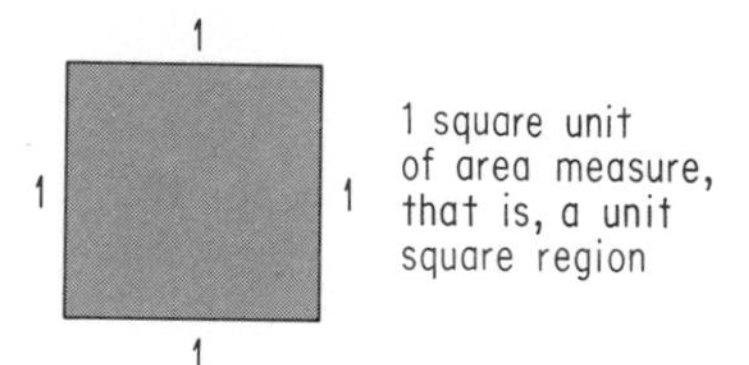

Consider a rectangular region $ABCD$ with $m(\overline{AB}) = b$ and $m(\overline{BC}) = h$. When b and h are whole numbers, the rectangular region has the same area measure as bh unit square regions. In the second figure $b = 3$, $h = 2$, and the rectangular region $PQRS$ has area measure 6; that is, the area of rectangular region $PQRS$ is 6 square units.

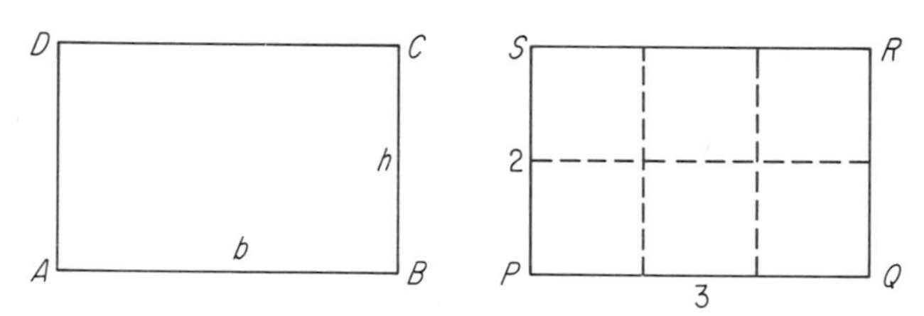

We use the symbol $\square ABCD$ for rectangle $ABCD$ and follow the custom of speaking of the area of the rectangle when we mean the rectangular region. Thus we write $\mathcal{A}(\square ABCD)$ for the **area measure** of rectangle $ABCD$. Then

$$\mathcal{A}(\square PQRS) = 6$$

and, in general,

$$\mathcal{A}(\square ABCD) = bh$$

where $m(\overline{AB}) = b$ and $m(\overline{BC}) = h$. We have illustrated the area formula for whole numbers b and h. We *define* the area measure of the rectangle to be bh in square units for all linear measures b and h.

A square is a special case of a rectangle, $b = h$. Thus the area of a square region with an edge e units long is e^2 square units. As in the case of rectangles we often speak of the areas of squares or other polygons and mean thereby the areas of the polygonal regions.

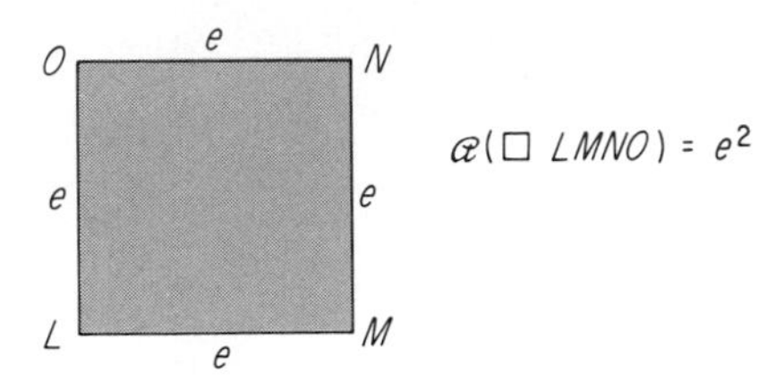

A point has area measure 0 and a line segment has area measure 0. If two plane regions have area measures and the area measure of their intersection is 0, then the area measure of the union of the two regions is equal to the sum of their area measures.

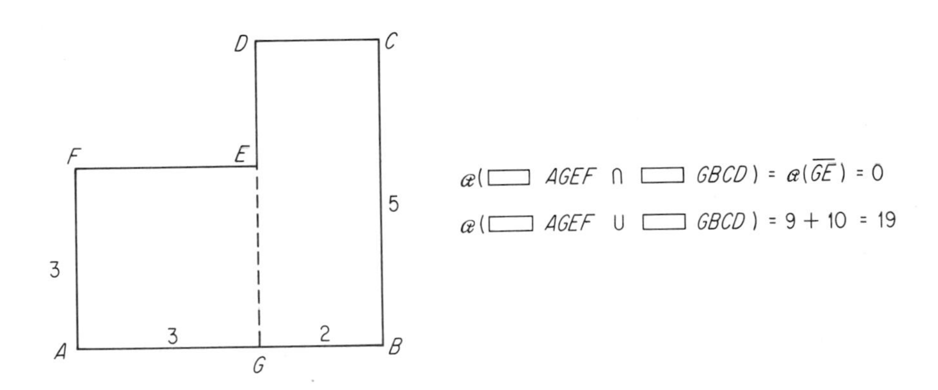

In order to find the area measures of triangular regions, we assume that the triangular regions determined by any rectangle and one of its diagonals have the same area measure. Then the area measure of a right triangle with legs of length b and h is $\frac{1}{2}bh$, as indicated in the next figure.

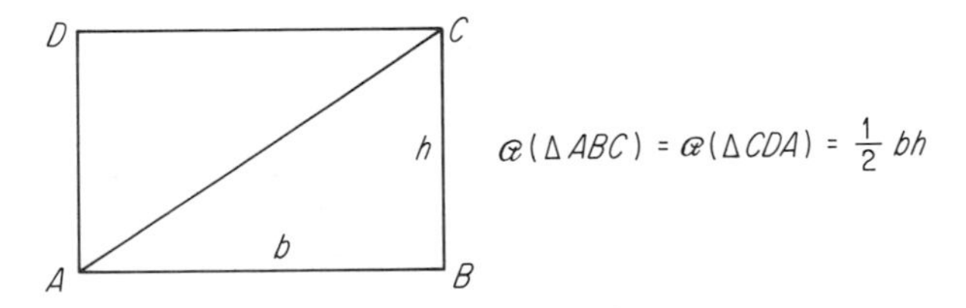

For any triangle ABC a side, such as $\overline{AB}$, may be selected as base. Then the line that contains C and is perpendicular to $\overleftrightarrow{AB}$ intersects $\overleftrightarrow{AB}$ at a point D. There are three cases to consider according as D is an interior point of $\overline{AB}$, an endpoint of $\overline{AB}$, or an exterior point of $\overline{AB}$.

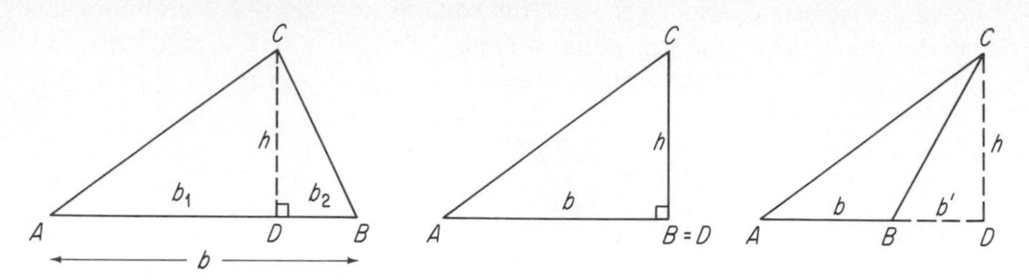

In the figure on the left D is an interior point of $\overline{AB}$, $m(\overline{AB}) = b$, $m(\overline{AD}) = b_1$, $m(\overline{DB}) = b_2$, $b_1 + b_2 = b$, and $m(\overline{CD}) = h$. Therefore $\mathcal{A}(\triangle ADC) = \frac{1}{2}b_1h$, $\mathcal{A}(\triangle DBC) = \frac{1}{2}b_2h$, and

$$\begin{aligned}\mathcal{A}(\triangle ABC) &= \mathcal{A}(\triangle ADC) + \mathcal{A}(\triangle DBC)\\ &= \tfrac{1}{2}b_1h + \tfrac{1}{2}b_2h\\ &= \tfrac{1}{2}(b_1 + b_2)h = \tfrac{1}{2}bh\end{aligned}$$

In the center figure $B = D$, $m(\overline{AB}) = b$, $m(\overline{CD}) = h$. Therefore $\mathcal{A}(\triangle ABC) = \frac{1}{2}bh$.

In the figure on the right D is an exterior point of $\overline{AB}$, $m(\overline{AB}) = b$, $m(\overline{BD}) = b'$, $m(\overline{AD}) = b + b'$, $m(\overline{CD}) = h$, and

$$\begin{aligned}\mathcal{A}(\triangle ABC) &= \mathcal{A}(\triangle ADC) - \mathcal{A}(\triangle BDC)\\ &= \tfrac{1}{2}(b + b')h - \tfrac{1}{2}b'h\\ &= \tfrac{1}{2}[(b + b') - b']h\\ &= \tfrac{1}{2}[b + (b' - b')]h = \tfrac{1}{2}bh\end{aligned}$$

In all cases, and thus for any triangle ABC, $\mathcal{A}(\triangle ABC) = \frac{1}{2}bh$.

Areas of polygonal regions may be found, as in the pedagogical explorations for this section, by subdividing the polygonal region into nonoverlapping rectangular or triangular regions, that is, regions with no common interior points. A **parallelogram** is a quadrilateral with its opposite sides parallel. Any parallelogram $ABCD$ with base b and height h has area bh.

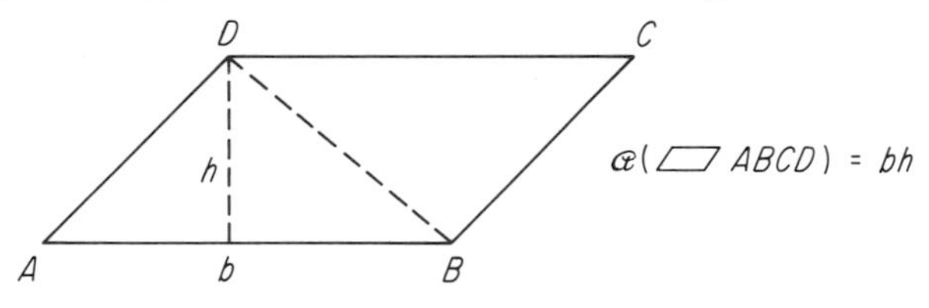

Example: Find the area of the given region.

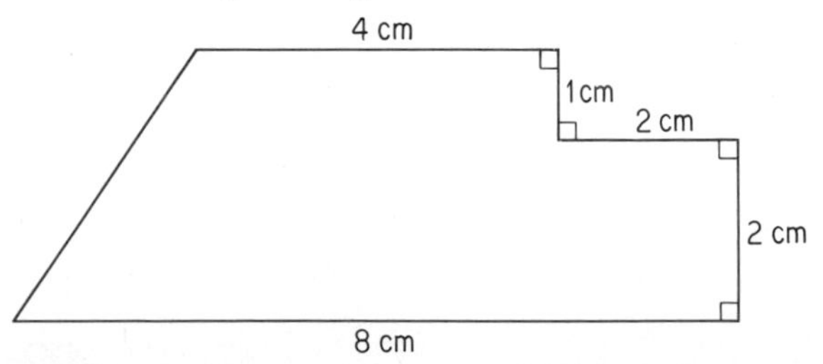

Solution: We subdivide the region into triangular and rectangular regions, as shown in the next figure. The triangular region has area measure of $\frac{1}{2} \times 2 \times 3 = 3$; the rectangular regions have area measures $3 \times 4 = 12$ and $2 \times 2 = 4$. Thus the area measure of the given region is 19 and the area is 19 square centimeters, that is, 19 cm².

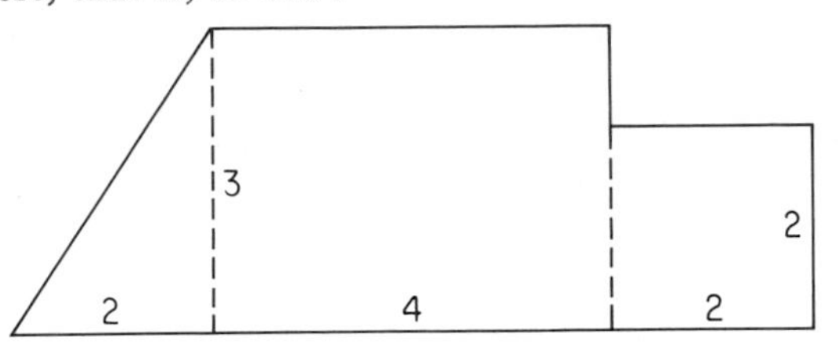

In the example notice that, as for linear measure and length, the area measure is a number; the area is that number of square units.

EXERCISES

Find the area measurement of each region.

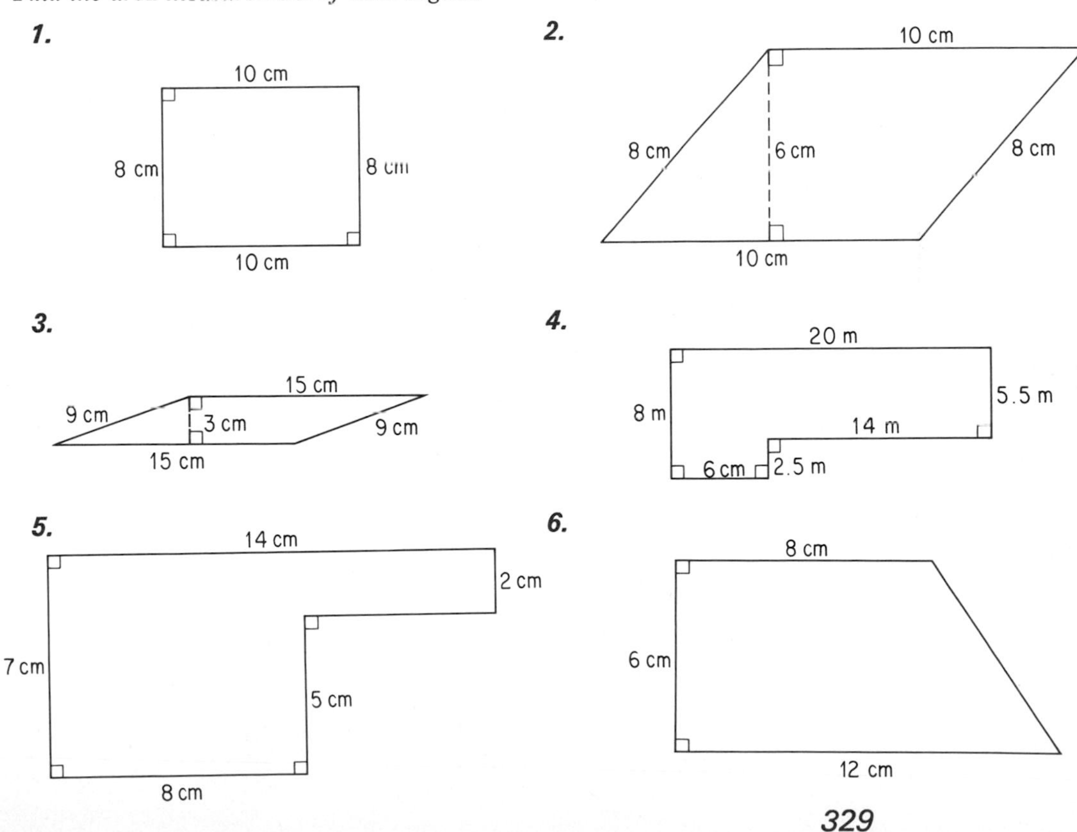

7.

8 cm
6 cm
3 cm
15 cm

8.

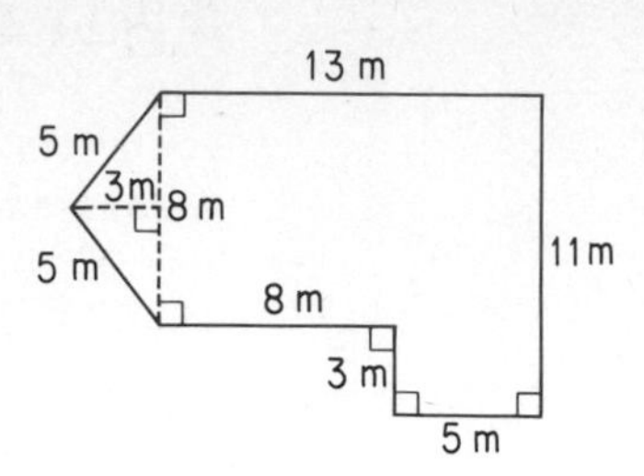

Describe the change in the area of:

9. A square when the measure of each of its edges is doubled.

10. A square when the measure of each of its edges is multiplied by **(a)** 3; **(b)** $\frac{1}{2}$; **(c)** a constant k.

11. A rectangle when the measure of each of its edges is multiplied by **(a)** 2; **(b)** 3; **(c)** $\frac{1}{2}$; **(d)** a constant k.

12. A triangle when the measure of each of its edges is multiplied by **(a)** 2; **(b)** 3; **(c)** $\frac{1}{2}$; **(d)** a constant k.

13. A polygon when the measures of its angles are not changed and the measure of each of its edges is multiplied by **(a)** 2; **(b)** a constant k.

The formula for the area of a circle is given by the formula $A = \pi r^2$, where r is the radius. Approximate π as 3.14 and find the area of the circle with the given radius.

14. $r = 3$ cm

15. $r = 10$ cm

16. $r = 9$ mm

17. $r = 5$ mm

Any trapezoid $PQRS$ can be represented as in the figure. Let $m(\overline{PQ}) = b_1$, $m(\overline{SR}) = b_2$, $\overleftrightarrow{PQ} \| \overleftrightarrow{SR}$, $\overleftrightarrow{ST}$ be perpendicular to $\overleftrightarrow{PQ}$ with T on $\overleftrightarrow{PQ}$, $\overleftrightarrow{QU}$ be perpendicular to $\overleftrightarrow{SR}$ with U on $\overleftrightarrow{SR}$, and $m(\overline{ST}) = h$. Give a reason for each of statements 18 through 20.

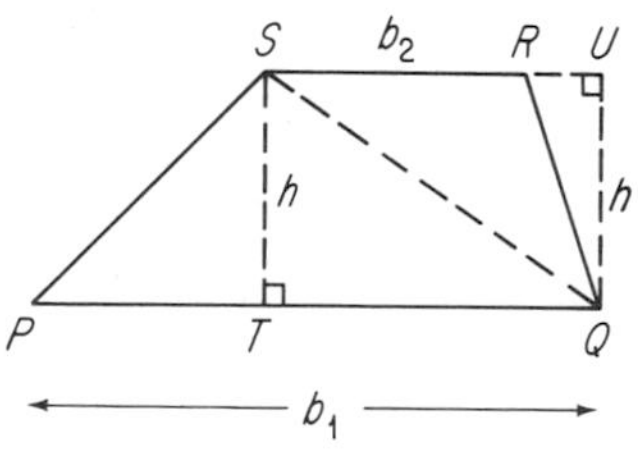

18. $\mathcal{A}(\triangle PQS) = \frac{1}{2}b_1h$

19. $\mathcal{A}(\triangle QRS) = \frac{1}{2}b_2h$

20. $\mathcal{A}(\square PQRS) = \frac{1}{2}(b_1 + b_2)h$

21. Find a formula for the radius of a circle in terms of the area. That is, solve $A = \pi r^2$ for r.

PEDAGOGICAL EXPLORATIONS

A **geoboard** is a square array of nails or pegs around which rubber bands may be stretched. Effective use can be made of clear plastic models on the overhead projector for demonstration purposes. If there are an insufficient number of geoboards available for classroom use, students may use dot paper or graph paper. There is a formula, known as **Pick's formula**, that relates the area of an enclosed region to the number of points involved. See if you can discover the formula from the following explorations.

1. For each of the following figures, count the number of points b in the boundary of the figure. Then compute the area A in terms of square units as in the given table.

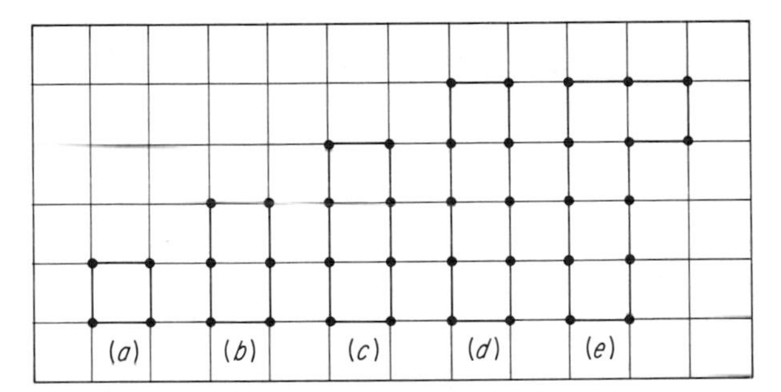

	(a)	(b)	(c)	(d)	(e)
Number (b) of boundary points	4	6	8	10	12
Area (A)	1	2			

2. Conjecture a formula that relates A in terms of b as in the table for Exploration 1.
3. Each of the figures that follow have boundary points b and interior points i. Compute the area A of each figure in terms of square units as in the given table.

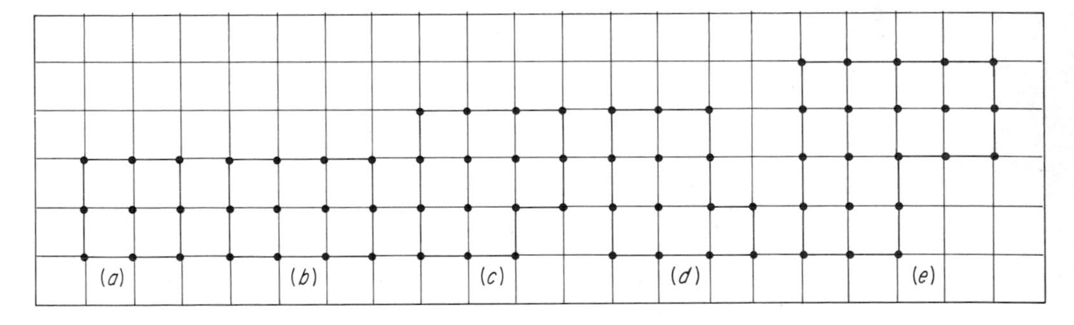

	(a)	(b)	(c)	(d)	(e)
Number (b) of boundary points	8	10	12		
Number (i) of interior points	1	2	3		
Area (A)	4	6			

4. Conjecture a formula that relates A in terms of b and i as in the table for Exploration 3.

8-6 volume measures

We now define a number of three-dimensional figures so that we may consider measures of their volumes. On a plane a polygon is a simple, closed union of line segments, that is, a union of line segments (sides of the polygon) such that exactly two sides have each endpoint of a line segment (vertex of the polygon) in common and no side contains an interior point of another side. In space a **polyhedron** (plural **polyhedra**) is a simple, closed union of polygonal regions (**faces** of the polyhedron) such that exactly two faces have each side of a polygonal region (**edge** of the polyhedron) in common and no face contains an interior point of another face. The vertices of the polygonal regions are also called **vertices** of the polyhedron. Any two polyhedra are *congruent* if they are of the same shape and of the same size. Any two congruent polyhedra have the same volume measure and the same surface measure.

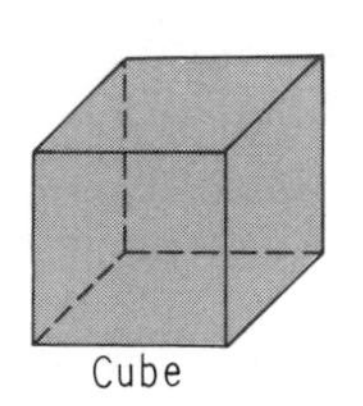
Cube

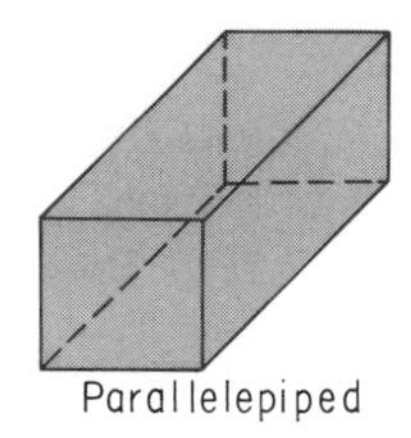
Parallelepiped

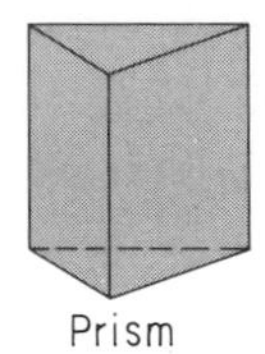
Prism

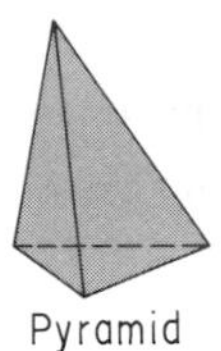
Pyramid

Each of these figures is a polyhedron. The cube has square faces; the rectangular box (parallelepiped) has rectangular faces. A polyhedron with all of its vertices in two parallel planes is a **prismatoid.** If the edges that are not in the parallel planes are parallel edges, then the polyhedron is a **prism.** The faces of the prism in the two parallel planes are the **bases** of the prism. The parallel edges joining the two bases are **lateral edges.** In more advanced courses it can be shown that the bases of a prism are congruent polygons. Notice in the sketches that the "hidden" edges are indicated by dotted line segments.

A polyhedron with all but one of its vertices on a plane is a **pyramid.** The vertices on a plane are the vertices of the **base** of the pyramid. Prisms and pyramids are often classified as triangular, rectangular, square, and so forth according to their bases.

Volume measure is a space measure and is used for cubic and other regions. As for area measure it is customary to speak of the volume of a cube or other polyhedron and to mean thereby the volume of the polyhedral region. The unit of volume measure is the measure of a cubic region having an edge with length 1 unit of linear measure. For example, 1 cubic inch, 1 cubic centimeter, and 1 cubic decimeter (1 liter) may each be used as units of volume measure.

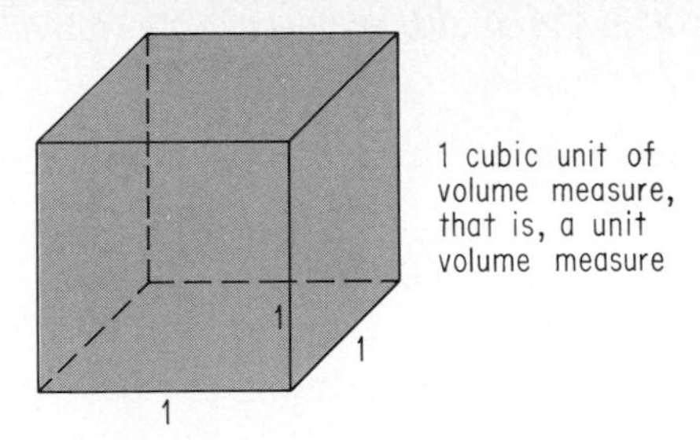

Volumes are considered in a manner very similar to areas. We define any rectangular parallelepiped with edges of linear measures a, b, and c on a common endpoint to have volume measure abc.

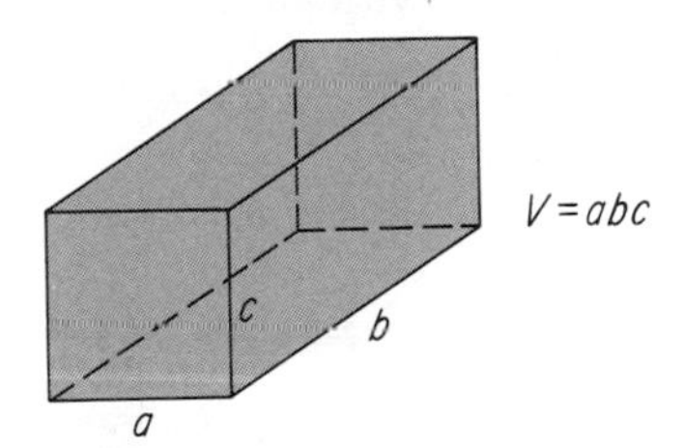

Points, line segments, and plane regions have volume measure 0. Any pyramid with a base of area B square units and height to that base h linear units, can be proved in more advanced courses to have volume measure $\frac{1}{3}Bh$.

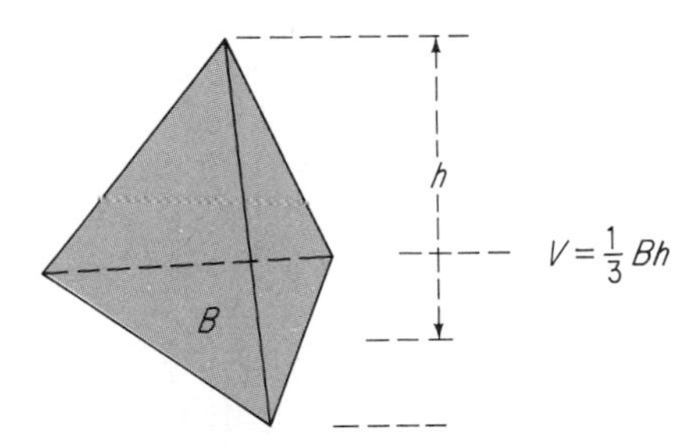

Volumes of many other regions may be obtained by subdividing the region into pyramids or rectangular parallelepipeds, or by using the prismoidal formula (Exercise 14). As in the case of length and area, the volume measure is a number; the volume is that number of cubic units. The notation for the units of a volume in cubic centimeters is cm^3; in cubic meters, m^3.

EXERCISES

Find **(a)** *the surface area, and* **(b)** *the volume, of a cube with edge of length:*

1. 5 centimeters ***2.*** 8 decimeters ***3.*** 7 centimeters ***4.*** 2 meters

Find (**a**) *the surface area, and* (**b**) *the volume, of each rectangular parallelepiped.*

5. **6.**

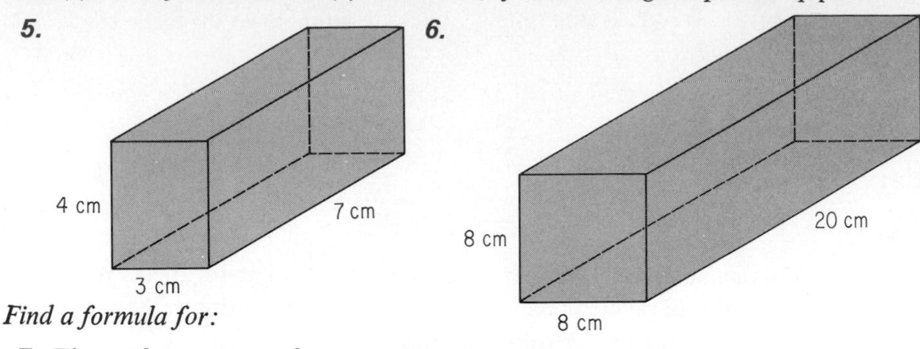

Find a formula for:

7. The surface area S of a cube with edge of linear measure e.

8. The surface area S of a rectangular parallelepiped with edges of linear measure b, d, and h.

Find the volume of each pyramid.

9. **10.**

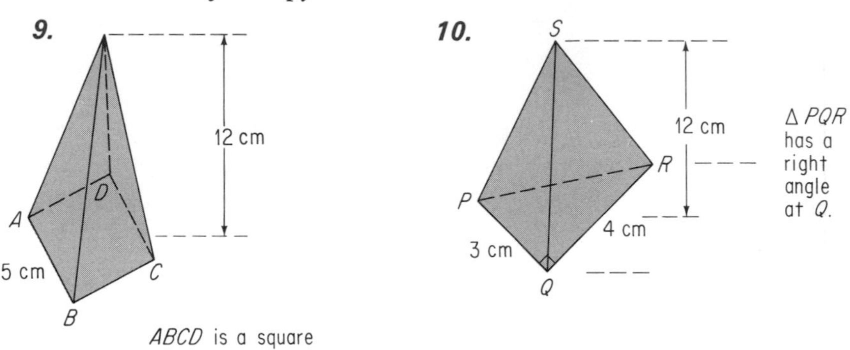

In Exercises 11 through 13 describe the change in the volume of:

11. A cube when the measure of each of its edges is doubled.

12. A cube when the measure of each of its edges is multiplied by (**a**) 3; (**b**) $\frac{1}{2}$; (**c**) a constant k.

13. A rectangular parallelepiped when the measure of each of its edges is multiplied by (**a**) 2; (**b**) 3; (**c**) $\frac{1}{2}$; (**d**) a constant k.

14. The most general common formula for volume measures is the **prismoidal formula**

$$V = \frac{h}{6}(B_1 + 4M + B_2)$$

This formula may be used, for example, for any prismatoid where B_1 and B_2 are the area measures of the bases, M is the area measure of the intersection of the prismatoid with the plane halfway between the bases, and the bases are h units apart; that is, the altitude to the bases has linear measure h. The plane section halfway between the bases is often called the **midsection.** Then M is the area measure of the midsection.

For a cube with an edge of linear measure e, $B_1 = e^2$, $M = e^2$, and $B_2 = e^2$. Show that the prismoidal formula gives the usual expression for the volume.

***15.** Think of a sphere as in the figure and use the prismoidal formula to find an expression for the volume of a sphere of radius r. (The area measure of a circle of radius r is πr^2.)

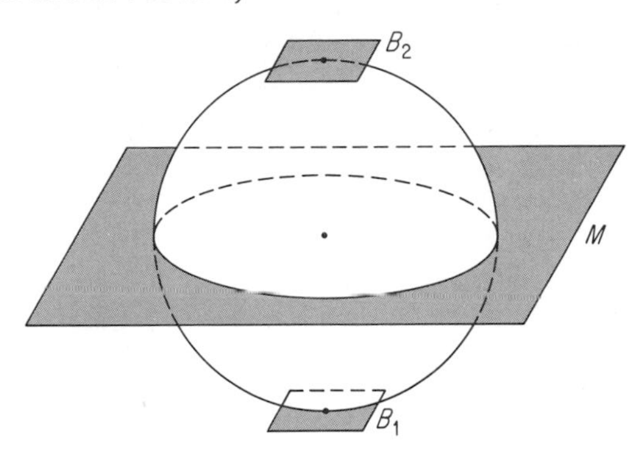

***16.** For a square pyramid with height h and a base with an edge of linear measure e the midsection is a square with an edge $\frac{1}{2}e$. Use the prismoidal formula to find an expression for the volume of the square pyramid.

***17.** Use the prismoidal formula to find an expression for the volume of a circular cone with height h if the base has radius r and the midsection has radius $\frac{1}{2}r$.

PEDAGOGICAL EXPLORATIONS

1. Examine a set of elementary mathematics textbooks for grades 1 through 6 and list all of the formulas presented therein for the volumes of geometric solids.
2. An interesting way to introduce elementary school students to the concept of volume is to ask such questions as "Do you think that it is possible to fit one million pennies into this room?" "Can one million ping-pong balls fit into this room?" Answer these questions for one of your classrooms, and then develop some procedure to help you verify your answers.
3. A contest to guess the number of beans, or other such similar objects, in a jar is another interesting approach to the concept of volume for elementary school students. Try to list five specific suggestions for activities that help develop the idea of volume.
4. Find a formula for the volume of a right circular cylinder, such as an ordinary fruit juice can. Prepare a short lesson plan to illustrate how

this formula might be presented to upper elementary school students, using discovery or laboratory-type procedures.

5. Graph paper is an excellent aid for sketching space figures on a plane. Consider the three steps shown below for sketching a cube. In step 1 we draw the top face, with opposite sides parallel and equal in length. In step 2 draw four edges of the cube as shown, parallel and of equal length. Note that the "hidden" edge is shown as a dashed line. The figure is completed in step 3, again showing "hidden" edges with dashed lines.

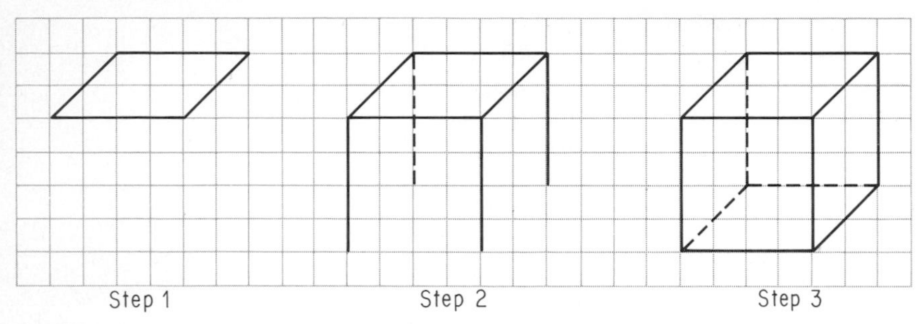

Use graph paper, follow these suggested steps, and draw a number of sketches of your own of prisms, parallelepipeds, and pyramids.

6. The "pictures" (sketches, drawings) of space figures on a sheet of paper never completely represent the figure and often are hard to visualize. Consider the accompanying patterns and make a model of (**a**) a tetrahedron (triangular pyramid), and (**b**) a cube.

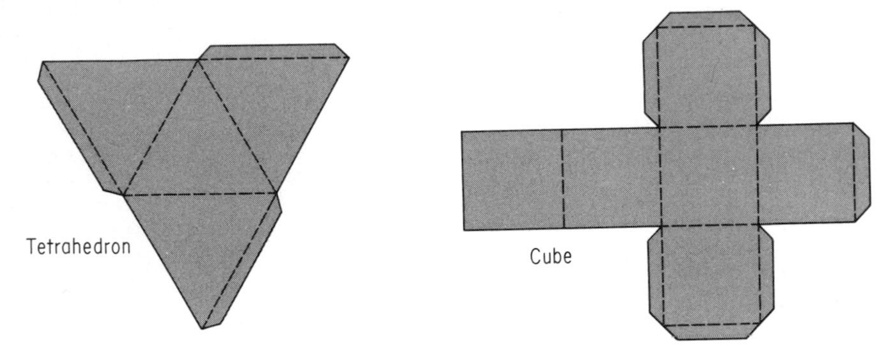

CHAPTER TEST

Change each of the following as indicated.

1. 3 meters to centimeters

2. 5 kilometers to meters

3. 2 centimeters to millimeters

4. 3.75 meters to centimeters

5. 5000 meters to kilometers

6. 3500 milligrams to grams

7. 12 grams to milligrams

8. 1500 milliliters to liters

Select the best answer.

9. A distance of 500 kilometers is about how many miles?
   (a) 100 (b) 200 (c) 300 (d) 400 (e) 500

10. A weight of 90 kilograms is about how many pounds?
   (a) 50 (b) 100 (c) 150 (d) 200 (e) 250

11. A capacity of 20 liters is about how many gallons?
   (a) 5 (b) 10 (c) 20 (d) 40 (e) 80

12. Write 35 degrees Celsius in its equivalent in degrees Fahrenheit.

13. Find the radian measure for an angle of 120°.

14. Find the degree measure for an angle of $\pi/6$ radians.

On a number line find the linear measure of:

15. $\overline{P_2P_{-7}}$

16. $\overline{P_{-8}P_{-3}}$

17. Copy the given figure and label each angle with its measurement.

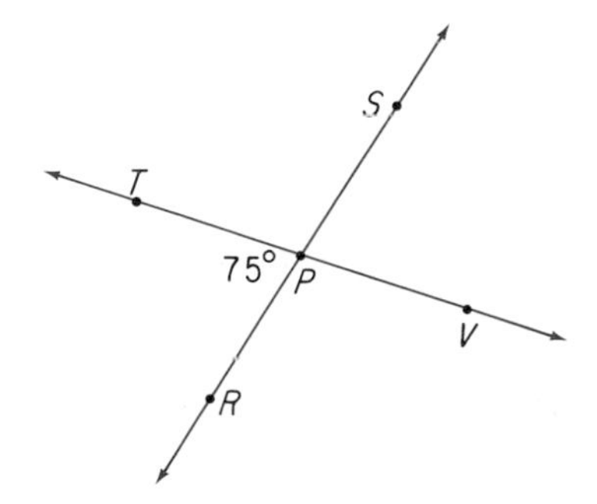

Find the area measurement of each figure.

18.

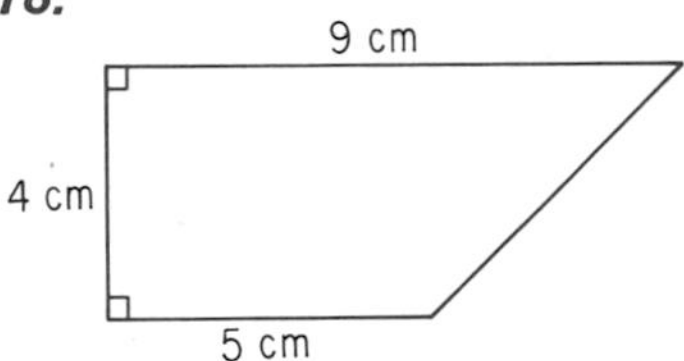

19.

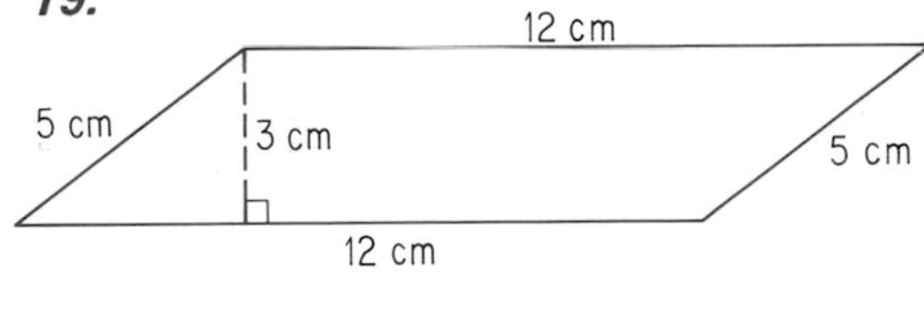

20. Find the surface area of a cube with an edge of length 3 meters.

21. Find the surface area of a rectangular parallelepiped with edges of lengths 3 meters, 5 meters, and 7 meters.

22. Find the volume of a cube with an edge of length 6 centimeters.

23. Find the volume of a rectangular parallelepiped with edges of lengths 3 centimeters, 4 centimeters, and 5 centimeters.

24. Describe the change in the surface area of a cube when the measure of each of its edges is multiplied by 5.

25. Repeat Exercise 24 for the change in the volume.

READINGS AND PROJECTS

1. Prepare a report on ways to implement the metric system in the elementary school curriculum, with suggestions for grade placement. Use as your sources examination of recently published elementary textbooks, interviews with teachers in local school systems, and the report of the NCTM Metric Implementation Committee. This report can be found in the May 1974 issue of *The Arithmetic Teacher,* and is reprinted in *A Metric Handbook for Teachers,* a 1974 publication of the National Council of Teachers of Mathematics.
2. Write to the Metric Information Office, National Bureau of Standards, Washington, D.C., 20234 for a list of available materials on the metric system. Prepare a set of ten activities suitable for use in introducing metric concepts to elementary school students.
3. Elementary school students can be introduced to the concept of the approximate nature of measurement through the use of arbitrary units of measure. For example, they can be asked to measure various line segments using lima beans as the unit of measure. Then consider other possible units, such as postage stamps. Prepare a 30-minute lesson plan on this topic for a particular grade level, using a laboratory approach.
4. Read Chapter 7 (Concepts of Measurement) of the twenty-seventh yearbook of the National Council of Teachers of Mathematics. Then complete as many of the exercises given at the end of the chapter as you can.
5. Read Booklet No. 15 (Measurement) of the thirtieth yearbook of the National Council of Teachers of Mathematics. From your reading report on the meaning of the *precision* of a measurement.
6. The Greek mathematician Archimedes discovered a relationship between the volume of a cylinder and that of a cone with the same base and height. See if you can find this relationship, and then try to demonstrate it through the use of physical models.
7. Examine a set of recently published elementary textbooks and determine the grade placement for the various formulas for area and volume of geometric figures.
8. Read Chapter 8 (Using Models as Instructional Aids) of the thirty-fourth yearbook of the National Council of Teachers of Mathematics. In particular, examine the models suggested for teaching concepts of measurement and prepare a collection of these that can be used for this purpose.

9. Read "A Move to Metric: Some Considerations," by Donald H. Firl on pages 581 through 584 of the November 1974 issue of *The Mathematics Teacher*. In this same issue you can find "A Metric Bibliography" by Stuart A. Choate on pages 586 and 587, a good source for further references.
10. For an interesting research study read "Measurement Concepts of First- and Second-Grade Students" by Thomas P. Carpenter on pages 3 through 13 of the January 1975 issue of the *Journal for Research in Mathematics Education*, a publication of the National Council of Teachers of Mathematics. Report on the author's findings as he compares them with those of Piaget.
11. Read Chapter 8 (A Laboratory Approach) of the thirty-fifth yearbook of the National Council of Teachers of Mathematics. Prepare a class demonstration of some of the uses of the geoboard described by the author.
12. Read "Let's Do It! Organizing a Metric Center in Your Classroom" by James V. Bruni and Helene J. Silverman on pages 80 through 87 of the February 1976 issue of *The Arithmetic Teacher*. Prepare at least one additional activity for measures of length, area, volume, and mass.
13. Read "Introducing Decimal Fractions with the Meterstick" by Robert B. Ashlock on pages 201 through 206 of the March 1976 issue of *The Arithmetic Teacher*. Prepare at least three additional activities similar to the ones suggested in the article.
14. Read Chapter 10 (Measurement) of the thirty-seventh yearbook of the National Council of Teachers of Mathematics. Suggest at least three additional activities to develop the basic concepts of measurement.

9

Sentences in One Variable

A symbol that may be replaced by any member of a set of elements is a variable. *Variables* such as $\square$, n, x, and t are often used in arithmetic. However, the study of sentences involving variables is a part of algebra. Thus our present study of sentences in one variable represents a shift of emphasis from arithmetic to algebra. This is part of the shift from teaching arithmetic in elementary school to teaching mathematics in elementary school.

Algebra is an extension of the arithmetic of integers, rational numbers, and real numbers that we considered in Chapters 5 through 7. Arithmetic sentences involve only numbers and operations on numbers. Each such sentence can be identified as true or false. Algebraic sentences may involve variables and be true for some replacements of the variable and false for others.

In this chapter we use number concepts and the geometric representation of real numbers on a number line to explore and to explain algebraic concepts. We shall solve statements of equality (equations) and statements of inequality that contain only one *real variable*, a variable whose replacement set is the set of real numbers. Traditionally a secondary school topic, this material is now informally introduced at a very early stage in most programs of elementary school mathematics.

9-1 sentences and statements

As in §1-8, sentences that can be identified as true or identified as false are called *statements*. Each of the following is a statement:

$7 + 3 = 3 + 7$ (a true statement of equality)
$7 - 3 = 3 - 7$ (a false statement of equality)

Statements of equality, whether true or false, are called **equations.** We may also write *statements of inequality* such as the following:

$7 + 2 > 7 - 2$ (a true statement of inequality)
$5 + 3 \neq 3 + 5$ (a false statement of inequality)
$8 - 2 < 8$ (a true statement of inequality)

We formally define the **order** of any two real numbers a and b, where a is less than b, as follows:

> $a < b$ (and $b > a$) if and only if there is a positive number $\mathbf{c}$ such that $a + \mathbf{c} = b$.

For example, $2 < 5$ (or $5 > 2$) and there is a positive number 3 such that $2 + 3 = 5$. The *trichotomy law*, cited for counting numbers in Exercise 14 of §4-7, holds for all real numbers. If a and b are any two real numbers, then exactly one of these relations must hold:

$$a < b, \qquad a = b, \qquad a > b$$

Let us now consider the statement of inequality $x - 1 < 3$. The symbol x is called a *variable*, and we need to determine the set of replacements that can be used for the variable, that is, the *replacement set* of the variable. In each case the set of numbers that make the sentence true is called the *solution set* or the *truth set* of the given sentence for that replacement set. Recall that the elements of a set are generally listed within a pair of braces as shown in the array.

Sentence	*Replacement Set*	*Solution Set*
$x - 1 < 3$	Counting numbers	$\{1, 2, 3\}$
$x - 1 < 3$	Whole numbers	$\{0, 1, 2, 3\}$
$x - 1 < 3$	Integers	$\{\ldots, -2, -1, 0, 1, 2, 3\}$
$x - 1 < 3$	Real numbers	All real numbers less than 4

It is *hereafter assumed* that, unless otherwise specified, the replacement set of each variable is the set of real numbers.

Example 1: Find the solution set for $x + 1 > 4$ when x is a whole number.

Solution: When $x = 3$, $3 + 1 = 4$. The solution set S consists of all whole numbers greater than 3; that is,

$$S = \{4, 5, 6, \ldots\}$$

Example 2: Find the solution set for $n + 2 = 2 + n$ when n is a real number.

Solution: This sentence is a true statement of equality for *all* possible replacements of the variable, since it is an application of the commutative property of addition. The solution set is the set of real numbers.

An equation that is true for all possible replacements of the variable (as in Example 2) is called an **identity.** A statement of inequality, such as $n + 2 > n$, may also be classified as an identity.

Example 3: For what values of x is the sentence $x + 2 = x$ true for integers x?

Solution: Regardless of the integer selected as the replacement for x, the sentence $x + 2 = x$ is *always* false. The solution set is the empty set, $\varnothing$.

Statements of inequality may also have the empty set as their solution set. For example, there is no replacement for x that will make the sentence $x + 2 < x$ true; the solution set is the empty set.

Example 4: For what values of x is the sentence $x^2 = 2$ true for rational numbers x?

Solution: There is no rational number x for which $x^2 = 2$ since $x = \sqrt{2}$ is not a rational number. (See §7-2.) The solution set is the empty set.

As in §1-5, the solution set of any open sentence in one variable may be graphed on a number line. The following examples illustrate various types of graphs as well as the relationship between algebraic sentences and geometric figures. In each example we are to find and graph the solution set of the given *open sentence* for the given replacement set. The *graph* of the sentence is the graph of its solution set.

Example 5: Graph the sentence $x + 3 = 5$ for integers x.

Solution: The solution set consists of a single element, {2}. We graph this solution set on a number line by drawing a solid dot at 2. The graph consists of this single *point.*

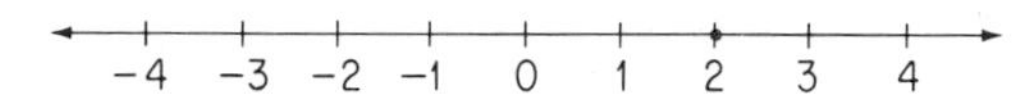

Example 6: Graph the sentence $x + 3 > 5$ for real numbers x.

Solution: The solution set consists of all real numbers greater than 2. The graph of the solution set is drawn by first placing a hollow dot at 2 on the number line to indicate that this point is not a member of the solution set. We then draw a heavily shaded arrow to show that all numbers greater than 2 satisfy the given inequality.

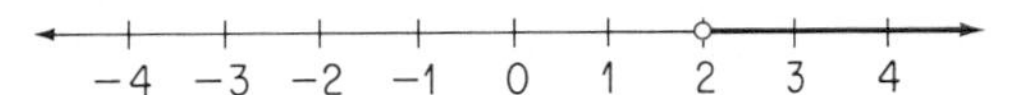

For any real number b the figure formed by the graph of an inequality of the form $x > b$ (see Example 6) is called a **half-line.** (The figure formed by the graph of an inequality of the form $x < b$ is also a half-line.)

Example 7: Graph the sentence $x + 3 \geq 5$ for real numbers x.

Solution: This sentence is read "$x + 3$ is greater than or equal to 5." Therefore, it is true when $x = 2$, and it is true when $x > 2$.

For any real number b the figure formed by the graph of an inequality of the form $x \geq b$ (see Example 7) is called a *ray*. (The figure formed by the graph of an inequality of the form $x \leq b$ is also a ray. The sentence $x \leq b$ is read "x is less than or equal to b.")

Example 8: Graph the sentence $x + 3 \not< 5$ for real numbers x.

Solution: This sentence is read "$x + 3$ is not less than 5." This is equivalent to saying that $x + 3$ is greater than or equal to 5; $x + 3 \geq 5$. Thus the solution set is the same as for Example 7.

Example 9: Graph the sentence $-1 \leq x \leq 3$ for real numbers x.

Solution: This sentence is true when x is replaced by any real number -1 through 3.

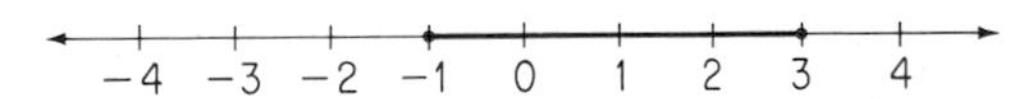

For any real numbers a and b where $a < b$ the graph of the sentence $a \leq x \leq b$ (see Example 9) is called a *line segment*.

Example 10: Graph the sentence $-1 < x < 3$ for real numbers x.

Solution: The solution set consists of all the real numbers *between* -1 and 3.

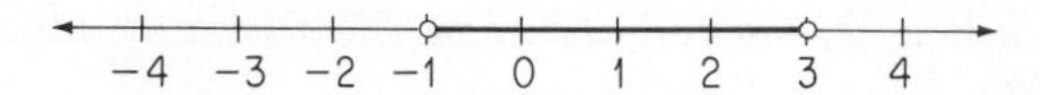

The points with coordinates -1 and 3 are the *endpoints* of the line segment graphed in Example 9. Notice that the graph in Example 10 can be obtained from the graph in Example 9 by removing the endpoints of the line segment. When we wish names for each of these graphs, a line segment with its endpoints is called a **closed line segment;** a line segment without either endpoint is an **open line segment;** a line segment with one endpoint but not both is neither closed nor open. For example, the following graph of $-1 \leq x < 3$ is neither closed nor open.

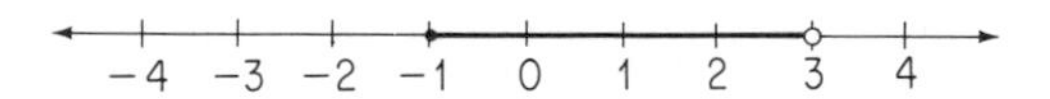

Example 11: Graph the sentence $x + 2 = 2 + x$ for real numbers x.

Solution: This sentence is true for all replacements of x; it is an identity. The solution set is the set of all real numbers. Thus the graph is the entire number line.

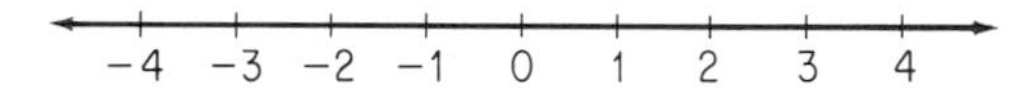

Example 12: Graph the sentence $x + 3 \leq 5$ for whole numbers x.

Solution: The solution set is $\{0, 1, 2\}$ and is graphed as a set of three points.

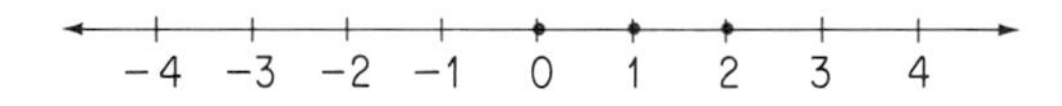

Example 13: Graph the sentence $x + 3 > 5$ for whole numbers x.

Solution: The solution set is $\{3, 4, 5, \ldots\}$, an infinite set of whole numbers. The sentence may be graphed as follows, where the heavily shaded arrowhead is used to indicate that the graph continues in the manner shown. The solution set consists of the set of all whole numbers greater than or equal to 3.

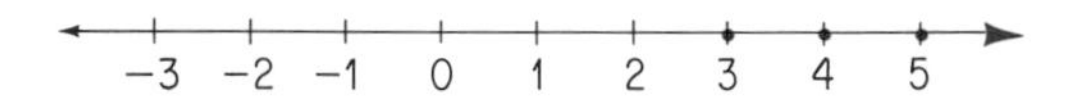

EXERCISES

For which of these sentences is the solution set always the empty set? Which sentences are identities?

1. $x + 3 > x$

2. $x + 2 \neq x$

3. $x + 2 \neq 2 + x$

4. $x > x + 3$

5. $x < x + 1$

6. $x + 1 < x + 2$

7. $x + 2 < x + 1$

8. $x > x - 1$

Find the solution set for each sentence when the replacement set is the set of whole numbers.

9. $x + 3 = 8$

10. $x - 3 = 2$

11. $x + 2 < 6$

12. $x - 2 < 6$

13. $x - 2 < 5$

14. $x - 2 > 5$

Find the solution set for each sentence when the replacement set is the set of integers.

15. $x + 1 < 5$

16. $x - 1 < 5$

17. $x - 2 < 1$

18. $x - 2 > 3$

Describe the solution set for each sentence when the replacement set is the set of real numbers.

19. $x + 1 < 3$

20. $x + 1 > 3$

21. $x - 2 < 4$

22. $x - 2 > 4$

23. $x + 2 \neq 7$

24. $x + 3 \neq 3$

Identify the graph of the solution set of each sentence as a point, a half-line, a ray, a line segment, or a line.

25. $x + 3 = 7$

26. $x - 2 > 7$

27. $x + 2 \geq 5$

28. $x < x + 3$

29. $-2 \leq x \leq 5$

30. $x + 3 \leq 5$

31. $x + 1 > x$

32. $x - 2 = 7$

33. $x - 1 < 5$

34. $-3 \leq x \leq 0$

35. $x + 2 \not> 5$

36. $x + 2 \not\leq x + 1$

Graph each sentence for real numbers x.

37. $x + 1 > 3$

38. $x - 2 \leq 5$

39. $-3 \leq x \leq 4$

40. $-2 < x < 2$

41. $x + 3 \geq 5$

42. $x - 2 < 4$

43. $x + 2 > 4$

44. $x + 2 \not< 4$

45. $x + 3 > x$

46. $x + 3 = x$

47. $x + 2 \not> 5$

48. $x + 3 \neq 5$

49. $x - 3 < 1$

50. $x - 1 < 3$

Graph each set for real numbers.

**51.* $\{x | 3x - 2 \geq -8\}$

**52.* $\{x | 2x + 5 \leq 1\}$

**53.* $\{x | x^2 = 36\}$

**54.* $\{x | x^2 + 3 = 28\}$

**55.* $\{x | x^2 \leq 9\}$

**56.* $\{x | x^2 > 16\}$

PEDAGOGICAL EXPLORATIONS

1. Elementary school students work with sentences and equations at various levels of abstraction. For example, each of the following is a sample of a type of question asked at an early grade level:

   The sum of 3 and some number is 5. What is the number?
   3 + □ = 5; what is the replacement for □ to obtain a true statement?
   3 + ? = 5; replace ? by a number to form a true sentence.
   3 + __ = 5; fill in the blank to form a true sentence.
   3 + *n* = 5; find *n*.
   3 + *x* = 5; solve for *x*.

   Review a recently published series of textbooks for grades 1, 2, and 3 and collect examples that illustrate this early introduction of the concept of a variable.

2. Do you believe that there is a specific age at which a child is first able to comprehend the concept of using a letter as a placeholder? That is, when does a child first understand the meaning of a sentence such as 3 + *x* = 5? Explore this concept with several preschool children of various ages. Use blocks or other aids and try to build up such an abstraction. Repeat this procedure with children in the early elementary grades and discuss your results in class.

3. About 1650 B.C., an Egyptian scribe named Ahmes copied an earlier manuscript which he described as "the entrance into knowledge of all existing things and all obscure secrets." Ahmes' copy is often called the Rhind Papyrus. It contains 85 problems. Here is one of those problems.

   A number and its one-fourth added together become 15.
   What is the number?

   Variables were not known 3000 years ago. Algebra did not exist. Problems were solved by arithmetic with a procedure that we now call the *method of false position*. This method can be used for some problems but is not useful for many others. For the preceding problem, we note that we need to take one-fourth of the number, so we try 4.

$$4 + \frac{1}{4}(4) = 4 + 1 = 5$$

If we try 4, we get 5. But we need 15, that is, 3×5. Therefore the answer is 3×4, that is, 12.

Try the method of false position for each of these problems from the Rhind Papyrus.

(a) A number and its one-fifth added together become 21. What is the number?

(b) A number, its one-third, and its one-quarter added together become 2. What is the number?

(c) If a number and its two-thirds are added together and from the sum one-third of the sum is subtracted, then 10 remains. What is the number?

9-2 compound sentences

Let us now turn our attention to compound sentences in algebra. For example, consider this compound sentence:

$$x + 1 > 2 \quad \text{and} \quad x - 2 < 1$$

Since no replacement set for x is specified, we assume that the replacement set is the set of real numbers. The sentence $x + 1 > 2$ is true for all x greater than 1; the sentence $x - 2 < 1$ is true for all x less than 3. Recall (§1-8 and §2-2) that a compound sentence of the form $p \wedge q$ (p and q) is true only when both parts are true. Thus the given compound sentence is true for the set of elements in the *intersection* of the two sets. Graphically we can show this as follows:

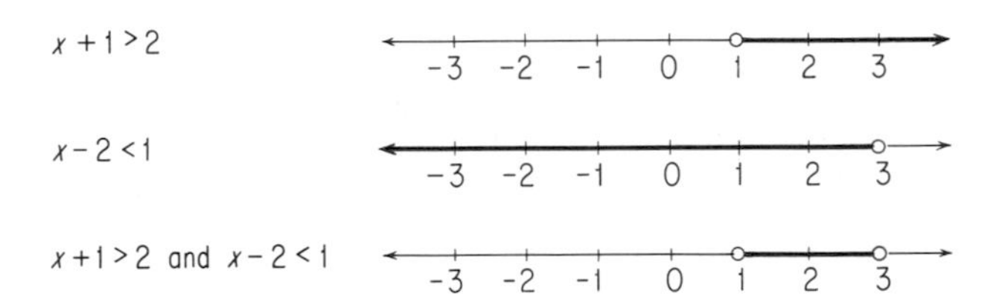

The graph of the compound sentence consists of an open line segment and can be described as

$$1 < x < 3$$

The solution set of this compound sentence can be written in *set-builder notation* as

$$\{x \mid x > 1\} \cap \{x \mid x < 3\}$$

This is read as "the intersection of the set of all x such that x is greater than 1 and the set of all x such that x is less than 3."

Example 1: Find the solution set for integers x:

$$x \geq -2 \quad \text{and} \quad x + 1 \leq 4$$

Solution: Here we want the set of integers that are greater than or equal to -2 but at the same time are less than or equal to 3. (If $x + 1 \leq 4$, then $x \leq 3$.) The solution set is $\{-2, -1, 0, 1, 2, 3\}$.

Example 2: Find the solution set for real numbers x:

$$x + 3 < 5 \quad \text{and} \quad 3 < 1$$

Solution: You will note that the second part of this sentence $(3 < 1)$ is false. If part of a sentence of the form $p \wedge q$ is false, then, as you may recall, the entire sentence is false. Thus the solution set is the empty set.

Next we consider a compound sentence involving the connective "or":

$$x + 1 < 2 \quad \text{or} \quad x - 2 > 1$$

The sentence $x + 1 < 2$ is true for all x less than 1; the sentence $x - 2 > 1$ is true for all x greater than 3. Recall that a sentence of the form $p \vee q$ (p or q) is true unless both parts are false. Thus the given compound sentence is true for the set of elements in the *union* of the two sets. Graphically we have the following:

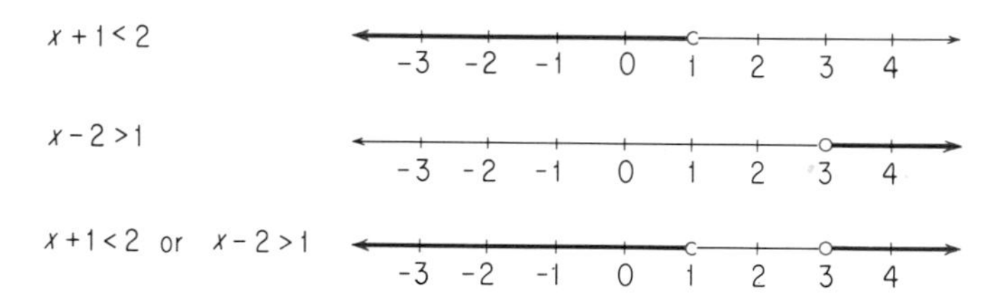

The graph of the compound sentence consists of the union of two half-lines. The solution set can be written in set-builder notation as

$$\{x \mid x < 1\} \cup \{x \mid x > 3\}$$

This is read as "the union of the set of all x such that x is less than 1 and the set of all x such that x is greater than 3."

Example 3: Find the solution set for real numbers x:

$$x + 5 > x \quad \text{or} \quad x + 2 < 5$$

Solution: As noted in §1-8 and §2-1, the compound sentence $p \vee q$ is true if at least one of the parts is true. Since the first part of the given sentence is always true, the whole sentence is true. That is, the compound sentence is true for *all* real numbers x; the solution set is the entire set of real numbers.

Example 4: Graph the solution set

$$x \leq -1 \quad \text{or} \quad x \geq 2$$

Solution: The graph is the union of two rays.

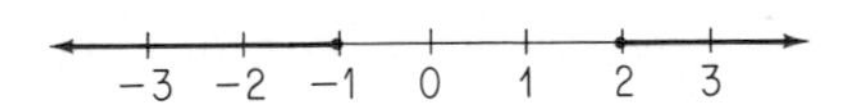

Example 5: Graph the sentence

$$x = -2 \quad \text{or} \quad x \geq 1$$

Solution:

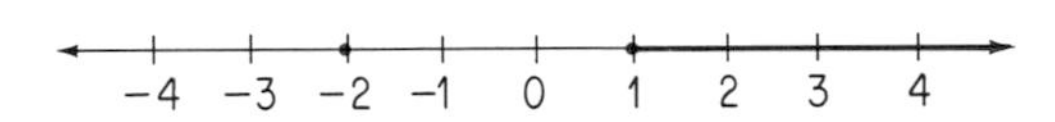

Example 6: Find the solution set

$$x + 2 \neq 2 + x \quad \text{or} \quad x + 2 < x$$

Solution: Both parts of the sentence are always false; the solution set is the empty set.

Example 7: Find the solution set

$$\{x \mid (x - 2)(x + 1) = 0\}.$$

Solution: The product of two numbers is zero if and only if one or the other or both of the numbers is zero. Therefore, if $(x - 2)(x + 1) = 0$, then $x - 2 = 0$ or $x + 1 = 0$. Thus $x = 2$ or $x = -1$; the solution set is $\{2, -1\}$.

EXERCISES

Give the solution set for integers x.

1. $x \geq 1$ and $x + 1 \leq 6$

2. $x \leq -2$ and $x > -5$

3. $x > 1$ and $x < 5$

4. $x \geq 2$ or $x \leq 1$

Graph each sentence for real numbers x.

5. $x \geq 2$ and $x \leq 5$

6. $x \leq 0$ or $x \geq 3$

7. $x + 2 < 5$ and $x \geq 0$

8. $x - 2 > 5$ or $x < 0$

9. $x + 3 \geq 5$ or $x - 1 < 0$

10. $x + 1 \geq 3$ or $x = -2$

11. $x + 3 \geq 5$ and $x - 1 < 0$

12. $x + 2 \geq 2$ and $x - 1 < 3$

13. $x - 1 > x$ and $x + 2 = 7$

14. $x < x + 1$ and $x - 3 > 5$

15. $x + 2 \neq x$ or $x + 2 \leq 5$

16. $x + 1 = x$ or $|x| < 0$

Graph each set for real numbers x.

17. $\{x \mid x > 0\} \cap \{x \mid x < 3\}$

18. $\{x \mid x \leq -2\} \cup \{x \mid x \geq 3\}$

19. $\{x \mid x + 2 < 4\} \cup \{x \mid x \geq 4\}$

20. $\{x \mid x - 1 < 3\} \cap \{x \mid x \geq -3\}$

21. $\{x \mid x(x + 3) = 0\}$

22. $\{x \mid (x - 2)(x + 5) = 0\}$

23. $\{x \mid (x + 2)(x + 5) = 0\}$

24. $\{x \mid (x - 3)(x - 1) = 0\}$

***25.** $\{x \mid x(x + 3) \geq 0\}$

***26.** $\{x \mid (x - 3)(x + 5) \leq 0\}$

***27.** $\{x \mid (x + 2)(x + 5) < 0\}$

***28.** $\{x \mid (x - 5)(x - 3) > 0\}$

PEDAGOGICAL EXPLORATIONS

1. Prepare a set of transparencies suitable for demonstrating graphs of compound sentences to an upper elementary or junior high school mathematics class. Use a number line as a base, and then two other transparencies in different colors for the graphs of two other sentences. For example, one transparency could have the graph of $x \geq 3$ and the other one the graph of $x \leq 5$. Each of these can be shown separately, and then superimposed on the base to show their union and their intersection.
2. Compound sentences in various forms are common at many grade levels, as in the following examples. Find the solution set of each sentence where $n \in \{1, 2, 3, 4, 5, \ldots, 25\}$.
   (a) n is a multiple of 5 and an odd number.
   (b) n is less than 17 but greater than 11.
   (c) n is a multiple of 3 and $2n$ is a multiple of 12.
   (d) n is divisible by 3 and by 2.
   (e) n is divisible by 3 or by 11.

9-3 sentences of the first degree

So far we have been solving sentences by informal methods. For example, if we are given that

$$x - 3 = 7$$

we recognize that x may be obtained as $(x - 3) + 3$; that is,

$$(x - 3) + 3 = x + (-3 + 3) = x + 0 = x$$

Therefore, we may solve the given equation as follows:

$$\begin{aligned} x - 3 &= 7 \\ (x - 3) + 3 &= 7 + 3 \\ x &= 10 \end{aligned}$$

We have added 3 to both members of the equation to obtain an **equivalent sentence,** that is, a sentence with the same solution set as the given sentence. Formally, we have used a property of equations. Here are some of the common properties of statements of equality involving real numbers a, b, c, and d:

Reflexive, =: $a = a$.

Addition, =: If $a = b$ and $c = d$, then $a + c = b + d$.

Multiplication, =: If $a = b$ and $c = d$, then $ac = bd$.

Uses of these properties together with the field properties of §6-1 and §7-4 are shown in the examples that follow.

Example 1: Solve $x + 3 = 7$, and explain each step.

Solution:

Statements	*Reasons*
$x + 3 = 7$	Given.
$(x + 3) + (-3) = 7 + (-3)$	Addition, =.
$x + [3 + (-3)] = 7 + (-3)$	Associative, +.
$x + 0 = 4$	Addition.
$x = 4$	Zero, +. (Addition property of zero.)

Example 2: Solve $2x - 3 = 7$, and explain each step.

Solution:

Statements	*Reasons*
$2x - 3 = 7$	Given.
$2x + (-3) = 7$	Definition, −.
$[2x + (-3)] + 3 = 7 + 3$	Addition, =.
$2x + [(-3) + 3] = 7 + 3$	Associative, +.
$2x + 0 = 10$	Addition.
$2x = 10$	Zero, +.
$\frac{1}{2}(2x) = \frac{1}{2} \times 10$	Multiplication, =.
$(\frac{1}{2} \times 2) \cdot x = \frac{1}{2} \times 10$	Associative, ×.
$1x = 5$	Multiplication.
$x = 5$	One, ×. (Multiplication property of one.)

In the preceding examples, the steps were listed in order to demonstrate the use of the field properties in the solution of equations. Normally, however, one completes most of these steps mentally. Thus the solution for Example 2, in condensed form, might appear as follows:

$$\begin{aligned} 2x - 3 &= 7 \\ 2x &= 10 \\ x &= 5 \end{aligned}$$

Notice that subtraction is not needed when solving equations since

$$a - b = a + (-b)$$

Similarly, division is not needed since

$$a \div b = a \times \frac{1}{b}$$

The methods used in Examples 1 and 2 may be used to solve any sentence that is an equation of the **first degree in one variable** x, that is, any sentence that can be expressed in the form

$$ax + b = 0, \qquad a \neq 0$$

The expression $ax + b$ is called an **expression of the first degree** in x.

Sentences involving inequalities may also be solved by making use of certain basic properties. First we list some of the common properties of order relations among real numbers a, b, c, and d.

Addition, <: If $a < b$ and $c = d$, then $a + c < b + d$.

Addition, >: If $a > b$ and $c = d$, then $a + c > b + d$.

Multiplication, <: If $a < b$, then $ac < bc$ if $c > 0$.
If $a < b$, then $ac > bc$ if $c < 0$.

Multiplication, >: If $a > b$, then $ac > bc$ if $c > 0$.
If $a > b$, then $ac < bc$ if $c < 0$.

In general, if the same number is added to both members of a statement of inequality, the "sense" of the inequality is preserved. Thus in the following examples the first sentence listed is true, and the second one must also be true.

$$\begin{array}{rr} 2 < 5 & \qquad 2 + 3 < 5 + 3 \\ 8 > 3 & \qquad 8 + 5 > 3 + 5 \\ -3 < 5 & \qquad -3 + 1 < 5 + 1 \end{array}$$

If both members of a true statement of inequality are multiplied by the same positive number, the resulting sentence is also true. If both members of an inequality are multiplied by a negative number, it is necessary to reverse the sense of the inequality to obtain an equivalent sentence. Thus in the following examples each member of the first inequality is multiplied by the same positive number, and the order of the inequality is maintained.

$$\begin{array}{ll} 3 < 7 & \qquad 2 \times 3 < 2 \times 7 \\ 5 > -1 & \qquad 3 \times 5 > 3 \times -1 \end{array}$$

In the next two examples, each member of the first inequality is multiplied by the same negative number, and the order of the inequality is reversed.

$$2 < 8 \qquad -3 \times 2 > -3 \times 8$$
$$3 > -1 \qquad -2 \times 3 < -2 \times (-1)$$

The uses of properties of order relations are shown in the examples that follow. Again, many steps are indicated although with practice one normally completes some of these steps mentally.

Example 3: Solve $\frac{1}{2}x - 2 > 5$ and explain each step.

Solution:

Statements	*Reasons*
$\frac{1}{2}x - 2 > 5$	Given.
$\frac{1}{2}x + (-2) > 5$	Definition, −.
$[\frac{1}{2}x + (-2)] + 2 > 5 + 2$	Addition, >.
$\frac{1}{2}x + [(-2) + 2] > 5 + 2$	Associative, +.
$\frac{1}{2}x + 0 > 7$	Addition.
$\frac{1}{2}x > 7$	Zero, +.
$2(\frac{1}{2}x) > 2 \times 7$	Multiplication, >.
$(2 \times \frac{1}{2})x > 2 \times 7$	Associative, ×.
$1x > 14$	Multiplication.
$x > 14$	One, ×.

Example 4: Solve $-2x + 3 < 7$ and explain each step.

Solution:

Statements	*Reasons*
$-2x + 3 < 7$	Given.
$(-2x + 3) + (-3) < 7 - (-3)$	Addition, <.
$-2x + [3 + (-3)] < 7 + (-3)$	Associative, +.
$-2x + 0 < 4$	Addition.
$-2x < 4$	Zero, +.
$-\frac{1}{2}(-2x) > (-\frac{1}{2})4$	Multiplication, <.
$(-\frac{1}{2} \times -2)x > (-\frac{1}{2})4$	Associative, ×.
$1x > -2$	Multiplication.
$x > -2$	One, ×.

Note the change in the sense of the inequality from $<$ to $>$ when both members were multiplied by the negative number, $-\frac{1}{2}$. Also note that both members were multiplied by $-\frac{1}{2}$ inasmuch as the product $(-\frac{1}{2} \times -2)x = 1x$, thus allowing us to solve for x as required.

EXERCISES

Explain each step in the following solutions.

1. $x + 2 = 7$

(a) $(x + 2) + (-2) = 7 + (-2)$
(b) $x + [2 + (-2)] = 7 + (-2)$
(c) $x + 0 = 5$
(d) $x = 5$

2. $x + 2 < 7$

(a) $(x + 2) + (-2) < 7 + (-2)$
(b) $x + [2 + (-2)] < 7 + (-2)$
(c) $x + 0 < 5$
(d) $x < 5$

3. $2x + 3 = 7$

(a) $(2x + 3) + (-3) = 7 + (-3)$
(b) $2x + [3 + (-3)] = 7 + (-3)$
(c) $2x + 0 = 4$
(d) $2x = 4$
(e) $\frac{1}{2}(2x) = \frac{1}{2}(4)$
(f) $(\frac{1}{2} \times 2)x = \frac{1}{2}(4)$
(g) $1x = 2$
(h) $x = 2$

4. $-\frac{2}{3}x + 2 < 8$

(a) $(-\frac{2}{3}x + 2) + (-2) < 8 + (-2)$
(b) $-\frac{2}{3}x + [2 + (-2)] < 8 + (-2)$
(c) $-\frac{2}{3}x + 0 < 6$
(d) $-\frac{2}{3}x < 6$
(e) $(-\frac{3}{2})(-\frac{2}{3}x) > (-\frac{3}{2}) \times 6$
(f) $[(-\frac{3}{2})(-\frac{2}{3})] \times x > (-\frac{3}{2}) \times 6$
(g) $1x > -9$
(h) $x > -9$

Solve for x.

5. $3x + 2 = 14$

6. $2x - 5 = 9$

7. $-3x + 1 = 10$

8. $-2x + 3 = 7$

9. $\frac{1}{2}x - 3 = 8$

10. $\frac{2}{3}x - 7 = 5$

11. $2x + 3 < 9$

12. $3x - 2 < 10$

13. $3x - 1 > 8$

14. $2x + 1 > 7$

15. $-2x + 1 < 9$

16. $-3x + 2 < 8$

17. $-\frac{1}{2}x + 3 < 7$

18. $-\frac{1}{3}x + 1 > 3$

19. $-\frac{3}{4}x - 2 > 10$

20. $-\frac{2}{3}x - 4 < 8$

21. $7x + 5 = 3x - 7$

22. $2(x + 1) = 2x - 3$

23. $3x - 1 > x + 5$

24. $3x + 5 > 3(x + 1)$

PEDAGOGICAL EXPLORATIONS

1. Many slow learners are helped to write and solve equations through the use of flow charts. For example, note the equations that the following flow charts describe.

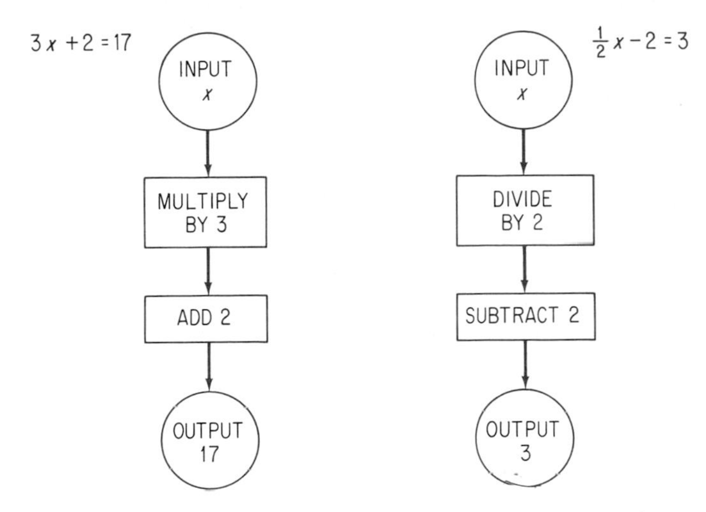

Draw flow charts to describe each of the following equations.

(a) $2x + 3 = 7$ **(b)** $3x - 5 = 10$
(c) $\frac{1}{2}x - 1 = 7$ **(d)** $\frac{1}{4}x + 2 = 6$

2. Students can be aided in their attempts to solve equations by drawing reverse flow charts. Thus they work backwards, starting with the output and using inverse operations as in these examples.

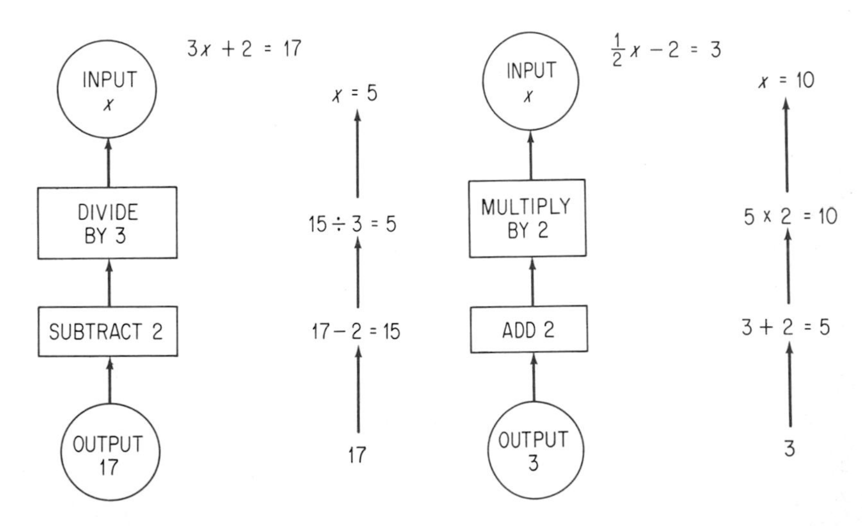

Draw reverse flow charts to solve each of the equations in Exploration 1 above.

3. Many mathematical "tricks" can be explained and developed through the use of algebraic techniques. Consider, for example, this set of directions.

   Think of a number.
   Multiply your number by 2.
   Add 9.
   Subtract 3.
   Divide by 2.
   Subtract the number you started with.
   Your final answer will always be 3.

   Use n to represent the original number, and write a mathematical phrase to represent each step in the set of directions given above. For example, the first three steps would give n, $2n$, and $2n + 9$. From this show why the result is always 3, regardless of the number one starts with.

4. Make up several similar "tricks" of your own corresponding to the one in Exploration 3.

9-4 sentences involving absolute value

The absolute value of a number x was defined on the number line in §5-4 as the distance from the origin of the point with coordinate x. Thus $|3| = 3$, $|-3| = 3$, and in general for any real number x:

$$|x| = x, \quad \text{if } x \text{ is positive or zero.}$$
$$|x| = -x, \quad \text{if } x \text{ is negative.}$$

Note that if x is negative, then $-x$ is the opposite of x and therefore is a positive number. Thus if $x = -3$, $|-3| = -(-3) = 3$. Accordingly, the absolute value of any real number different from zero is a positive number; $|0| = 0$.

Here is a flow chart for finding the absolute value of a number x.

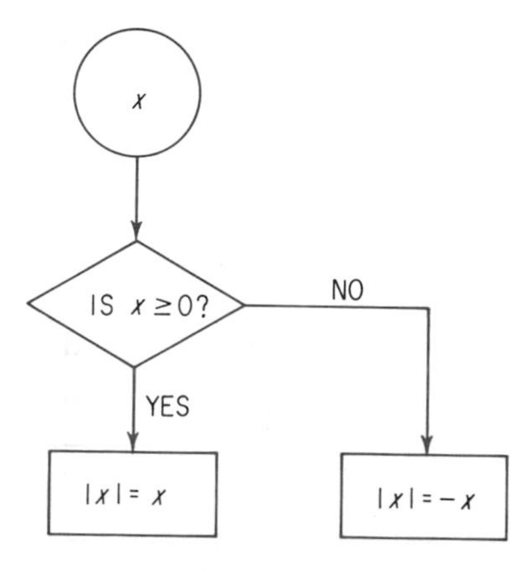

Example 1: Find the solution set for $|x| = 3$.

Solution: On a number line, the points with coordinates 3 and -3 are each three units away from the origin. Thus there are two numbers whose absolute value is 3, namely, 3 and -3. The solution set is $\{-3, 3\}$. This solution set can be described by the compound sentence

$$x = -3 \quad \text{or} \quad x = 3$$

The instruction to find the solution set of a sentence may be indicated briefly as "**solve**". Similarly, the instruction to graph a sentence may be abbreviated as "**graph.**"

Example 2: Graph $|x| \geq 3$.

Solution: We are to find all points on the number line that are at a distance of three units or more than three units from the origin.

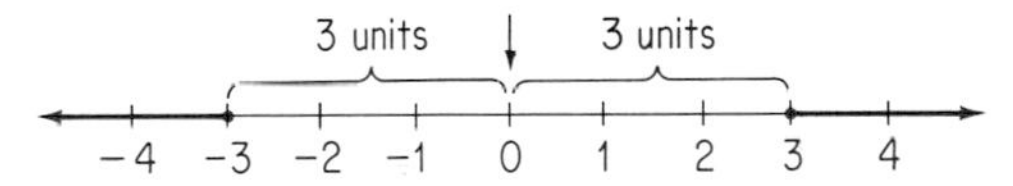

Note that the solution set of Example 2 can be described by the compound sentence

$$x \leq \;\; 3 \quad \text{or} \quad x \geq 3$$

In set-builder notation we may write this as

$$\{x \mid x \leq -3\} \cup \{x \mid x \geq 3\}$$

We may generalize the results of Example 2 to provide this definition of $|x| \geq k$ for any real number k:

$$\{\mathbf{x} \mid |\mathbf{x}| \geq \mathbf{k}\} = \{\mathbf{x} \mid \mathbf{x} \leq -\mathbf{k}\} \cup \{\mathbf{x} \mid \mathbf{x} \geq \mathbf{k}\}$$

Example 3: Graph $|x| \leq 3$.

Solution: We are to find all points on the number line that are at a distance of three units or less than three units from the origin; these are the points that are "at most three units" from the origin.

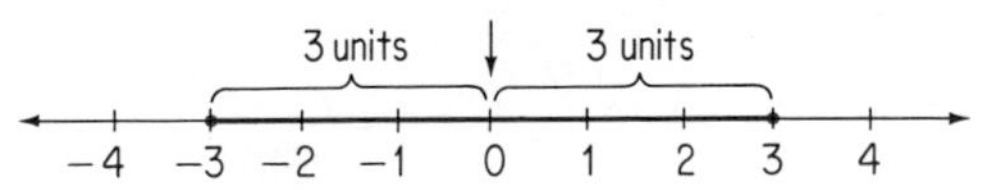

The solution set of Example 3 can be described in each of these ways:

$$x \geq -3 \quad \text{and} \quad x \leq 3$$

$$\{x \mid x \geq -3\} \cap \{x \mid x \leq 3\}$$

$$\{x \mid -3 \leq x \leq 3\}$$

We may generalize the results of Example 3 to provide this definition for $|x| \leq k$ for any nonnegative number k:

$$\{x \mid |x| \leq k\} = \{x \mid x \geq -k\} \cap \{x \mid x \leq k\}$$
$$= \{x \mid -k \leq x \leq k\}$$

Example 4: Solve $|x - 1| = 3$.

Solution: As in Example 1, if the absolute value of a number is 3, then that number is 3 or -3. If $x - 1 = 3$, then $x = 4$; if $x - 1 = -3$, then $x = -2$. Thus the solution set is $\{-2, 4\}$.

The points with coordinates -2 and 4 of the solution set for Example 4 are each at a distance of three units from the point with coordinate 1. Thus the given sentence

$$|x - 1| = 3$$

may be interpreted in terms of distance on a number line. The distance between the point with coordinate x and the point with coordinate 1 is 3. This interpretation is illustrated in the following figure.

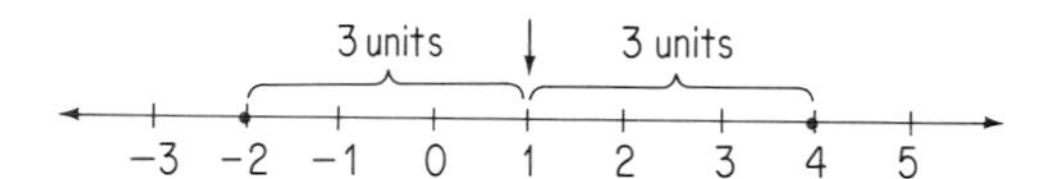

The absolute value notation may be used to indicate the distance between any two points on the number line. Thus $|x - h|$ represents the distance between the point with coordinate x and the point with coordinate h. Notice that the distance between the points with coordinates 2 and 5 may be denoted either as $|5 - 2|$ or as $|2 - 5|$, since $|5 - 2| = |3| = 3$ and $|2 - 5| = |-3| = 3$. In general, $|x - h| = |h - x|$ and the distance between any two distinct points is a positive number.

If we are told that $|x + 1| = 3$, we may consider this sentence in the form $|x - (-1)| = 3$; that is, the distance between the point with coordinate x and the point with coordinate -1 is 3. Thus, as shown in the figure, $x = -4$ or $x = 2$.

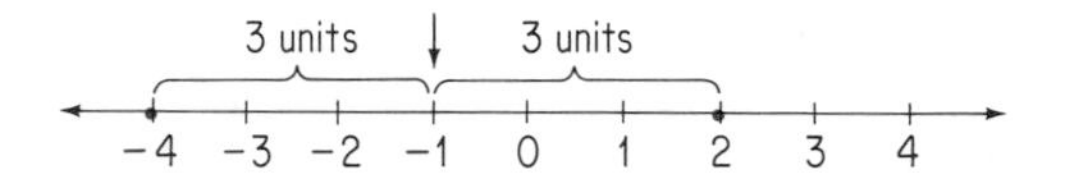

The problem $|x+1|=3$ can also be solved by using the fact that if the absolute value of a number is 3, then that number must be 3 or -3. If $x+1=3$, then $x=2$; if $x+1=-3$, then $x=-4$. Thus if $|x+1|=3$, then $x=-4$ or $x=2$.

Example 5: Graph $\{x \mid |x-1| \geq 3|$.

Solution: Consider the definition for $|x| \geq k$, and replace x by $x-1$:

$$\{x \mid |x-1| \geq 3\} = \{x \mid x-1 \leq -3\} \cup \{x \mid x-1 \geq 3\}$$

If $x-1 \leq -3$, then $x \leq -2$; if $x-1 \geq 3$, then $x \geq 4$. Thus the solution set is $\{x \mid x \leq -2 \text{ or } x \geq 4\}$; that is,

$$\{x \mid x \leq -2\} \cup \{x \mid x \geq 4\}.$$

The graph of the solution set is:

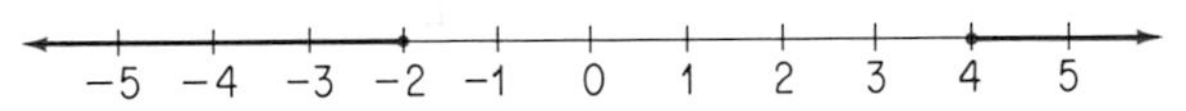

The graph of $|x-1| \geq 3$ is the union of two rays and consists of the points on the number line that are at a distance of three units or more than three units from the point with coordinate 1.

Example 6: Graph $|x-1| \leq 3$.

Solution: The graph of $|x-1| \leq 3$ consists of the points of the number line that are at a distance of at most three units from the point with coordinate 1. Thus the graph is a line segment.

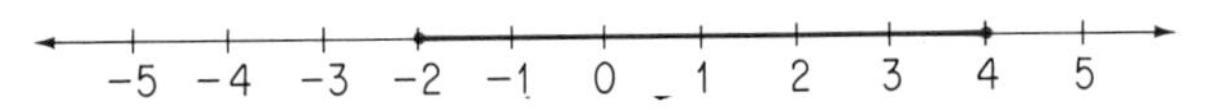

Note that the sentence in Example 6 can be solved algebraically by applying the definition for $|x| \leq k$, and replacing x by $x-1$:

$$\{x \mid |x-1| \leq 3\} = \{x \mid x-1 \geq -3\} \cap \{x \mid x-1 \leq 3\}$$

If $x-1 \geq -3$, then $x \geq -2$; if $x-1 \leq 3$, then $x \leq 4$. Thus the solution set is:

$$\{x \mid x \geq -2 \quad \text{and} \quad x \leq 4\} = \{x \mid x \geq -2\} \cap \{x \mid x \leq 4\}$$
$$= \{x \mid -2 \leq x \leq 4\}$$

EXERCISES

Give the set of integers for which each sentence is true.

1. $|x| = 5$

2. $|x| \leq 2$

3. $|x| < 3$

4. $|x| = -3$

5. $|x-1|=4$

6. $|x+1|=6$

7. $|x+2|=2$

8. $|x-2|=5$

Evaluate.

9. $|-9|$

10. $|-3|+|-5|$

11. $|(-3)+(-5)|$

12. $(|-6|)^2$

13. $|(-6)^2|$

14. $|-5|\times|-9|$

Identify the graph of each sentence on a real number line as two points, a line segment, or the union of two rays.

15. $|x|\leq 7$

16. $|x-2|=5$

17. $|x|\geq 1$

18. $|x+3|\leq 1$

Graph each sentence on a real number line.

19. $|x|\leq 3$

20. $|x|\geq 2$

21. $|x|>2$

22. $|x|<3$

23. $|x-1|\geq 3$

24. $|x+1|\leq 2$

25. $|x|=-2$

26. $|x|\geq 0$

27. $|x+2|\geq 3$

28. $|x-3|\leq 1$

29. $|x-3|<1$

30. $|x+2|>3$

*31. $2\leq|x|\leq 4$

*32. $2\leq|x-1|\leq 5$

*33. $|x|+2=|x+2|$

*34. $|x|=-x$

*35. $|x|+|x-3|=3$

*36. $|x^2-10|\leq 6$

PEDAGOGICAL EXPLORATIONS

1. Assume that integers have been studied and graphed on a number line. Prepare a brief introduction of absolute value for use with elementary school students with this background. Include in your plan suggestions for the use of visual aids to help get this concept across.
2. In a first course in algebra the following definition is usually made:

$$\sqrt{x^2}=|x|$$

For example, as a result of this definition we may write

$$\sqrt{(3)^2}=|3|=3 \quad \text{and} \quad \sqrt{(-3)^2}=|-3|=3.$$

Many students carelessly and incorrectly write $\sqrt{x^2}=x$. Show how this incorrect statement can contradict the use of the radical symbol to denote the nonnegative square root of a number. (See §7-5.)

CHAPTER TEST

Which sentences have the empty set as their solution set? Which sentences are identities?

1. (a) $x + 1 = 1 + x$ (b) $x + 3 \neq x$

2. (a) $x > x + 1$ (b) $x - 1 < x$

Describe the solution set for each sentence in terms of a specific geometric figure.

3. (a) $x - 3 \leq 5$ (b) $x + 2 = 2 + x$

4. (a) $2 \leq x \leq 5$ (b) $x + 1 > 5$

List the elements in the solution set of each sentence for whole numbers x.

5. $x + 3 \leq 6$

6. $0 < x \leq 3$

7. $x - 2 > 4$

8. $|x| \leq 2$

Classify as true or false.

9. $3 > 1$ and $-2 < -3$

10. $3 < 7$ and $|-2| = 2$

11. $-1 > 0$ or $0 > -2$

12. $2 < 0$ or $|5| = -5$

List the set of integers for which each sentence is true.

13. $|x - 2| = 3$

14. $|x + 3| = 2$

Graph the solution set for each sentence for real numbers x.

15. $x + 2 > 5$

16. $x - 2 \leq 5$

17. $x \leq 3$ or $x + 1 > 6$

18. $x - 1 \leq 3$ and $x > 0$

19. $|x| \leq 6$

20. $|x + 2| \geq 6$

21. Graph the solution set for $A \cup B$, where sets A and B are defined for real numbers x as follows:

$$A = \{x | x + 2 \leq 8\}, \quad B = \{x | x - 1 > 3\}$$

Solve for x.

22. $3x + 1 = 10$

23. $-\frac{3}{4}x + 4 > 16$

24. $5 - 3x < 8$

25. $|3 - 2x| = 1$

READINGS AND PROJECTS

1. Read Booklet No. 8 (Number Sentences) of the twenty-ninth yearbook of the National Council of Teachers of Mathematics. Complete at least three of the exercise sets given there.

2. Prepare a 20-item multiple-choice test that evaluates the basic concepts of this chapter.
3. Read the article (" 'Equation' Means 'Equal' ") by Barbara E. Bernstein on pages 697 and 698 in the December 1974 issue of *The Arithmetic Teacher.* Give your reactions to her suggested procedures.
4. Read Booklet No. 17 (Hints for Problem Solving) of the thirtieth yearbook of the National Council of Teachers of Mathematics. Summarize the suggestions given for effective problem solving.
5. Read Chapter 11 (Skills) of the thirty-third yearbook of the National Council of Teachers of Mathematics. Prepare a report on the specific suggestions offered for the effective development of skills.
6. Prepare a visual aid that provides a geometrical demonstration for the algebraic identity $(a + b)^2 = a^2 + 2ab + b^2$. To do so, prepare a square sheet of cardboard to represent $(a + b)^2$. Then use scotch tape and have four smaller pieces that can be folded over to show the regions with areas a^2, ab, ab, and b^2 as in the figure.

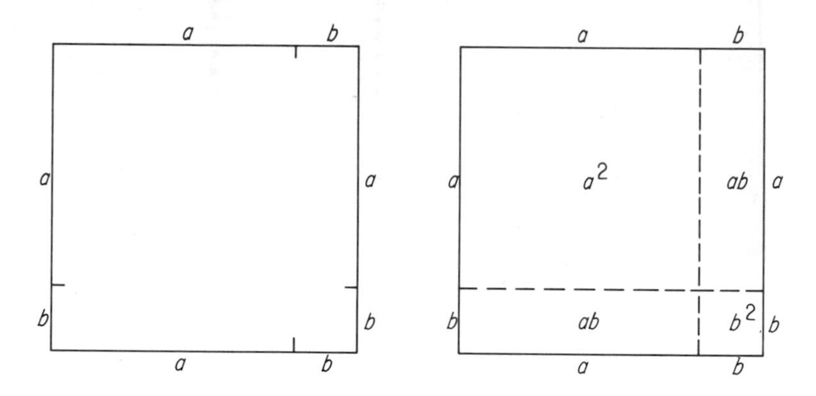

7. Try to prepare a visual aid that demonstrates the algebraic identity $(a - b)^2 = a^2 - 2ab + b^2$.
8. Begin a collection of mathematical tricks that can be explained through the use of algebraic techniques, as shown in the Pedagogical Explorations for §9-3. Here is another sample of a such a trick. Select any three-digit number with all digits the same, such as 444. Then find the sum of the digits. (In this case, $4 + 4 + 4 = 12$.) Finally, divide the original number by this sum ($444 \div 12 = 37$). The quotient will *always* be 37. Try to prove that this is so. (*Hint:* Represent a three-digit number as $100h + 10t + u$, where $h = t = u$.)
9. Read "What We Can Do With $a \times b = c$" by Ruth Erckmann on pages 181 through 183 of the March 1976 issue of *The Arithmetic Teacher.* Suggest at least two additional examples of the sentence $a \times b = c$ from the areas of arithmetic, geometry, algebra, and science.

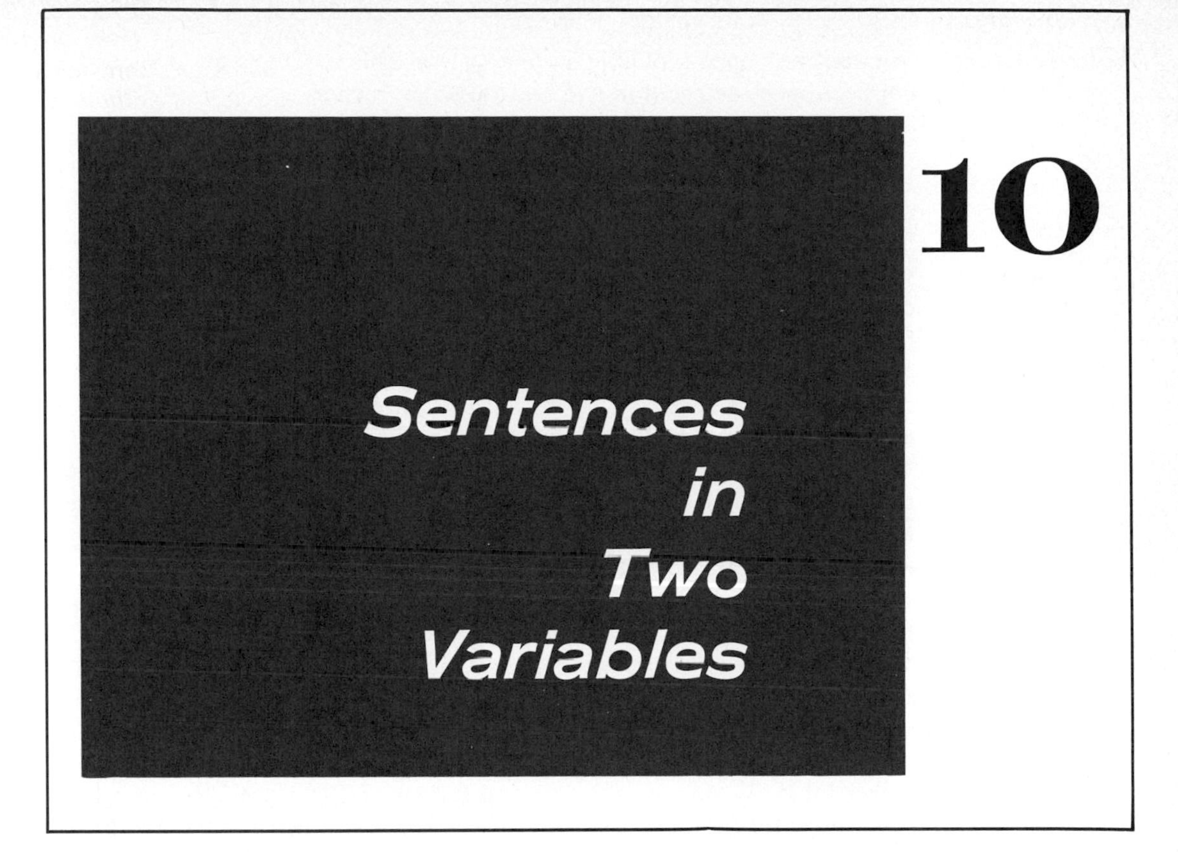

Sentences in two variables arise in many ways. For example, the sentence $d = 80h$ may be used to express the distance d in kilometers that John drives in h hours at 80 kilometers per hour. The sentence $p = 4s$ may be used to express the perimeter p of a square in terms of the side s. The sentence $I = 0.05P$ may be used to express the simple interest I on P dollars for one year at 5 per cent per year. The sentence $\mathrm{C} = \frac{5}{9}(\mathrm{F} - 32)$ may be used to express the temperature C in degrees Celsius in terms of the temperature F in degrees Fahrenheit.

In the seventeenth century the French mathematician René Descartes (1596–1650) made the initial discoveries of the branch of mathematics known as analytic geometry. Basically he provided a geometric interpretation for algebraic sentences in two variables. In the present chapter we shall have the opportunity to study such sentences and their graphs in light of Descartes' geometry.

Open sentences in one real variable have been considered in Chapter 9. As in that chapter we now use number concepts and geometric representations to explore and to explain open sentences in two real variables. The geometric representation of each value of one real variable is a point on the number line. The geometric representation of each ordered pair of real numbers is a point on a coordinate plane.

10-1 solution sets

We discussed open sentences in one real variable in Chapter 9. Now let us consider open sentences in two real variables, such as

$$x + y = 8$$

If a replacement is made for only one of the variables, the sentence $x + y = 8$ remains an open sentence. For example, if x is replaced by 5, the sentence becomes $5 + y = 8$, which is neither true nor false, since no value has been specified for the real number y. Thus, a *pair* of replacements is needed for an open sentence in two variables before we can determine whether it is true or false for these replacements.

If x is replaced by 5 and y is replaced by 3, the sentence $x + y = 8$ is true. Usually we think of the variables as an *ordered pair* (x, y) and speak of the replacements as an ordered pair of numbers (5, 3). By convention the variable x is assumed to be the first element of the ordered pair and is called the **independent variable** since we may replace it by any arbitrarily selected member of the replacement set. The first number in the ordered pair of numbers is the replacement for x, that is, the replacement for the independent variable. The variable y is called the **dependent variable** since the value that may be used to replace y depends upon the value selected for x. Thus the second element of the ordered pair is the replacement for y, that is, the replacement for the dependent variable. For the ordered pair (7, 1) we have $x = 7$ and $y = 1$; for the ordered pair (1, 7) we have $x = 1$ and $y = 7$. The sentence $x + y = 8$ is true or false for specified ordered pairs of numbers. It is true for (5, 3), since $5 + 3 = 8$; it is false for (2, 7), since $2 + 7 \neq 8$. The solution set for the sentence $x + y = 8$ is the set of ordered pairs of numbers for which the sentence is true. Indeed, the solution set for any sentence in two variables is a set of ordered pairs.

Just as with sentences in one variable, the solution sets of sentences in two variables depend upon the replacement sets of the variables used. For example, if the replacement set for x and y is the set of counting numbers, then the solution set for $x + y = 8$ is

$$\{(1, 7),(2, 6), (3, 5),(4, 4),(5, 3),(6, 2),(7, 1)\}$$

Example 1: Find the solution set for the sentence $x + y < 3$ if the replacement set for x and y is the set of whole numbers.

Solution: When $x = 0$, y must be less than 3. Thus the ordered pairs of numbers (0, 0), (0, 1), and (0, 2) make the sentence true. Similarly, when $x = 1$, y must be less than 2; when $x = 2$, y must be less than 1. Is there a whole number y that satisfies the inequality when $x = 3$? When $x > 3$? The solution set is

$$\{(0, 0), (0, 1), (0, 2), (1, 0), (1, 1), (2, 0)\}$$

Example 2: Find the solution set for the sentence $x + y \leq 3$ when x and y are counting numbers.

Solution: {(1, 1), (1, 2), (2, 1)}.

EXERCISES

Find the solution set for each sentence for the replacement set $\{1, 2, 3, 4\}$.

1. $x + y = 4$ **2.** $2x + y \leq 4$

3. $x + y \leq 3$ **4.** $y = x + 1$

5. $x + y < 2$ **6.** $y < x$

Find the solution set for each sentence, using the set of counting numbers as the replacement set for x *and* y.

7. $x + y = 6$ **8.** $x + y < 5$

9. $y \leq 4 - x$ **10.** $y = 5 - x$

For the replacement set $\{-3, -2, -1, 0, 1, 2, 3\}$ *find the solution set for each sentence.*

11. $y = |x|$ **12.** $y = |-x|$

13. $y = -|x|$ **14.** $y = |x| - 2$

15. $y = x^2 + 1$ **16.** $y = x^2 - 3$

PEDAGOGICAL EXPLORATIONS

Flow charts are an effective device for visualizing open sentences in two variables. Consider, for example, the following:

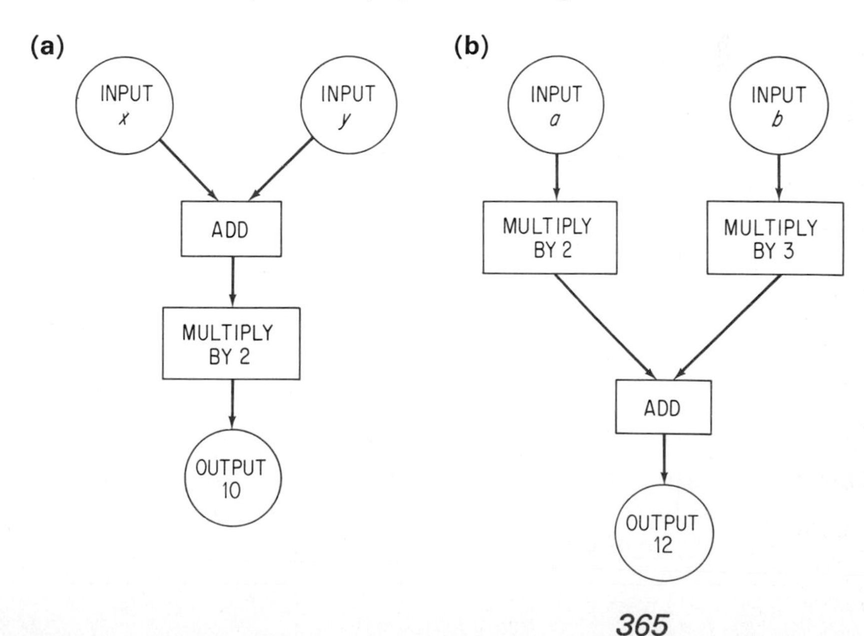

In (**a**) the equation developed is $2(x + y) = 10$. Following the steps indicated in (**b**) gives $2a + 3b = 12$. Draw a flow chart to show each of the following sentences:

1. $3x + 5y = 30$

2. $3(x + y) = 12$

3. $\frac{1}{2}x + 3y = 12$

4. $2x + \frac{1}{4}y = 16$

5. $3(2x + 3y) = 18$

6. $2(3x + y) = 24$

10-2 ordered pairs

A set of ordered pairs may be obtained from any given set of numbers. For example, consider the set of numbers

$$\mathcal{U} = \{1, 2, 3\}$$

We may form a set of all ordered pairs of numbers whose members are both elements of $\mathcal{U}$. A **tree diagram** is helpful in showing the possible pairs of numbers involved.

First element	*Second element*	*Ordered pairs*
1	1	(1, 1)
	2	(1, 2)
	3	(1, 3)
2	1	(2, 1)
	2	(2, 2)
	3	(2, 3)
3	1	(3, 1)
	2	(3, 2)
	3	(3, 3)

Each element of $\mathcal{U}$ may serve as the first member of an ordered pair and may be matched with each of the elements of $\mathcal{U}$ as the second member. As in §1-3 this set of ordered pairs is called the *Cartesian product* of $\mathcal{U}$ and $\mathcal{U}$, is written as $\mathcal{U} \times \mathcal{U}$, and is read "$\mathcal{U}$ cross $\mathcal{U}$." The set of all ordered pairs whose coordinates belong to the given set $\mathcal{U}$ is $\mathcal{U} \times \mathcal{U}$:

$$\mathcal{U} \times \mathcal{U} = \{(1, 1), (1, 2), (1, 3), (2, 1), (2, 2), (2, 3), (3, 1), (3, 2), (3, 3)\}$$

We shall refer to a set $\mathcal{U}$ as the **universal set** for a given problem.

Example 1: Let $\mathcal{U} = \{1, 2\}$ and give the set of elements in $\mathcal{U} \times \mathcal{U}$.

Solution: $\{(1, 1), (1, 2), (2, 1), (2, 2)\}$.

Example 2: Give the solution set for $x + y = 4$ if $\mathcal{U} = \{1, 2, 3, 4\}$.

Solution: We are permitted to replace x and y only by elements of $\mathcal{U}$. Thus for $x = 1, y = 3$; for $x = 2$; $y = 2$; for $x = 3$, $y = 1$. The solution set is $\{(1, 3), (2, 2), (3, 1)\}$. Why can we not let $x = 4$?

The Cartesian product $A \times B$ is the set of all ordered pairs with first elements from A and second elements from B.

sec. 10-2
ordered pairs

Example 3: Let $A = \{1, 2\}$ and $B = \{3, 4, 5\}$. Then give the set of elements in $A \times B$.

Solution: $\{(1, 3), (1, 4), (1, 5), (2, 3), (2, 4), (2, 5)\}$.

The universal set may include negative numbers. For example, if $\mathcal{U} = \{-3, -2, -1, 0, 1, 2, 3\}$, then $\mathcal{U} \times \mathcal{U}$ consists of a set of 49 ordered pairs of numbers. The graph of $\{(x, y) | y = x\}$ for this universal set may be shown as follows:

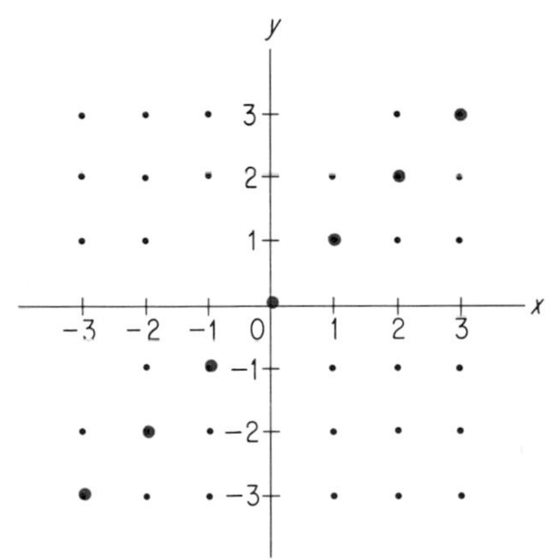

Example 4: Graph $\{(x, y) | y \geq x + 2\}$ for the universal set

$$\mathcal{U} = \{-3, -2, -1, 0, 1, 2, 3\}$$

Solution: For the corresponding statement of equality, we have

$$\{(x, y) | y = x + 2\} = \{(-3, -1), (-2, 0), (-1, 1), (0, 2), (1, 3)\}.$$

The graph of $y \geq x + 2$ consists of the points that are solutions of $y = x + 2$, and also all the points in $\mathcal{U} \times \mathcal{U}$ that are above (since $y > x + 2$) the points of the graph of $y = x + 2$.

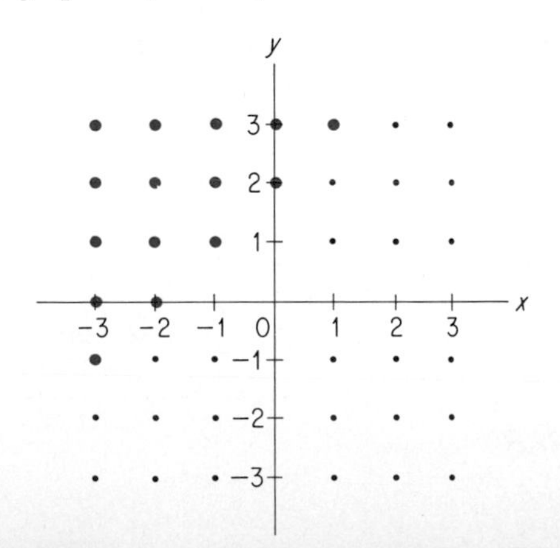

EXERCISES

1. Let $\mathcal{U} = \{1\}$ and give the set of elements in $\mathcal{U} \times \mathcal{U}$.

2. Let $A = \{1, 2\}$ and $B = \{3, 4\}$. Give the set of elements in $A \times B$.

Graph each set for $\mathcal{U} = \{-3, -2, -1, 0, 1, 2, 3\}$.

3. $\{(x, y) | y = x + 2\}$

4. $\{(x, y) | y = x - 2\}$

5. $\{(x, y) | y \geq x\}$

6. $\{(x, y) | y < x\}$

7. $\{(x, y) | y \leq x - 1\}$

8. $\{(x, y) | x + y \geq 3\}$

9. $\{(x, y) | y = x - 3\}$

10. $\{(x, y) | y = x + 3\}$

11. $\{(x, y) | y \geq x - 3\}$

12. $\{(x, y) | y \leq x + 3\}$

List the elements in each solution set for $\mathcal{U} = \{1, 2, 3, 4\}$.

13. $x + y = 4$

14. $y \leq x$

15. $y \geq x + 2$

16. $x + y < 5$

**17.* An *ordered triple* of numbers is a set of three numbers in order, such as (1, 1, 2). Give all possible ordered triples that can be formed from the set $\mathcal{U} = \{1, 2\}$; that is, give the set of elements of $\mathcal{U} \times \mathcal{U} \times \mathcal{U}$.

PEDAGOGICAL EXPLORATIONS

Ordered pairs may be used to show patterns. Some of these patterns may be expressed by algebraic sentences in two variables. Elementary school students usually enjoy the game of "trees" (originally developed by the University of Illinois Committee on School Mathematics). The teacher (or a selected student) thinks of a sentence in two variables. Initially two or three ordered pairs of integers that satisfy the sentence are stated and graphed. If a pair (x, y) satisfies the sentence, then "x trees y." If a pair does not satisfy the sentence, then "x does not tree y." Other pairs are suggested by members of the class as potential pairs for satisfying the sentence that is not known to the class. If for a suggested pair "x trees y," then the pair is graphed. Each student is encouraged to guess privately the sentence (or an equivalent) that is under discussion. To test a student who may know the sentence, a question such as the following is asked:

Does −5 tree 3?

Or, an instruction such as this may be given:

Let x be 7 and give a value for y.

After many students think that they know the sentence, the sentence is discussed openly.

Try this game of "trees" with some of your classmates using sentences such as:

1. $x + y = 6$	**2.** $y = x + 1$
3. $y = -x$	**4.** $y = 2 - x$
5. $y > x$	**6.** $y \leq x + 2$
7. $y = \|x\|$	**8.** $y > \|x\|$
9. $y \leq 1 + \|x\|$	**10.** $y < 2 + \|x\|$

10-3 relations and functions

A *relation* may be defined as any subset of $\mathcal{U} \times \mathcal{U}$; thus a relation is a set of ordered pairs of numbers. A relation may be represented by an equation (rule), a graph, or a table of values for the two variables. A *function* is a set of ordered pairs (x, y) such that for each value of x there is at most one value of y; that is, no first element appears with more than one second element. You may think of the first element as the *independent variable* and the second element as the *dependent variable*. Each variable has a set of possible values. The set of all first elements of the ordered pairs of numbers is called the **domain** of the function; the set of all second elements is called the **range** of the function. In terms of its graph, any vertical line drawn intersects the graph of a function in at most one point.

Here are the graphs of two relations that are also functions.

Notice that no vertical line can intersect the graph in more than one point.

Here are the graphs of two relations that are not functions.

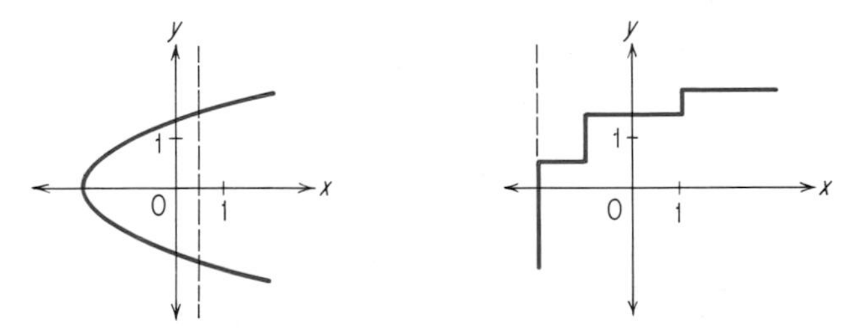

Notice that in each case there exists at least one vertical line that intersects the graph in two or more points.

A formula such as $y = x^2$ defines a function but the formula is not itself a function. We have defined a function to be a set of ordered pairs (x, y), such as those obtained from the formula $y = x^2$ for the real variable x. Thus, a formula may provide a rule by which the function may be determined. In other words, a formula may provide a means for associating a unique element in the range with each element in the domain. As soon as such distinctions are clearly understood it is customary to speak of "the function $y = x^2$" and mean "the function represented, or determined, by the formula $y = x^2$."

When a formula such as $y = x^2 - 2x + 3$ is used to define a function, it is customary to think of y as *a function of x* and to write the formula in **functional notation:**

$$y = f(x)$$

where

$$f(x) = x^2 - 2x + 3$$

Then the value of y for any value b of x may be expressed as $f(b)$. For example,

$$f(2) = 2^2 - 2(2) + 3 = 3$$
$$f(-1) = (-1)^2 - 2(-1) + 3 = 6$$
$$f(0) = 0^2 - 2(0) + 3 = 3$$
$$f(1) = 1^2 - 2(1) + 3 = 2$$

Notice that for the function $f(x) = (x - 1)^2 + 2$, the domain of the function is the set of all real numbers, and the range of the function is the set of real numbers greater than or equal to 2. The range may be determined from the graph of the function or from the observation that since $(x - 1)^2 \geq 0$ we must have $(x - 1)^2 + 2 \geq 2$. Other letters, as in $g(x)$ and $h(y)$, may be used in designating functions.

Example 1: Graph the function $y = x^2 - 1$ and identify the domain and range of this function.

Solution: The variable x may assume any real number as a value. Thus the domain is the set of real numbers. The range is the set of real numbers greater than or equal to -1, that is, $y \geq -1$, as in the graph.

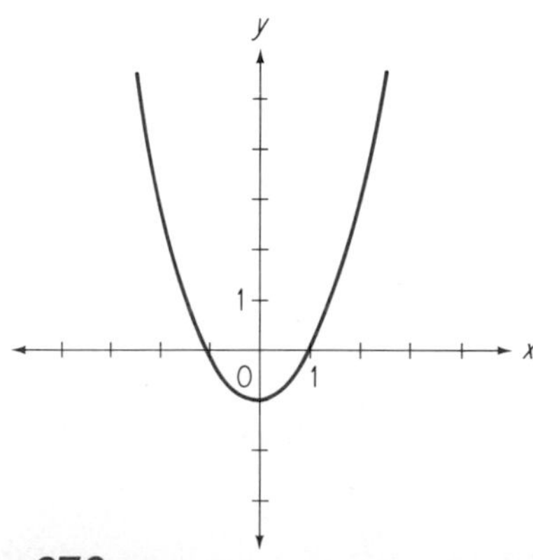

Example 2: Find the domain and range of $y = |x|$.

Solution: Domain: set of real numbers. Range: set of nonnegative real numbers, that is, $y \geq 0$.

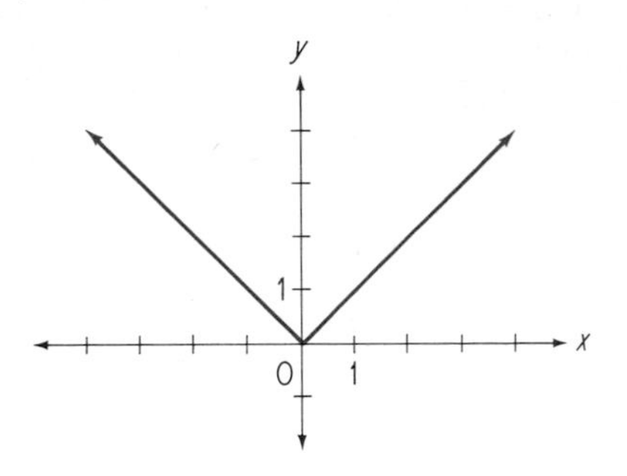

Example 3: Which of these formulas define functions? (*a*) $y = x^3$; (*b*) $x = y^2$; (*c*) $x^2 + y^2 = 9$; (d) $y = (x + 1)^2$.

Solution: Note that the formulas in (a) and (d) define functions.

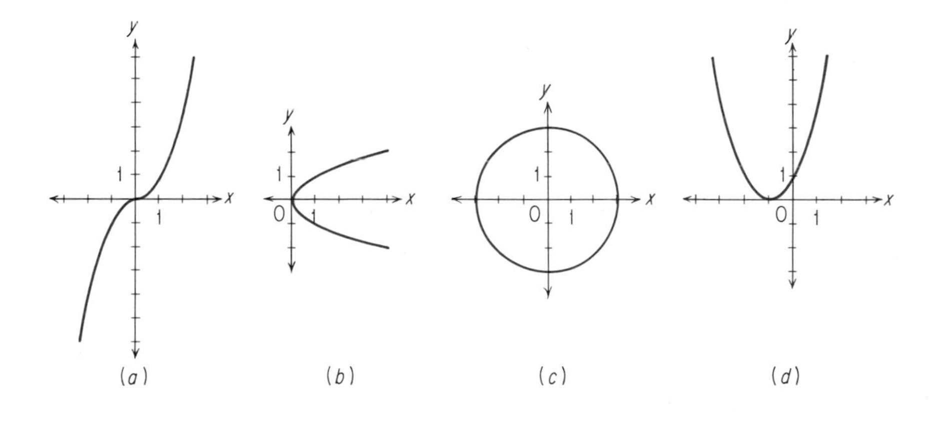

EXERCISES

1. If $f(x) = x - 3$, find **(a)** $f(1)$; **(b)** $f(3)$; **(c)** $f(10)$.

2. If $f(x) = x^2 - 4x + 9$, find **(a)** $f(0)$; **(b)** $f(2)$; **(c)** $f(-2)$.

3. If $g(x) = x^3 - 5$, find **(a)** $g(2)$; **(b)** $g(-2)$; **(c)** $g(-3)$.

4. If $f(x) = |x| + x$, find **(a)** $f(-3)$; **(b)** $f(0)$; **(c)** $f(5)$.

In Exercises 5 through 8, graph each relation, identify the relations that are functions, and state the domain and range for each one that is a function. In each case $\mathcal{U} = \{1, 2, 3\}$.

5. $y = x$ 6. $y = x + 1$

7. $y \leq x$ 8. $y > x + 1$

Proceed as in Exercises 5 through 8 for the universe of real numbers.

9. $y = x + 2$ 10. $y = x - 2$

11. $y = |x - 2|$ 12. $y = |x + 2|$

13. $y = |x| + 2$ 14. $y = |x| - 2$

15. $y \geq -x$ 16. $y \leq x - 2$

17. $y = -x^2$ 18. $y = (x + 2)^2$

PEDAGOGICAL EXPLORATIONS

1. Many elementary texts now introduce the concept of function (mapping) and relation on an intuitive basis through the use of arrow diagrams. For example, each of the following shows a correspondence (relation) between two sets. In each case the arrow indicates which members of the second set (range) correspond to the given elements of the first set (domain). Which of the relations are functions?

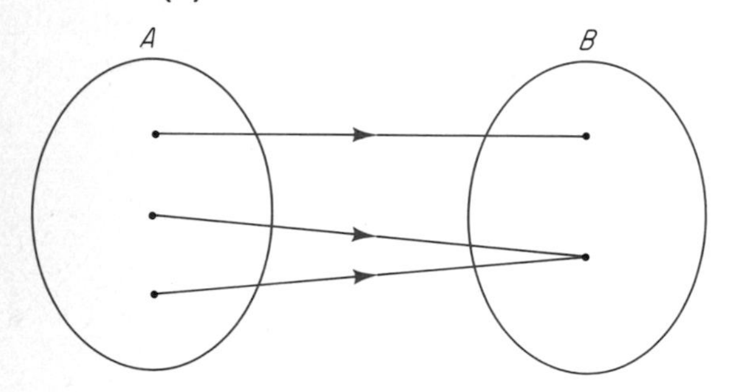

(b)

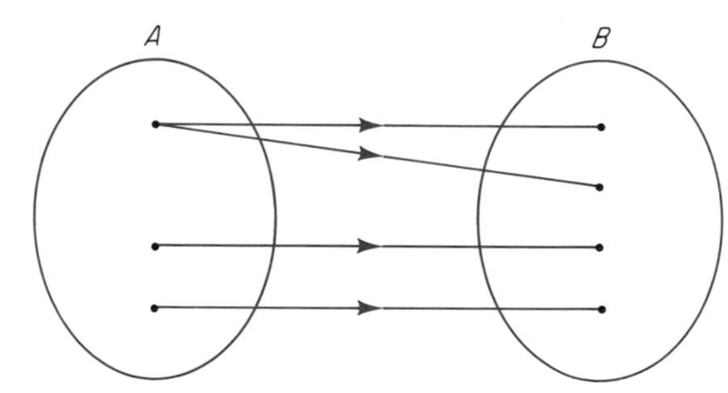

(c)

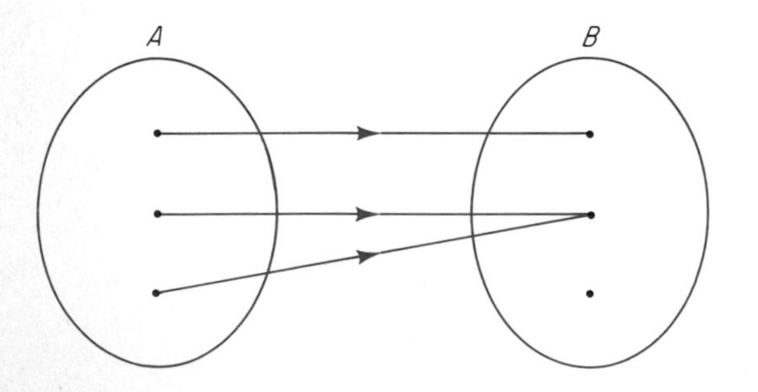

(d)

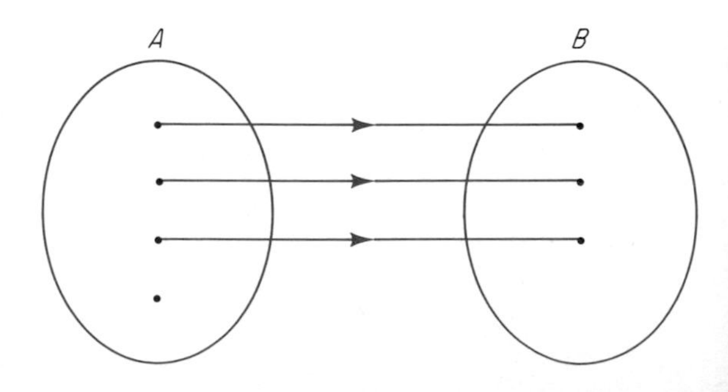

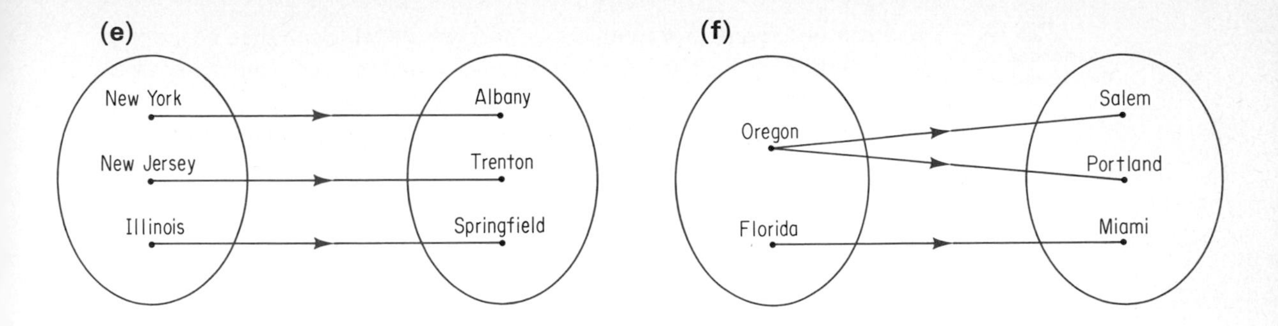

2. A mapping of a set A *onto* another set B is a correspondence where each element of A corresponds to a unique element of B, and each element of B is the image of at least one element in A. Draw several different arrow diagrams that illustrate such a mapping.
3. A *one-to-one* mapping of a set A onto a set B is one where each element of set A corresponds to a unique element of set B, and each element of set B is the image of a unique element in set A. Draw an arrow diagram that illustrates such a mapping.

10-4 graphs on a plane

In §10-2 we graphed the solution sets of sentences using finite coordinate systems. Now we shall extend these concepts and consider $\mathcal{U}$ as the set of real numbers. Then $\mathcal{U} \times \mathcal{U}$ becomes an infinite collection of ordered pairs of numbers and its graph consists of an entire plane. In honor of René Descartes, such a plane is frequently referred to as the **Cartesian plane.** The numbers of each ordered pair of real numbers are the **Cartesian coordinates** of a point of the plane. Each point of a Cartesian plane can be represented by (has as its *coordinates*) an ordered pair of real numbers, and each ordered pair of real numbers can be used to identify (locate) a unique point of the plane. We follow the custom of speaking of "the point with coordinates (x, y)" as "the point (x, y)."

Let us now discuss the graphs of sentences for the universe of real numbers. Consider, for example, the sentence $y = x + 2$. We may list in a table of values several ordered pairs of numbers that are solutions of the sentence. Thus for $x = -1$, we have $y = -1 + 2 = 1$ and for $x = 2$, we have $y = 2 + 2 = 4$. Confirm the other entries given in this table:

$y = x + 2$:

x	-3	-2	-1	0	1	2	3
y	-1	0	1	2	3	4	5

Each ordered pair of numbers (x, y) from the table can then be graphed. These points may be connected in the order of the x-coordinates and the graph of $y = x + 2$ obtained as in the following figure.

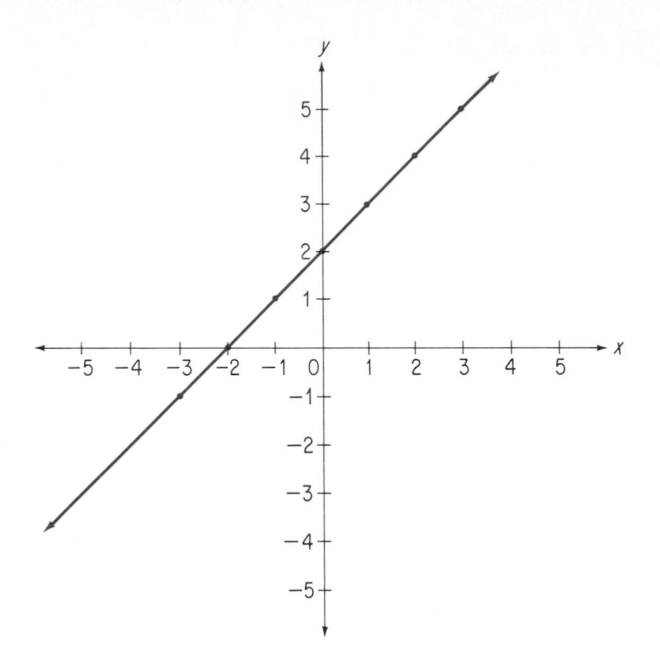

The graph of $y = x + 2$ is a straight line and extends indefinitely as indicated by the arrowheads. Often these arrowheads are omitted. In general, for the universe of real numbers, any equation that can be expressed in the form

$$ax + by + c = 0$$

where a and b are not both zero, is said to be a **linear equation** and has a straight line as its graph.

Inasmuch as a straight line is determined by two points, we can graph a linear equation after locating two of its points. When $a \neq 0$ and $b \neq 0$ the most convenient points to locate are the point $(0, y)$ at which the graph crosses the y-axis and the point $(x, 0)$ at which the graph crosses the x-axis. The graph of $y = x + 2$ crosses the y-axis when $x = 0$, that is, at the point A: $(0, 2)$, where 2 is the **y-intercept** of the graph. The graph of $y = x + 2$ crosses the x-axis when $y = 0$, that is, at the point B: $(-2, 0)$, where -2 is the **x-intercept** of the graph.

Example 1: Graph $\{(x, y) \mid x - 2y = 4\}$.

Solution: The equation may be expressed as $x - 2y - 4 = 0$ and thus has a line as its graph. For $x = 0$, $-2y = 4$ and $y = -2$. For $y = 0$, $x = 4$. The x-intercept 4 and the y-intercept -2 determine the line in the next graph.

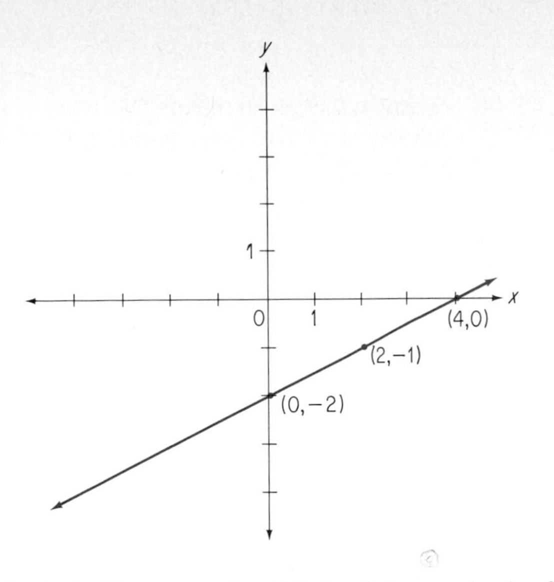

Usually it is desirable to graph a third point as a check of your work. For example, when $x = 2$ we have $2 - 2y = 4$, $-2y = 2$, and $y = -1$. Thus $(2, -1)$ may be graphed and the point should be on the line.

Statements of inequality can also be graphed for the universe of real numbers. Consider the sentence $y \leq x + 2$. Here the graph consists of all the points on the line $y = x + 2$, as well as the points in the **half-plane** below the line as indicated by the shaded portion of the graph. Note that for any value b of x the point $(b, b + 2)$ is on the line $y = x + 2$, and for any value of y less than $b + 2$ the point (b, y) is below the line. Thus the graph of $y < x + 2$ is the union of a line and a half-plane.

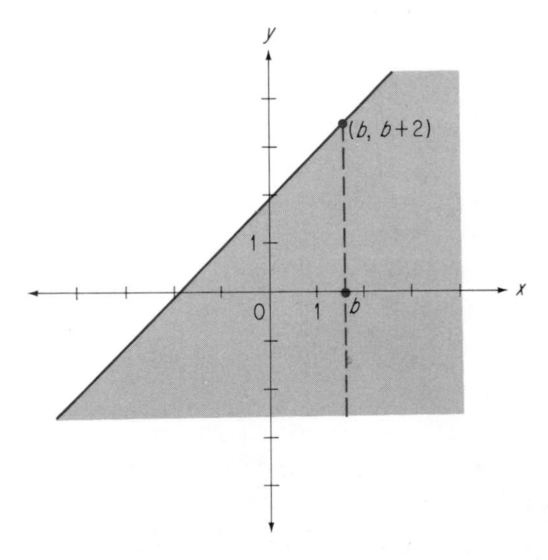

In the preceding figure, the line $y = x + 2$ divided the plane into two half-planes. One of these half-planes was shown as the desired graph and satisfied the given inequality. One way to determine which half-plane is needed is to try some points in the given inequality. For example, (0, 0) satisfies the inequality ($y < x + 2$) and therefore is part of the desired half-planc. If any point satisfies the inequality, the point is on the desired half-plane. If any point, such as (0, 3) does not satisfy the given inequality, the point is on the opposite side of the line from the desired half-plane.

Example 2: Graph $y > x - 1$.

Solution: It is helpful to draw first the corresponding statement of equality, $y = x - 1$, as a dotted line. This line separates the plane into two half-planes.

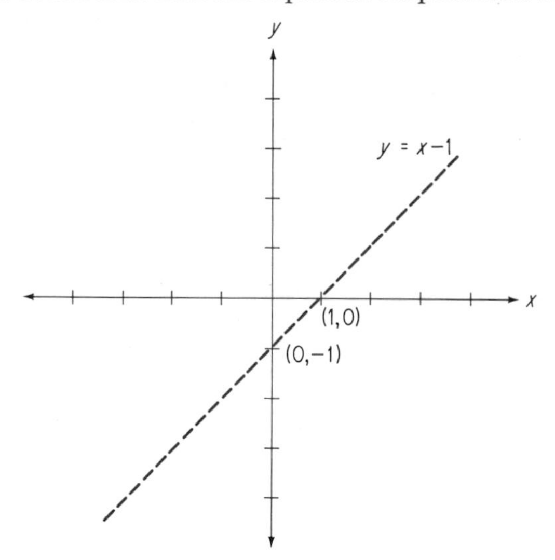

To determine which half-plane to shade, we note that the problem requires all the values of *y greater than* $x - 1$. Thus we shade the half-plane above the line to represent the solution set.

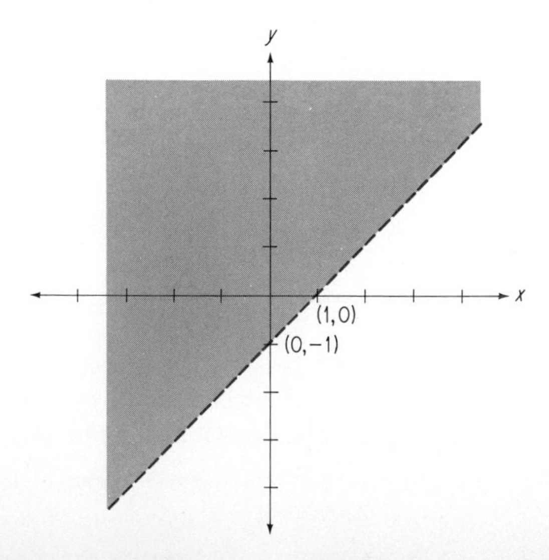

Note that the point (0, 0) satisfies the given inequality and is part of the desired half-plane. The point (2, 0) does not satisfy the given inequality and is on the opposite side of the line from the desired half-plane.

The line in the graph for Example 2 is dotted, since the points of the line *are not* points of the graph. The line in the graph for Example 3 is solid, since the points of the line *are* points of the graph.

Example 3: Graph $x - 2y - 4 \geq 0$.

Solution: This sentence is equivalent to $x - 2y \geq 4$. From Example 1 we have the graph of the line $x - 2y = 4$. To determine which half-plane to shade we may proceed in one of two ways. We can place the given sentence in ***y*-form** as follows:

$$x - 2y \geq 4$$
$$-2y \geq -x + 4 \qquad \text{Why?}$$
$$y \leq \tfrac{1}{2}x - 2 \qquad \text{Why?}$$

This tells us that the half-plane *below* the line is included in the graph of the solution set. An alternate plan for determining which half-planc is included

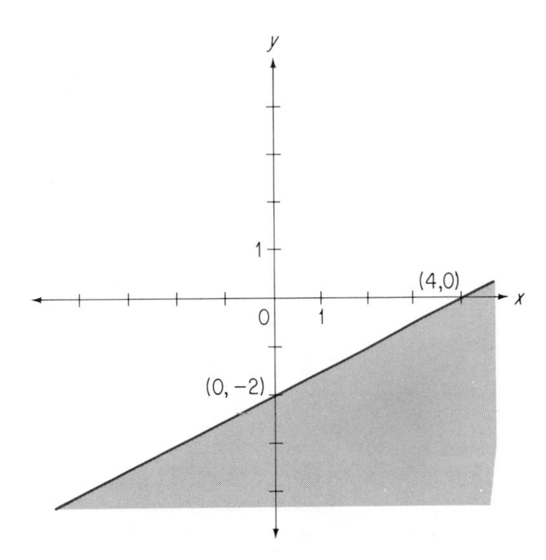

in the solution set is to select a particular point of one half-plane and see whether or not that point belongs in the graph. The origin, (0, 0), is a convenient point to consider. When $x = 0$ and $y = 0$, we have the false statement $0 - 0 \geq 4$. Thus the points of the half-plane that contains (0, 0) are not in the solution set and again we conclude that the graph consists of the line and the half-plane below the line.

EXERCISES

For the graph of each sentence, find (a) the x-intercept; (b) the y-intercept.

1. $x + y = 8$
2. $x - y = 3$
3. $2x - 3y = 12$
4. $3x - 2y = 12$
5. $y = 2x + 5$
6. $y = 3x + 5$
7. $2x - y + 8 = 0$
8. $x + 3y - 6 = 0$
9. $5x + 4y = 10$

*10. $ax + by + c = 0$

Write each sentence in y-form, that is, solve for y.

11. $2x + y = 9$
12. $4x - 2y = 12$
13. $2x + y \geq 8$
14. $x + 3y \leq 6$
15. $2x - y \leq 8$
16. $3x - 2y \geq 12$

Graph each set for the universe of real numbers.

17. $\{(x, y) | y = x + 3\}$
18. $\{(x, y) | y = x - 1\}$
19. $\{(x, y) | x + y = 2\}$
20. $\{(x, y) | x - y = 5\}$
21. $\{(x, y) | 2x + 3y + 6 = 0\}$
22. $\{(x, y) | 3x + 2y - 6 = 0\}$
23. $\{(x, y) | x = 3\}$
24. $\{(x, y) | y = -2\}$
25. $\{(x, y) | y \leq x\}$
26. $\{(x, y) | y \geq x - 2\}$
27. $\{(x, y) | 2x + y > 6\}$
28. $\{(x, y) | x - 2y < 4\}$
29. $\{(x, y) | x + y - 1 \leq 0\}$
30. $\{(x, y) | 3x - 4y + 12 \geq 0\}$

PEDAGOGICAL EXPLORATIONS

1. When junior high school students are first introduced to the concept of rectangular coordinates, they enjoy an exercise that involves plotting points that form figures. Thus a set of ordered pairs of numbers are given that are to be located, then connected in order. The outcome is some figure, such as an animal, a Christmas tree, etc. Prepare a set of such ordered pairs that gives rise to some figure.
2. An interesting exercise for upper elementary grade students is to have them draw an object in the first (upper right) quadrant of a Cartesian plane. Then they must attempt to draw the reflection of this object in each of the other quadrants, using the x-axis and y-axis as axes of symmetry. For example, the next figure shows the reflection of the letter F in each quadrant:

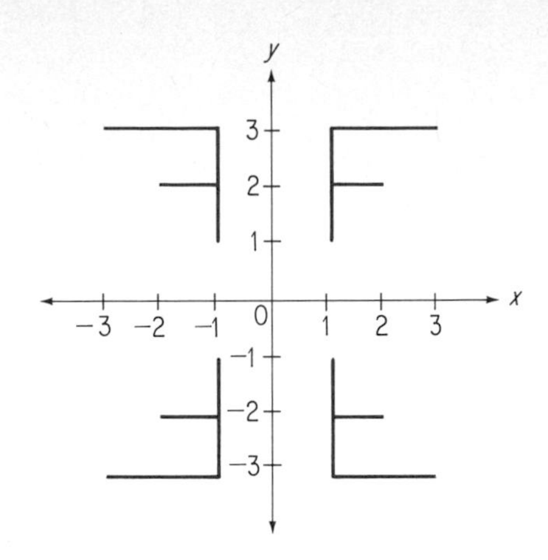

Prepare a similar set of drawings for the letters P and R. Are there any letters that will look the same in all four quadrants?

3. Assign coordinates to each of the vertices of F in the first quadrant in the preceding figure. Then label each vertex of the images of F with its coordinates.
4. Try to write the word SEVEN as it would appear if held up to a mirror. Then use a mirror to check your result. Such activities help elementary school students develop concepts of reflection and symmetry.
5. Make a list of the number of places that you encounter mirror images in daily life. This list may range from such activities as simply combing your hair while looking in a mirror, to seeing the word "ambulance" printed on the front of an ambulance so that it reads correctly when you see it in your rear view mirror.

10-5 graphs involving absolute value

Not all sentences with two variables have graphs that are straight lines—only those sentences that can be written in the form $ax + by + c = 0$. In §9-4, where we considered graphs of sentences in one variable, we graphed sentences involving absolute value. We next explore sentences that include two variables and involve absolute value; for example, $y = |x|$. As an aid to graphing this sentence for the universe of real numbers we first consider the graph for $\mathcal{U} = \{-3, -2, -1, 0, 1, 2, 3\}$. The values to be graphed are summarized in the table of values below.

Recall the definition of $|x|$:

$$|x| = x \text{ for } x \geq 0$$
$$|x| = -x \text{ for } x < 0$$

x	−3	−2	−1	0	1	2	3
y	3	2	1	0	1	2	3

These ordered pairs (x, y) such that $y = |x|$ can be graphed on $\mathcal{U} \times \mathcal{U}$ as follows:

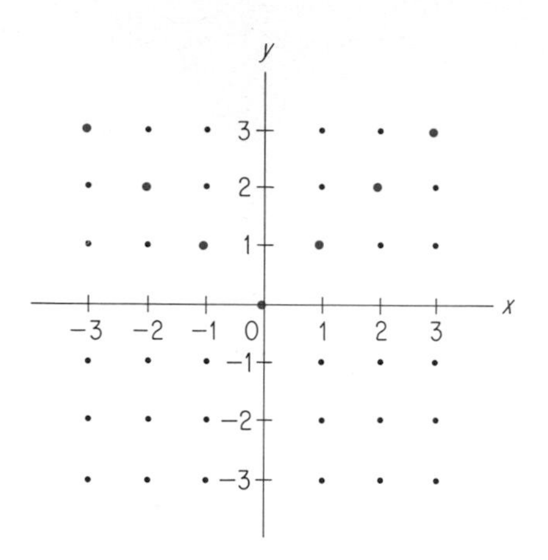

If $\mathcal{U}$ is the set of real numbers, our graph consists of an infinite collection of points. Using the preceding graph as a guide, we are now able to graph $y = |x|$ for the set of real numbers. This next graph may also be obtained directly from the definition of $|x|$.

$\{(x, y) | y = |x|\} = \{(x, y) | y = x \text{ and } x \geq 0\} \cup \{(x, y) | y = -x \text{ and } x < 0\}$.

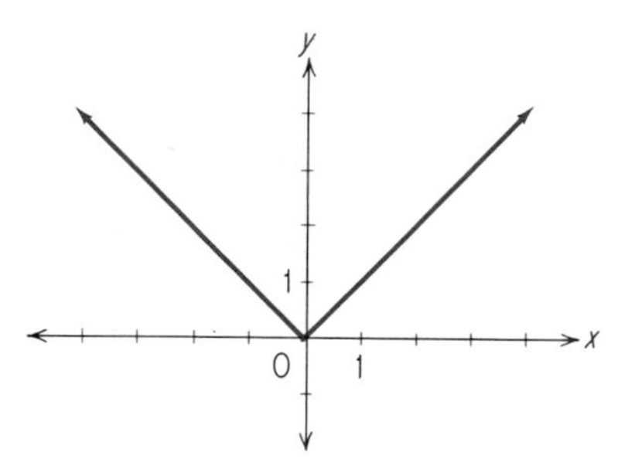

The graph of $y = |x|$ separates the plane into two regions. These regions are the graphs of $y < |x|$ and $y > |x|$. The desired region may be determined intuitively or by testing points as in the previous section. The graph of $y \geq |x|$ is completed by shading in the portion of the plane above the graph, as shown in the figure at the top of the next page.

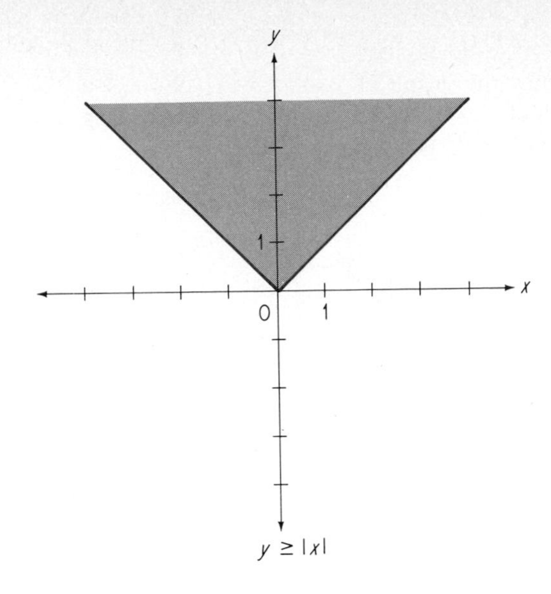

$y \geq |x|$

Example 1: Graph $\{(x, y) | y = |x + 1|\}$.

Solution: Recall that we assume the universal set to be the set of real numbers unless otherwise specified.

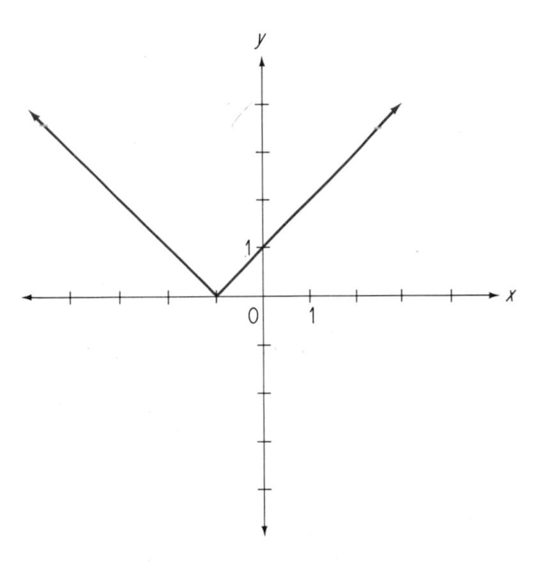

Notice that if $x = -1$, then $y = 0$; if $x = 0$ or -2, then $y = 1$; if $x = 1$ or -3, then $y = 2$; and so forth.

Example 2: Graph $\{(x, y) | y \geq |x + 1|\}$.

Solution: From Example 1 we know the graph of $y = |x + 1|$. We test a point such as (0, 0). Then the graph of $y \geq |x + 1|$ is completed by shading in the portion of the plane above the graph as shown.

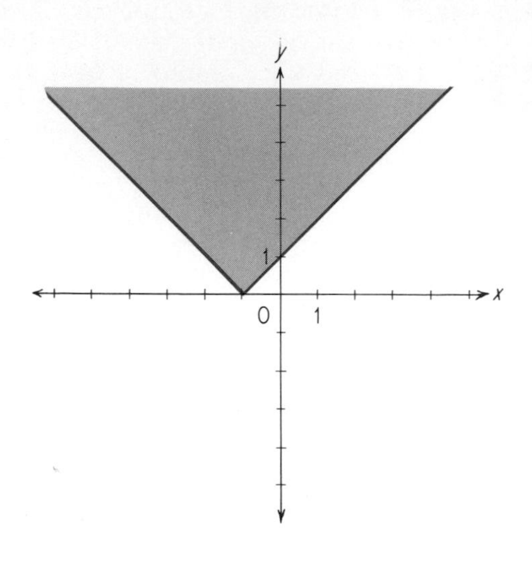

EXERCISES

Graph the solution set for $\mathcal{U} = \{-3, -2, -1, 0, 1, 2, 3\}$.

1. $y = |x + 1|$

2. $y = |x - 1|$

3. $y = -|x|$

4. $y = -|x + 1|$

5. $y = |x| + 2$

6. $y = |x| - 2$

7. $y \geq |x| + 1$

8. $y \leq |x| - 1$

Graph each set for the universe of real numbers.

9. $\{(x, y) | y = |x| + 2\}$

10. $\{(x, y) | y = |x| - 2\}$

11. $\{(x, y) | y = |x + 2|\}$

12. $\{(x, y) | y = |x - 2|\}$

13. $\{(x, y) | y > |x + 3|\}$

14. $\{(x, y) | y < |x - 3|\}$

***15.** $\{(x, y) | |x| \leq 3\}$

***16.** $\{(x, y) | |y| \geq 2\}$

***17.** $\{(x, y) | |x + y| = 2\}$

***18.** $\{(x, y) | |x - y| = 3\}$

PEDAGOGICAL EXPLORATIONS

1. The overhead projector is an effective device for showing graphs on a plane. Prepare a set of coordinate axes and grid lines on a sheet of acetate. On a second sheet of acetate draw the graph of $y = |x|$. Demonstrate each of the following graphs by an appropriate translation (sliding) of the second sheet of acetate while keeping the first sheet (the base) fixed.

(a) $y = |x| + 1$
Shift curve one unit up.

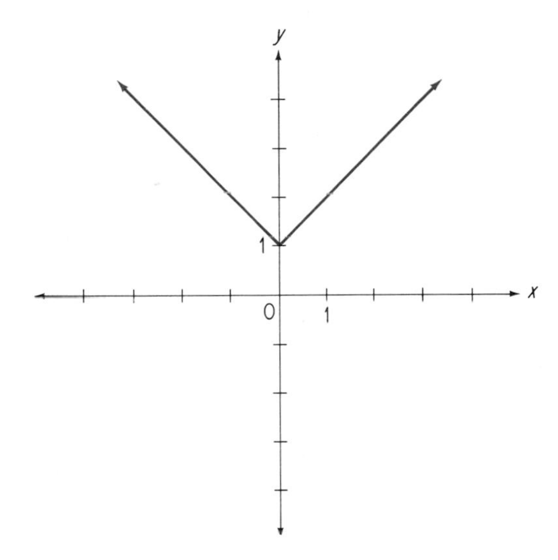

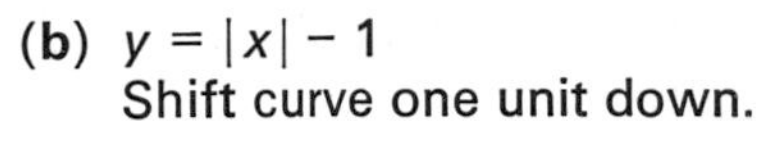
(b) $y = |x| - 1$
Shift curve one unit down.

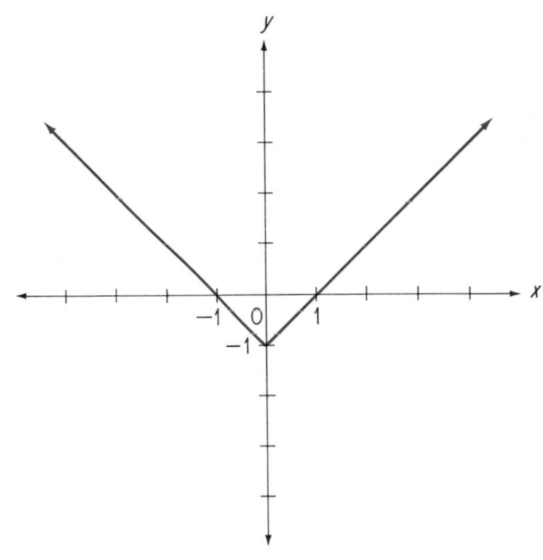

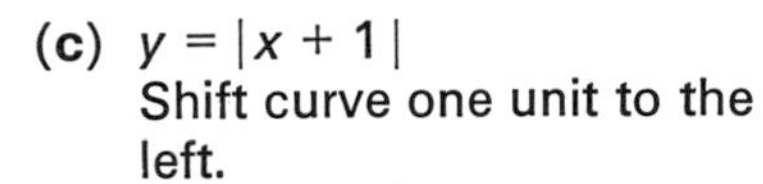
(c) $y = |x + 1|$
Shift curve one unit to the left.

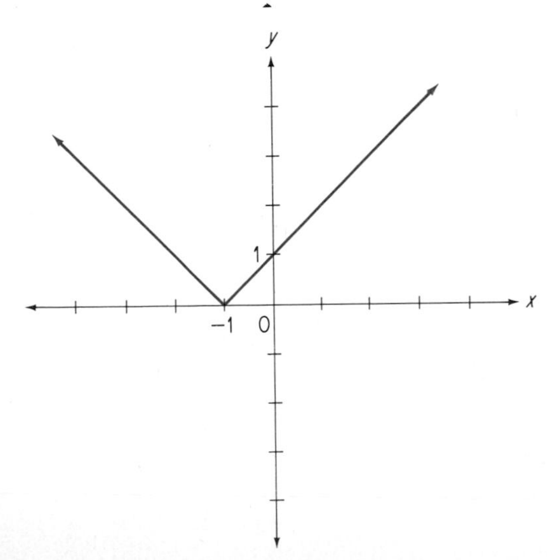

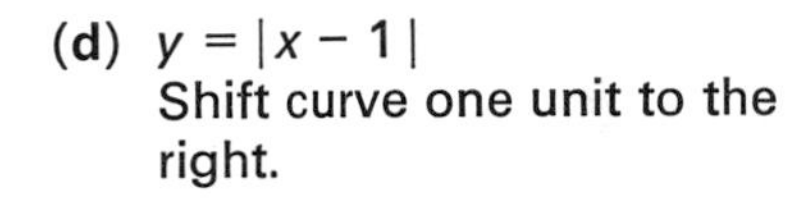
(d) $y = |x - 1|$
Shift curve one unit to the right.

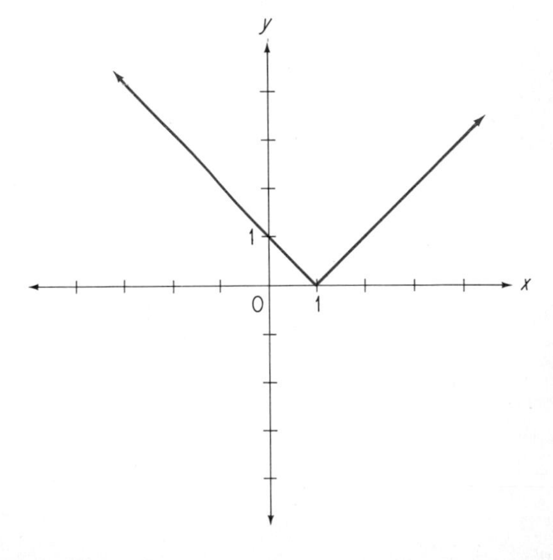

2. A 180° rotation of the graph of $y = |x|$ about the x-axis will give the graph of $y = -|x|$. Use this idea to show each of the following by appropriate rotations and translations:

(a) $y = -|x|$
(b) $y = -|x| + 1$
(c) $y = -|x| - 1$
(d) $y = -|x + 1|$

10-6 linear systems

In Chapter 9 we considered compound sentences that involved one variable. Now we shall turn our attention to compound sentences involving two variables. Consider, for example, the compound sentence

$$x + y - 3 = 0 \quad \text{and} \quad x - y - 1 = 0$$

This is frequently written in the following form:

$$\begin{cases} x + y - 3 = 0 \\ x - y - 1 = 0 \end{cases}$$

and is referred to as a **system of linear equations** or as a set of **simultaneous linear equations.** To solve a set of two simultaneous equations we find the set of ordered pairs that are solutions of both equations. Our approach in this section will be a graphical one rather than an algebraic one; that is, we draw the graph of each sentence and identify the point of intersection, if any, of the two lines. For the given system of equations we have this graph:

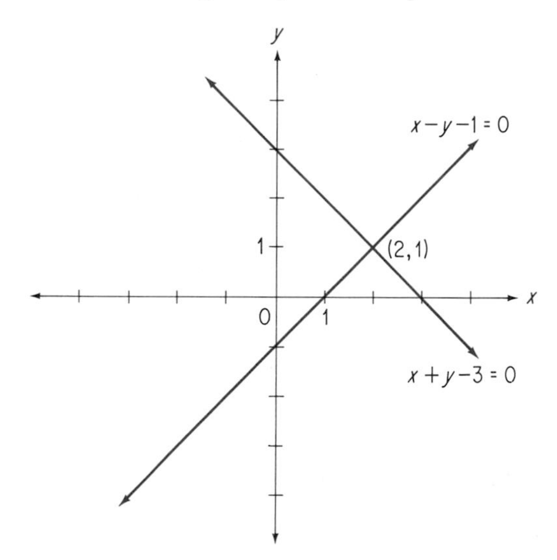

The point located at (2, 1) is on both lines and (2, 1) is the solution of the given system of equations. We may express this fact as

$$\{(x, y) \mid x + y - 3 = 0\} \cap \{(x, y) \mid x - y - 1 = 0\} = \{(2, 1)\}$$

Consider next the compound sentence

$$x + y - 3 = 0 \quad \text{or} \quad x - y - 1 = 0$$

This word "or" indicates that we are to find the set of ordered pairs of numbers that are solutions of either one or of both of the given equations; that is,

$$\{(x, y) | x + y - 3 = 0\} \cup \{(x, y) | x - y - 1 = 0\}$$

The graph of the solution set consists of all the points that are on at least one of the two lines; that is, the graph is the union of the points of the two lines drawn in the preceding figure.

Example 1: Graph $x - y - 1 = 0$ or $x - y + 2 = 0$.

Solution: The graph consists of the union of the points of the two lines.

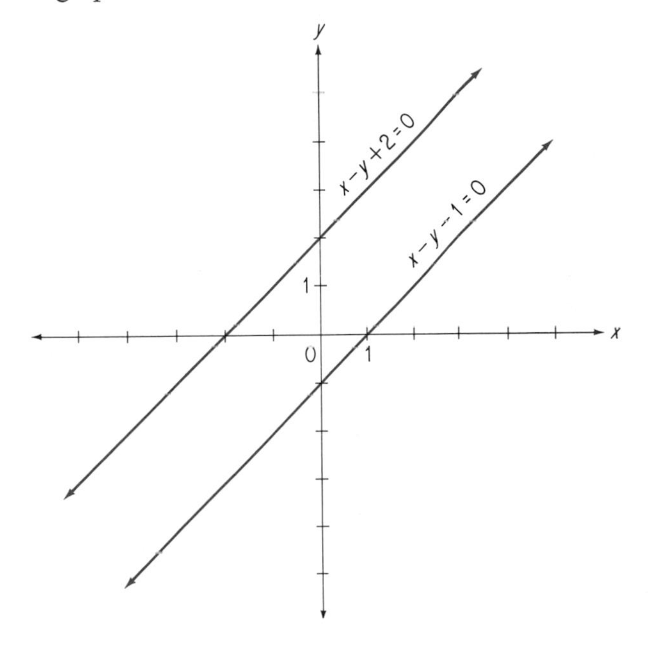

Notice that the two lines in the solution for Example 1 are parallel. The solution set of the sentence "$x - y - 1 = 0$ *and* $x - y + 2 = 0$" would be the empty set.

Example 2: Graph $\{(x, y) | (x - y - 1)(x - y + 2) = 0\}$.

Solution: First we note that the product $a \times b = 0$ implies that $a = 0$ or $b = 0$. (Recall that this use of the word "or" also includes the case that $a = 0$ *and* $b = 0$.) Thus the given sentence can be written as the equivalent sentence $x - y - 1 = 0$ or $x - y + 2 = 0$. But this is precisely the problem in Example 1. Thus the graph of the solution set consists of the two lines drawn in the figure given there.

Systems of inequalities can be solved graphically as the union or intersection of half-planes. Consider the system

$$\begin{cases} x + y - 3 > 0 \\ x - y - 1 > 0 \end{cases}$$

The corresponding statements of equality have been graphed earlier in this section. The graphs of the inequalities are shown in these figures.

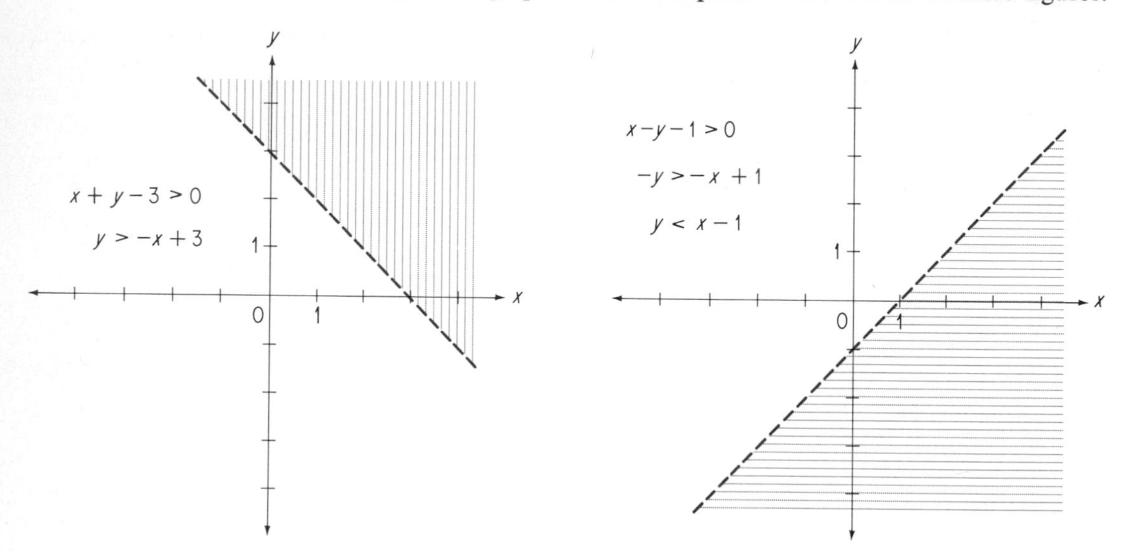

Note that the corresponding lines are dotted since the points of the lines are *not* included in the graphs. Also one graph has vertical shading and the other graph has horizontal shading. The graph of the system consists of the intersection of these two graphs; that is, the points of the region that is shaded both vertically and horizontally when the graphs of the two inequalities are drawn on the same coordinate plane.

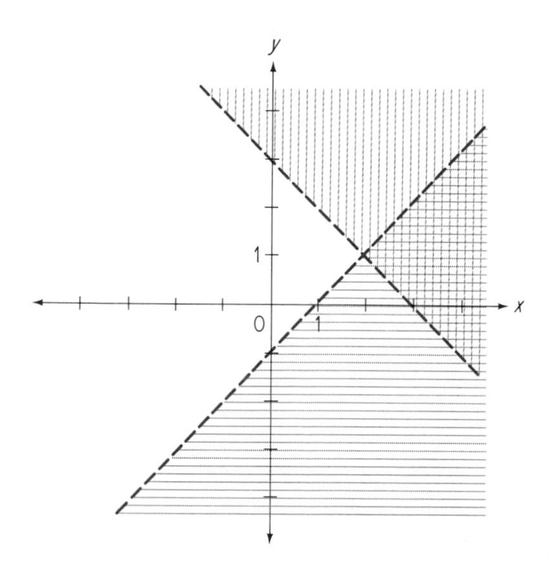

We can also determine from this figure the graph of the sentence $x + y - 3 > 0$ or $x - y - 1 > 0$, namely the union of the points of the two shaded regions. The unshaded region (including the dotted lines) represents the graph of the solution set of the system

$$\begin{cases} x + y - 3 \leq 0 \\ x - y - 1 \leq 0 \end{cases}$$

Example 3: Graph

$$\{(x, y) | x - y + 2 \leq 0\} \cap \{(x, y) | 2x + y - 4 \geq 0\}$$

Solution: The graph of $x - y + 2 \leq 0$ has horizontal shading in the figure; the graph of $2x + y - 4 \geq 0$ has vertical shading. The graph of the solution set consists of the points in the region with both horizontal and vertical shading and includes the points of the two rays that serve as the boundary of this region.

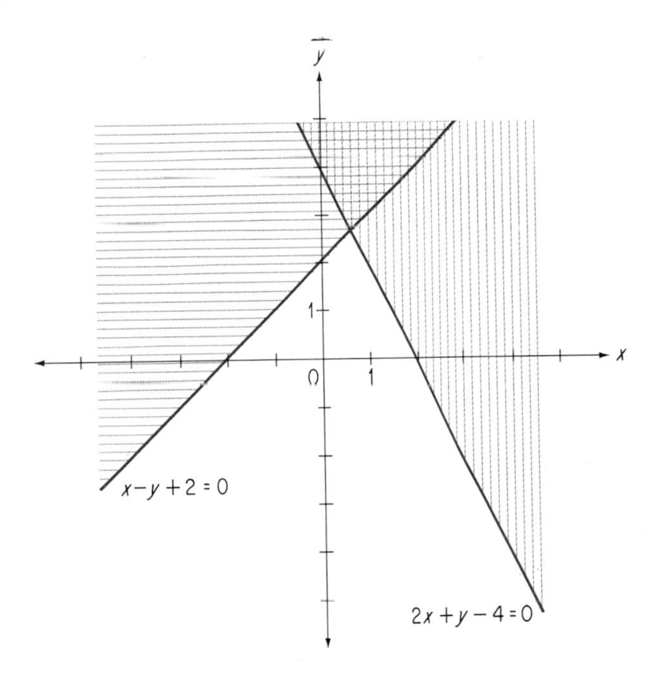

Example 4: Graph

$$\{(x, y) | x - y + 2 \leq 0\} \cup \{(x, y) | 2x + y - 4 \geq 0\}$$

Solution: Consider the graph for Example 3. The solution set in Example 4 consists of all the points in the regions that are shaded in any way, that is, all points are included *except* those in the unshaded region.

EXERCISES

Graph the solution set for each system.

1. $\begin{cases} x + y - 6 = 0 \\ x - y + 4 = 0 \end{cases}$

2. $\begin{cases} x + y - 5 = 0 \\ x - y - 3 = 0 \end{cases}$

3. $\begin{cases} x + y = 2 \\ x + 2y = 4 \end{cases}$

4. $\begin{cases} 2x - y = 8 \\ x + 2y = -4 \end{cases}$

5. $\{(x, y) | x + y - 3 = 0\} \cup \{(x, y) | x + y + 4 = 0\}$

6. $\{(x, y) | x + 2y - 4 = 0\} \cup \{(x, y) | 2x + 4y - 8 = 0\}$

7. $\{(x, y) | 2x - y - 2 = 0\} \cap \{(x, y) | 4x - 2y + 4 = 0\}$

8. $\{(x, y) | (x - y + 2)(2x + 3y + 6) = 0\}$

9. $\begin{cases} x - y + 3 > 0 \\ x + y - 3 > 0 \end{cases}$

10. $\begin{cases} 2x + y - 6 < 0 \\ x - 2y + 6 < 0 \end{cases}$

11. $\begin{cases} 3x - 2y - 6 \leq 0 \\ 2x + 3y + 6 \geq 0 \end{cases}$

12. $\begin{cases} x + 2y - 6 \leq 0 \\ 2x - y + 4 \geq 0 \end{cases}$

13. $\{(x, y) | x - y - 5 \leq 0\} \cap \{(x, y) | x + y - 5 \leq 0\}$

14. $\{(x, y) | x + 2y - 4 \geq 0\} \cap \{(x, y) | 2x - y + 6 \leq 0\}$

15. $\{(x, y) | 3x - y + 6 \leq 0\} \cup \{(x, y) | 3x + 4y + 12 \geq 0\}$

16. $\{(x, y) | x + 4y - 8 \geq 0\} \cup \{(x, y) | 4x - 2y + 8 \geq 0\}$

*17. $\{(x, y) | (2x - 3y + 6)(x + 2y - 4) \leq 0\}$

*18. $\{(x, y) | y \geq |x + 2|\} \cap \{(x, y) | y \leq 5\}$

*19. $\{(x, y) | y \geq |x| - 2\} \cap \{(x, y) | y \leq -|x| + 2\}$

*20. $\{(x, y) | |x + y| \leq 3\} \cap \{(x, y) | |x| \leq 3\}$

*21. $\{(x, y) | |x| + |y| = 2\}$

*22. $\{(x, y) | |x| + |y| \leq 2\}$

*23. $\{(x, y) | |x| - |y| = 2\}$

PEDAGOGICAL EXPLORATIONS

1. A student teacher ran into trouble when trying to teach a ninth grade algebra class how to draw the graph of the solution set for the inequality $2x - 3y + 6 > 0$. First he constructed the graph of the corresponding equality as in the figure at the top of page 389. He then tested the point (0, 0) in the upper half-plane, which gives this true statement of inequality:

$$0 - 0 + 6 > 0$$

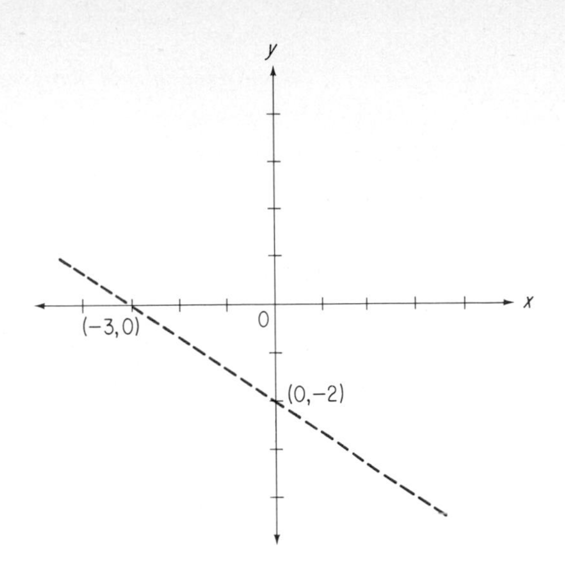

Thus he concluded that the required half-plane includes the origin, and must be *above* the line. As a check, he then went on to write the given inequality in y-form:

$$2x - 3y + 6 > 0, \qquad -3y > -2x - 6, \qquad y < \tfrac{2}{3}x + 2$$

This last statement indicates that the required half-plane must be *below* the line! Where did he make the error?

2. Having made the error described above, discuss different procedures the student teacher could have followed in terms of teaching techniques so as to capitalize on the error.

CHAPTER TEST

Find the solution set for the given universes.

1. $y < x - 2$, $\mathcal{U} = \{-3, -2, -1, 0, 1, 2, 3\}$

2. $y = -|x| + 1$, $\mathcal{U} = \{-3, -2, -1, 0, 1, 2, 3\}$

Graph each set for $\mathcal{U} = \{-3, -2, -1, 0, 1, 2, 3\}$.

3. $\{(x, y) \mid y = -x\}$

4. $\{(x, y) \mid y > x + 2\}$

Find the x-intercept and the y-intercept for the graph of each sentence.

5. $2x - y = 7$

6. $3x - 2y + 18 = 0$

Solve for y.

7. $x + 3y = 12$

8. $2x - y \geq 4$

Graph for the universe of real numbers.

9. $\{(x, y) | y = x - 2\}$

10. $\{(x, y) | x - y = 3\}$

11. $\{(x, y) | x + 2y \geq 6\}$

12. $\{(x, y) | y \geq |x| + 1\}$

13. $\{(x, y) | |x| \geq 2\}$

14. $\{(x, y) | |y| < 1\}$

15. $\{(x, y) | x + y - 3 \geq 0\} \cap \{(x, y) | x - y + 2 \leq 0\}$

16. $\{(x, y) | x + y - 2 \leq 0\} \cup \{(x, y) | x - y + 1 \geq 0\}$

17. If $f(x) = x^2 + 3x - 5$, find **(a)** $f(-3)$; **(b)** $f(2)$.

18. If $g(x) = x^3 - |x|$, find **(a)** $f(-1)$; **(b)** $f(\frac{1}{2})$.

Graph each function and state the domain and range for the universe of real numbers.

19. $y = 2x - 6$

20. $y = |x + 1|$

READINGS AND PROJECTS

1. Prepare a report on the life of René Descartes, stressing his contributions in the field of analytic geometry. See Chapter 3 of E. T. Bell's *Men of Mathematics*, Dover Publications, Inc., 1937.
2. Examine a recently published elementary textbook series to discover where the concept of an ordered pair of numbers is first introduced. Then trace the development of this concept to determine the extent to which it is used in the first six elementary grades.
3. Prepare a set of transparencies suitable for use on an overhead projector to demonstrate the idea of a system of inequalities. This is best done by having a base transparency of the coordinate plane, and then two overlays that each show a half-plane shaded in different colors.
4. The function concept is often cited as one of the unifying themes of mathematics. Examine a recently published elementary textbook series to determine how extensively this concept is used.
5. Prepare a plan for a 20-minute lesson that serves to introduce the concept of rectangular coordinates to an elementary mathematics class. Search for an innovative strategy for your approach. For example, consider the use of the game of tic-tac-toe to introduce coordinates as described by Max A. Sobel and Evan M. Maletsky on pages 65 through 67 of *Teaching Mathematics*, Prentice-Hall, Inc., 1975.
6. Read Booklet No. 13 (Graphs, Relations, and Functions) of the thirtieth yearbook of the National Council of Teachers of Mathematics. Complete at least three of the exercise sets given there.
7. Read Chapter 3 (Relations and Functions) of the twenty-fourth yearbook of the National Council of Teachers of Mathematics.

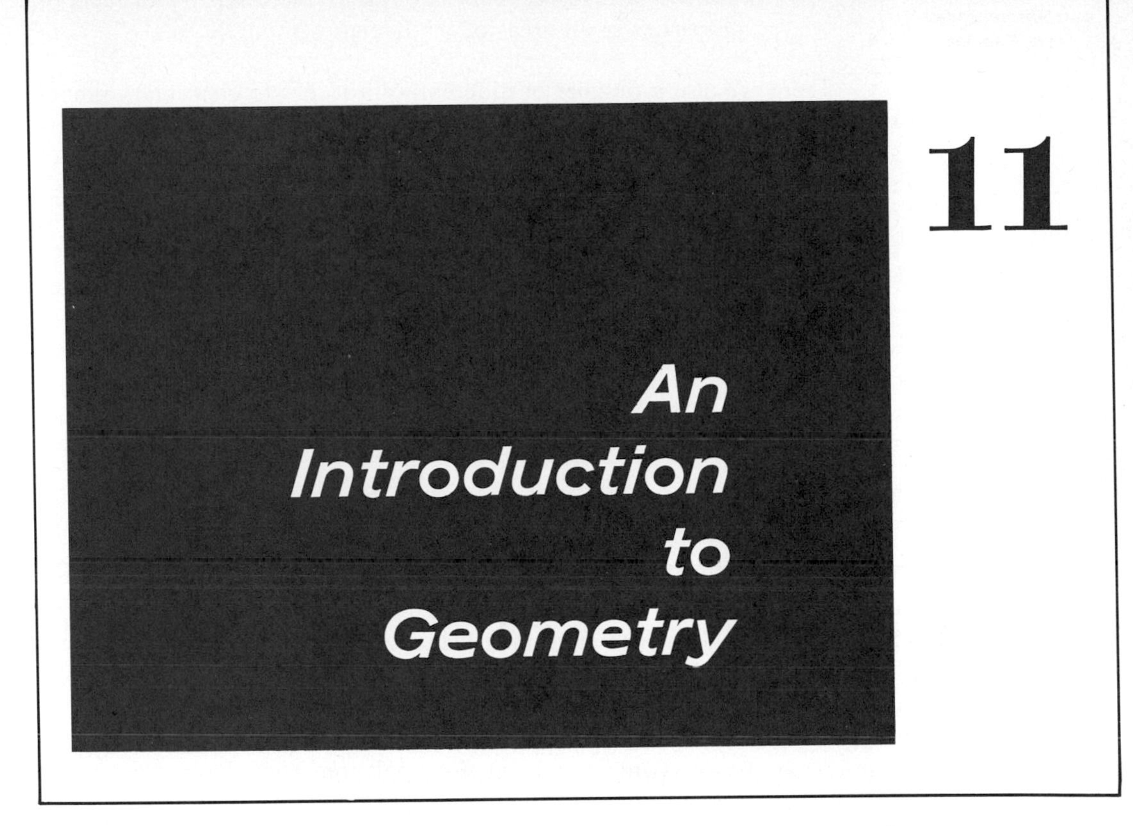

Geometry has evolved from a concern for earth measure (geo-metry), through the use of line segments and other figures to represent physical magnitudes, to a study of properties of sets of geometric elements. The geometric figures serve as the elements of geometry. Relations among these elements and proofs of their properties from given sets of postulates are considered in more advanced courses.

All geometric figures are usually considered to be sets of points where points are accepted as undefined elements. Although other basic elements are considered in abstract geometry, we shall restrict our consideration to points. This is not a serious restriction, since points may be *interpreted* in many ways. For example, we usually think of a point as a position on a line, on a plane, or in space. We may also think of a point on a number scale in terms of its coordinate. We may even think of cities as points on a map, with the air routes joining them considered as lines. This freedom to interpret points in a variety of ways provides a basis for more abstract geometries.

Elementary school geometry includes many of the same words as secondary school geometry but the approach is entirely different. The objectives in elementary school geometry include:

> The recognition of figures to establish a mathematical vocabulary.
> The informal visualization of relations among geometric figures.

The use of geometric figures as aids to the discovery of facts or properties in all areas of mathematics.

There are also a number of different approaches to elementary school geometry. Geometric figures may be studied in terms of unions and intersections of sets of points, lines, and planes (*nonmetric geometry*), in terms of measures of geometric figures (*metric geometry*), or in terms of coordinates (*coordinate geometry*). Our introduction to geometry includes all three of these approaches.

Measures of geometric figures have been considered in Chapter 8. Uses of geometric figures as aids to the discovery of facts and properties in other areas of mathematics have been common throughout this book. Thus this chapter serves as an extension of previously considered geometric concepts and a summary of the geometric concepts used in contemporary elementary school mathematics.

11-1 geometric figures

The language of mathematics includes the names of many geometric figures. Line segments, closed curves, and simple plane curves were introduced in §1-5. These introductions were primarily by examples of the figures, as is done in elementary schools. In the pedagogical explorations of that section a union of line segments that forms a simple, closed, plane curve was identified as a *polygon*. The line segments are *sides* of the polygon; the endpoints of the line segments are *vertices* (singular *vertex*) of the polygon. Then the common polygons were classified in terms of the number of their sides.

As in the classic story of Humpty Dumpty and Alice (§1-5) words are sometimes used in different ways in different series of elementary school textbooks. This situation arises relative to the word *quadrilateral*. All quadrilaterals are four-sides polygons but in some texts the polygons must be *convex* while in other texts this is not necessary. Similar distinctions are then made for polygons of more than four sides. A polygon is a **convex polygon** if every line segment PQ with points of the polygon as endpoints is either a subset of a side of the polygon or has only the points P and Q in common with the polygon. A polygon that is not a convex polygon is a **concave polygon.**

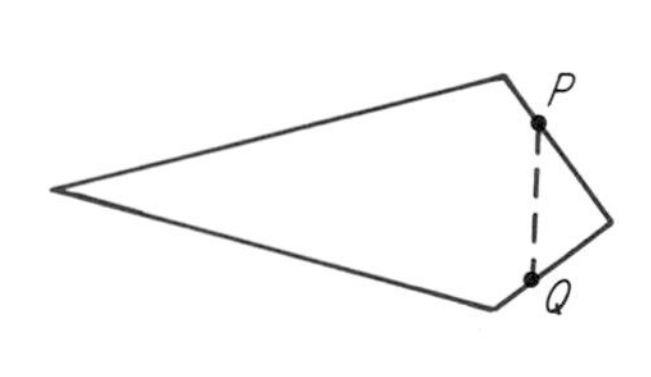

Convex polygon

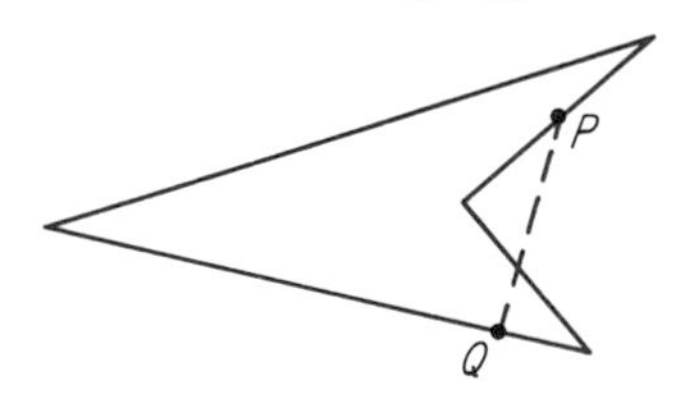

Concave polygon

Any line segment PQ has the points P and Q as *endpoints*; all other points of the line segment are **interior points**. If P and Q are points of a convex polygon and $\overline{PQ}$ is not a subset of a side of the polygon, then the interior points of $\overline{PQ}$ are **interior points** of the convex polygon. The union of the points of a convex polygon and its interior points is a **polygonal region.** On the plane of a convex polygon those points that are not points of the polygonal region are **exterior points** of the polygon.

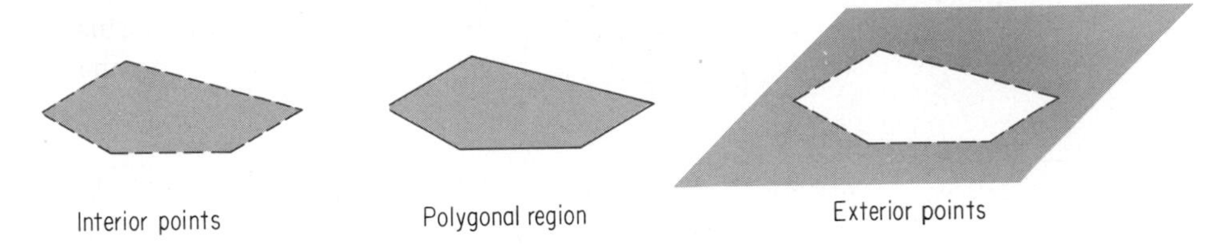

The linear measures and angular measures introduced in Chapter 8 enable us to classify triangles either in terms of the measures of their sides or in terms of the measures of their angles. A triangle may be classified according to the lengths of its sides as:

A **scalene triangle** if no two sides are congruent.
An **isosceles triangle** if at least two sides are congruent.
An **equilateral triangle** if all three sides are congruent.

Note that every equilateral triangle is an isosceles triangle, but not every isosceles triangle is an equilateral triangle. A triangle may be classified according to the measures of its angles as:

An **acute triangle** if all three angles are acute angles.
A **right triangle** if one angle is a right angle.
An **obtuse triangle** if one angle is an obtuse angle.
An **equiangular triangle** if all three angles are congruent.

On an ordinary plane any two distinct lines either intersect (have exactly one point in common) or are *parallel lines*. Quadrilaterals are usually classified in terms of the concepts of parallel lines, **consecutive sides** (sides with a common endpoint), and **opposite sides** (sides that do not have a common endpoint. Then a quadrilateral is:

A **trapezoid** if at least one pair of its opposite sides are parallel.
A **parallelogram** if each of its pairs of opposite sides are parallel.

A parallelogram is:

A **rectangle** if at least one of its angles is a right angle.
A **rhombus** if at least one pair of its consecutive sides are congruent.

Note that if at least one angle of a parallelogram is a right angle, then we can show that all four angles are right angles. The definition only requires that at least one angle be a right angle since we want explanations (proofs) of why certain parallelograms are rectangles to be as simple as possible.

Similarly, if at least one pair of consecutive sides of a parallelogram are congruent, then it can be proved that all four sides are congruent. A rectangle is:

A **square** if at least one pair of its consecutive sides are congruent.

A rhombus is:

A *square* if at least one of its angles is a right angle.

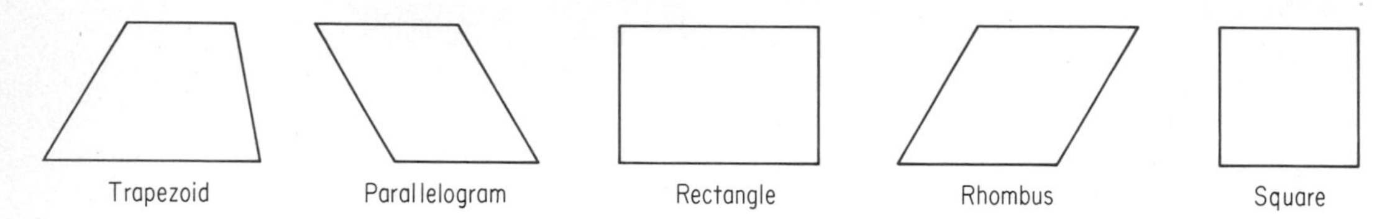

An equilateral triangle is also equiangular. Thus each of its sides is congruent to any one of the others and each of its angles is congruent to any one of the others. Any polygon such that each of its sides is congruent to any one of the others and each of its angles is congruent to any one of the others is a **regular polygon**. Then a regular polygon of three sides is an equilateral triangle; a regular polygon of four sides is a square. Several regular polygonal regions are often used for covering surfaces, for example, for tiling a floor.

Space figures are also considered in elementary school mathematics. Spheres, cubes, cylinders, pyramids, and triangular pyramids have been mentioned in §1-5. Volumes of *polyhedra* (plural of polyhedron) such as cubes, parallelepipeds, prismatoids, and prisms were considered in §8-6. Recognition of space figures and some of their measures is an important part of elementary school mathematics. Elementary school teachers also need to be able to draw reasonable pictures of such space figures. In making such drawings the basic principle is to dash (or dot) the edges (or parts of edges) that are hidden by other parts of the figure. For example, see the figures in §8-6. This principle is followed throughout this text. Elementary school students usually find space figures easier to discuss and less abstract than line or plane figures. Models of space figures can be made and examined in detail. We live in a three-dimensional world.

EXERCISES

Sketch a figure that is a union of line segments and is:

1. Not simple and not closed.
2. Simple but not a polygon.
3. A convex polygon of seven sides.
4. A concave polygon of seven sides.

If possible sketch the requested figure. If not possible, explain why it is not possible.

5. A scalene triangle.
6. An isosceles triangle.
7. An acute scalene triangle.
8. A right scalene triangle.
9. An obtuse scalene triangle.
10. An equiangular scalene triangle.
11. An acute isosceles triangle.
12. A right isosceles triangle.
13. An obtuse isosceles triangle.
14. An equiangular isosceles triangle.
15. An acute equilateral triangle.
16. An obtuse equilateral triangle.
17. A parallelogram that is not a rectangle.
18. A parallelogram that is not a rhombus.
19. A rhombus that is not a square.
20. A rhombus that is not a rectangle.
21. A rhombus that is not a parallelogram.
22. A trapezoid that is not a parallelogram.
23. A trapezoid that is not a rectangle.
24. A rectangle that is not a trapezoid.

*25. A cube.

*26. A rectangular parallelepiped that is not a cube.

*27. A parallelepiped that is not rectangular.

*28. A pyramid with a square base; that is, a square pyramid.

*29. A sphere.

PEDAGOGICAL EXPLORATIONS

Construct a triangle *ABC* with sides of lengths 3 cm, 4 cm, and 5 cm. Then construct a triangle *PQR* with sides of lengths 6 cm, 8 cm, and 10 cm, respectively.

1. Explain the relationship between the perimeters of the two triangles.
2. Explain the relationship between the areas of the two triangular regions.

3. Let a new triangle be constructed with sides respectively k times the lengths of the sides of a given triangle.
   (a) What is the ratio of the perimeter of the new triangle to that of the given triangle?
   (b) What is the ratio of the area of the new triangular region to that of the given triangle?
4. Select at least three different values of k for Exploration 3, construct appropriate figures, make necessary measurements, and check your answers for that exploration.

Construct a model of a cube with edges of length 6 cm. Then construct a model of a cube with edges of length 3 cm.

5. Find the sum of the lengths of the edges of each cube and compare these sums.
6. Repeat Exploration 5 for the surface areas of the cubes.
7. Repeat Exploration 5 for the volumes of the cubes.
8. If an edge of one cube is k times an edge of another cube, what is the relation between:
   (a) The sum of the lengths of the edges of the cubes?
   (b) The surface areas of the cubes?
   (c) The volumes of the cubes?

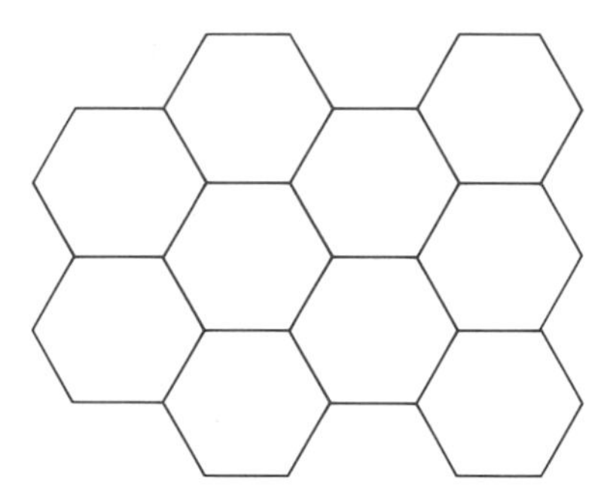

Suppose that you needed to make tile for a floor or patio and wanted all the tiles to be of the same size and shape. Suppose also that in preparation for this task you make some paper (or cardboard) model tiles in the shapes of regular polygons. One way to draw a regular polygon of n sides ($n \geq 3$) is to follow these steps:

Draw a circle, mark its center O, and draw one radius $\overline{OA}$.
Since one revolution about the center is 360 degrees, use a protractor and draw radii at intervals of $(360/n)^\circ$ from $\overline{OA}$. (After n steps you should be back at $\overline{OA}$.)
For each of the radii just drawn mark its endpoint on the circle.
Then join A and these endpoints in order around the circle with line segments returning to A to complete the regular polygon with n sides.

9. Make a regular triangle and use it to cut out at least ten triangular regions. Could you use regular triangular regions to tile a floor? Use your models to demonstrate your answer.
10. Repeat Exploration 9 for squares.
11. Repeat Exploration 9 for regular five-sided polygons (pentagons).
12. Repeat Exploration 9 for regular six-sided polygons (hexagons).
13. Repeat Exploration 9 for regular seven-sided polygons.

Observe the angles of the regular polygons made for Explorations 9 through 13.

14. (a) As the number of sides of the polygons increases what happens to the size of the angles of the polygons?
    (b) Explain your answer for part (a) using the measures of the angles.
15. What is the smallest number of tiles that may share a common vertex when you tile a floor?
16. Explain why you do or do not expect that regular polygons of more than six sides can be used as tile for a floor.

11-2 constructions of geometric figures

Two figures are *equal* if and only if they consist of the same points, that is, they are the same figure, usually named in different ways. For example, $\triangle ABC = \triangle BCA$; also in $\triangle ABC$ we have $\angle A = \angle BAC = \angle CAB$. Two figures are *congruent* if and only if under some matching of parts they have the same measures. For example, two line segments are congruent if they have the same linear measure; two angles are congruent if they have the same angular measure. Two circles are congruent if their *radii* (the distances from the centers to points of the circles) are equal. Actually, the word *radius* is used for both the line segment and the distance. Thus for congruent circles the line segments are congruent and the distances are equal. If two circles are congruent, then they have equal circumferences and equal areas.

Congruent line segments, angles, and circles are used in the constructions of many common figures. When we construct a figure congruent to a given figure, we *copy* the given figure. The usual "tools" for making constructions are a straightedge for drawing lines and a compass for drawing circles. The straightedge may be an unmarked ruler or simply the crease formed when a piece of paper is folded. The compass may be set for any given distance (radius) and used with any given point as center.

An instruction such as "sketch" or "draw" is used instead of "construct" whenever a protractor is to be used for measuring angles or a ruler for measuring distances. Only a straightedge and a compass may be used in the classical constructions of geometric figures.

Any triangle has three vertices and three sides. For any triangle ABC it is customary to name the angle at each vertex by the capital letter that names that vertex. Then the length of the side determined by the other two vertices (the *side opposite* the given vertex) is named by the corresponding small (lower-case) letter.

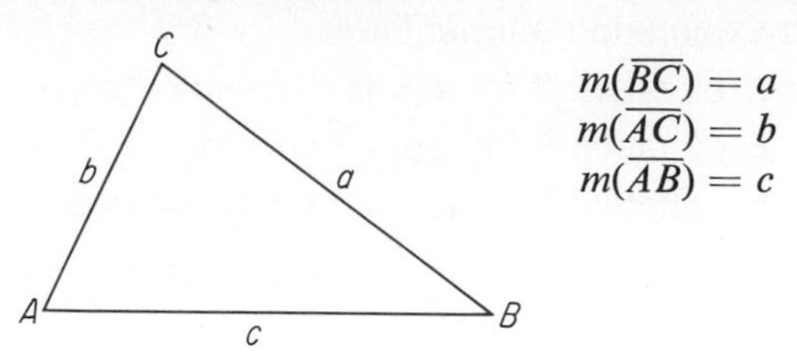

$m(\overline{BC}) = a$
$m(\overline{AC}) = b$
$m(\overline{AB}) = c$

Example 1: Copy the given line segment AB onto the given line DE such that A is copied onto D.

Solution: The given line segment may be copied onto either the ray DE or the ray DH *opposite* the ray DE. Select one of these rays. Set the compass for radius AB. Use D as center and draw a circle of radius AB or an arc of the circle so that the arc intersects the selected ray. Label the point of intersection F; $\overline{AB} \cong \overline{DF}$.

A line segment AB may be copied onto (is congruent with) exactly one line segment DF on a given ray DE.

Example 2: Copy $\angle ABC$ onto the given ray DE so that B corresponds to D.

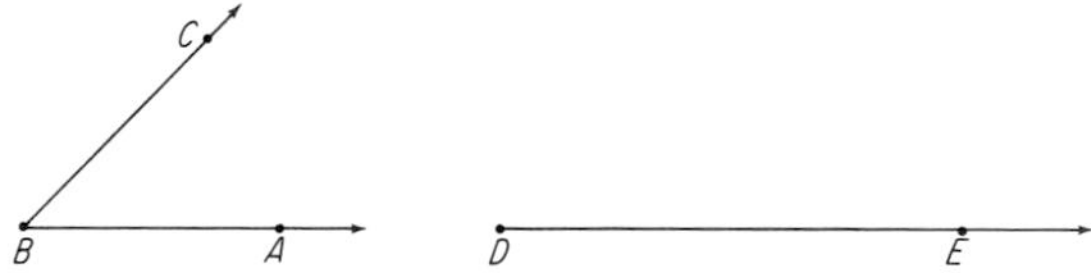

Solution: Use a compass and draw any two congruent circles with centers B and D, or sufficiently large arcs of these circles.

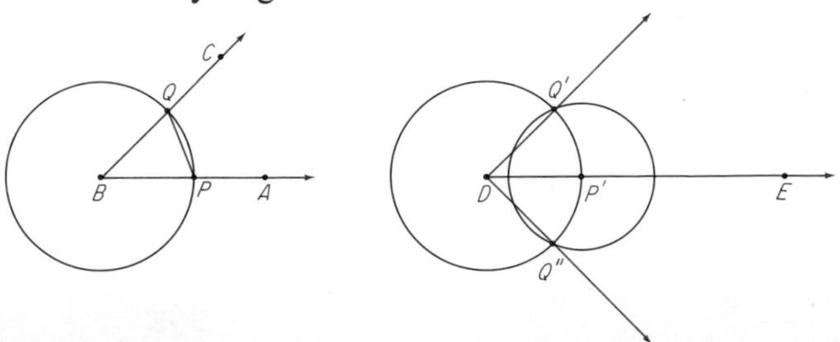

Label the intersections of the circle with center B and the rays BA and BC as P and Q respectively. The line segment PQ has its endpoints on the circle and is a **chord** of the circle. Then label the intersection of the ray DE and the circle with center D as P'. Since the two circles are congruent circles, $\overline{DP'} \cong \overline{BP}$. Next draw a circle with center P' and $m(\overline{PQ})$ as radius. Label the intersections of the circles with centers D and P' as Q' and Q''. Draw rays DQ' and DQ''. Both $\angle EDQ'$ and $\angle EDQ''$ satisfy the given requirements; $\angle EDQ' \cong \angle ABC$ and $\angle EDQ'' \cong \angle ABC$.

In Example 2 note that a given angle ABC may be copied onto a given ray DE with D as the new vertex in two ways; that is, using points in either the half-plane above the line DE or the half-plane below the line DE. An angle ABC may be copied onto (is congruent with) exactly one angle EDQ where the ray DE is given and Q is in a given half-plane with respect to the line DE.

Example 3: Use straightedge and compass to construct the midpoint of a given line segment AB.

Solution: Select a radius r greater than one-half $m(\overline{AB})$. Then draw circles of radius r and centers A and B. These circles intersect at two points P and Q. Draw $\overleftrightarrow{PQ}$ and label its intersection with $\overline{AB}$ as M. Then M is the midpoint of $\overline{AB}$.

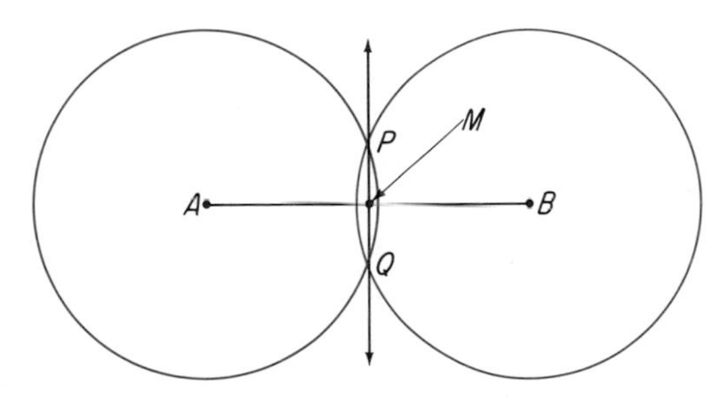

In Example 3 the line $\overleftrightarrow{PQ}$ is also *perpendicular* to $\overleftrightarrow{AB}$ at M, that is, $\angle AMP$, $\angle AMQ$, $\angle BMP$, and $\angle BMQ$ are all right angles.

Example 4: Given $\triangle ABC$ and line segment DE construct $\triangle DEF$ such that $\angle A \cong \angle D$ and $\angle B \cong \angle E$.

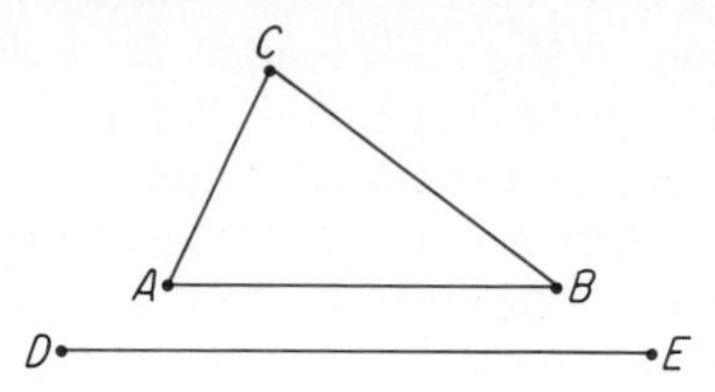

Solution: We select a half-plane with respect to $\overleftrightarrow{DE}$, such as the half-plane above $\overleftrightarrow{DE}$. Then as in Example 2 we copy $\angle BAC$ such that $\angle EDQ \cong \angle BAC$, copy $\angle ABC$ such that $\angle DEP \cong \angle ABC$, and let $\overrightarrow{DQ} \cap \overrightarrow{EP} = F$.

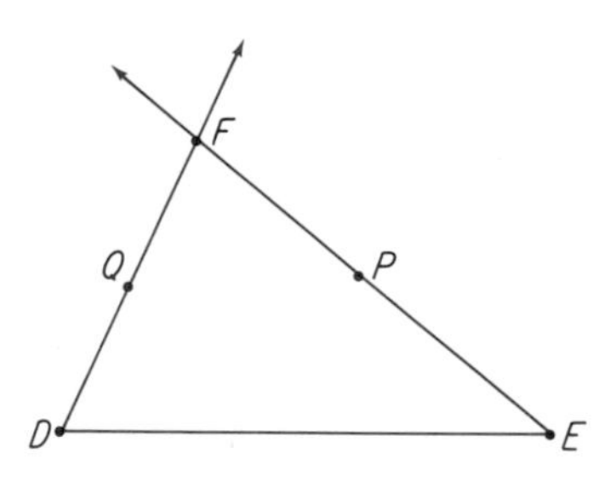

In Example 4 use a protractor and measure $\angle C$ and $\angle F$; they should have the same measures within the accuracy of your ability to measure them. This illustrates the fact that if two angles of a triangle are congruent to two angles of another triangle, then their third angles are also congruent. The triangles ABC and DEF appear to have the same *shape*; they are **similar triangles**, written

$$\triangle ABC \sim \triangle DEF$$

In the symbolic statement the triangles are named by listing the vertices in order so that the angles at corresponding vertices are congruent; that is,

$$\angle A \cong \angle D, \qquad \angle B \cong \angle E, \qquad \angle C \cong \angle F$$

Example 4 indicates that a triangle may be constructed similar to any given triangle and with any given line segment as a side. The most familiar properties of similar figures were considered in Pedagogical Explorations 1 through 8 of §11-1. Similar figures provide the basis for scale drawings, blueprints, and maps.

Suppose that you wish to copy a triangle so that the new triangle has the same *size and shape* as the original triangle. Then the two triangles will be **congruent triangles**, written $\triangle ABC \cong \triangle PQR$

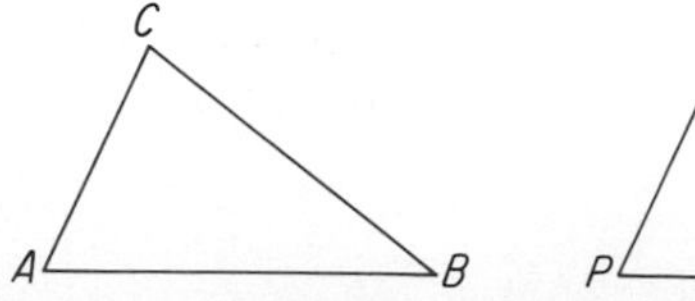

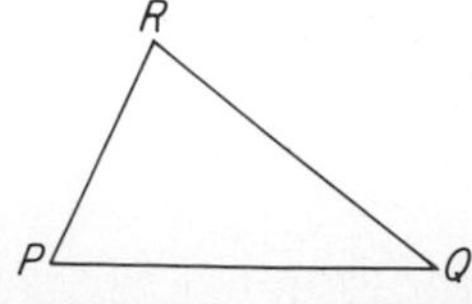

where the notation implies that

$$\angle A \cong \angle P, \qquad \overline{AB} \cong \overline{PQ}$$
$$\angle B \cong \angle Q, \qquad \overline{BC} \cong \overline{QR}$$
$$\angle C \cong \angle R, \qquad \overline{AC} \cong \overline{PR}$$

Note that the angles are congruent in the order of their listed vertices.

When we constructed a triangle similar to a given triangle, we found that if we copied two of the angles, then the third angles were automatically congruent. Fortunately with six desired congruences for congruent triangles, we do not need to use all six congruences to copy a triangle. For example, it is sufficient to copy the three sides as in Example 5.

Example 5: Use the sides of triangle ABC and construct $\triangle DEF \cong \triangle ABC$.

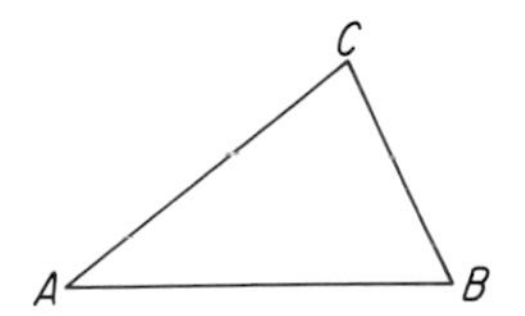

Draw any line DH and select any side of the line (either half-plane) for the third vertex of the triangle. Copy $\overline{AB}$ onto the ray DH so that $\overline{AB} \cong \overline{DE}$. Draw intersecting arcs of a circle with radius AC and center D and a circle with radius BC and center E. Label the intersection F. Draw $\overline{DF}$ and $\overline{EF}$. Then $\overline{DF} \cong \overline{AC}$, $\overline{EF} \cong \overline{BC}$, and $\triangle DEF \cong \triangle ABC$.

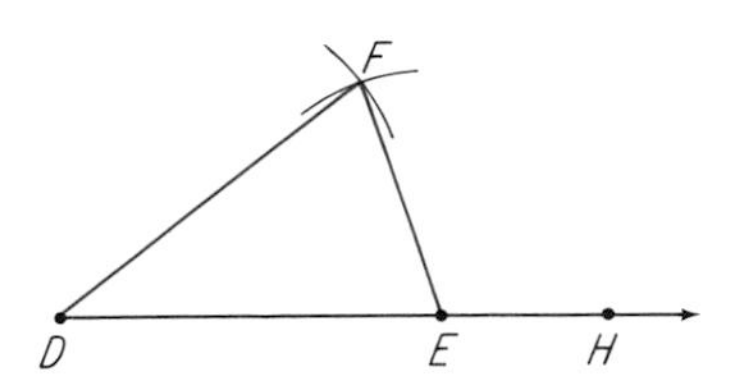

Example 5 illustrates the fact that if three sides of one triangle are congruent to three sides of another triangle, then the triangles are congruent (s.s.s. $\cong$ s.s.s.). If the congruent triangular regions were cut out, it would be found that the triangles were of exactly the same shape and size (area). Other selections of the parts to be copied to obtain congruent triangles are considered in most elementary school textbooks.

Relationships among polygons may be considered in terms of those between corresponding triangles. For example, quadrilateral $ABCD$ is similar to quadrilateral $MNOP$ if $\triangle ABC \sim \triangle MNO$, $\triangle ACD \sim \triangle MOP$, the triangular regions in $ABCD$ have only the line segment AC in common, and the triangular regions in $MNOP$ have only the line segment MO in common.

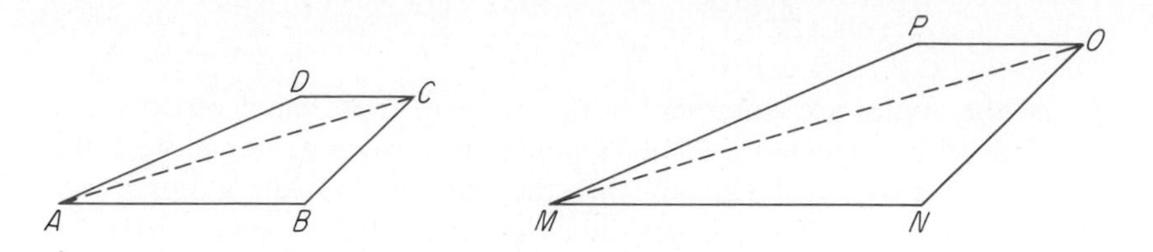

The need for the statements about intersecting triangular regions may be seen in the next two figures. Such restrictions were probably covered in your high school geometry course by saying that the triangles ACD and MOP were "similar and similarly placed."

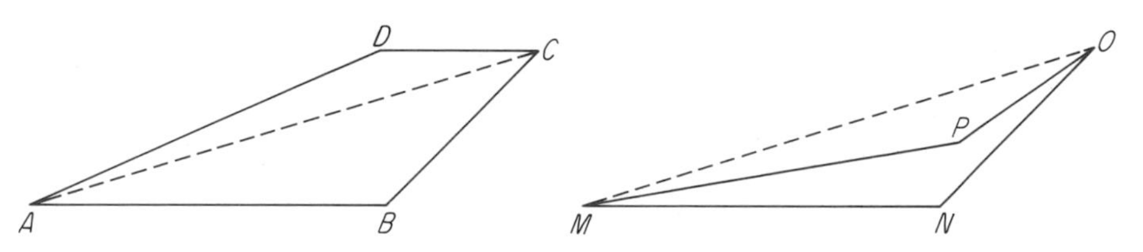

Two distinct points A and B determine a unique line and the line may be named by any two of its distinct points (§1-5). Similarly, three points that are not on the same line determine a unique plane and the plane may be named by any three of its noncollinear points. For example, the plane of the quadrilateral $ABCD$ may be identified by any one of the names ABC, ABD, ACD, and BCD. Also, in any of these names the letters may be in any order. For example, ABC, ACB, BCA, BAC, CBA, and CAB are all names of the same plane.

EXERCISES

The given figure is a rectangle. Each of its angles is a right angle. Its opposite sides are congruent.

1. Write two statements of:
- **(a)** Equality of angles.
- **(b)** Congruence of angles that are not equal.
- **(c)** Equality of sides.
- **(d)** Congruence of sides that are not equal.
- **(e)** Equality of rectangles.

2. Copy the figure given in Exercise 1, draw line segments QS and TR, and let $\overline{QS} \cap \overline{TR} = P$.
(a) Name a triangle that appears to be (or is) equal to triangle QRS.
(b) Name a triangle that appears to be (or is) congruent but not equal to triangle QRS.
(c) Write a statement of the congruence for part (b) to show the correspondence of the vertices.

3. Repeat Exercise 2 for triangle PQR.

4. Repeat Exercise 2 for triangle QPT.

In each exercise sketch in the construction marks if you do not have access to a compass. Also be prepared to explain the steps that you have used.

5. Draw a line segment PQ and a line RS.
(a) Copy $\overline{PQ}$ onto the ray $\overrightarrow{RS}$.
(b) Show that $\overline{PQ}$ can be copied onto $\overleftrightarrow{RS}$ so that P corresponds to R and Q corresponds to either of two points T or U. Give an expression for the length of $\overline{TU}$ in terms of the length of $\overline{PQ}$.

6. Draw an acute angle RST and a ray $\overrightarrow{XY}$.
(a) Copy $\angle RST$ so that $\overrightarrow{XY}$ is a side of the new angle.
(b) Show that part (a) can be done in two ways.

Use straightedge and compass (or sketch markings), construct, and label:

7. A straight angle ABC.

8. A right angle DEF.

9. A line GH that intersects a given line IJ at a given point K so that the two lines are perpendicular.

Construct each specified figure. In each case describe the steps used and explain whether or not you and a friend can both do the exercise, obtain different appearing figures, and both be correct.

10. A square with edges 4 cm long.

11. A rectangle with edges 2 cm and 5 cm long.

12. A rhombus with edges 3 cm long.

13. A parallelogram with edges 3 cm and 5 cm long.

If possible construct a triangle with sides of the specified numbers of units of length. You may select any convenient size for the units. If you consider the construction impossible, explain your reasoning.

14. 7, 5, 4

15. 20, 30, 40

16. 0.08, 0.11, 0.15

17. 2, 4, 6

18. 25, 15, 10

19. 6, 2, 3

20. 8, 13, 15

21. $\frac{1}{4}, \frac{1}{3}, \frac{1}{5}$

22. $\frac{1}{2}, \frac{1}{3}, \frac{1}{6}$

23. $\sqrt{2}, \sqrt{3}, \sqrt{5}$

PEDAGOGICAL EXPLORATIONS

Consider cubes and their interior points.

1. A cube has 2-centimeter edges and is painted red. Think of this 2-centimeter cube as cut into eight 1-centimeter cubes using three planes, one parallel to each pair of parallel faces of the cube. How many of the 1-centimeter cubes have six painted faces? Five painted faces? Four painted faces? Three painted faces? Two painted faces? One painted face? No painted faces?
2. Repeat Exploration 1 for a painted 3-centimeter cube divided into 27 1-centimeter cubes.
3. Repeat Exploration 1 for a painted 4-centimeter cube divided into 64 1-centimeter cubes.
4. Repeat Exploration 1 for a painted 5-centimeter cube divided into 125 1-centimeter cubes.
5. Repeat Exploration 1 for a painted n-centimeter cube, $n \geq 2$, divided into n^3 1-centimeter cubes.

Prepare for use with a specified elementary school grade:

6. A lesson plan for discovering that the sum of the measures of the angles of any triangle is the same as the sum for any other triangle. Consider paper folding as well as other approaches.
7. A set of transparencies to be used one on top of the other and showing each step of the construction of the perpendicular bisector of a given line segment.
8. A lesson plan for discovering the number of diagonals for any convex polygon.
9. A display of pictures of common applications of similar figures and of congruent figures.

11-3 relations among geometric figures

The pedagogical explorations of the previous section are similar to some of the activities in contemporary elementary school classes. Relations among figures are explored and illustrated until they appear familiar and reasonable to the student. Sometimes the relationships are not pursued to the stage of formal verbalization. At other times the relationships are summarized in words and in symbols. The examples and exercises of §11-2 probably reminded you of some of the following theorems (statements) that you assumed or proved in secondary school geometry:

The sum of the measures, in degrees, of the angles of a triangle is 180.

If two angles of one triangle are congruent to two angles of another triangle, then the triangles are similar triangles.

If two triangles are similar, then their corresponding sides are in the same ratio (proportional).

If the corresponding sides of two triangles are in the same ratio, then the triangles are similar triangles.

Any two plane triangles are congruent if there is a matching of the vertices of the triangles such that at least one of the following is true:

- The three sides of one triangle are respectively congruent to the three sides of the other triangle (s.s.s. $\cong$ s.s.s.).
- Two sides and the included angle of one triangle are respectively congruent to the corresponding parts of the other triangle (s.a.s. $\cong$ s.a.s.).
- Two angles and the included side of one triangle are respectively congruent to the corresponding parts of the other triangle (a.s.a. $\cong$ a.s.a.).

Two right triangles are congruent if there is a matching of the vertices such that either of the following is true:

- The hypotenuse and a leg of one triangle are congruent to the corresponding parts of the other triangle.
- The hypotenuse and an acute angle of one triangle are congruent to the corresponding parts of the other triangle.

These ways of determining that triangles are congruent can be based upon ways of constructing a triangular region that can be cut out and fitted onto another triangular region. The word "superposition" formerly was used to describe this process. However, "superposition" has connotations that are contrary to contemporary secondary school geometry. Thus the word "superposition" has become nearly obsolete. Congruent triangles are used extensively in secondary school mathematics and in applications of mathematics. Readiness for these uses of congruent triangles is developed through experiences in elementary school mathematics.

The heavy emphasis upon congruence that you may have experienced in your own school years should not be allowed to exclude the recognition of other relations among figures. Similarity of triangles and other polygons has already been noted. There are also several other important relations, often defined only for certain types of figures.

Coincidence of geometric figures is another name for equality; that is, two figures coincide if they consist of the same points.

Intersections of geometric figures have the same meaning, and indeed are, intersections of sets of points.

Separation has many applications, some of which we have already considered informally. Two points A and C are separated on a line by a point B if $B \in \overline{AC}$. On a line two points A and C are not separated by a point D

if $D \notin \overline{AC}$. Any point B of a line AC separates the line into three sets. If $B \in \overline{AC}$, then the three sets are:

The point B.
The half-line BC.
The half-line BA.

On a plane two points A and C are separated by a line b if $A \notin b$, $C \notin b$, and $\overline{AC} \cap b \neq \varnothing$. Two points A and D are not separated by a line b if $\overline{AD} \cap b = \varnothing$. Any line b of the plane separates the plane into three sets. If A and C are points of the plane and the line b intersects $\overline{AC}$, then the three sets are:

The line b.
The points on the same side of b as A, that is, the half-plane of A with respect to b.
The points on the opposite side of b from A, that is, the half-plane of C with respect to b.

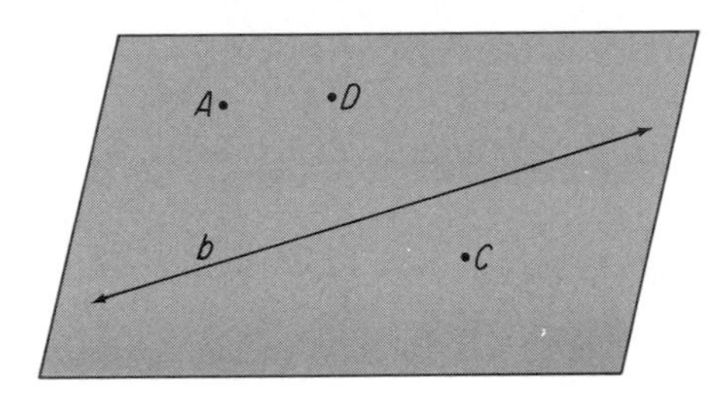

In ordinary space two points A and C are separated by a plane β if $A \notin \beta$, $C \notin \beta$, and $\overline{AC} \cap \beta \neq \varnothing$. Two points A and D are not separated by a plane β if $\overline{AD} \cap \beta = \varnothing$. Any plane β separates space into three sets. If β intersects $\overline{AC}$, then the three sets are:

The plane β.
The points on the same side of β as A, that is, the *half-space* of A with respect to β.
The points on the opposite side of β from A, that is, the half-space of C with respect to β.

In advanced mathematics we say that *any simple closed curve in a plane separates the plane into two regions.* This is the **Jordan curve theorem.** The curve is the common boundary of the two regions, and one cannot cross from one region to the other without crossing the curve. The Jordan curve theorem is a very powerful theorem and yet a very simple one. Notice that it is independent of the size or shape of the curve.

How can such a simple theorem have any significance? It provides a basis for Euclid's assumption that any line segment joining the center of a circle to a point outside the circle must contain a point of the circle. In more advanced courses it provides a basis for the existence of at least one zero of a polynomial in any interval on which the polynomial changes sign. It provides a basis for Euler diagrams and Venn diagrams in any two-valued logic. In §1-6 we assumed that all points of the universal set were either points of a circular region A or points of A', the complement of A.

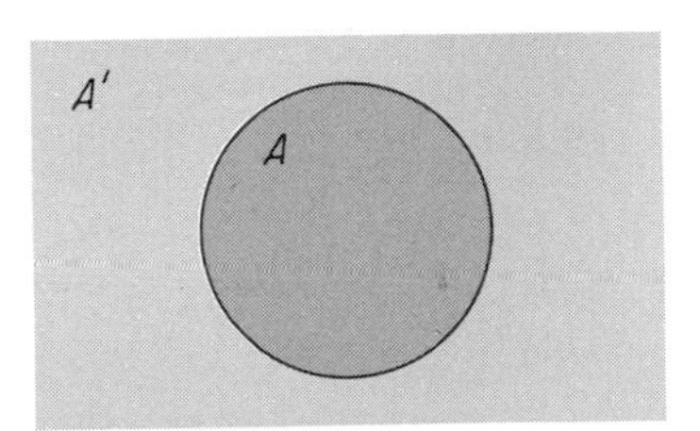

In §2-2 we considered statements p that are either true or false (in which case $\sim p$ is true) but not both true and false. Because of the Jordan curve theorem we were able to represent the situations under which statements were true (or false) by Euler diagrams and by Venn diagrams.

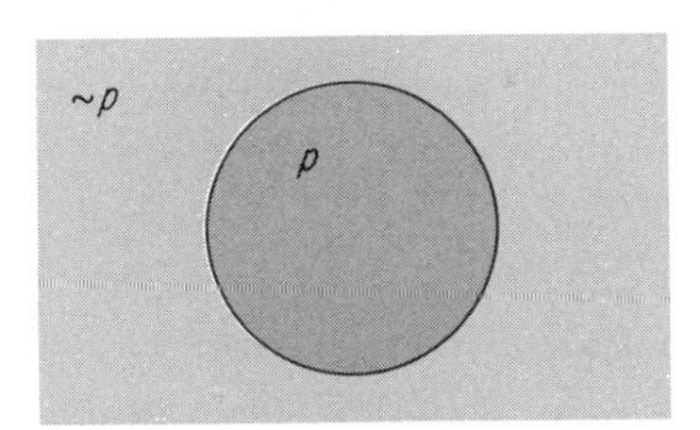

This simple theorem regarding the existence of an inside and an outside of any simple closed curve may also be used to answer questions raised by the next problem. This problem is a popular one, and many people have spent hours working on it. Consider three houses (×) in a row and three utilities (∘) in a second row on a plane surface. The problem is to join each house to each utility by an arc on the plane in such a way that no two arcs cross or pass through houses or utilities except at their endpoints. As in the figure on the left it is easy to designate paths from one house to each of the

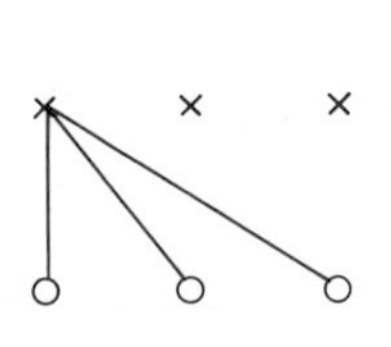

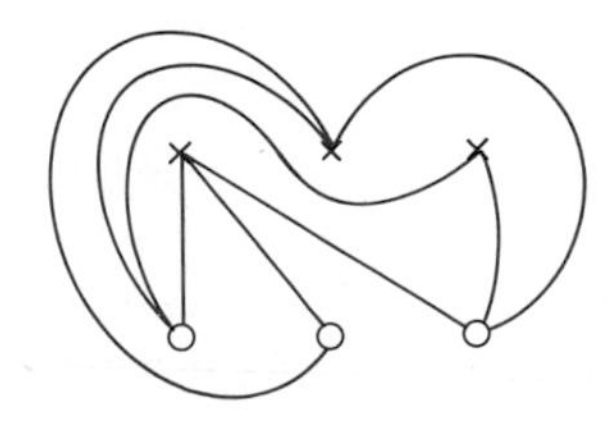

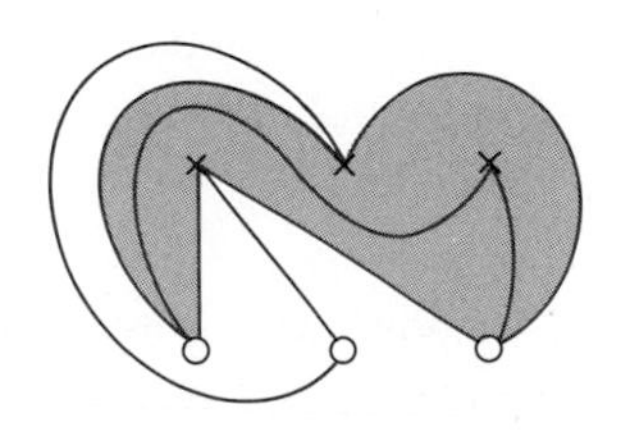

three utilities. One can also designate paths to each utility from the second house as in the second figure. Then one can designate two of the paths from the third house, but it is not possible on an ordinary plane to draw the path from the remaining house to the remaining utility. This assertion is based upon the fact that the simple closed curve indicated in the third figure divides the plane into two regions; the third house is inside the curve (shaded region), the remaining utility is outside the curve, and the two cannot be joined without crossing the curve.

A very useful relation among figures is *perpendicularity* (intersection at right angles). This relation is defined for:

Line to line: coplanar lines that intersect at right angles.

Line to plane and plane to line: a line m that intersects a plane in a point P and is perpendicular to at least two (actually all) of the lines that are on the plane and contain the point P.

Plane to plane: two planes such that either plane contains at least one line that is perpendicular to the other plane.

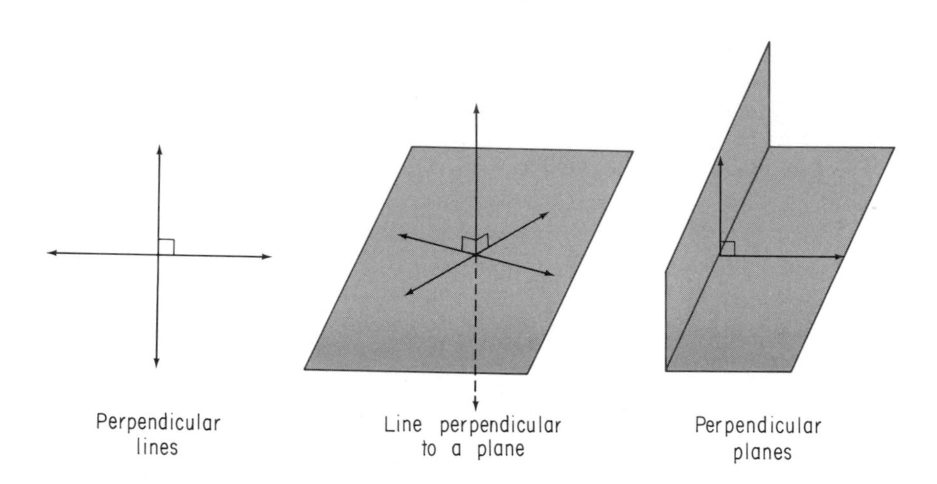

Still another useful relation among geometric figures is *parallelism.* This relation is defined for:

Line to line: coplanar lines that do not have exactly one point in common. (This definition allows a line to be parallel to itself; however, books differ on this.)

Line to plane and plane to line: a line and a plane that do not have exactly one point in common. (This definition allows a line on a plane to be parallel to that plane; however, books differ on this.)

Plane to plane: two planes that coincide (books differ on this) or do not have any points in common.

Relations among lines in space may be observed in terms of the intersections of the walls, ceiling, and floor of a traditional classroom (rectangular parallelepiped). On any given wall the line along the ceiling and the line along the floor do not appear to intersect, that is, do not have any point in common no matter how far they are extended. The line along the ceiling and one wall also does not appear to intersect the line of intersection of two other walls. Any distinction between these two situations must take into consideration the fact that in the first case the two lines were on the same wall (plane), whereas in the second case there could not be a single plane containing both lines. In general, two lines that are on the same plane and do not have any point in common are also said to be *parallel lines*; two lines that are not on the same plane are called **skew lines;** two distinct lines that have a point in common are called **intersecting lines;** lines that have all their points in common may be visualized as two names for the same line and are called *coincident lines.*

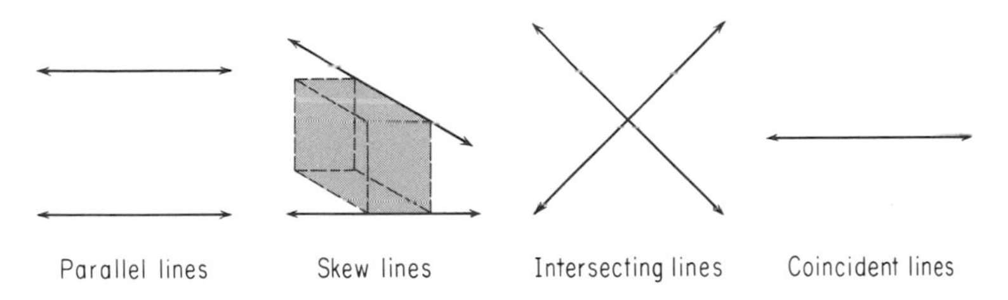

Note that the arrows indicate that the lines may be extended indefinitely, that is, that the figure involves representations for lines rather than actual lines. These representations are on the printed page and thus on a plane. Thus the figure for skew lines must be visualized as a "picture" on a plane of skew lines in space.

Points, lines, and planes are the building blocks of geometry. We use these elements and relations such as intersections, parallelism, coincidence, skewness, and separation as we define other geometric figures.

EXERCISES

1. Label three points A, B, and C that are not on the same line. Draw and name each of the lines that are determined by these points.
2. Repeat Exercise 1 for four points A, B, C, and D such that no three of the points are on the same line.
3. Assume that the four points in Exercise 2 are not on the same plane. Name each of the planes that is determined by these points.
4. Assume that all of the given points are on the same plane and no three

of the given points are on the same line. State how many lines are determined by the given number of points: **(a)** 2; **(b)** 3; **(c)** 4; **(d)** 5; **(e)** 6; ***(f)** 10; ***(g)** n.

5. Assume that all of the given lines are on the same plane, that each line intersects each of the other lines, and that no three of the given lines are on the same point. State how many points are determined by the given number of lines: **(a)** 2; **(b)** 3; **(c)** 4; **(d)** 5; **(e)** 6; ***(f)** 10; ***(g)** n.

6. Consider the lines along the edges of the rectangular solid shown in the given figure and state whether the specified lines are parallel lines, skew lines, or intersecting lines:

(a) $\overleftrightarrow{AB}$ and $\overleftrightarrow{CD}$
(b) $\overleftrightarrow{AB}$ and $\overleftrightarrow{EF}$
(c) $\overleftrightarrow{AB}$ and $\overleftrightarrow{EH}$
(d) $\overleftrightarrow{CD}$ and $\overleftrightarrow{FG}$

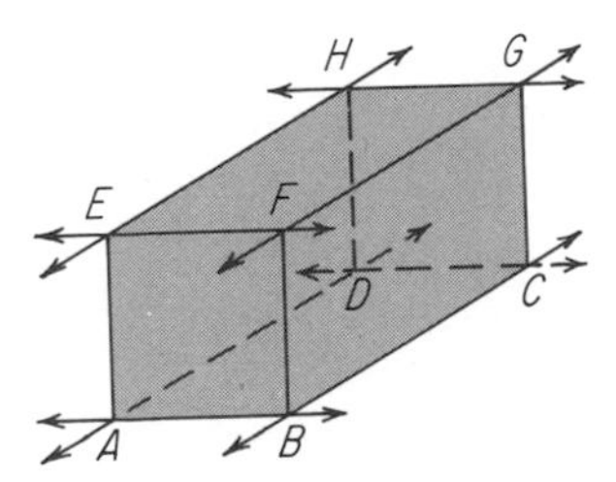

Use pencils, table tops, etc. to represent lines and planes and determine whether each statement appears to be true or false in Euclidean geometry.

7. Given any line m and any point P that is not a point of m, there is exactly one line t that is parallel to m and contains P.

8. Given any line m and any point P that is not a point of m, there is exactly one line q such that q contains P and m and q are intersecting lines.

9. Given any line m and any point P that is not a point of m, there is exactly one line s such that s contains P and s and m are skew lines.

10. Given any line m and any point P that is not a point of m, there is exactly one plane that is parallel to m and contains P.

11. Given any plane ABC and any point P that is not a point of ABC, there is exactly one plane that contains P and intersects ABC.

12. Given any plane ABC and any point P that is not a point of ABC, there is exactly one plane that contains P and is parallel to ABC.

13. Given any plane ABC and any point P that is not a point of ABC, there is exactly one line that contains P and intersects ABC.

14. Given any plane ABC and any point P that is not a point of ABC, there is exactly one line that contains P and is parallel to ABC.

15. Given any plane ABC and any line m that is parallel to ABC, there is exactly one plane that contains m and intersects ABC.

16. Given any plane ABC and any line m that is parallel to ABC, there is exactly one plane that contains m and is parallel to ABC.

17. Given any plane ABC and any line m that is parallel to ABC, there is for any given point P that is not on m and not on ABC exactly one line that is parallel to m and also parallel to ABC.

Many interesting figures may be obtained by paper folding. Wax paper is especially well suited for these exercises.

18. Select two points P and Q. Fold the point P onto the point Q and form the crease for the fold. What relation does the line represented by the crease have to the line segment PQ?

19. Describe and demonstrate the formation of a rectangle by paper folding.

20. Describe and demonstrate the formation of a square by paper folding.

21. Draw a circle and select about twenty points somewhat equally spaced around the circle. Fold each of these points of the circle to the center of the circle, form the creases, and describe the resulting figure.

PEDAGOGICAL EXPLORATIONS

Find the numbers V of vertices, A of arcs (including line segments), and R of regions for each figure.

1. (a) 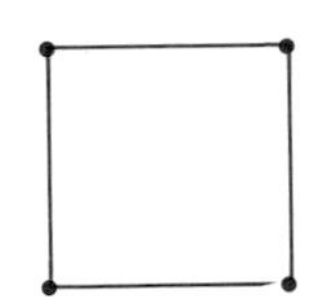 (b) 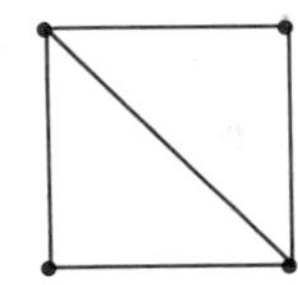

2. (a) 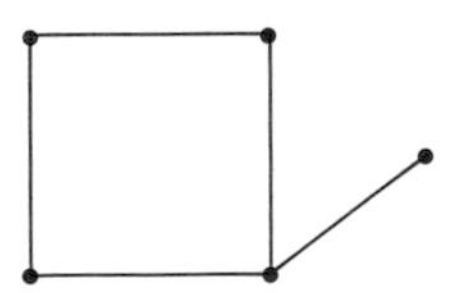 (b) 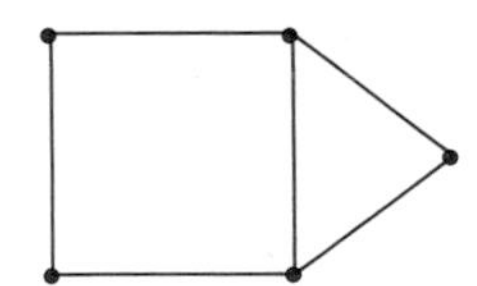

3. (a)  (b) 

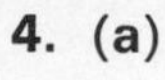

4. (a)

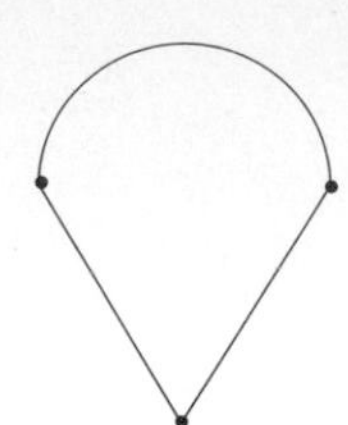

(b)

5. (a)

(b)

6. (a)

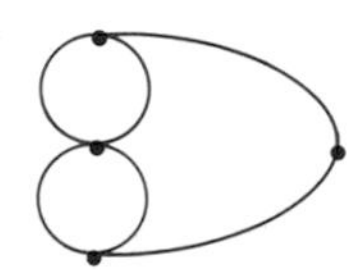

(b)

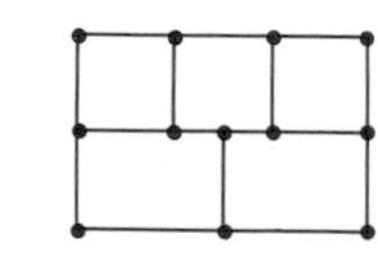

7. Study the values of V, A, and R obtained in Explorations 1 through 6 and find an expression for A in terms of V and R. You should obtain a form of the *Euler formula for networks.*
8. Let V be the number of vertices, E the number of edges, and F the number of faces. Then make a table of values of these numbers for each of the following figures:
   (a) A triangular pyramid.
   (b) A square pyramid.
   (c) A cube.
   (d) A triangular prism.
   (e) A pyramid with a pentagon as a base.
   (f) A prism with a pentagon as a base.
   (g) A pyramid with a hexagon as a base.
9. Compare the value of E with the value of $V + F$ for each of the figures considered in Exploration 8. Then conjecture a formula for E in terms of V and F. You should obtain the famous *Euler formula for polyhedra.*

11-4 networks

Any set of line segments (or arcs) forms a **network.** If the network can be drawn by tracing each line segment exactly once without removing the point of the pencil from the paper, the network is **traversable.** The study of the traversability of networks probably stemmed from a problem concerning the bridges in the city of Königsberg. There was a river flowing through the city, two islands in the river, and seven bridges as in the figure. The people of Königsberg loved a Sunday stroll and thought it would be nice to take a

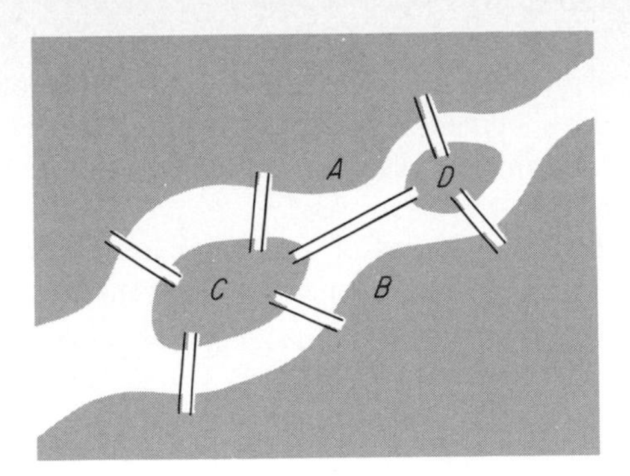

walk that would take them across each bridge exactly once. They found that no matter where they started or what route they tried, they could not cross each bridge exactly once. Gradually it was observed that the basic problem was concerned with paths between the two sides of the river A, B, and the two islands C, D as in the figure. With this representation of the problem by

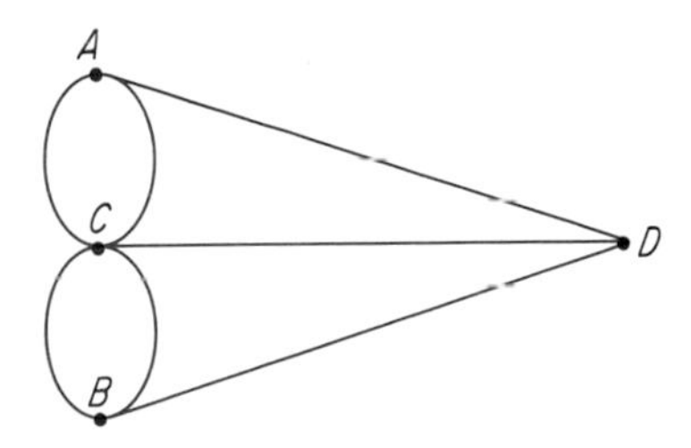

a network, it was no longer necessary to discuss the problem in terms of walking across the bridges. Instead one could discuss whether or not the network associated with the problem was traversable. The problem was solvable if and only if its network was traversable.

When is a network traversable? One can walk completely around any ordinary city block, and it is not necessary to start at any particular point to do so. In general, one may traverse any simple closed broken line in a single trip. We next consider walking around two blocks and down the street $\overline{BE}$ separating them. This problem is a bit more interesting in that it is necessary to start at B or E. Furthermore, if one starts at B, one ends at E, and conversely.

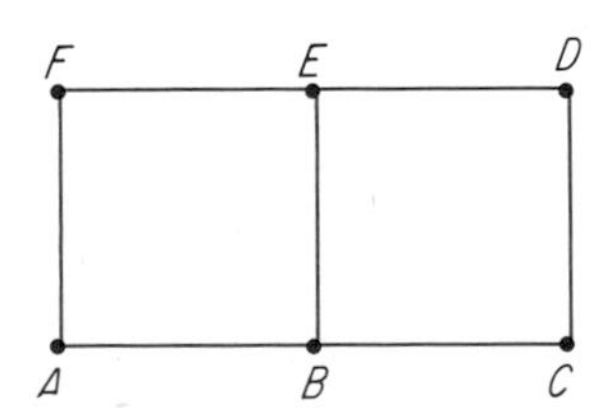

Note that it is permissible to pass through a vertex several times, but one may traverse a line segment only once. The peculiar property of the vertices B and E is based upon the fact that each of these endpoints is an endpoint of three line segments, whereas each of the other vertices is an endpoint of two line segments. A similar observation led a famous mathematician named Leonhard Euler (1707–1783) to devise a complete theory for traversable networks.

Euler classified the vertices of a network as odd or even. For example, in the given figure the vertex A is an endpoint of three arcs $\overline{AB}$, $\overline{AC}$, and $\overline{AD}$ and thus is an odd vertex; B is an endpoint of two arcs $\overline{BA}$ and $\overline{BC}$ and is an even vertex; C is an endpoint of four arcs $\overline{CB}$, $\overline{CA}$, $\overline{CD}$, and $\overline{CE}$ and is an even vertex; D is an endpoint of two arcs $\overline{DA}$ and $\overline{DC}$ and is an even vertex; E is an endpoint of one arc $\overline{EC}$ and is an odd vertex. Thus the figure has two odd vertices A and E and three even vertices B, C, and D. For any network

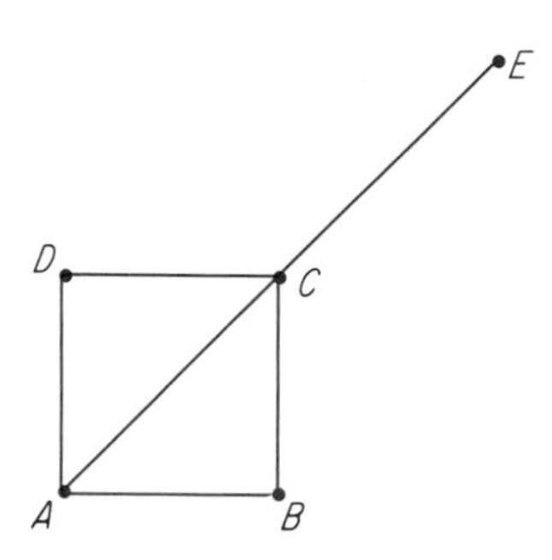

a vertex that is an endpoint of an odd number of arcs is an **odd vertex;** a vertex that is an endpoint of an even number of arcs is an **even vertex.** Since each arc has two endpoints, there must be an even number of odd vertices in any network.

Any network, such as a square, that has only even vertices is traversable, and the trip may be started at any vertex. Furthermore, the trip will terminate at its starting point. If a network has exactly two odd vertices, it is traversable, but the trip must start at one of the odd vertices and will terminate at the other. If a network has more than two odd vertices, it is not traversable. In general, a network with $2k$ odd vertices may be traversed in k distinct trips. The network for the Königsberg bridge problem has four odd vertices and thus is not traversable in a single trip. Notice that the Königsberg bridge problem is independent of the size and shape of the river, bridges, or islands.

The study of simple closed curves, networks, and other plane curves that we have considered is a part of a particular geometry called **topology.**

EXERCISES

In Exercises 1 through 8 identify **(a)** *the number of even vertices;* **(b)** *the number of odd vertices;* **(c)** *whether or not the network is traversable and, if it is traversable, the vertices that are possible starting points.*

1.

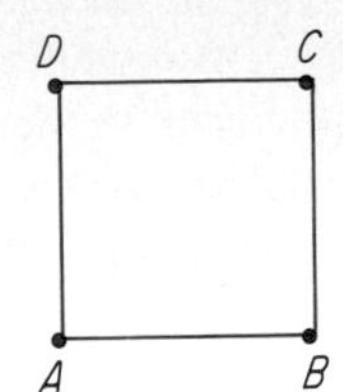

2.

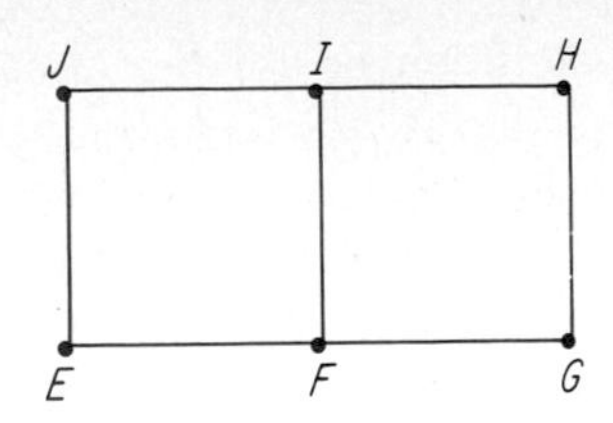

3.

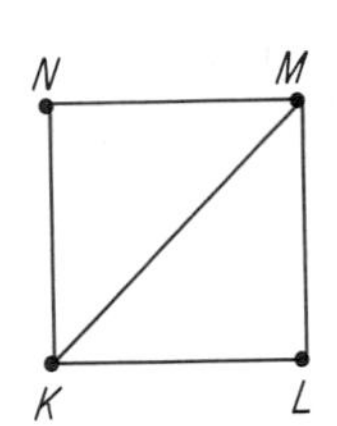

4.

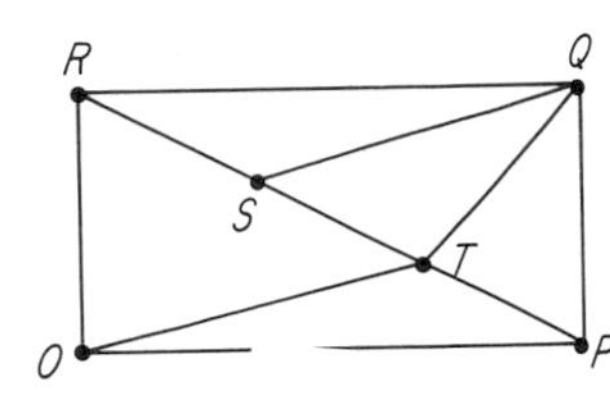

5.

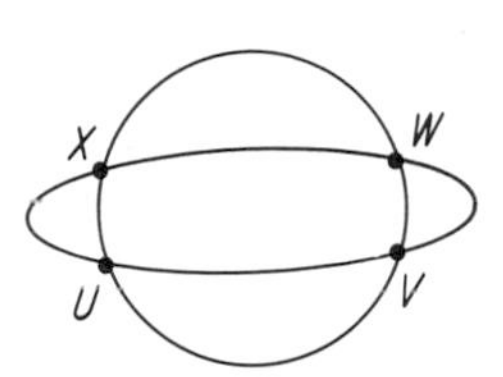

6.

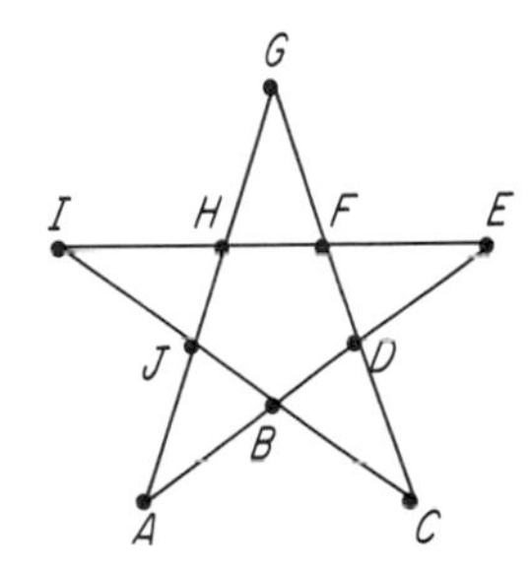

7. The network formed by the edges of a triangular pyramid.

8. The network formed by the edges of a cube.

9. Explain how a highway inspector can use a network to determine whether or not he can inspect each section of a highway without traversing the same section of the highway twice.

10. Use a related network to obtain and explain your answer to the following question: Is it possible to take a trip through a house with the floor plan indicated in the figure and pass through each doorway exactly once?

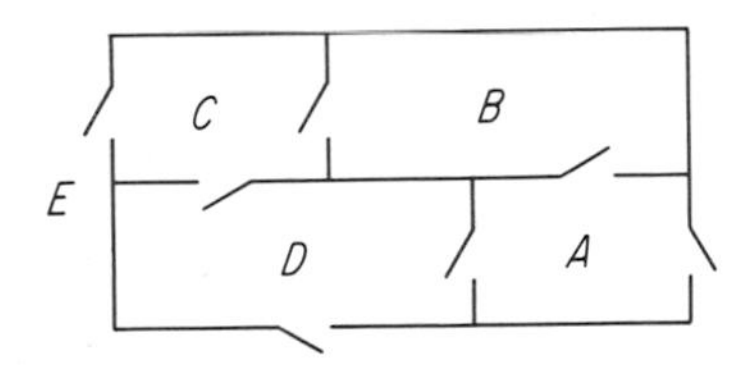

11. Explain whether or not it is possible to draw a simple connected broken line cutting each line segment of the given figure exactly once.

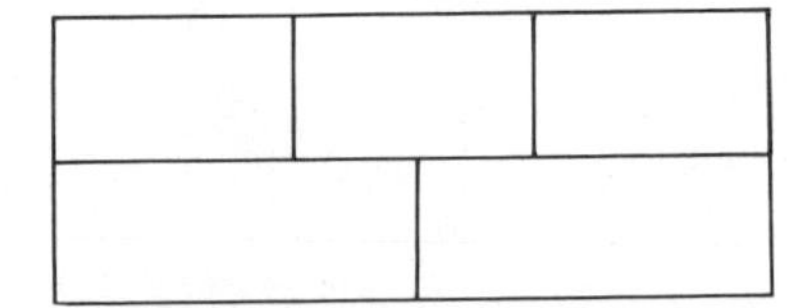

12. Consider the use of highway networks by a salesman who wants to visit each town exactly once. (**a**) Does the salesman always need to travel each highway at least once? (**b**) Consider each of the networks in Exercises 1 through 4 and indicate whether or not the salesman can visit each town exactly once without retracing any highway. (**c**) On the basis of your answer for part (b), does traversability of the network appear to be of interest to the salesman?

***13.** Describe a modification that would make the Königsberg bridge problem possible.

***14.** Describe a modification of the network for Exercise 11 such that the possibility of the desired construction is changed.

PEDAGOGICAL EXPLORATIONS

1. Find a way of describing to third graders that the above curve is traversable. (*Hint:* Consider ways of drawing a "likeness" that is obviously traversable.)

There are many figures in topology that have very unusual properties. One of these is a surface that is one-sided. A fly can walk from any point on this surface to any other point without crossing an edge. Unlike a table top or a wall, it does not have a top and a bottom or a front and a back. This surface is called a **Möbius strip**, and it may be very easily constructed from a rectangular piece of paper such as a strip of gummed tape. Size is theoretically unimportant, but a strip an inch or two wide and about a foot long is easy to handle. We may construct a Möbius strip by twisting the strip of gummed tape just enough (one half-twist) to stick the gummed edge of one end to the gummed edge of the other end. If we cut across this strip, we again get a single strip similar to the one we started with. But if we start with a rectangular strip and cut around the center of the Möbius

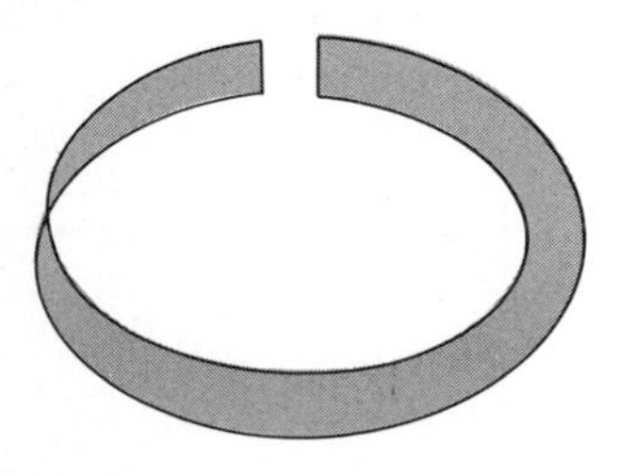

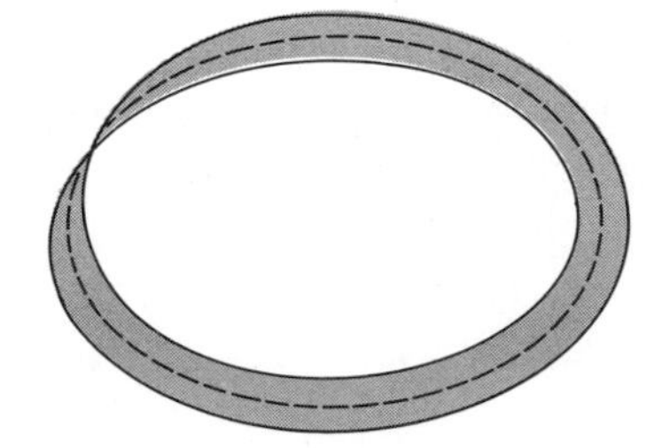

strip (see the dotted line in the second figure), we do not get two strips. Rather, we get one strip with two half-twists in it.

On one occasion one-sided surfaces of this sort were used as place cards at a seven-year-old's birthday party. While waiting for dessert, the youngsters were encouraged to cut the strip down the middle while guessing what the result would be. They were suitably impressed when they found only one piece, and were anxious to cut it again. Once more they were impressed when they found two pieces linked together. Almost a year after the party, one of the boys asked about the piece of paper that was in only one piece after it was cut in two.

When confronted with unusual properties such as those of the Möbius strip, both children and adults may ask questions that the nearest teacher cannot answer and that many college mathematics professors cannot answer. This is good for all concerned, since it impresses upon them that there is more to mathematics than formal algebraic manipulations and classical geometric constructions.

2. Construct a Möbius strip and cut around the center to obtain one strip with two half-twists in it.

3. Repeat Exploration 2 and then cut around the center again.

4. Construct a Möbius strip and cut along a path that is about one-third of the distance from one edge to the other.

5. Construct a Möbius strip, mark a point A on it, and draw an arc from A around the strip until you return to the point A.

6. Explain why a Möbius strip is called a one-sided surface.

7. Does a Möbius strip have one or two simple closed curves as its edge? Explain the reason for your answer.

8. Be on the alert for applications of Möbius strips as you go to repair shops and other places where machines are operated by belts and pulleys. Describe to your class any applications that you observe and explain their usefulness.

11-5 figures on a coordinate plane

The use of coordinates as an aid to the study of geometry is becoming common in elementary schools. Coordinate geometry is also important for prospective elementary school teachers as a means of increasing their own understanding of the geometric figures considered in elementary schools. Teachers of grades 4 through 8 will find that many of the concepts considered in this section and the next are already in the curriculum for their students.

Any plane may be considered as a coordinate plane as in §10-4. Any two points A: (x_1, y_1) and B: (x_2, y_2) determine a line AB.

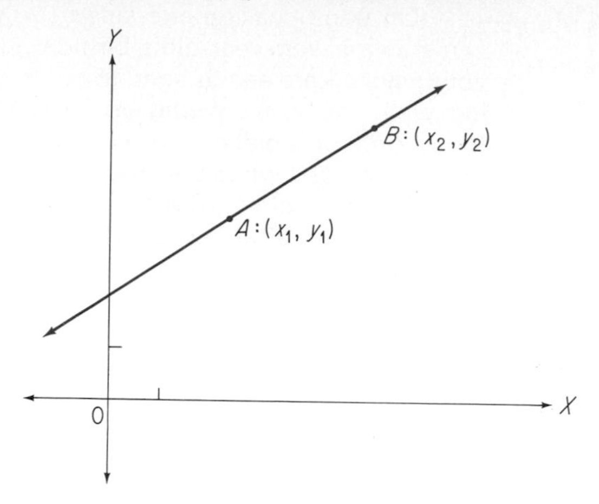

Example 1: Graph A: (2, 1) and B: (2, 5). Draw the line AB and find the length of the line segment AB.

Solution: The given points may be graphed as in the figure. The line AB is parallel to the y-axis and has equation $x = 2$ since A and B both have x-coordinate 2.

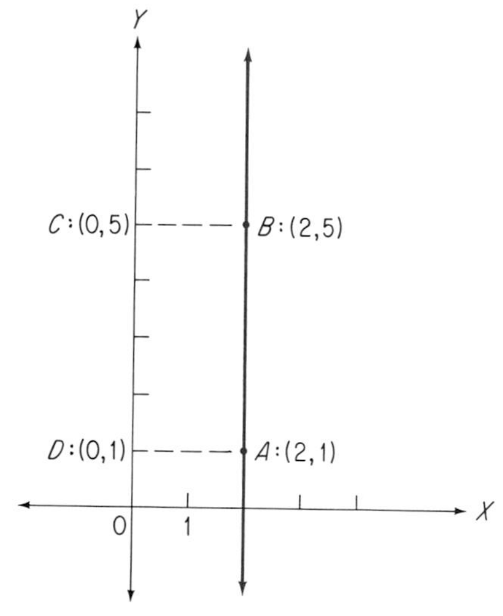

Because $ABCD$ is a rectangle, the opposite sides $\overline{AB}$ and $\overline{CD}$ must be congruent. Therefore to find the length of $\overline{AB}$, we may consider the two points D: (0, 1) and C: (0, 5) on the y-axis and note that $m(\overline{CD}) = 4$. Then the length of $\overline{AB}$ is 4 units.

For any points A: (x_1, y_1) and B: (x_1, y_2) the line AB is parallel to the y-axis and the linear measure of the line segment AB is $|y_2 - y_1|$. Similarly, for any points A: (x_1, y_1) and C: (x_2, y_1) the line AC is parallel to the x-axis and $m(\overline{AC}) = |x_2 - x_1|$. In general, the linear measure of any line segment is a number; the length of the line segment is that number of linear units.

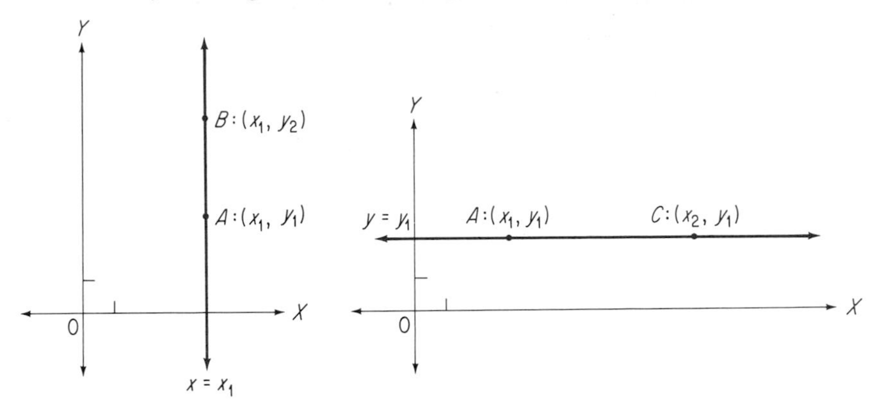

Consider the points P_x with coordinates x on a number line. We define the **directed distance** from P_r to P_s to be $s - r$. Then the directed distance from P_1 to P_4 is $4 - 1$, that is, 3; the directed distance from P_4 to P_1 is $1 - 4$, that is, -3; the directed distance from P_{-1} to P_1 is $1 - (-1)$, that is, 2. We use directed distances to develop a formula for the coordinate of the midpoint of a line segment. To find the midpoint M of a line segment P_rP_s we start at P_r and proceed halfway from P_r to P_s. Therefore, the coordinate of M is $r + \frac{1}{2}(s - r)$, that is, $\frac{1}{2}(r + s)$.

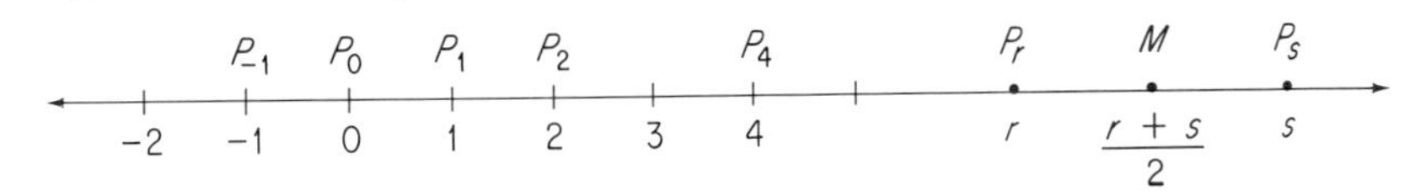

Example 2: Find the linear measure and the coordinates of the midpoint of $\overline{AB}$ for A: (1, 5) and B: (7, 5).

Solution: Consider the points C: (1, 0) and D: (7, 0) and the rectangle $CDBA$ as in the figure. Then

$$m(\overline{AB}) = m(\overline{CD}) = |7 - 1| = 6$$

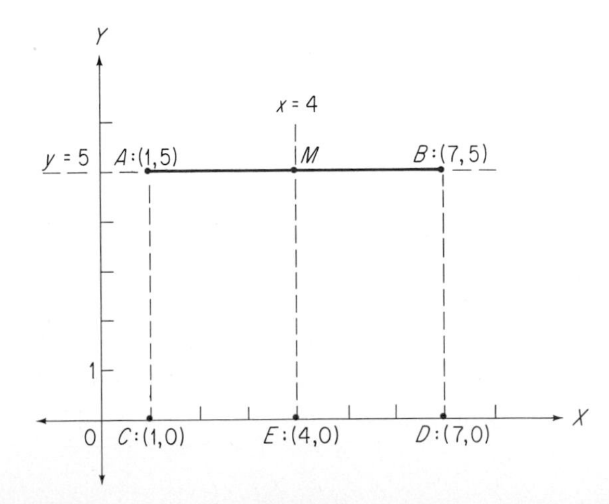

The y-coordinate of the midpoint M of $\overline{AB}$ must be 5, since every point of $\overleftrightarrow{AB}$ has y-coordinate 5. The midpoint E of $\overline{DC}$ has x-coordinate $\frac{1}{2}(1 + 7)$, that is, 4. The lines $\overleftrightarrow{AC}$, $x = 4$, and $\overleftrightarrow{BD}$ are parallel and intercept (cut off) congruent segments $\overline{CE}$ and $\overline{ED}$ on the x-axis. Therefore these lines intercept congruent segments on $\overline{AB}$, M must be on the line $x = 4$, and M has coordinates (4, 5).

In general, any two distinct points A: (x_1, y_1) and C: (x_2, y_1) on a line $y = y_1$ parallel to the x-axis are endpoints of a line segment AC with linear measure $|x_2 - x_1|$ and midpoint $(\frac{1}{2}(x_1 + x_2), y_1)$. Similarly, any two distinct points A: (x_1, y_1) and B: (x_1, y_2) on a line $x = x_1$ parallel to the y-axis are endpoints of a line segment AB with linear measure $|y_2 - y_1|$ and midpoint $(x_1, \frac{1}{2}(y_1 + y_2))$.

Example 3: Find the coordinates of the midpoints of the line segments with endpoints (a) (1, 0) and (−3, 0); (b) (2, 1) and (2, 7).

Solution:

(a) $\left(\frac{1 + (-3)}{2},\ 0\right) = (-1, 0)$

(b) $\left(2,\ \frac{1 + 7}{2}\right) = (2, 4)$

For two points A: (x_1, y_1) and B: (x_2, y_2) such that $\overline{AB}$ is not parallel to a coordinate axis we consider also the point C: (x_2, y_1). Then the midpoints of $\overline{AC}$ and $\overline{AB}$ are on the line $x = \frac{1}{2}(x_1 + x_2)$, and the midpoints of $\overline{CB}$ and $\overline{AB}$ are on the line $y = \frac{1}{2}(y_1 + y_2)$. Thus, as in the figure at the top of the next page, the midpoint M of $\overline{AB}$ has coordinates

$$\left(\frac{x_1 + x_2}{2},\ \frac{y_1 + y_2}{2}\right)$$

This is the **midpoint formula** for points on a coordinate plane. Notice that the formula holds whether or not $\overline{AB}$ is parallel to a coordinate axis.

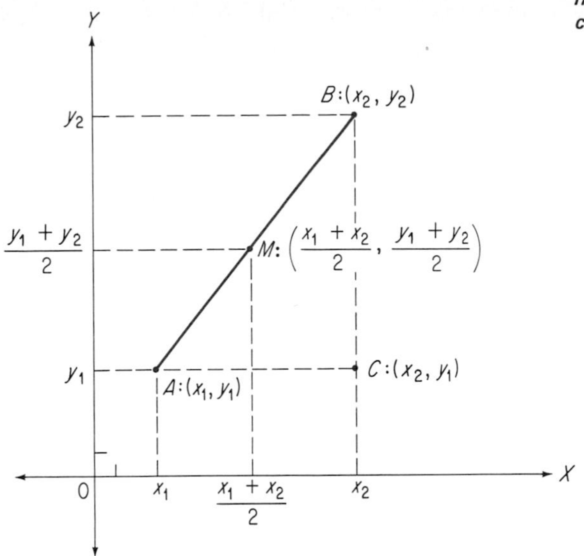

Example 4: Find the midpoint of $\overline{AB}$ for (a) A: (6, 1) and B: (—2, 5); (b) A: (—3, 4) and B: (1, —2).

Solution:

(a) $\left(\frac{6+(-2)}{2}, \frac{1+5}{2}\right) = (2, 3)$

(b) $\left(\frac{-3+1}{2}, \frac{4+(-2)}{2}\right) = (-1, 1)$

The Pythagorean theorem may be used to obtain a general formula for the linear measure of any line segment on a coordinate plane. Let the endpoints of the line segment be A: (x_1, y_1) and B: (x_2, y_2). Either the line segment AB is parallel to a coordinate axis, or a right triangle may be formed with the vertex of the right angle at C: (x_2, y_1). The linear measures of the legs of the right triangle are $|x_2 - x_1|$ and $|y_2 - y_1|$. By the Pythagorean theorem the linear measure of the hypotenuse is

$$m(\overline{AB}) = \sqrt{(x_2 - x_1)^2 + (y_2 - y_1)^2}$$

This is the **distance formula** on a plane. The formula holds whether or not the line segment is parallel to a coordinate axis.

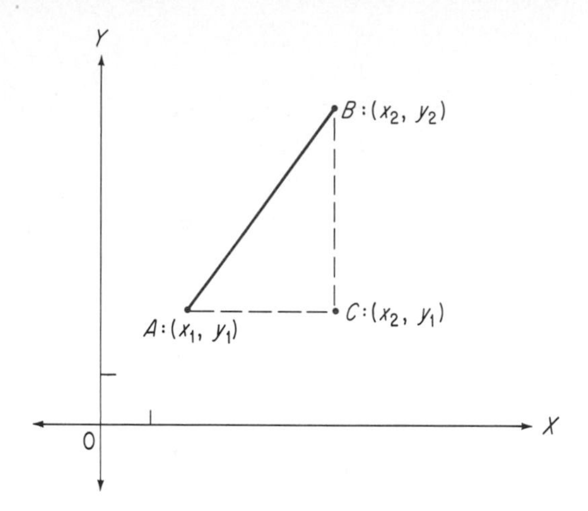

Example 5: Find the linear measure of the line segment with endpoints A: (2, —3) and B: (5, 1).

Solution:

$$m(\overline{AB}) = \sqrt{(5-2)^2 + [1-(-3)]^2} = \sqrt{3^2 + 4^2} = 5$$

On a coordinate plane the circle with center at C: (h, k) and radius r is the set of points at a distance r from the point C. The distance of any point P: (x, y) from C: (h, k) is

$$\sqrt{(x-h)^2 + (y-k)^2}$$

Therefore, the circle with center C: (h, k) and radius r has equation

$$(x-h)^2 + (y-k)^2 = r^2$$

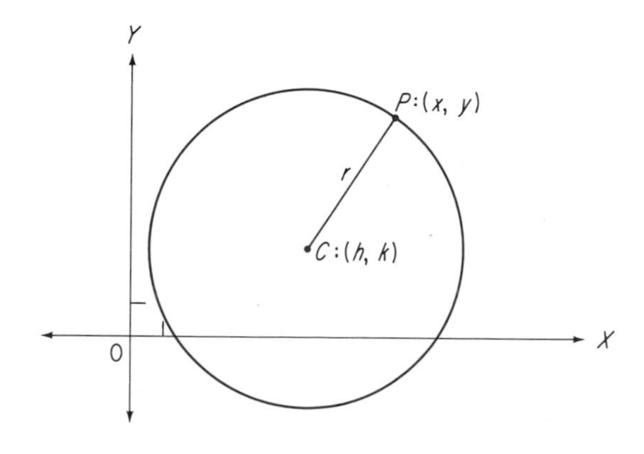

Example 6: Describe the set of points (x, y) such that
(a) $x^2 + y^2 = 4$
(b) $(x - 1)^2 + (y - 3)^2 = 1$
(c) $(x + 2)^2 + (y + 1)^2 = 9$

Solution: (a) Since $x = x - 0$ and $y = y - 0$, the given equation can be expressed as

$$(x - 0)^2 + (y - 0)^2 = 2^2$$

Each point of the set is 2 units from the origin. Thus the set of points is a circle with center (0, 0) and radius 2.

(b) The set of points is a circle with center (1, 3) and radius 1.

(c) Since $x + 2 = x - (-2)$ and $y + 1 = y - (-1)$, the given equation can be expressed as

$$[x - (-2)]^2 + [y - (-1)]^2 = 3^2$$

The set of points is a circle with center (−2, −1) and radius 3.

Example 7: Give an algebraic representation for the points that are on a coordinate plane and are at most 2 units from (3, −5).

Solution: The points form a **circular region** (the points of a circle and the interior points of the circle) with center (3, −5) and radius 2. Thus we have

$$(x - 3)^2 + (y + 5)^2 \leq 4$$

EXERCISES

For each line segment identified by its endpoints find its linear measure; the coordinates of its midpoint.

1. **(a)** (0, 5) and (7, 5) **(b)** (3, 1) and (8, 1)

2. **(a)** (−1, 5) and (−1, −3) **(b)** (5, −3) and (5, 11)

Find expressions for the lengths of line segments with endpoints:

3. **(a)** $(b, 0)$ and $(0, 0)$ **(b)** $(0, k)$ and $(0, 0)$

4. **(a)** (x_1, y_1) and (x_2, y_1) **(b)** (x_1, y_1) and (x_1, y_2)

Find linear measures of the line segments with endpoints:

5. A: (2, 3) and B: (5, 7)

6. C: (−2, 5) and D: (3, −7)

7. E: (1, 15) and F: (8, −9)

8. G: (−1, 6) and H: (8, 12)

9. I: (−2, −4) and J: (1, −1)

10. K: (5, −4) and L: (−6, −10)

Find the coordinates of:

11. The midpoint of $\overline{AB}$ for **(a)** A: (1, −6) and B: (7, 2); **(b)** A: (5, 11) and B: (−1, −3); **(c)** A: (−2, −7) and B: (−4, −5).

12. The endpoint B if A: (1, 3) and C: (2, 5) is the midpoint of $\overline{AB}$.

13. The endpoint B if A: (3, 0) and C: (−1, 4) is the midpoint of $\overline{AB}$.

14. The endpoint A if B: (−5, 6) and C: (−3, −1) is the midpoint of $\overline{AB}$.

15. The vertex C of a square $ABCD$ with A: (0, 0), B: $(a, 0)$, and D: $(0, a)$.

16. The vertex S of a rectangle $QRST$ with Q: (0, 0), R: $(a, 0)$, and T: $(0, b)$.

***17.** The vertex S of an isosceles trapezoid $QRST$ with Q: (0, 0), R: $(a, 0)$, and T: (b, c), where $2b < a$.

Draw coordinate axes, sketch each set of points, and label each line with its equation.

18. The points 5 units from the y-axis.

19. The points 2 units from the x-axis.

20. The points 3 units from the line $x = 2$.

21. The points 4 units from the line $y = -1$.

Describe the set of points (x, y) *such that:*

22. $x^2 + y^2 = 6$

23. $x^2 + y^2 \leq 25$

24. $(x - 1)^2 + (y - 2)^2 \leq 9$

25. $(x + 3)^2 + (y - 1)^2 < 4$

26. $(x + 1)^2 + (y + 2)^2 \leq 25$

27. $(x + 3)^2 + (y + 1)^2 > 9$

Consider the points on a coordinate plane and give an algebraic representation for:

28. The points of the circle with center (−3, 4) and radius 3.

29. The points of the circular region with center (−3, −1) and radius 5.

30. The interior points of the circle with center (6, −4) and radius 6.

31. The exterior points of the circle with center (−4, 3) and radius 7.

32. The exterior points of the circle with center (−3, −2) and radius 4.

PEDAGOGICAL EXPLORATIONS

In Explorations 1 through 3 use a geoboard (§8-5, Pedagogical Explorations) or use the intersections of the "rules" on a sheet of graph (coordinate) paper as the pegs (or nails) of a geoboard.

1. Prepare a lesson plan on the representation of at least five different types of triangles on a geoboard.

2. Repeat Exploration 1 for quadrilaterals.
3. Repeat Exploration 1 for lines.
4. We expect the area covered by polygonal regions to be independent of the arrangement of the regions.
   (a) Use graph paper and cut out a square with 16 units on each side. Note that the area of this square is 256 square units.
   (b) Mark and cut the square into four pieces as indicated in the figure.

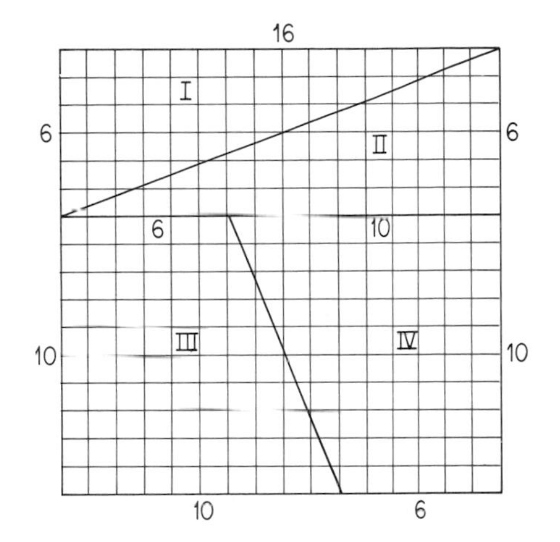

   (c) Rearrange the pieces as in this figure.

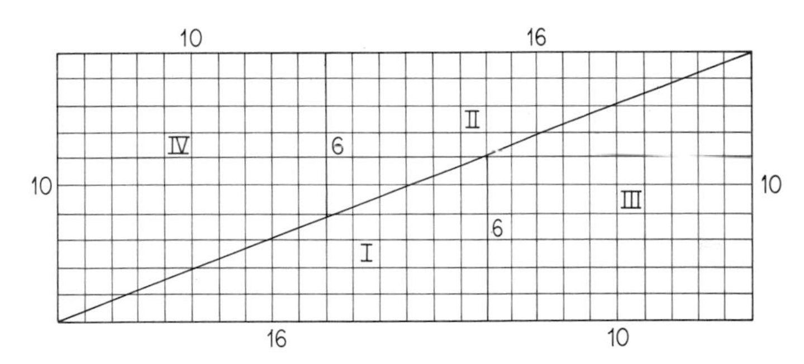

   (d) Note that a rectangular region 10 by 26 units has area 260 square units. Explain the increase of 4 square units.

Explorations (proofs) of properties of geometric figures are often relatively easy on a coordinate plane. Use properties of a coordinate plane and prove:

5. If A: (0, 0), B: (a, 0), and C: (b, c) are the vertices of a triangle, then the midpoints of $\overline{AC}$ and $\overline{BC}$ are endpoints of a line segment that is parallel to $\overline{AB}$ and half as long as $\overline{AB}$.
6. The line segments with the midpoints of opposite sides of a quadrilateral as endpoints bisect each other. [*Hint:* Use vertices A: (0, 0), B: (a, 0), C: (b, c), and D: (d, e).]

11-6 figures on a coordinate space

We may think of the xy-plane in space with a line OZ perpendicular to the xy-plane at the origin. It is customary to select the coordinate axes $\overleftrightarrow{OX}$, $\overleftrightarrow{OY}$, and $\overleftrightarrow{OZ}$ as in the figure, where the tick marks indicate the unit points on each axis.

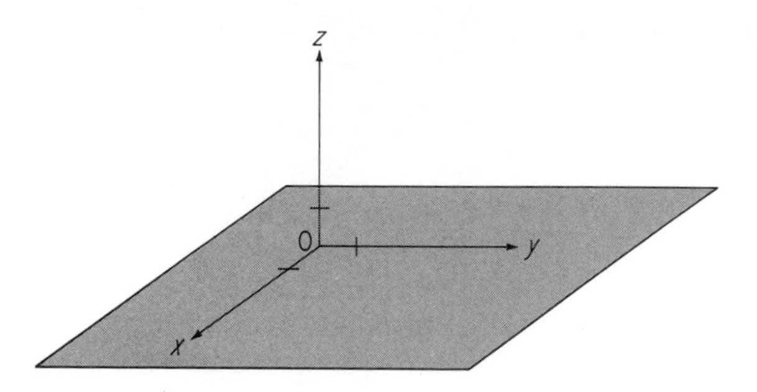

In space the locus of points 1 unit from the xy-plane consists of two planes; that is, the plane $z = 1$ one unit above the xy-plane and the plane $z = -1$ one unit below the xy-plane. Any plane parallel to the xy-plane has an equation of the form

$$z = t$$

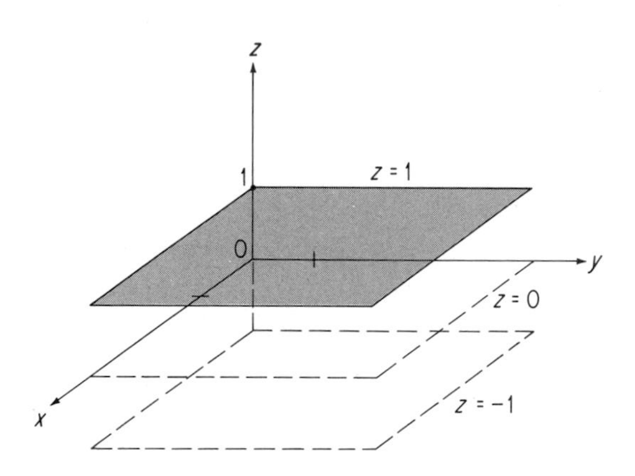

Similarly, any plane parallel to the xz-plane has an equation of the form

$$y = n$$

Any plane parallel to the yz-plane has an equation of the form

$$x = k$$

The location of any point (k, n, t) may be determined as the intersection of the three planes $x = k$, $y = n$, and $z = t$. By convention the x-coordinate is always listed first, the y-coordinate second, and the z-coordinate third in specifying the coordinates of a point in space as an **ordered triple** of real numbers (k, n, t).

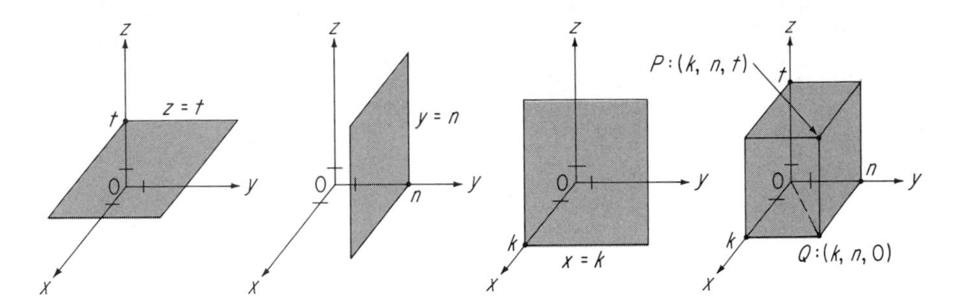

Each point in space has a unique ordered triple of real numbers as its *coordinates* with respect to a given set of coordinate axes; each ordered triple of real numbers has a unique point in space as its *graph*. When k, n, and t are each different from zero, we frequently can visualize the position of the point P: (k, n, t) after sketching a rectangular parallelepiped with a vertex at the origin, three edges on the coordinate axes, and a vertex at P. For any point P: (k, n, t) that is in a coordinate space but not on a coordinate plane, we consider also Q: $(k, n, 0)$. Then for R: $(k, 0, 0)$ there is a right triangle ORQ and by the Pythagorean theorem

$$m(\overline{OQ}) = \sqrt{k^2 + n^2}$$

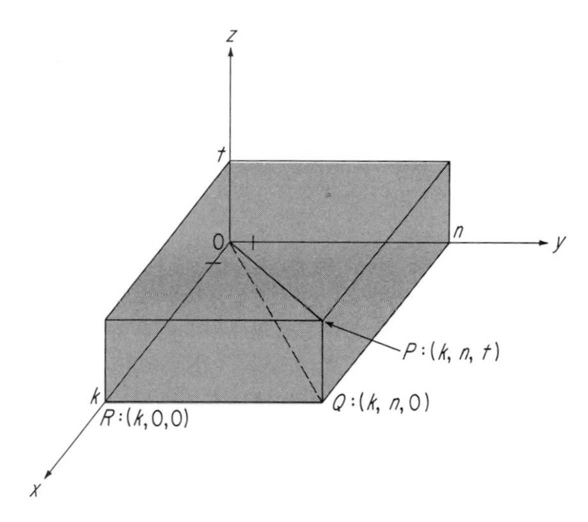

Also there is a right triangle OQP and by the Pythagorean theorem

$$\mathbf{m(\overline{OP})} = \sqrt{[m(\overline{OQ})]^2 + t^2} = \sqrt{\boldsymbol{k^2 + n^2 + t^2}}$$

This formula holds whether or not P is on a coordinate plane.

Example 1: Find $m(\overline{OP})$ for P: (1, 5, 7).

Solution:

$$m(\overline{OP}) = \sqrt{1^2 + 5^2 + 7^2} = \sqrt{75} = 5\sqrt{3}$$

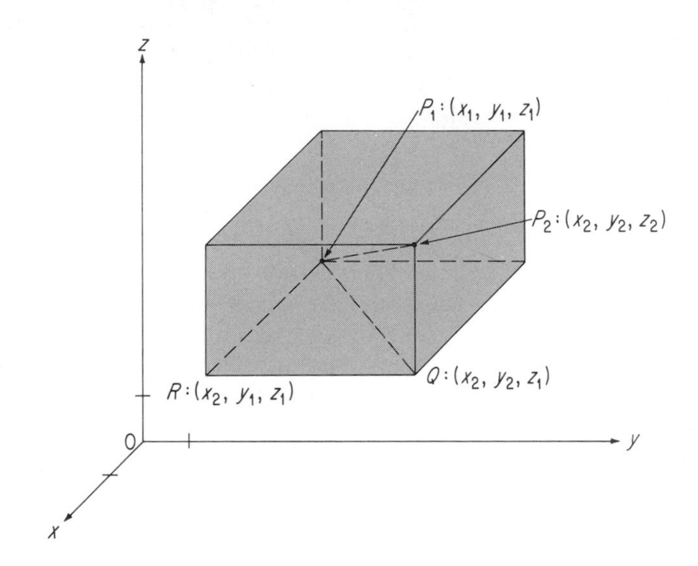

Let P_1: (x_1, y_1, z_1) and P_2: (x_2, y_2, z_2) be any two points that are not on a plane that is parallel to a coordinate plane. Then, as in the figure, we can identify the vertices of a rectangular parallelepiped with P_1 and P_2 as vertices and with its edges parallel to the coordinate axes. For Q: (x_2, y_2, z_1) we have

$$m(\overline{P_1Q}) = \sqrt{(x_2 - x_1)^2 + (y_2 - y_1)^2}$$

and from the right triangle P_1QP_2

$$m(\overline{P_1P_2}) = \sqrt{[m(\overline{P_1Q})]^2 + [m(\overline{P_2Q})]^2}$$

$$\mathbf{m(\overline{P_1P_2}) = \sqrt{(x_2 - x_1)^2 + (y_2 - y_1)^2 + (z_2 - z_1)^2}}$$

This **distance formula** holds for any points P_1 and P_2 in space.

Example 2: Find $m(\overline{AB})$ for A: (2, −3, 5) and B: (7, 1, 11).

Solution:

$$\begin{aligned} m(\overline{AB}) &= \sqrt{(7-2)^2 + [1-(-3)]^2 + (11-5)^2} \\ &= \sqrt{5^2 + 4^2 + 6^2} \\ &= \sqrt{25 + 16 + 36} = \sqrt{77} \end{aligned}$$

The sphere with center (a, b, c) and radius r is the set of points at a distance of r from the point (a, b, c). The distance of any point P: (x, y, z) from the point (a, b, c) is

sec. 11-6 figures on a coordinate space

$$\sqrt{(x-a)^2+(y-b)^2+(z-c)^2}$$

Therefore, the sphere with center (a, b, c) and radius r has equation

$$(x-a)^2+(y-b)^2+(z-c)^2=r^2$$

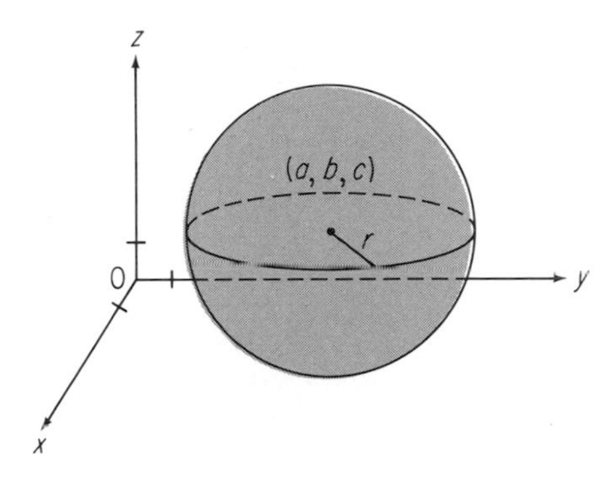

Example 3: Find an equation for the sphere with center at the origin and radius 3.

Solution: From the formula, $(x-0)^2+(y-0)^2+(z-0)^2=9$, that is, $x^2+y^2+z^2=9$.

Example 4: Find an equation for the sphere with center $(5, -7, 11)$ and radius 4.

Solution:

$$(x-5)^2+(y+7)^2+(z-11)^2=16$$

Example 5: Describe the set of points at most 2 units from the point $(1, -2, 3)$ by an inequality.

Solution:

$$(x-1)^2+(y+2)^2+(z-3)^2\leq 4$$

EXERCISES

Graph each equation in coordinate space.

1. $z=4$

2. $x=6$

3. $y=-2$

4. $z=-3$

Graph each point.

5. (3, 4, 1)

6. (2, 4, 3)

7. (1, 4, 2)

8. (3, −2, 2)

Describe the set of points in coordinate space such that:

9. $x = 0$

10. $y = 0$

11. $z = 0$

12. $x = a$

13. $y = b$

14. $z = c$

15. $\begin{cases} x = 0 \\ y = 0 \end{cases}$

16. $\begin{cases} y = 0 \\ z = 0 \end{cases}$

17. $\begin{cases} x = 0 \\ z = 0 \end{cases}$

18. $\begin{cases} x = 1 \\ z = 0 \end{cases}$

A rectangular box (parallelepiped) may be sketched in coordinate space to assist in the identification of the position of the given point P: (k, n, t). The rectangular box in the figure has $m(\overline{AB}) = 2$, $m(\overline{AF}) = 4$, *and* $m(\overline{AD}) = 3$.

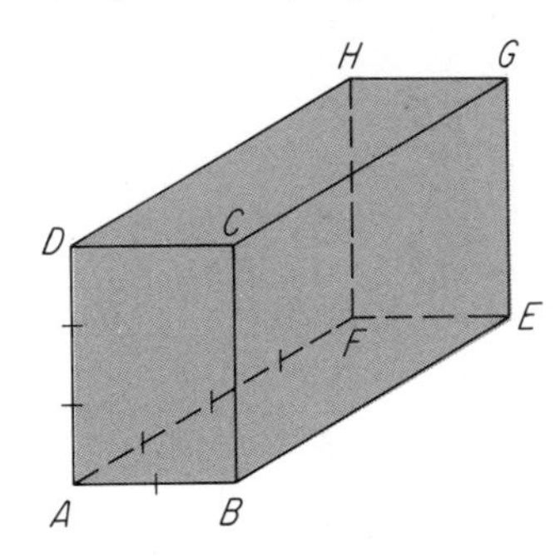

For Exercises 19 through 22 copy the box ABCDEFGH and give the coordinates of each vertex if:

19. The origin is at F with $\overrightarrow{OX} = \overrightarrow{FA}$, $\overrightarrow{OY} = \overrightarrow{FE}$, and $\overrightarrow{OZ} = \overrightarrow{FH}$.

20. The origin is at A with $\overrightarrow{OX}$ opposite $\overrightarrow{AF}$, $\overrightarrow{OY} = \overrightarrow{AB}$, and $\overrightarrow{OZ} = \overrightarrow{AD}$.

21. The origin is at E with $\overrightarrow{OX} = \overrightarrow{EB}$, $\overrightarrow{OY}$ opposite $\overrightarrow{EF}$, and $\overrightarrow{OZ} = \overrightarrow{EG}$.

22. The origin is at H with $\overrightarrow{OX} = \overrightarrow{HD}$, $\overrightarrow{OY} = \overrightarrow{HG}$, and $\overrightarrow{OZ}$ opposite $\overrightarrow{HF}$.

23–26. Repeat Exercises 19 through 22 for the next box with $m(\overline{AB}) = 5$, $m(\overline{AF}) = 7$, and $m(\overline{AD}) = 2$.

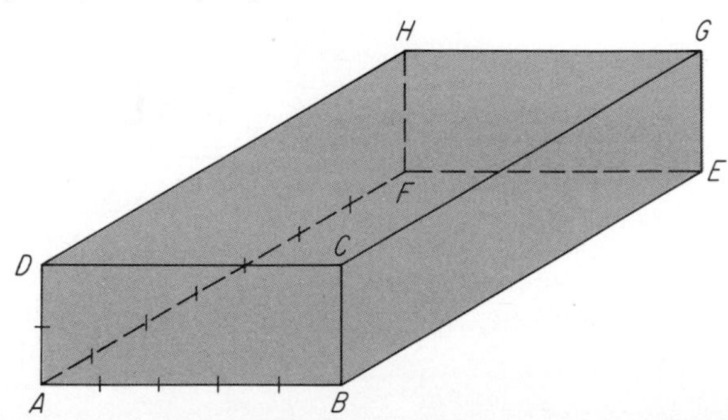

Find the linear measure of $\overline{AB}$ for:

27. A: (1, 0, 2) and B: (1, 4, 5).

28. A: (2, 7, −3) and B: (1, 5, 0).

29. A: (2, −1, 0) and B: (3, 1, 2).

30. A: (1, 5, −6) and B: (−2, 3, −1).

Find an equation for the sphere with:

31. Center A: (1, −2, 5) and radius 3.

32. Center B: (2, 3, −4) and radius 4.

33. Center C: (−1, −2, 3) and radius 5.

34. Center D: (5, −7, 11) and radius 8.

Describe the set of points such that:

35. $x^2 + y^2 + z^2 = 4$

36. $(x - 1)^2 + (y - 2)^2 + (z - 3)^2 = 16$

37. $(x - 2)^2 + (y + 1)^2 + (z + 2)^2 = 25$

38. $(x + 3)^2 + (y - 1)^2 + (z + 1)^2 = 9$

Describe each set of points in space by an equation, inequality, or system of such statements.

39. The points 3 units above the xy-plane.

40. The points above the xy-plane.

41. The points 3 units from (1, −2, 3).

42. The points at most 4 units from (2, 3, 4).

43. The points at least 4 units from (2, 3, 4).

44. The points at most 5 units from (1, 2, −3).

45. The points at least 3 units above the xy-plane.

46. The points at least 2 and at most 3 units above the xy-plane.

***47.** The points less than 4 units from the yz-plane.

***48.** The points less than 5 units from the xz-plane.

***49.** The points at least 3 and at most 5 units from (1, 6, 4).

***50.** The points at least 4 and at most 7 units from (2, −3, 5).

PEDAGOGICAL EXPLORATIONS

Each of the following explorations involves the determination of a shortest path a spider has to crawl along the outside of a box to reach a fly F that is assumed to stay in one place. As in the next figure, we consider a closed rectangular box 12 inches wide, 18 inches long, and 8 inches tall.

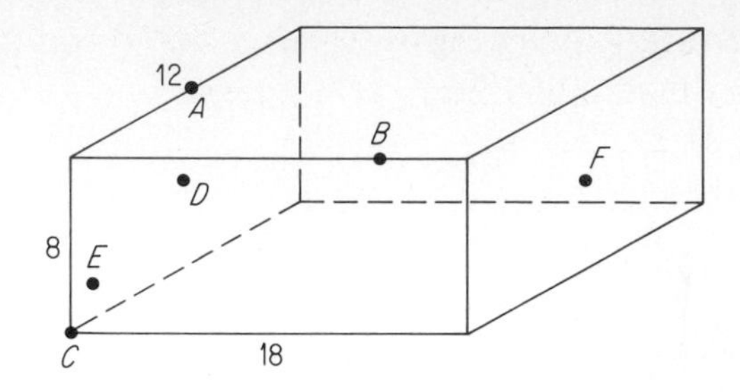

For each exploration assume that the fly is at the center *F* (intersection of the diagonals) of an end of the box. Copy the figure, draw the shortest path(s) for the spider, and find the length of the shortest path. If there are two or more paths of minimal length, draw them all.

1. The spider is at position *A* in the center of the top edge of the opposite end of the box.
2. The spider is at position *B* at the top of an adjoining side and 4 inches from the end containing the fly. (*Hint:* Think of the box as flattened out in some way.)
3. The spider is in the corner *C* at the bottom of the opposite end of the box.
4. The spider is at the middle *D* of the end of the box opposite the fly.
5. The spider is at position *E* one inch from the vertical edge over *C* and two inches from the bottom of the box.

CHAPTER TEST

In Exercises 1 through 9 sketch:

1. A convex polygon with five sides.
2. A network with two odd vertices and two even vertices.
3. A regular polygon that is not a square.
4. An obtuse scalene triangle.
5. A rhombus that is not a square.
6. Two similar triangles that are not congruent.
7. Two congruent quadrilaterals.
8. A figure that includes a line perpendicular to a plane (in the figure identify such a line and plane).
9. A figure that includes two perpendicular planes (in the figure identify two such planes).

10. Consider the six lines along the edges of a triangular pyramid $ABCD$ as in the figure. Identify the lines that (a) intersect $\overleftrightarrow{AB}$ in a single point; (b) form skew lines with $\overleftrightarrow{AB}$.

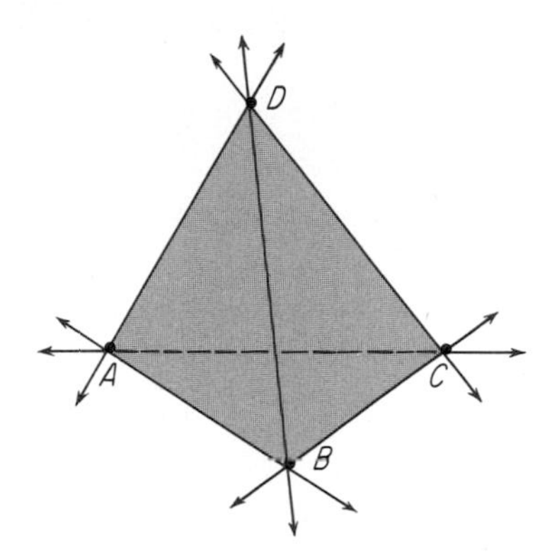

Consider each figure and state whether or not the network is traversable. If a network is traversable, identify the vertices that are possible starting points.

11. (a)

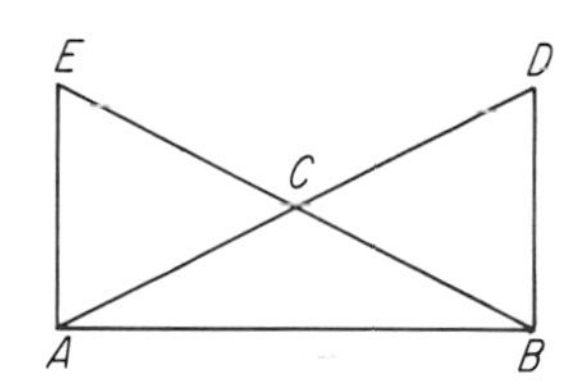

(b)

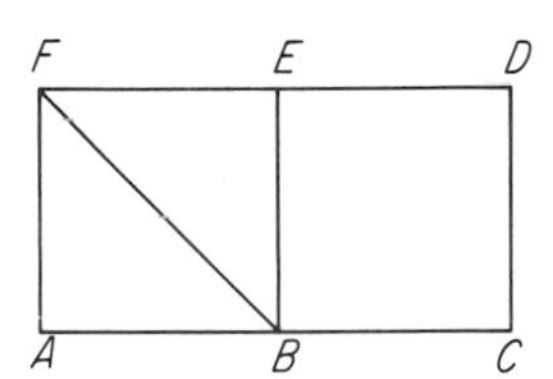

12. (a)

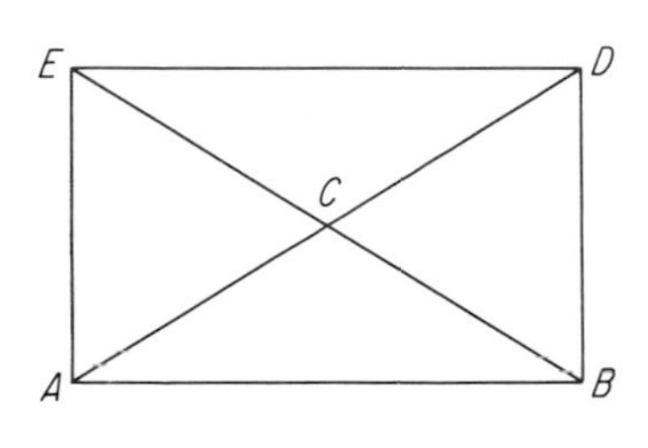

(b)

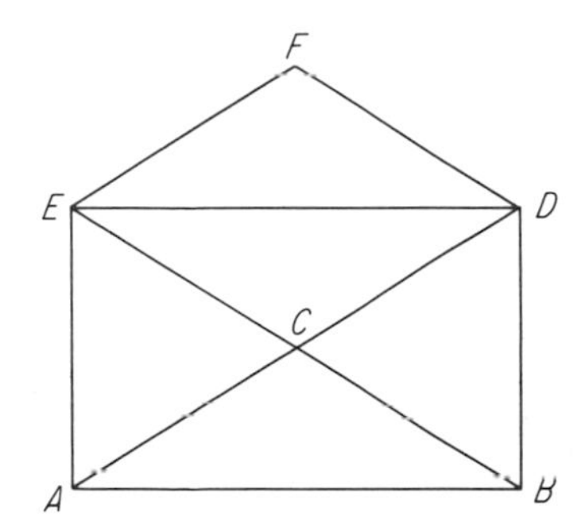

Identify as true (T, always true) or false (F, not always true).

13. (a) There is one and only one line that is perpendicular to a given plane and contains a given point.
(b) There is one and only one plane that is perpendicular to a given plane and contains a given line.

14. (a) Any two lines intersect or are parallel.
(b) Any two planes intersect or are parallel.

Sketch and describe the construction of:

15. A triangle with sides of length 6, 8, and 10 units.

16. A rectangle that is not a square.

On a coordinate plane:

17. Find the linear measure and the coordinates of the midpoint of the line segment with endpoints (2, —5) and (7, 7).

18. Find the coordinates of the endpoint B of a line segment AB with endpoint A: (2, 6) and midpoint C: (5, 1).

19. Describe the set of points such that:
(a) $x^2 + y^2 \leq 81$ (b) $(x - 1)^2 + (y + 2)^2 > 4$

20. Find an algebraic representation for the circle with center (5, —2) and radius 3.

21. Find an algebraic representation for the points that are equidistant from the point (—3, 0) and the point (5, 0).

In coordinate space:

22. Give an algebraic representation for the y-axis.

23. Find $m(\overline{AB})$ for A: (1, 5, 3) and B: (7, 1, 5).

24. Find an equation for a sphere with center (2, —3, 1) and radius 5.

25. Describe the set of points such that $x^2 + y^2 + (z + 2)^2 = 4$.

READINGS AND PROJECTS

1. Read, and write a summary of, Chapter 2 (Informal Geometry in Grades K–6) of the thirty-sixth yearbook of the National Council of Teachers of Mathematics.
2. Read, and try several of the activities in, Chapter 7 (Aids and Activities) and Chapter 8 (A Laboratory Approach) of the thirty-fifth yearbook of the National Council of Teachers of Mathematics.
3. Read Chapter 10 (Mathematics Projects, Exhibits and Fairs, Games, Puzzles, and Contests) of the thirty-fourth yearbook of the National Council of Teachers of Mathematics. Prepare a summary of the aspects of this chapter that are applicable to elementary school mathematics.
4. Read Booklet No. 14 (Informal Geometry) of the thirtieth yearbook of the National Council of Teachers of Mathematics. Complete at least three of the exercise sets given there.
5. Read Booklet No. 18 (Symmetry, Congruence, and Similarity) of the thirtieth yearbook of the National Council of Teachers of Mathematics. Complete at least three of the exercise sets given there.

6. Prepare a report for your college class on Chapter 9 (Geometry in the Grades) and Chapter 10 (Topology) of the twenty-seventh yearbook of the National Council of Teachers of Mathematics.
7. Read "Graphs in the Primary Grades" by Morris Pincus and Frances Morgenstern on pages 499 through 501 of the October 1970 issue of *The Arithmetic Teacher*. Prepare a lesson plan for a primary grade class and make use of some of this and related material.
8. Read "Dressing Up Mathematics" by Jean J. Pedersen on pages 118 through 122 of the February 1968 issue of *The Mathematics Teacher*. Make a pattern for at least one of the items that she describes.
9. Read "Activities: Tessellations" by L. Carey Bolster on pages 339 through 342 of the April 1973 issue of *The Mathematics Teacher*. Based upon this and related material prepare a lesson plan suitable for use with a specified elementary school grade level.
10. Read Chapter 9 (Geometry) of the thirty-seventh yearbook of the National Council of Teachers of Mathematics. Summarize the long-range mathematical and pedagogical goals of the study of geometry as discussed by the author.
11. Read "Let's Do It! From Shadows To Mathematics" by James V. Bruni and Helene J. Silverman on pages 232 through 239 of the April 1976 issue of *The Arithmetic Teacher*. Suggest additional activities with shadows that can be used to develop geometric concepts.
12. Read "Minimal Surfaces Rediscovered" by Sister Rita M. Ehrmann on pages 146 through 152 of the February 1976 issue of *The Mathematics Teacher*. Attempt to reproduce at least one of the experiments described in the article and demonstrate this to your class.

See also the following publications of the National Council of Teachers of Mathematics:

13. *Mathematics through Paper Folding* by Alton T. Olson. Prepare a lesson plan based upon using paper folding with a specified elementary school grade.
14. *Polyhedron Models for the Classroom* by Magnus J. Wenninger. Construct several of the models.
15. *Boxes, Squares, and Other Things: A Teacher's Guide for a Unit in Informal Geometry* by Marion I. Walter.
16. *The Pythagorean Proposition* by Elisha S. Loomis. Prepare to present at least three of the 366 proofs of the proposition to your class.
17. *Teacher-made Aids for Elementary School Mathematics: Readings from the Arithmetic Teacher.* Make and demonstrate several of the aids described.

12 An Introduction to Probability

We make frequent reference to probability in everyday language. For example, we frequently say, hear, or read such comments as the following:

> It probably will rain.
> The odds are in his favor.
> His chances are 50-50.

Ever since the seventeenth century, mathematicians have been exploring the topic of probability. Interestingly enough, the subject is said to have had its foundation in the realm of gambling and to have arisen from a discussion of the distribution of stakes in an unfinished game. Probabilities are still used to understand games of chance such as the tossing of coins, throwing of dice, or drawing of lottery tickets. Probabilities are also used extensively by persons who are concerned with the cost of insurance (based upon rates of mortality), the construction of various polls such as the Gallup Poll to appraise public opinion, and numerous other types of statistical studies. The topics in this chapter provide a basis for understanding probabilities and their applications to real life situations. As you read this chapter, look for other applications, if possible among your own activities, of the concepts considered.

12-1 counting problems

Many problems depend for their solution upon an enumeration of all possible outcomes. Thus, the simple task of counting becomes an important one in the study of probability. To illustrate various problems in this chapter, we shall invent a fictitious club consisting of a set M of members:

$$M = \{\text{Betty, Doris, Ellen, John, Tom}\}$$

Let us form a committee that is to consist of one boy and one girl, each selected from the set M of club members. How many such committees are possible? One way to answer this question is by means of a *tree diagram*, which helps list each of the possibilities.

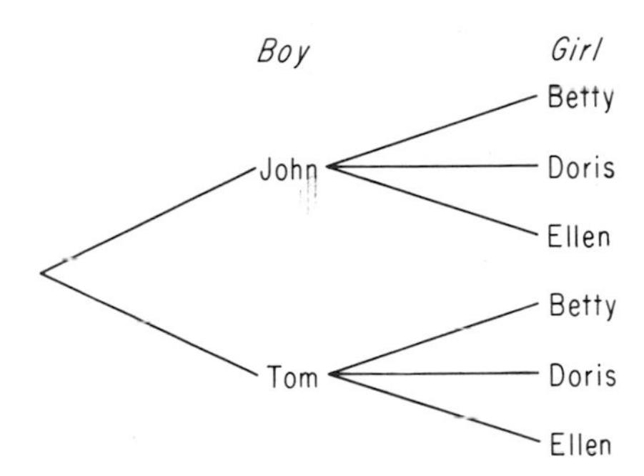

For each of the two possible choices of a boy, there are three possible choices of a girl. Thus the following six distinct possible committees can be formed and can be read from the tree diagram:

John-Betty Tom-Betty
John-Doris Tom-Doris
John-Ellen Tom-Ellen

Suppose that we had selected a girl first. Then the tree diagram would be as shown in the next figure and there would still be six possibilities, the same six committees as before.

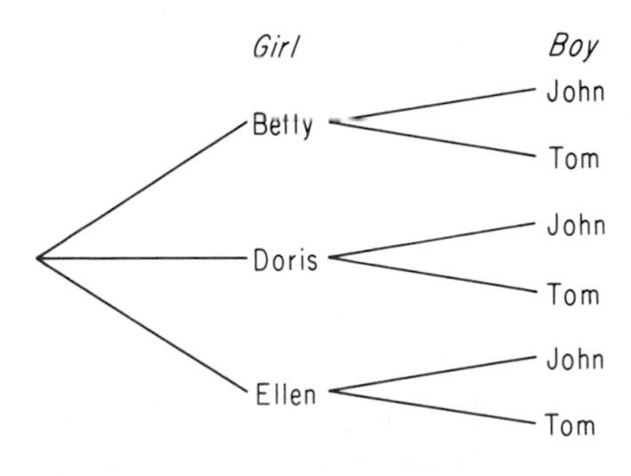

Example 1: How many different selections of two officers, a president and a vice-president, can be elected from the set M of club members?

Solution: Let us select the officers in two stages. There are five possible choices for the office of president. Each of these five selections may be paired with any one of the remaining four members. Thus there are 20 possible choices in all which can be read from the following diagram.

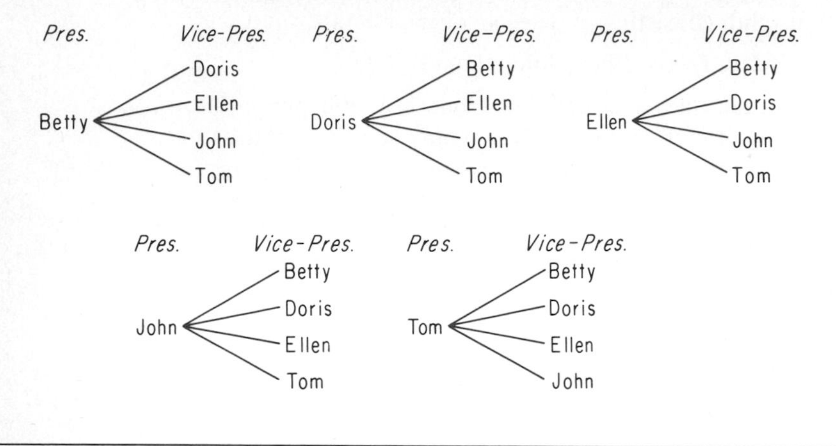

In general, if one task can be performed in m different ways and a second task can be performed in n different ways, then the first and second tasks together can be performed in $m \times n$ different ways. This **general principle of counting** can be extended if there are additional tasks:

$$m \times n \times r \times \cdots \times t$$

Example 2: The club members M must send a delegate to a meeting tomorrow and also a delegate to a different meeting next week. How many different selections of these delegates may be made if any member of the club may serve as a delegate to each meeting?

Solution: There are five possible choices of a delegate to the first meeting. Since no restriction is made, we assume that the same member may attend each of the two meetings. Thus, there are five choices for the delegate to next week's meeting. In all, there are 5×5, that is, 25 choices.

Example 3: How many three-letter "words" may be formed from the set of vowels $V = \{a, e, i, o, u\}$ if no letter may be used more than once? (A word in this sense is any arrangement of three letters, such as aeo, iou, etc.)

Solution: There are five choices for the first letter, four for the second, and three for the third. In all, there are $5 \times 4 \times 3$, that is, 60 possible "words." Notice that 125 words are possible if repetitions of letters are permitted.

---

1. How many three-letter "words" may be formed from the given set $R = \{m, a, t, h\}$ if (**a**) no letter may be used more than once? (**b**) repetitions of letters are permitted?
2. Repeat Exercise 1 for four-letter words.
3. Repeat Exercise 1 for the set $S = \{m, e, t, r, i, c\}$.
4. Repeat Exercise 2 for the set $S = \{m, e, t, r, i, c\}$.
5. Bob has five sport shirts and three pairs of slacks. Assuming that he can wear any combination of these, how many different outfits can he assemble?
6. Jane has six dresses, three hats, and four pairs of shoes. Assuming that she can wear any combination of these, how many different outfits can she assemble?
7. A baseball team has six pitchers and four catchers. How many different batteries consisting of a pitcher and a catcher can they form?
8. Show on a tree diagram the number of different methods of travel or means of transportation that are possible from New York to Los Angles, via Chicago, if you can go from New York to Chicago by one train or one plane, and from Chicago to Los Angeles by one train, one plane, or one bus.
9. (**a**) How many different two-digit numbers can be formed from the set of digits $D = \{1, 2, 3, 4, 5\}$ if repetitions of digits are allowed? (**b**) How many numbers of two different digits can be formed from the set D?
10. Repeat Exercise 9 for $D = \{1, 2, 3, 4, 5, 6, 7, 8\}$.
11. The Jonesboro Swim Club has 15 members. How many different sets of officers consisting of a president, a vice-president, and a secretary-treasurer can it form? No person can hold more than one of these three offices at a time.
12. The Portland Swim Club has 260 members with 200 of these members eligible to hold any office. Repeat Exercise 11 for the Portland Swim Club.
13. How many five-digit numbers can be formed using the ten decimal digits without repetitions if zero is not to be used as the first digit?
14. How many two-digit even numbers can be formed without repetitions from the set $I = \{1, 2, 3, \ldots, 9\}$?
15. Find the number of "words" of three different letters that may be formed from the set $V = \{a, e, i, o, u\}$ if (**a**) the first letter must be i; (**b**) the first letter must be e and the last letter must be i.
16. (**a**) How many different appearing license plates can be made using a letter from our alphabet followed by three decimal digits if the first digit

must not be zero? (**b**) How many are possible if the first digit must not be zero and no digit may be used more than once?

***17.** Find the number of three-digit numbers that may be formed from the set $W = \{0, 1, 2, \ldots, 9\}$ if zero is not an acceptable first digit and (**a**) the number must be even; (**b**) the number must be divisible by 5; (**c**) no digit may be used more than once; (**d**) the number must be odd and less than 500.

***18.** In a certain so-called "combination" lock, there are 50 different positions. To open the lock you move to a certain number in one direction, then to a different number in the opposite direction, and finally to a third number in the original direction. What is the total number of such "combinations":

(**a**) If the first turn must be clockwise?

(**b**) If the first turn may be either clockwise or counterclockwise?

***19.** Find the number of different possible license plates if each one is to consist of two letters of the alphabet followed by four decimal digits, the first digit may not be zero, and no repetitions of letters or numbers are permitted.

***20.** Repeat Exercise 19 for license plates consisting of four consonants followed by three decimal digits.

PEDAGOGICAL EXPLORATIONS

Find two boxes to use as (or make) two cubes. Cover each cube with paper so that the cubes are of different colors, such as red and green. To introduce the vocabulary of dice games think of one cube as a red die and the other cube as a green die. Number the faces of each die 1 through 6. Have two students hold the "dice" so that the rest of the class can see only one number on each die.

1. If the number on the green die is 5, can a number be selected for showing on the red die so that the sum of the two numbers is:

(a) 5? (b) 6? (c) 7?
(d) 11? (e) 3? (f) 12?

2. If the number on the green die is 2, what is the set of numbers that can be obtained as sums by suitable selections of numbers on the red die?

3. What is the set of numbers that can be obtained as sums by suitable selections of a number on each of the two dice?

4. Which of the numbers obtained in Exploration 3 can be obtained in only one way? List this way for each of the numbers.

5. Which of the numbers obtained in Exploration 3 can be obtained in only two ways? List the ways for each of the numbers.
6. Which of the numbers obtained in Exploration 3 can be obtained in the largest number of ways? List these ways.
7. Start a collection of puzzle problems that may be used in an elementary school class and that involve counting problems. Present at least three such problems in your class.

12-2 definition of probability

When a **normal** coin is tossed, we know that there are two distinct and **equally likely** ways in which it may land, heads or tails. We say that the probability of getting a head is one out of two, or simply $\frac{1}{2}$.

In rolling one of a pair of **normal** dice, there are six equally likely ways in which the die may land. We say that the probability of rolling a 5 on one throw of a die is one out of six, or $\frac{1}{6}$.

In each of these two examples, the **events** that may occur are said to be **mutually exclusive.** That is, one and only one of the events can occur at any given time. When a coin is tossed, there are two possible events (heads and tails); one and only one of these may occur. When a single die is rolled, there are six events $\{1, 2, 3, 4, 5, 6\}$; one and only one of these may occur. Informally we define the probability of success as the ratio of the number of successes of an event to the number of possible outcomes of that event. More generally we define probability as follows:

> If an event can occur in any one of n mutually exclusive and equally likely ways, and if m of these ways are considered favorable, then the **probability** $P(A)$ that a favorable event A will occur is given by the formula
>
> $$P(A) = \frac{m}{n}$$

The probability m/n satisfies the relation $0 \leq m/n \leq 1$, since m and n are integers and $m \leq n$. When success is inevitable, $m = n$ and the probability is 1; when an event cannot possibly succeed, $m = 0$ and the probability is 0. For example, the probability of getting either a head or a tail on a single toss of a coin is 1, assuming that the coin does not land on an edge. The probability of tossing a sum of 13 with a single toss of a pair of normal dice is 0. (Here, and in all future work, assume that normal dice are used unless otherwise instructed.)

The sum of the probability of an event's occurring and the probability of that same event's not occurring is 1.

$$\text{If} \quad P(A) = \frac{m}{n}, \qquad \text{then} \quad P(\text{not } A) = 1 - \frac{m}{n}.$$

Example 1: A single card is selected from a deck of 52 bridge cards. What is the probability that it is a spade? What is the probability that it is not a spade? What is the probability that it is an ace or a spade?

Solution: Of the 52 cards, 13 are spades. Therefore, the probability of selecting a spade is $\frac{13}{52}$, that is, $\frac{1}{4}$. The probability that the card selected is not a spade is $1 - \frac{1}{4}$, that is, $\frac{3}{4}$. There are 4 aces and 12 spades besides the ace of spades. Therefore, the probability that the card selected is an ace or a spade is $(4 + 12)/52$; that is, $\frac{16}{52}$, which we express as $\frac{4}{13}$.

It is very important that only equally likely events be considered when the probability formula is applied; otherwise faulty reasoning can occur. Consider again the first question asked in Example 1. One might reason that any single card drawn from a deck of cards is either a spade or is not a spade; thus there are two possible outcomes, and the probability of drawing a spade must therefore be $\frac{1}{2}$. It is correct to say that there are these two possible outcomes, but of course they are *not* equally likely since there are 13 spades in a deck of cards and 39 cards that are not spades.

Example 2: A committee of two is to be selected from the set

$$M = \{\text{Betty, Doris, Ellen, John, Tom}\}$$

by drawing names out of a hat. What is the probability that both members of the committee will be girls?

Solution: We solve this problem by first listing all of the possible committees of two that can be formed from the set M.

[Betty – Doris]	Doris – John
[Betty – Ellen]	Doris – Tom
Betty – John	Ellen – John
Betty – Tom	Ellen – Tom
[Doris – Ellen]	John – Tom

Of the ten possible committees, there are three (those boxed) that consist of two girls. Thus the probability that both members selected are girls is $\frac{3}{10}$. What is the probability of selecting a committee to consist of two boys?

EXERCISES

What is the probability of tossing on a single toss of one die:

1. An even number?

2. An odd number?

3. A number greater than 2?

4. A number less than 4?

5. A number different from 4?

6. A number different from 0?

7. The number 0?

8. A number less than 7?

In Exercises 9 through 12 what is the probability of drawing on a single draw from an ordinary deck of 52 bridge cards:

9. An ace?

10. A king?

11. A spade?

12. A red card?

In Exercises 13 through 16, we will consider naming a committee of two from the set $N = \{Alice, Bob, Carolyn, Doug, Ellen\}$.

13. How many different committees of two can be formed?

14. What is the probability that a committee of two will consist of two boys?

15. What is the probability that a committee of two will consist of two girls?

16. What is the probability that a committee of two will consist of one boy and one girl?

17. What is the probability that the first-named author of this text was born in the month of December?

18. What is the probability that your instructor's telephone number has a 7 as its final digit?

19. The probability of obtaining all heads in a single toss of three coins is $\frac{1}{8}$. What is the probability that not all three coins are heads on such a toss?

20. What is the probability that the next person you meet was not born on a Sunday?

21. Two distinct numbers between 1 and 7 are selected at random (each possible number is equally likely). What is the probability that **(a)** the first number selected is even? **(b)** both numbers are even?

22. Repeat Exercise 21 for numbers between 1 and 13.

PEDAGOGICAL EXPLORATIONS

1. Students enjoy cartoons that deal with mathematical ideas. Start a collection of such cartoons.

NANCY® **By Ernie Bushmiller**

2. Prepare a bulletin board display showing the use of different concepts of probability that appear in newspaper and magazine articles.
3. Identify at least one item that appears regularly in the daily newspaper and is always based upon probability.

If you are enrolled in a class of 30 students, what would you guess is the probability that there are two members of the group that celebrate their birthdays on the same date? Normally one would reason that the probability is low, inasmuch as there are 365 different days of the year on which an individual's birth date could occur, and 30 is small relative to 365.

Although the mathematical explanation is beyond our scope at this time, it can be shown that the actual probability of at least two birth dates occurring on the same day of the year in a random group of 30 students is 0.71. That is, you can safely predict that 71% of the time such an occurrence will take place. Or, alternatively, if you bet that at least two members of the group have the same birth date, then you can expect to win such a bet approximately 71 times out of 100.

The graph at the top of the next page shows the probability of two common birth dates for people in groups of different sizes. Note that the probability in a group of 23 is approximately 0.50. That is, for a group of 23 individuals, there is a 50% chance of having such an occurrence take place. For a group of 60 the probability is 0.99, or almost certainty!

4. Use the given graph to determine the probability of two or more common birth dates for the members of a class of (**a**) 20 students; (**b**) 40 students.
5. Use the members of your class to test the results shown in the graph.

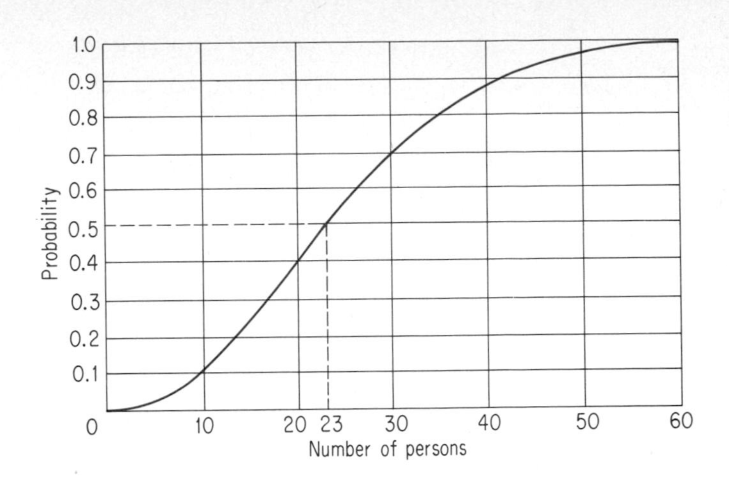

6. Test the results shown in the graph by going to an encyclopedia and using the dates of birth of 30 individuals selected at random.
7. Observe and list in order the last two digits of 20 license plates on automobiles in any large parking lot. Repeat this procedure for at least five sets of 20 two-digit numbers. For each set of 20 numbers note how often you find a repetition of pairs of digits (the same two-digit number appearing at least twice).
8. Repeat the preceding exploration by opening a telephone book and selecting 20 telephone numbers at random. Record the last two digits only, and note the frequency with which one finds a repetition of pairs of digits within a set of 20 two-digit numbers.

12-3 sample spaces

It is often convenient to solve problems of probability by making a list of all possible outcomes. Such a listing is called a **sample space.** Consider first the problem of tossing two coins. The sample space for this problem is given by the following set of all possible outcomes:

$$\{HH, HT, TH, TT\}$$

These four possible outcomes may be obtained by using a tree diagram to list all possible choices or by listing the possibilities as in the following chart.

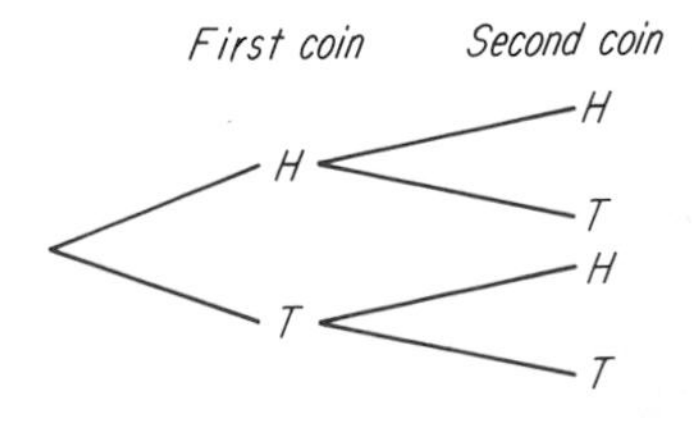

First coin	*Second coin*
H	*H*
H	*T*
T	*H*
T	*T*

From the chart we may observe that:

Two heads occur in one event.
One head and one tail occur in two events.
No heads, that is, two tails, occur in one event.

Then we may list the various probabilities regarding the tossing of two unbiased coins as follows:

Event	Probability
2 heads	$\frac{1}{4}$
1 head	$\frac{2}{4}$
0 heads	$\frac{1}{4}$

Since all possibilities have been considered, the sum of the probabilities should be 1; $\frac{1}{4} + \frac{2}{4} + \frac{1}{4} = 1$. This provides a check on our computation. The list of probabilities is sometimes called a **probability distribution.**

For the case of three coins, the following tree diagram and array may be made.

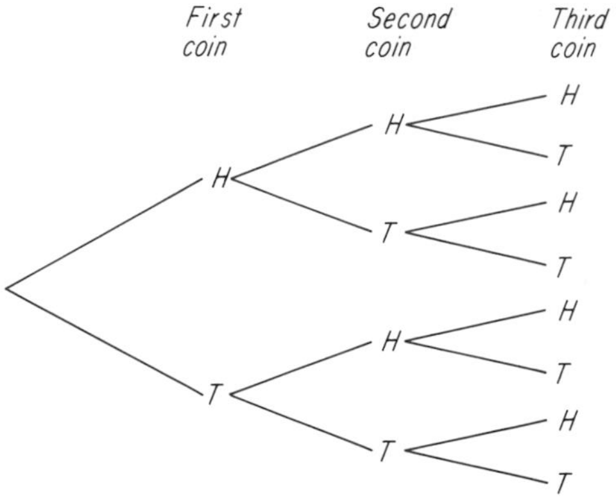

First coin	Second coin	Third coin	First coin	Second coin	Third coin
H	H	H	H	H	H
		T	H	H	T
	T	H	H	T	H
		T	H	T	T
T	H	H	T	H	H
		T	T	H	T
	T	H	T	T	H
		T	T	T	T

From the tree diagram we have the following sample space for the tossing of three coins:

$$\{HHH, HHT, HTH, HTT, THH, THT, TTH, TTT\}$$

We may also list the probabilities of specific numbers of heads.

Event	Probability
0 heads	$\frac{1}{8}$
1 head	$\frac{3}{8}$
2 heads	$\frac{3}{8}$
3 heads	$\frac{1}{8}$

Both for two coins and for three coins the sum of the probabilities is 1; that is, all possible events have been listed and these events are mutually exclusive. Note also that for two coins there were four possible outcomes and for three coins there were eight possible outcomes. For n coins there would be 2^n possible outcomes.

Example: A box contains two red and three white balls. Two balls are drawn in succession without replacement. List a sample space for this experiment.

Solution: To identify individual balls, we denote the red balls as R_1 and R_2; the white balls as W_1, W_2, W_3. Then the sample space is as follows:

R_1R_2	R_2R_1	W_1R_1	W_2R_1	W_3R_1
R_1W_1	R_2W_1	W_1R_2	W_2R_2	W_3R_2
R_1W_2	R_2W_2	W_1W_2	W_2W_1	W_3W_1
R_1W_3	R_2W_3	W_1W_3	W_2W_3	W_3W_2

EXERCISES

Use the sample space of the above Example to find the probability that:

1. Both balls are red.
2. Both balls are white.
3. The first ball is red.
4. The first ball is red and the second ball is white.
5. One ball is red and the other is white.

In Exercises 6 through 8 use the sample space for the outcomes when three coins are tossed to find the probability that:

6. All three coins are heads.
7. At least two coins are heads.
8. At most one coin is tails.

9. Make a tree diagram and give the sample space for the tossing of four coins.

In Exercises 10 through 12 use the sample space for the outcomes when four coins are tossed (Exercise 9) to find the probability that:

10. **(a)** No coins are heads.
    (b) All four coins are tails.

11. (a) At least three coins are heads.
(b) At most one coin is tails.

12. (a) At most two coins are tails.
(b) At least two coins are heads.

13. Make a tree diagram and give a sample space for the tossing of a pair of dice. Represent each outcome by an ordered pair of numbers. For example, let (1, 3) represent a 1 on the first die and a 3 on the second die.

In Exercises 14 through 20 use the sample space for the outcomes when a pair of dice are tossed (Exercise 13) to find the probability that:

14. (a) The number on the first die is 2.
(b) The number on the first die is not 2.

15. (a) The number 1 is on both dice.
(b) It is not true that the number 1 is on both dice.

16. (a) The same number is on both dice.
(b) There are different numbers on the two dice.

17. (a) The sum of the numbers obtained is 11.
(b) The sum of the numbers obtained is not 11.

18. (a) The sum of the numbers obtained is 7.
(b) The sum of the numbers obtained is not 7.

19. (a) The number on the second die is twice the number on the first die.
(b) The number on one die is twice the number on the other die.

20. (a) The number on one die is three more than the number on the other die.
(b) The number on one die is two less than the number on the other die.

21. A box contains two red balls R_1 and R_2 and two white balls W_1 and W_2. List a sample space for the outcomes when two balls are drawn in succession without replacement. Find the probability that both balls are red.

22. Repeat Exercise 21 for the case in which the first ball is replaced before the second ball is drawn.

23. Repeat Exercise 21 for a box that contains three red balls and two white balls.

PEDAGOGICAL EXPLORATIONS

Tree diagrams may be used to solve many problems that are of interest to elementary school children.

1. Suppose that two evenly-matched teams, the Bees and the Hornets, play a single game.

(a) What is the probability of the Bees winning?
(b) What is the probability of the Hornets winning?
(c) In the language of probability what is the meaning of "evenly-matched"?

2. Suppose that the evenly-matched Bees and Hornets are school teams participating in a playoff that is determined by the best two out of three games.
   (a) Make a tree diagram of the possible outcomes. Label the outcome of each game with its probability and be sure that after each game the sum of the probabilities is 1.
   (b) What is the probability of the Bees winning the first two games?
   (c) What is the probability of the Bees winning the playoff?

3. Repeat Exploration 2 if the Bees had a 60% chance of winning each game.

4. Suppose that the Hornets are one of two evenly-matched teams that are entered in a regional tournament with the best three out of five games needed for the winner.
   (a) Make a tree diagram for the Hornets' progress based upon winning or not winning each game. Label each outcome with its probability.
   (b) What is the probability that the Hornets will win the first three games?
   (c) What is the probability that the Hornets will win three of the first four games?
   (d) What is the probability that the Hornets will win the tournament?

There are many interesting probability questions whose answers are not intuitively obvious. Here are two popular ones that are best solved by means of a sample space.

5. Three cards are in a box. One is red on both sides, one is white on both sides, and one is red on one side and white on the other. A card is drawn at random and placed on a table. It has a red side showing. What is the probability that the side not showing is also red? (Contrary to popular belief, the answer is not $\frac{1}{2}$.)

As an aid to the solution of this problem, let us identify the three cards as in this diagram:

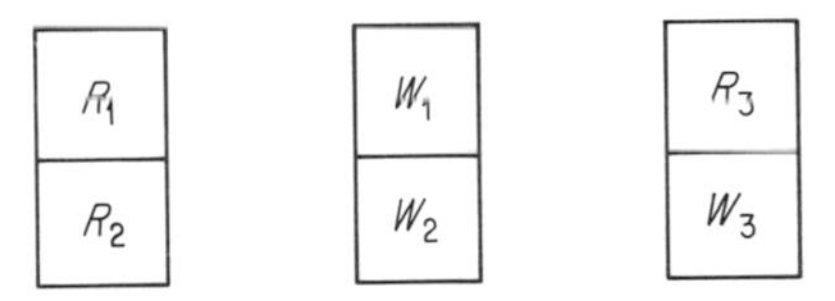

Now consider the set of possible outcomes when a card is drawn at random and placed on the table. In each of the following ordered pairs, the first side noted represents the one placed face up on the table whereas the second side indicates the face hidden from view.

$$\{R_1R_2, \quad R_2R_1, \quad R_3W_3, \quad W_1W_2, \quad W_2W_1, \quad W_3R_3\}$$

Inasmuch as we are told that a red side is showing, we may narrow down the sample space to the first three pairs only. Of these three possibilities, if the first side is red can you now tell the probability that the second side is also red?

6. One of the authors of this book has two children. One of these children is a boy. What is the probability that they are both boys? Believe it or not, the answer is *not* $\frac{1}{2}$!

12-4 computation of probabilities

If A and B represent two mutually exclusive events, then

$$P(A \text{ or } B) = P(A) + P(B)$$

That is, the probability that one event or the other will occur is the sum of the individual probabilities. Consider, for example, the probability of drawing an ace or a picture card (a jack, queen, or king) from an ordinary deck of 52 bridge cards.

The probability of drawing an ace, $P(A)$, is $\frac{4}{52}$.
The probability of drawing a picture card, $P(B)$, is $\frac{12}{52}$.
Then $P(A \text{ or } B) = \frac{4}{52} + \frac{12}{52} = \frac{16}{52} = \frac{4}{13}$.

Example 1: A bag contains three red, two black, and five yellow balls. Find the probability that a ball drawn at random will be red or black.

Solution: The probability of drawing a red ball, $P(R)$, is $\frac{3}{10}$. The probability of drawing a black ball, $P(B)$, is $\frac{2}{10}$. Then

$$P(R \text{ or } B) = P(R) + P(B) = \frac{5}{10} = \frac{1}{2}$$

This process can be extended to find the probability of any finite number of mutually exclusive events.

$$P(A_1 \text{ or } A_2 \text{ or } A_3 \text{ or } \cdots A_n) = P(A_1) + P(A_2) + P(A_3) + \cdots + P(A_n)$$

Example 2: A single die is tossed. What is the probability that either an odd number or a number greater than 3 appears?

Solution: There are three odd numbers possible, (1, 3, 5), so the probability of tossing an odd number is $\frac{3}{6}$. The probability of getting 4, 5, or 6, that is, a number greater than 3, is also $\frac{3}{6}$. Adding these probabilities gives $\frac{3}{6} + \frac{3}{6} = 1$. Something is obviously wrong, since a probability of 1 implies certainty and we can see that an outcome of 2 is neither odd nor greater than 3. The difficulty lies in the fact that the events are *not* mutually exclusive; a number may be both odd and also greater than 3. In particular, 5 is both odd and greater than 3 at the same time. Thus $P(5)$ has been included twice. Since $P(5) = \frac{1}{6}$, our answer should be $\frac{3}{6} + \frac{3}{6} - \frac{1}{6}$, that is, $\frac{5}{6}$.

Example 2 can be illustrated by means of the following Venn diagram that lists all possible outcomes when a single die is tossed. Note that the outcome 5 is listed in the intersection of the two sets since it is both odd and greater than 3. On the other hand, the outcome 2 fits into neither of these descriptions and is thus placed outside the circles.

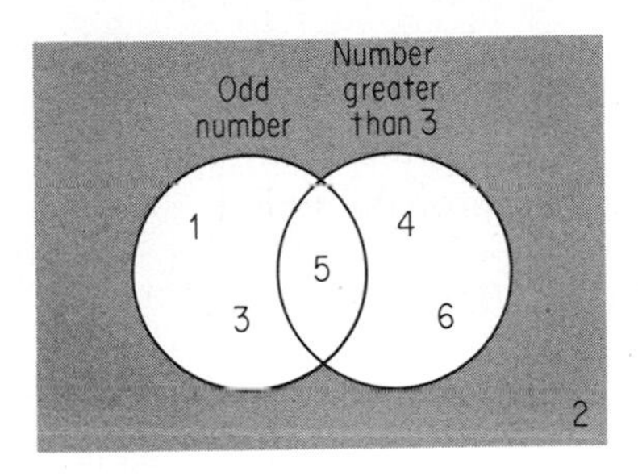

In situations like that of Example 2 we need to subtract the probability that both events occur at the same time. Thus, where A and B are not mutually exclusive events, we have

$$P(A \text{ or } B) = P(A) + P(B) - P(A \text{ and } B)$$

Note that in Example 2, we had the following probabilities:

$P(A)$, the probability of an odd number, is $\frac{3}{6}$.
$P(B)$, the probability of a number greater than 3, is $\frac{3}{6}$.
$P(A \text{ and } B)$, the probability that a number (in this case 5) is odd and greater than 3, is $\frac{1}{6}$.
$P(A \text{ or } B) = \frac{3}{6} + \frac{3}{6} - \frac{1}{6} = \frac{5}{6}$.

By an actual listing we can see that five of the six possible outcomes in Example 2 are either odd or greater than 3, namely 1, 3, 4, 5, and 6. The only "losing" number is 2. Thus the probability $P(A \text{ or } B)$ must be $\frac{5}{6}$.

In general, the two situations $P(A \text{ or } B)$ and $P(A \text{ and } B)$ that we have considered can be described by means of diagrams, where the points of the circular regions represent probabilities of events.

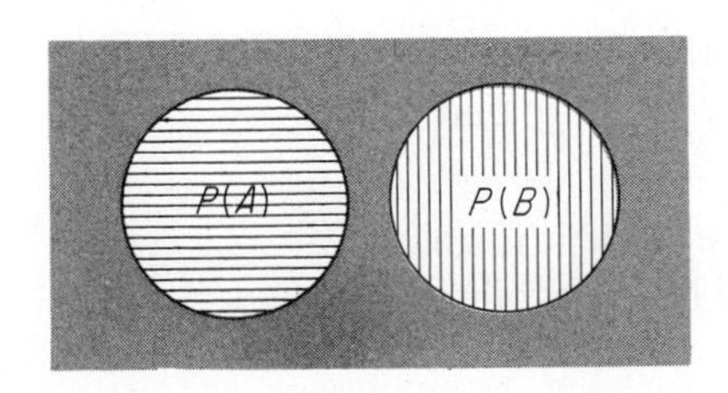

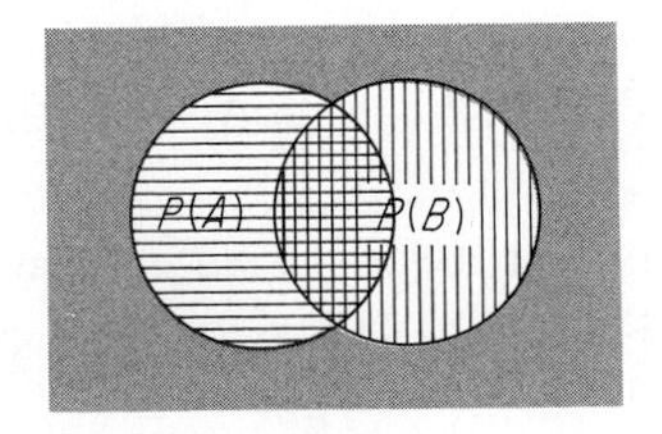

The first of the two preceding diagrams shows mutually exclusive events:

$$P(A \text{ or } B) = P(A) + P(B)$$

In the second figure, the region that is shaded both horizontally and vertically represents $P(A \text{ and } B)$, $P(A \text{ and } B) \neq 0$, the events are not mutually exclusive, and

$$P(A \text{ or } B) = P(A) + P(B) - P(A \text{ and } B)$$

We must subtract $P(A \text{ and } B)$, since we have counted it twice in the sum $P(A) + P(B)$.

Next we turn our attention to the probability that several events will occur, one after the other. Consider the probability of tossing a coin twice and obtaining heads on the first toss and tails on the second toss. From a sample space we see that the probability is $\frac{1}{4}$.

$$\{HH, \enclose{circle}{HT}, TH, TT\}$$

Furthermore, we see that the probability $P(A)$ that the first coin is heads is $\frac{1}{2}$. The probability $P(B)$ that the second coin is tails is $\frac{1}{2}$. Then $P(A \text{ and } B) = \frac{1}{2} \times \frac{1}{2} = \frac{1}{4}$. Note that these events are **independent;** that is, the outcome of the first toss does not affect the second toss.

In general, let the probability that an event A occurs be $P(A)$. Let the probability that a second event occurs, after A has occurred, be $P(B)$. Then

$$P(A \text{ and } B) = P(A) \times P(B)$$

This can be extended to any finite number of events.

$$P(A_1 \text{ and } A_2 \text{ and } A_3 \text{ and } \cdots A_n) = P(A_1) \times P(A_2) \times P(A_3) \times \cdots \times P(A_n)$$

Example 3: Urn A contains three white and five red balls. Urn B contains four white and three red balls. One ball is drawn from each urn. What is the probability that they are both red?

Solution: Let $P(A)$ be the probability of drawing a red ball from urn A and $P(B)$ be the probability of drawing a red ball from urn B. Then

$$P(A) = \tfrac{5}{8}, \qquad P(B) = \tfrac{3}{7}, \qquad P(A \text{ and } B) = \tfrac{5}{8} \times \tfrac{3}{7} = \tfrac{15}{56}.$$

Example 4: Two cards are selected in succession, without replacement, from an ordinary bridge deck of 52 cards. What is the probability that they are both aces?

Solution: The probability that the first card is an ace is $\frac{4}{52}$. If it is an ace, then the probability that the second card is an ace is $\frac{3}{51}$. The probability that both cards are aces is $\frac{4}{52} \times \frac{3}{51}$, that is, $\frac{1}{221}$.

What is the probability of tossing on a single toss of one die:

1. An even number or a number greater than 4?

2. An odd number or a number less than 4?

3. An odd number or a number greater than 4?

4. An even number or a number greater than 6?

5. An odd number or a number greater than 7?

6. An odd number or a number less than 6?

7. An even number or a number less than 7?

A single card is drawn from an ordinary deck of 52 bridge cards. In Exercises 8 through 15 find the probability that the card selected is:

8. An ace or a queen.

9. A spade or a diamond.

10. A spade or a queen.

11. A spade and an ace.

12. A spade and a queen.

13. A heart or a king or a queen.

14. A club or an ace or a king.

15. A club and a spade.

Two cards are drawn in succession from an ordinary deck of 52 bridge cards without the first card's being replaced. Find the probability that:

16. **(a)** Both cards are spades.
(b) Both cards are aces of spades.

17. **(a)** The first card is a spade and the second card is a heart.
(b) The first card is an ace and the second card is the king of hearts.

18. **(a)** The two cards are of the same suit.
(b) The two cards are of different suits.

Two cards are drawn in succession from an ordinary deck of 52 bridge cards with the first card replaced before the second card is drawn. In Exercises 19 and 20 find the probability that:

19. **(a)** Both cards are diamonds.
(b) Both cards are aces of diamonds.

20. **(a)** The first card is a heart and the second card is a club.
(b) The first card is an ace and the second card is the king of hearts.

21. A coin is tossed six times. What is the probability that all six tosses are heads?

22. A coin is tossed six times. What is the probability that at least one head is obtained? (*Hint:* First find the probability of getting no heads.)

23. A coin is tossed and then a die is rolled. Find the probability of obtaining **(a)** a head and a 3; **(b)** a head and an even number; **(c)** a head or a 3; **(d)** a head or an even number.

24. A bag contains four red balls and seven white balls. **(a)** If one ball is drawn at random, what is the probability that it is white? **(b)** If two balls are drawn at random, what is the probability that they are both white?

25. Five cards are drawn at random from an ordinary bridge deck of 52 cards. Find the probability that all five cards drawn are spades.

26. A box contains three red, four white, and six green balls. Three balls are drawn in succession, without replacement. Find the probability that **(a)** all three are red; **(b)** the first is red, the second is white, and the third is green; **(c)** none are green; ***(d)** all three are of the same color.

27. Repeat Exercise 26 if each ball is replaced after it is drawn.

28. A die is tossed three times. Find the probability that **(a)** a 6 is tossed on the first toss; **(b)** a 6 is tossed on the first two tosses; **(c)** a 6 is tossed on all three tosses; **(d)** a 6 is tossed on the first toss and not tossed on the second or third tosses.

29. A die is tossed three times. Find the probability that **(a)** an even number is tossed on all three tosses; **(b)** an even number is tossed on the first two tosses and an odd number on the third toss; **(c)** an odd number is tossed on the first toss and an even number on the second and third tosses; **(d)** exactly one even number is tossed.

***30.** A die is tossed three times. Find the probability that **(a)** at least one 6 is tossed; **(b)** exactly one 6 is tossed.

PEDAGOGICAL EXPLORATIONS

Many interesting class experiments can be performed that illustrate basic concepts of probability. Several such experiments are suggested here that you should attempt to complete with a class of students. Also make a collection of any other such experiments that you can find.

1. Consider the network of streets shown in the following figure.

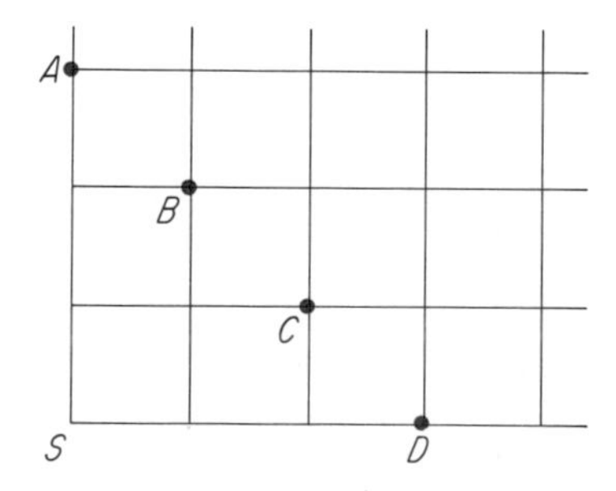

You are to start at the point S and move three "blocks." Each move is determined by tossing a coin. If the coin lands "tails," move one block to the right; for "heads," move one block up. Your terminal point will be at A, B, C, or D. Try to predict the number of times you will land at each point if the experiment is to be repeated 16 times. Then complete 16 trials, keeping a tally of the number of times you land at each point. Compare your actual results with your predictions as well as with the results obtained by your classmates. Finally, on the basis of your experimentation, revise your predictions if necessary.

2. Repeat Exercise 1 for the following network. This time you are to make four moves and will land at point A, B, C, D, or E.

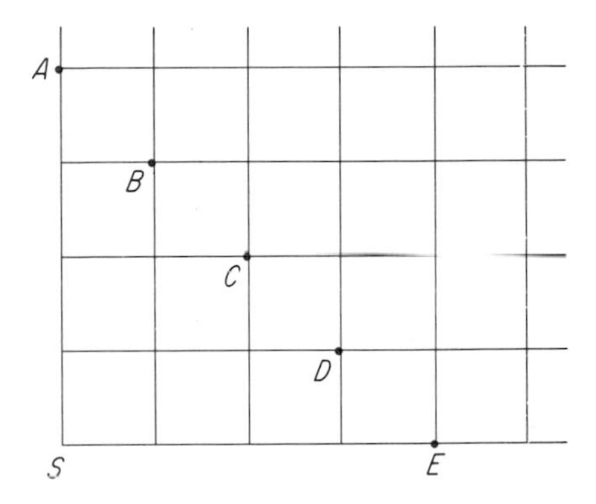

3. If a pair of normal dice are tossed repeatedly, one can theoretically expect a sum of 7 to occur in one out of six tosses. Thus if a pair of dice are tossed 36 times, on the average six of the tosses will give a sum of 7. Toss a pair of dice 36 times and count the frequency with which a sum of 7 appears. Compare your actual results with the theoretical probability of $\frac{1}{6}$. Collect this data for each member of your class and compare actual results with expected results.

4. Repeat Exploration 3 for 72 tosses of a pair of dice. Record each toss by placing an x over the numeral representing the sum, using graph paper. The final result should show a bar graph that is fairly "normal" in shape. The figure below shows the results obtained after one experiment of 18 tosses.

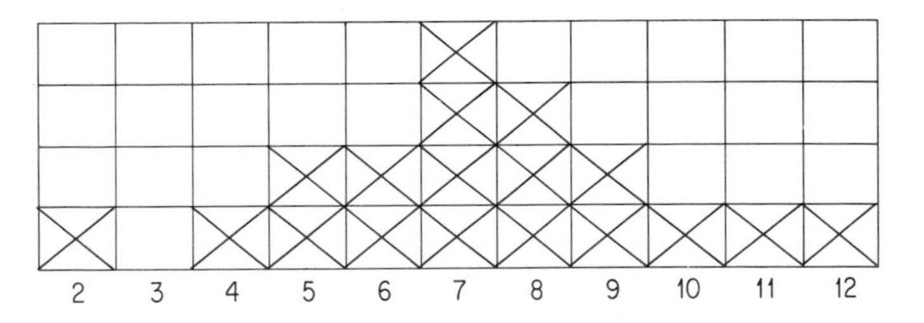

12-5 odds and mathematical expectation

The sports section of a newspaper often includes statements regarding the "odds" in favor of, or against, a particular team or individual's winning or losing some encounter. For example, we may read that the odds in favor of the Cardinals' winning the pennant are "4 to 1." In this section we shall attempt to discover just what such statements really mean.

Consider the problem of finding the odds against obtaining a 3 in one toss of a die. Since the probability of obtaining a 3 is known to be $\frac{1}{6}$, most people would say that the odds are therefore 6 to 1 against rolling a 3. This is not correct—because out of every six tosses of the die, in the long run, one expects to toss one 3. The other five tosses are not expected to be 3's. Therefore, the correct odds against rolling a 3 in one toss of a die are 5 to 1. The odds in favor of tossing a 3 are 1 to 5. Formally we define odds as follows:

> The **odds in favor** of an event are defined as the ratio of the probability that an event will occur to the probability that the event will not occur. The reciprocal of this ratio gives the **odds against** the occurrence of the event.

Thus the odds in favor of an event that may occur in several equally likely ways is the ratio of the number of favorable ways to the number of unfavorable ways.

Notice that odds and probabilities are very closely related. Indeed, if either the odds for or the probability of an event is known, then the other can be found. For example, if the odds for an event are 1 to 2, then we have the ratio

$$\frac{\text{number of favorable ways}}{\text{number of unfavorable ways}} = \frac{1}{2}$$

and the probability is the ratio

$$\frac{\text{number of favorable ways}}{\text{total number of all ways (favorable and unfavorable)}} = \frac{1}{3}$$

Similarly, if the probability is $\frac{2}{5}$, then the odds are 2 to 3.

Example 1: Find the *odds in favor* of drawing a spade from an ordinary deck of 52 bridge cards and the *odds against* drawing a spade from an ordinary deck of 52 bridge cards.

Solution: Since there are 13 spades in a deck of cards, the probability of drawing a spade is $\frac{13}{52} = \frac{1}{4}$. The probability of failing to draw a spade is $\frac{3}{4}$. The odds in favor of obtaining a spade are $\frac{1}{4} \div \frac{3}{4}$; that is, $\frac{1}{3}$.

The *odds in favor* of drawing a spade are stated as $\frac{1}{3}$. They may also be stated as "1 to 3" or as "1 : 3." Similarly, the *odds against* drawing a spade are $\frac{3}{1}$, which may be written as "3 to 1" or "3 : 1."

Mathematical expectation is closely related to odds and is defined as the product of the probability that an event will occur and the amount to be received upon such occurrence. Suppose that you are to receive $2.00 each time you obtain two heads on a single toss of two coins. You do not receive anything for any other outcome. Then your mathematical expectation will be one-fourth of $2.00; that is, $0.50. This means that you should be willing to pay $0.50 each time you toss the coins if the game is to be a fair one. *In the long run,* both you and the person who is running the game would break even. For example, if you played the game four times it would cost you 4 × $0.50; that is, $2.00. You expect to win, *in the long run,* once out of every four games played. Assuming that you do so, you will win $2.00 once every four times and thus be even.

If an event has several possible outcomes that occur with probabilities p_1, p_2, p_3, and so forth, and for each of these outcomes one may expect the amounts m_1, m_2, m_3, and so on, then the mathematical expectation E may be defined as

$$E = m_1p_1 + m_2p_2 + m_3p_3 + \cdots$$

Whenever this formula is used, it is worthwhile to check that all possible outcomes have been considered by checking that the sum of the probabilities is equal to 1.

Example 2: Suppose that you play a game wherein you are to toss a coin twice and are to receive 10 cents if two heads are obtained, 5 cents if one head is obtained, and nothing if both tosses produce tails. What is your expected value in this game?

Solution: The probabilities of obtaining two, one, and no heads, respectively, are $\frac{1}{4}$, $\frac{1}{2}$, and $\frac{1}{4}$. Therefore, the expected value E, in cents, is found to be

$$E = (10)(\tfrac{1}{4}) + (5)(\tfrac{1}{2}) + (0)(\tfrac{1}{4}) = 5$$

This solution may be interpreted in several ways. For one thing, it is the price you should be willing to pay for the privilege of playing this game. It may also be interpreted as the average amount of winnings per game that one may expect *when one is playing a large number of games.*

EXERCISES

What are the odds in favor of obtaining:

1. Two heads in a single toss of two coins?
2. At least two heads in a single toss of three coins?

3. Two heads when a single coin is tossed twice?

4. At least two heads when a single coin is tossed three times?

In Exercises 5 through 8 what are the odds against obtaining:

5. Two heads in a single toss of two coins?

6. An ace in a single draw from a deck of 52 bridge cards?

7. An ace or a king in a single draw from a deck of 52 bridge cards?

8. A 7 or an 11 in a single toss of a pair of dice?

9. For the event of rolling a 7 or an 11 in a single toss of a pair of dice, what are the odds in favor of the event?

10. One hundred tickets are sold for a lottery. The grand prize is $1000. What is your mathematical expectation if you are given a ticket?

11. Repeat Exercise 10 for the case in which 250 tickets are sold.

12. What is your mathematical expectation when you are given one of 300 tickets for a single prize worth $750?

13. What is your mathematical expectation in a game in which you will receive $10 if you toss a "double" (the same number on both dice) on a single toss of a pair of dice?

14. A box contains three dimes and two quarters. You are to reach in and select one coin, which you may then keep. Assuming that you are not able to determine which coin is which by its size, what would be a fair price for the privilege of playing this game?

15. There are three identical boxes on a table. One contains a five-dollar bill, one contains a one-dollar bill, and the third is empty. A man is permitted to select one of these boxes and to keep its contents. What is his expectation?

16. Three coins are tossed. What is the expected number of heads?

***17.** Two bills are to be drawn from a purse that contains three five-dollar bills and two ten-dollar bills. What is the mathematical expectation for this drawing?

18. If there are two pennies, a nickel, a dime, a quarter, and a half-dollar in a hat, what is the mathematical expectation of the value of a random selection of a coin from the hat?

***19.** Repeat Exercise 18 for the selection of two coins from the hat.

20. Suppose that the probability of your obtaining an A in this course is 0.3. What are the odds **(a)** in favor of your obtaining an A? **(b)** against your obtaining an A?

***21.** If the probability of an event is p, what are the odds **(a)** in favor of the event? **(b)** against the event?

***22.** Suppose that n tickets are sold at $1.00 each for a single prize of $1,000.00. Express the theoretically expected value of each ticket in terms of n. (*Hint:* This value is the algebraic sum of the expected loss and the expected gain.)

*23. Repeat Exercise 22 if there is a first prize of $5,000.00 and a second prize of $1,000.00.

PEDAGOGICAL EXPLORATIONS

See at least one recently published series of mathematical texts and summarize the use of probability, odds, and mathematical expectation in:

1. Grades K through 6.
2. Grades 7 and 8.

Games, newspaper clippings, and classroom experiments may be very effectively used for teaching concepts of probability, odds, and mathematical expectation.

3. Name at least three games that involve probability, that is, games that are "games of chance" rather than "games of skill."
4. Describe at least one classroom experiment that can be used effectively for one or more of these purposes.
5. Start a collection of newspaper clippings that can be used with an elementary school class for one or more of these purposes.

12-6 permutations

Suppose that three people, Ruth, Joan, and Debbie, are waiting to play singles at a tennis court. Two of the three can be selected in six different ways if the order in which they are named is significant, for example, if the first person named is to serve first. We may identify these six ways from a tree diagram.

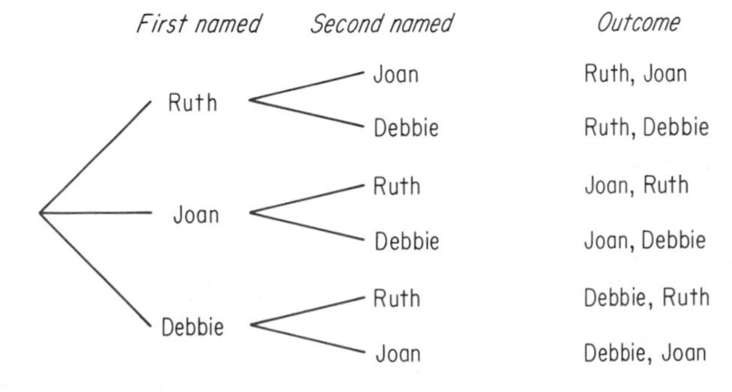

We say that there are 3×2 (that is, 6) *permutations* of the set of three people selected two at a time. In each case the *order* in which the two people are named is significant. A **permutation** of a set of elements is an *arrangement* of certain of these elements in a specified order. In the problem just discussed, the number of permutations of three things taken two at a time is 6. In symbols we write

$${}_3P_2 = 6$$ (read as "the number of permutations of three things taken two at a time is equal to six.")

To find a general formula for ${}_nP_r$, the number of permutations (arrangements) of n things taken r at a time, we use the general principle of counting (§12-1). Note that we can fill the first of the r positions in any one of n different ways. Then the second position can be filled in $n - 1$ different ways, and so on.

Position:	1	2	3	4	...	r
	↓	↓	↓	↓		↓
Number of choices:	n	$n-1$	$n-2$	$n-3$	...	$n-(r-1)$ (i.e., $n-r+1$)

The product of these r factors gives the number of different ways of arranging r elements selected from a set of n elements, that is, the permutations of n things taken r at a time.

$${}_nP_r = (n)(n-1)(n-2)\ldots(n-r+1)$$

where n and r are integers and $n \geq r$.

Example 1: Find ${}_8P_4$.

Solution: Here $n = 8$, $r = 4$, and $n - r + 1 = 5$. Thus

$${}_8P_4 = 8 \times 7 \times 6 \times 5 = 1680$$

Note that there are r, in this case 4, factors in the product.

Example 2: How many different three-letter "words" can be formed from the 26 letters of the alphabet if each letter may be used at most once?

Solution: We wish to find the number of permutations of 26 things taken three at a time.

$${}_{26}P_3 = 26 \times 25 \times 24 = 15{,}600$$

A special case of the permutation formula occurs when we consider the permutations of n things taken n at a time. For example, let us see in how many different ways we may arrange in a row the five members of a given set. Here we have the permutations of five things taken five at a time:

$${}_5P_5 = 5 \times 4 \times 3 \times 2 \times 1$$

In general, for n things n at a time, $n = n$, $r = n$, and $n - r + 1 = 1$;

$$_nP_n = (n)(n-1)(n-2)\ldots(3)(2)(1)$$

We use a special symbol, $n!$, read "n factorial," for this product of integers 1 through n. The following examples should illustrate the use of the new symbol.

$$
\begin{array}{ll}
1! = 1 & 5! = 5 \times 4 \times 3 \times 2 \times 1 \\
2! = 2 \times 1 & 6! = 6 \times 5 \times 4 \times 3 \times 2 \times 1 \\
3! = 3 \times 2 \times 1 & 7! = 7 \times 6 \times 5 \times 4 \times 3 \times 2 \times 1 \\
4! = 4 \times 3 \times 2 \times 1 & 8! = 8 \times 7 \times 6 \times 5 \times 4 \times 3 \times 2 \times 1
\end{array}
$$

Also, we *define* $0! = 1$ so that $(n - r)!$ may be used when $r = n$.

Using this **factorial notation,** we are now able to provide a different, but equivalent, formula for $_nP_r$:

$$_nP_r = n(n-1)(n-2)\ldots(n-r+1) \times \frac{(n-r)(n-r-1)(n-r-2)\ldots(3)(2)(1)}{(n-r)(n-r-1)(n-r-2)\ldots(3)(2)(1)} = \frac{n!}{(n-r)!}$$

Example 3: Evaluate $_7P_3$ in two different ways.

Solution:

(a) $_7P_3 = 7 \times 6 \times 5 = 210$

(b) $_7P_3 = \dfrac{7!}{4!} = \dfrac{7 \times 6 \times 5 \times 4 \times 3 \times 2 \times 1}{4 \times 3 \times 2 \times 1} = 7 \times 6 \times 5 = 210$

Example 4: A certain class consists of 10 boys and 12 girls. They wish to elect officers in such a way that the president and treasurer are boys and the vice-president and secretary are girls. In how many ways can this be done?

Solution: The number of different ways of selecting the president and treasurer is $_{10}P_2$. The number of ways of selecting the vice-president and secretary is $_{12}P_2$. The total number of ways of choosing officers is

$$(_{10}P_2) \times (_{12}P_2) = (10 \times 9) \times (12 \times 11) = 11{,}880$$

Example 5: (a) Find the number of three-digit numbers that can be formed using the digits 7, 8, 9 if no digit may be used more than once in a number. (b) How many of these numbers will be even? (c) What is the probability that the number will be odd?

Solution: (a) $_3P_3 = 3! = 6$

(b) If we select the units digit first, we have only one choice. Then using the general principle of counting, two of the numbers will be even; namely, 798 and 978.

(c) The probability that such a number will be odd (not even) is $(6 - 2)/6$; that is, $\frac{2}{3}$.

EXERCISES

1. Consider the set $S = \{\text{bat, ball}\}$ and list the permutations of the elements of S taken (**a**) one at a time; (**b**) two at a time.
2. Consider the set $R = \{\text{reading, writing, arithmetic}\}$ and list the permutations of the elements of R taken (**a**) one at a time; (**b**) two at a time; (**c**) three at a time.
3. Consider the set $T = \{A, B, C, D\}$ and list the permutations of the elements of T taken (**a**) one at a time; (**b**) two at a time; (**c**) three at a time; (**d**) four at a time.

Evaluate.

4. $5!$ **5.** $6!$ **6.** $\frac{8!}{6!}$

7. $\frac{11!}{7!}$ **8.** ${}_7P_2$ **9.** ${}_7P_3$

10. ${}_{10}P_1$ **11.** ${}_{10}P_{10}$ **12.** ${}_{12}P_{12}$

13. ${}_{12}P_3$ **14.** ${}_{10}P_3$ **15.** ${}_{10}P_7$

Express in terms of n.

16. ${}_nP_1$ **17.** ${}_nP_{n-1}$ **18.** ${}_nP_2$

19. ${}_nP_{n-2}$ **20.** ${}_nP_n$ **21.** ${}_nP_0$

In Exercises 22 through 27 solve for n.

22. ${}_nP_1 = 7$ **23.** ${}_nP_1 = 21$ **24.** ${}_nP_{n-1} = 6$

25. ${}_nP_{n-1} = 120$ **26.** ${}_nP_2 = 20$ **27.** ${}_nP_2 = 30$

28. Find the number of different arrangements of the set of letters $V = \{a, e, i, o, u\}$ if they are taken (**a**) two at a time; (**b**) five at a time.

29. (**a**) Find the number of four-digit numbers that can be formed using the digits 1, 2, 3, 4, 5 if no digit may be used more than once in a number.
(**b**) How many of these numbers will be even? (**c**) What is the probability that such a four-digit number will be even? (**d**) Odd?

30. Find the number of different signals that can be formed by running up three flags on a flagpole, one above the other, if seven different flags are available.

PEDAGOGICAL EXPLORATIONS

Make a worksheet of 20 problems for elementary school students to use:

1. For exploring the number of permutations of five things taken three at a time.

2. In developing an understanding of whether or not the order of the elements in a given specific set is important. (This is precisely the distinction between permutations and combinations.)

Permutations are assumed to be linear (along a line) unless otherwise specified. Suppose that we consider the permutations of seating people at a circular table. The various places at the table are assumed to be indistinguishable and only the relative positions of the people are considered. For such permutations (circular permutations) one person is seated as a reference point and then the various arrangements of the others with reference to this person are considered.

3. In how many ways can four people be seated at a circular table?
4. In how many ways can five people be seated at a circular table?
5. Explain the appropriateness of the expression $\frac{{}_nP_n}{n}$ for the number of permutations of n people in n seats around a circular table.

Unlike the seating of people around a table, a key ring can be turned over. Consider the effect of this phenomenon on the number of distinguishable permutations of keys on a circular key ring.

6. In how many ways can five keys be arranged on a circular key ring?
7. In how many ways can seven keys be arranged on a circular key ring?
8. Give an expression for the number of ways in which n keys can be arranged on a circular key ring.

12-7 combinations

A fictitious club with five members was considered in §12-1. The set of members is {Betty, Doris, Ellen, John, Tom}. The number of ways in which a president and a vice-president can be selected is essentially the number ${}_5P_2$ of permutations of five things taken two at a time. Here order is important in that Betty as president and Doris as vice-president is a different set of officers than Doris as president and Betty as vice-president.

Now, suppose we wish to select a committee of two members from set M without attaching any meaning to the order in which the members are selected. Then the committee consisting of Betty and Doris is certainly the same as the one consisting of Doris and Betty. In this case, we see that *order is not important*, and we call such an arrangement a **combination.** One way to determine the number of possible committees of two to be formed from the set M is by enumeration. There are ten possible committees.

Betty-Doris	Doris-John
Betty-Ellen	Doris-Tom
Betty-John	Ellen-John
Betty-Tom	Ellen-Tom
Doris-Ellen	John-Tom

We summarize this discussion by saying that the number of combinations of five things, taken two at a time, is 10. In symbols we write

$_5C_2 = 10$ (read as "the number of combinations of five things taken two at a time is equal to ten")

Note that $_5P_2 = 20$ and $_5C_2 = 10 = {_5P_2} \div 2$ since each combination of two elements such as {Betty, Doris} could have come from either of two permutations of those elements, in this case Betty-Doris or Doris-Betty. The basic distinction between permutations and combinations is:

A *permutation* is an *arrangement* of elements and *order is important.*
A *combination* is a *set* of elements and *order is not important.*

Consider a set of three elements; then

$$_3P_3 = 3 \times 2 \times 1 = 6 \quad \text{and} \quad {_3C_3} = 1$$

In general $_nP_n = n!$ but $_nC_n = 1$ for any counting number n. Also for n greater than or equal to 2

$$_nC_2 = \frac{_nP_2}{_2P_2} = \frac{n(n-1)}{2 \times 1}$$

where $_2P_2$ is the number of permutations (arrangements) associated with each combination of two elements.

To find $_5C_3$ consider again the specific problem of selecting committees of three from the set M. There are ten such possibilities, and we list them for $M = \{B, D, E, J, T\}$ using only the first initial of each name:

B, D, E   B, D, J   B, D, T   B, E, J   B, E, T
B, J, T   D, E, J   D, E, T   D, J, T   E, J, T

Note that selecting committees of three is equivalent to selecting sets of two to be omitted. That is, omitting J and T is the same as selecting B, D, and E. Therefore, we find that $_5C_3 = {_5C_2} = 10$.

Inasmuch as we wanted only committees, and assigned no particular jobs to the members of each committee, we see that order is not important. However, suppose that each committee is now to elect a chairman, secretary, and historian. In how many ways can this be done within each committee? This is clearly a problem in which order is important; we must therefore use permutations. The number of such possible arrangements within each committee is $_3P_3$; that is, 3!. For example, the committee consisting of B, D, and E can rearrange themselves as chairman, secretary, and historian, respectively, as follows:

B, D, E   B, E, D   D, E, B   D, B, E   E, B, D   E, D, B

All six of these permutations are associated with just one combination.

In general, each combination of three elements is associated with $_3P_3$, that is, 3! permutations of these elements;

$$_nC_3 = \frac{_nP_3}{_3P_3} = \frac{n(n-1)(n-2)}{3 \times 2 \times 1}$$

As in the cases of ${}_nC_2$ and ${}_nC_3$, the general form of ${}_nC_r$ may be expressed as a quotient and is frequently written in symbols in any one of these forms:

$$ {}_nC_r = \binom{n}{r} = \frac{{}_nP_r}{{}_rP_r} = \frac{{}_nP_r}{r!} $$

$$ = \frac{n(n-1)(n-2)(n-3)\dots(n-r-1)}{r(r-1)(r-2)(r-3)\dots 1} = \frac{n!}{r!\,(n-r)!} $$

Note that when ${}_nC_r$ was first expressed as a quotient of numbers, the denominator was $r!$, the product of r successive integers starting with r and decreasing; also the numerator was the product of r successive integers but in this case starting with n and decreasing.

Example 1: Evaluate ${}_7C_2$ in two ways.

Solution: Use ${}_nC_r = \frac{{}_nP_r}{r!}$ and ${}_nC_r = \frac{n!}{r!\,(n-r)!}$.

(a) $ {}_7C_2 = \frac{{}_7P_2}{2!} = \frac{7 \times 6}{2 \times 1} - 21 $

(b) $ {}_7C_2 = \frac{7!}{2!\,5!} = \frac{7 \times 6 \times 5 \times 4 \times 3 \times 2 \times 1}{2 \times 1 \times 5 \times 4 \times 3 \times 2 \times 1} = 21 $

Example 2: How many different possible hands of five cards each can be dealt from a deck of 52 cards? What is the probability that a particular hand contains four aces and the king of hearts?

Solution: The order of the five cards is unimportant, so this is a problem involving combinations.

$$ {}_{52}C_5 = \frac{52!}{5!\,47!} = \frac{52 \times 51 \times 50 \times 49 \times 48}{5!} \times \frac{(47!)}{(47!)} = 2{,}598{,}960 $$

The probability of obtaining any one particular hand, such as that containing the four aces and the king of hearts, is $\frac{1}{2{,}598{,}960}$.

Many problems in probability are most conveniently solved through the use of the concepts of combinations presented here. For example, let us consider again Example 4 of §12-4. There we were asked to find the probability that both cards would be aces if two cards are drawn from a deck of 52 cards. This problem can be solved by noting that ${}_{52}C_2$ is the total number of ways of selecting two cards from a deck of 52 cards. Also ${}_4C_2$ is the total number of ways of selecting two aces from the four aces in a deck. The required probability is then given as

$$ \frac{{}_4C_2}{{}_{52}C_2} = \frac{\dfrac{4!}{2!\,2!}}{\dfrac{52!}{2!\,50!}} = \frac{4 \times 3}{52 \times 51} = \frac{1}{221} $$

EXERCISES

In Exercises 1 through 6 evaluate:

1. $\frac{7!}{4!\,3!}$ **2.** $\frac{8!}{4!\,4!}$ **3.** $\frac{10!}{3!\,7!}$

4. ${}_9C_5$ **5.** ${}_{11}C_3$ **6.** ${}_{15}C_4$

7. List the ${}_3P_2$ permutations of the elements of the set $\{a, b, c\}$. Then identify the permutations for each combination and find ${}_3C_2$.

8. List the ${}_4P_3$ permutations of the elements of the set $\{p, q, r, s\}$. Then identify the permutations for each combination and find ${}_4C_3$.

9. List the elements of each combination of ${}_4C_3$ for the set $\{w, x, y, z\}$. Then match each of these combinations with a combination of ${}_4C_1$ and thereby illustrate the fact that ${}_4C_3 = {}_4C_1$.

10. Find a formula for ${}_nC_n$ for any positive integer n.

11. Find, and give an interpretation of, ${}_nC_0$ for any positive integer n.

12. Evaluate ${}_3C_0$, ${}_3C_1$, ${}_3C_2$, ${}_3C_3$ and check that the sum of these combinations is 2^3, the number of possible subsets that can be formed from a set of three elements.

13. Evaluate ${}_5C_0$, ${}_5C_1$, ${}_5C_2$, ${}_5C_3$, ${}_5C_4$, ${}_5C_5$ and check the sum as in Exercise 12.

14. Use the results obtained in Exercises 12 and 13 and conjecture a formula for any positive integer n for

$${}_nC_0 + {}_nC_1 + {}_nC_2 + {}_nC_3 + \cdots + {}_nC_{n-1} + {}_nC_n$$

15. How many sums of money (include the case of no money) can be selected from a set of coins consisting of a penny, a nickel, a dime, a quarter, and a half-dollar?

16. A man has a penny, a nickel, a dime, a quarter, and a half-dollar in his pocket. How many different amounts can he leave as a tip if he wishes to use exactly two coins?

17. A class consists of 8 boys and 12 girls. How many different committees of four can be selected from the class if each committee is to consist of two boys and two girls?

18. How many different hands of 13 cards each can be selected from a bridge deck of 52 cards?

19. How many choices of three books to read can be made from a set of seven books?

20. Explain why a so-called "combination lock" should really be called a permutation lock.

21. Urn A contains five balls and urn B contains ten balls. How many different selections of ten balls each can be made if three balls are to be selected from urn A and seven from urn B?

22. An urn contains three black balls and three white balls. (a) How many different selections of four balls each can be made from this urn? (b) How many of these selections will include exactly three black balls?

sec. 12-7 combinations

23. Repeat Exercise 22 for an urn that contains ten black balls and five white balls.

24. Three arbiters are to be chosen by lot from a panel of ten. What is the probability that a certain individual on the panel will be one of those chosen?

25. Five cards are drawn at random from an ordinary bridge deck of 52 cards. Find the probability that all four aces are among the cards drawn.

State whether each question involves a permutation, a combination, or a permutation and a combination. Then answer each question.

26. The twenty members of the Rochester Tennis Club are to play on a certain Saturday evening. How many different pairs can be selected for playing "singles"?

27. How many different selections of a set of four records can be made by a disc jockey who has ten records?

28. In how many ways can seven people line up at a single theater ticket window?

29. How many lines are determined by ten points if no three points are collinear?

30. How many selections of nine students to play baseball can be made from a class of nine students?

31. How many different hands can be dealt from a deck of 52 bridge cards if each hand contains (a) 4 cards? (b) 7 cards?

32. A class is to be divided into two committees of at least one student each. How many different pairs of committees are possible from a class of eight students?

*33. How many different pairs of committees of four students each can be formed from a class of eight students?

PEDAGOGICAL EXPLORATIONS

1. Suppose that the students in your class are to play games in which only two people participate, for example, checkers. How many different sets of two students are possible (how many different pairings of students are possible) if your class has
   (a) 8 members? (b) 12 members?
   (c) 16 members? (d) 20 members?
2. Repeat Exploration 1 for games in which four students play.

Recall that the notation ${}_nC_r$ may also be written in the form $\binom{n}{r}$. Consider the following array for sets of n elements where $n = 1, 2, 3, \ldots$:

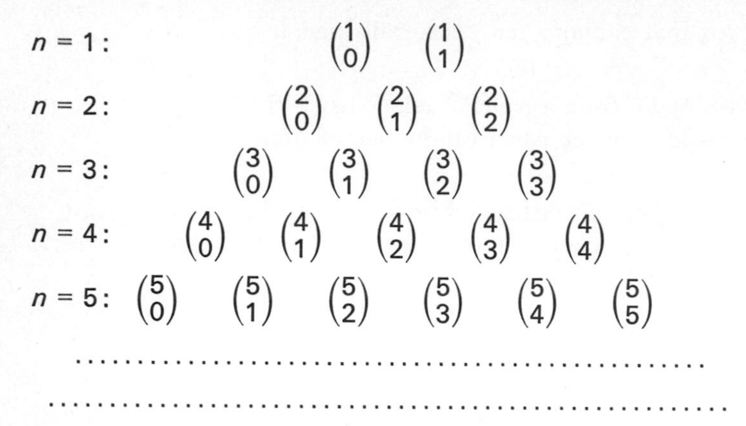

If we replace each symbol by its equivalent number, we may write the following array, known as **Pascal's triangle**. Generally ascribed to the French mathematician Blaise Pascal (1623–1662), this array of numbers is said to have been known to the Chinese in the early fourteenth century.

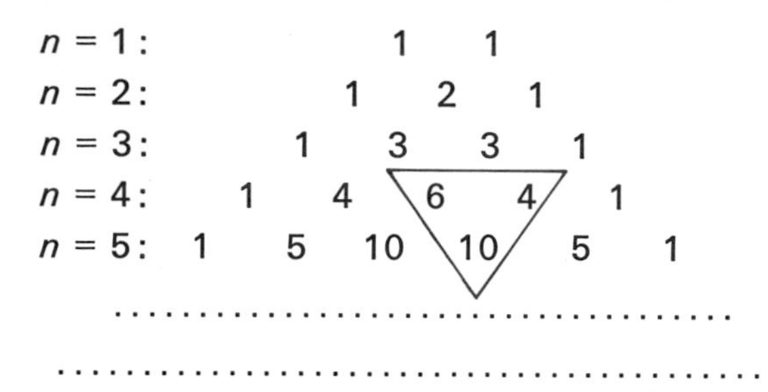

We read each row in this array by noting that the first entry in the nth row is $\binom{n}{0}$, the second is $\binom{n}{1}$, the third is $\binom{n}{2}$, and so on until the last entry, which is $\binom{n}{n}$. Since $\binom{n}{0} = \binom{n}{n} = 1$, each row begins and ends with 1.

There is a simple way to continue the array with very little computation. In each row the first number is 1 and the last number is 1. Each of the other numbers may be obtained as the sum of the two numbers appearing in the preceding row to the right and left of the position to be filled. Thus, to obtain the sixth row, begin with 1. Then fill the next position by adding 1 and 5 from the fifth row. Then add 5 and 10 to obtain 15, add 10 and 10 to obtain 20, and so forth as in this diagram:

$n = 5$: 1 5 10 10 5 1

$n = 6$: 1 6 15 20 15 6 1

Pascal's triangle may be used in a routine manner to compute probabilities as follows. The elements of the second row are the numerators for the probabilities when two coins are tossed; the elements of the third row are the numerators when three coins are tossed; and so on. The denominator in each case is found as the sum of the elements in the row used. For example, when three coins are tossed, we examine the third row (1, 3, 3, 1). The sum is 8. The probabilities of 0, 1, 2, and 3 heads are then given as

$$\frac{1}{8}, \quad \frac{3}{8}, \quad \frac{3}{8}, \quad \frac{1}{8}$$

Note that the sum of the entries in the second row is 4, the sum in the third row is 2^3 or 8, the sum in the fourth row is 2^4 or 16, and, in general, the sum in the nth row will be 2^n.

3. Use the fourth row of Pascal's triangle to find the probability of 0, 1, 2, 3, 4 heads in a single toss of four coins.
4. Construct Pascal's triangle for $n = 1, 2, 3, \ldots, 10$.
5. List the entries in the eleventh row of Pascal's triangle.
6. Repeat Exploration 5 for the twelfth row.
7. See how many different patterns of numbers you can find in Pascal's triangle. For example, do you see the sequence 1, 2, 3, 4, 5, . . .?

CHAPTER TEST

1. Evaluate: (a) $\frac{17!}{13!}$ (b) $\frac{101!}{99!}$
2. Evaluate: (a) ${}_{20}P_2$ (b) ${}_{20}C_2$
3. Consider the set $S = \{c, o, w\}$ and make a tree diagram to show the permutations of the letters of S taken two at a time.
4. How many different four-letter "words" can be formed from the set $B = \{p, e, n, c, i, l\}$ if no letter may be used more than once.
5. Repeat Exercise 4 if repetitions of letters are permitted.
6. A single card is selected from a deck of 52 bridge cards. What is the probability that the selected card is **(a)** not a spade? **(b)** a spade or a heart?
7. A certain florist has five different kinds of roses. How many different selections of three different kinds can be made?
8. A man has a nickel, a dime, a quarter, and a half-dollar. If he leaves a tip of exactly two coins, how many values are possible for such a tip?

9. Use a sample space to represent the possible outcomes described in Exercise 8 and find the probability that the tip is at least 30 cents when the selection of the two coins is at random.

10. What is the probability of tossing on a single toss of two dice (**a**) a 7? (**b**) a number greater than 7?

11. Two integers between 0 and 10 are selected at random. What is the probability that at least one of the numbers is odd?

12. Make a sample space for a toss of a coin followed by the toss of a single die.

A single card is drawn from an ordinary deck of 52 bridge cards. Find the probability that the card selected is:

13. (**a**) The ace of spades. (**b**) An ace or a spade.

14. (**a**) An ace or a red card. (**b**) An ace and a red card.

In Exercises 15 and 16 two cards are selected from an ordinary deck of 52 bridge cards. The first card is not replaced before the second is selected. Find the probability that the cards selected are:

15. (**a**) Both black cards. (**b**) Both black cards or both red.

16. (**a**) Both kings. (**b**) Both kings or both queens.

17. What are the odds in favor of obtaining an ace when one card is drawn from an ordinary deck of 52 bridge cards?

18. What is your mathematical expectation when you are given three tickets out of a total of 6000 tickets for a single prize worth $1000.00?

19. Four coins are tossed.
(**a**) What is the expected number of heads?
(**b**) What are the odds in favor of at least one head?

20. A bag contains five red balls and three green balls. Two balls are drawn in succession without replacement. Find the probability that (**a**) both balls are red; (**b**) the first ball is red and the second one is green.

21. Assume that n is greater than or equal to 5 and give an expression for:
(**a**) ${}_nP_5$ (**b**) ${}_nC_5$

22. Find and simplify a formula for:
(**a**) ${}_nP_0$ (**b**) $\frac{{}_nC_r}{{}_nP_r}$

23. Three coaches are to be selected by lot from a panel of ten for a particular game. What is the probability that a particular member of the panel will not be chosen?

24. Nine people draw lots for their positions (1 through 9) in a line for concert tickets. How many outcomes are possible?

25. A class of ten students is divided into pairs of committees of at least one student each. How many such pairs of committees are possible?

READINGS AND PROJECTS

1. Read, and write a summary of, Chapter 8 (Probability in the Elementary School) of the twenty-seventh yearbook of the National Council of Teachers of Mathematics.
2. Read Chapter 5 (Probability) of the twenty-fourth yearbook of the National Council of Teachers of Mathematics.
3. Read "Pascal's Triangle" by Nathan Hoffman on pages 190 through 198 of the March 1974 issue of *The Arithmetic Teacher* and the comments of W. A. Ewbank on page 710 of the December 1974 issue.
4. Read as much as feasible of Chapter 5 "Pascal's Triangle" by John D. Neff in *Topics for Mathematics Clubs* published by the National Council of Teachers of Mathematics and Mu Alpha Theta.
5. Discuss the use of concepts considered in the "Ideas" section on pages 686 through 696 of the December 1974 issue of *The Arithmetic Teacher* to introduce probability to elementary school students.
6. Read the section on probability, pages 127 through 134, in *Teaching Mathematics: A Sourcebook of Aids, Activities, and Strategies* by Max A. Sobel and Evan M. Maletsky, Prentice-Hall, Inc., 1975. Then develop a worksheet and plan for using one of the five experiments with a specified elementary school grade.
7. Find and prepare a report on an article concerning the teaching of probability in elementary schools.
8. For a specified elementary school grade prepare a 20-minute lesson based upon about twenty events selected to help students recognize that a few events are certain but many are probable with some uncertainty, usually due to unforeseen circumstances. Include a discussion of the probabilities of some of the events. For example, consider weather forecasts and the probability of the class meeting on the following day.
9. Read about the approximation of π using a probability experiment, usually called **Buffon's needle problem**. Among the many possible references is *Mathematics and the Imagination* by Edward Kasner and James Newman, Simon and Schuster, Inc., 1940. Try to repeat the experiment and report upon your results.
10. Ask each member of a class to select an integer between 1 and 10. Tabulate the number of times that each integer is selected. Discuss the results relative to a recent conjecture by several scholars that 7 is the most frequently selected number under such circumstances.
11. Read "Path Tracing And Vote Counting" by William L. Lepowsky on pages 22 through 26 of the January 1976 issue of *The Mathematics Teacher.* Describe a simple path tracing problem that would be appropriate for presentation to an upper elementary mathematics class.
12. Read "Extrasensory Probability" by John F. Loase on pages 116 through 118 of the February 1976 issue of *The Mathematics Teacher.* Try to reproduce the ESP experiment described in this article and report on your results.

13

An Introduction to Statistics

The average consumer is besieged daily with statistical data that are presented over radio and television, in newspapers, and in various other media. He is urged to buy a certain commodity because of statistical evidence presented to show its superiority over other brands. He is cautioned *not* to consume a particular item because of some other statistical study carried out to show its danger. He is told to watch certain programs, read certain magazines, see certain movies, and eat certain foods because of evidence produced to indicate the desirability of these acts as based upon data gathered concerning the habits of others.

Almost every issue of the daily newspaper presents data in graphical form to help persuade the consumer to follow certain courses of action. Thus we are told what the "average" citizen eats, what he reads, what he earns, and even what he does with his leisure time. Unfortunately, too many of us are impressed by statistical data regardless of their source.

Because of the widespread use of statistics for the consumer, we find that even elementary school children are ready to quote facts that they read or hear, often without real understanding. Unfortunately, this is equally true for the adult. Therefore in this chapter we shall make an effort to acquaint the reader with a sufficient number of basic concepts so that he may better understand and interpret statistical data.

13-1 uses and abuses of statistics

Statistical data and concepts affect each of us and often have a profound influence upon the opportunities available to us. For example, among the items considered when you applied for admission to college were:

> Your rank (first, second, third, . . .) in your high school class.
> Your verbal Scholastic Aptitude Test score.
> Your mathematics Scholastic Aptitude Test score.

In most states you are affected almost daily by sales taxes or restaurant food taxes. Probably you are also already affected by rates of interest whether it is on a mortgage, a loan, or a savings account. Statistical computations determine the cost of your life insurance, health insurance, and automobile insurance—to mention only a few such items.

It has become fashionable, as well as informative and impressive, to support all sorts of predications and assertions with statistical computations. Our main purposes in this chapter are to help you become aware of the extensive use of statistics and to recognize some of its abuses.

Many of the most blatant abuses of statistics occur in advertisements. For example, we frequently read or hear statements such as:

> Brush your teeth with GLUB and you will have fewer cavities.

The basic question that this statement fails to answer is: "Fewer cavities than what?" Regardless of the merits of GLUB, it is probably safe to say the following of every brand of toothpaste: "Brush your teeth with XXXX and you will have fewer cavities than *if you never brush your teeth at all.*" In other words, the example given is misleading in that it implies the superiority of GLUB without presenting sufficient data to warrant comparisons.

Many types of statements must be examined very carefully to separate their statistical implications from the impressions they are intended to make. Consider this statement:

> More people are killed in automobile accidents than in wars. Therefore, it is safer to be on the battlefield.

Although the first statement may be true, the second one does not necessarily follow. Inasmuch as there are generally many more people at home driving cars than there are in combat, the *per cent* of deaths on the battlefield is actually much higher than in automobiles.

Frequently graphs are presented in a style that misleads readers.

> Why not subscribe to MATH magazine? The next graph shows our growth of new subscribers over the past five years.

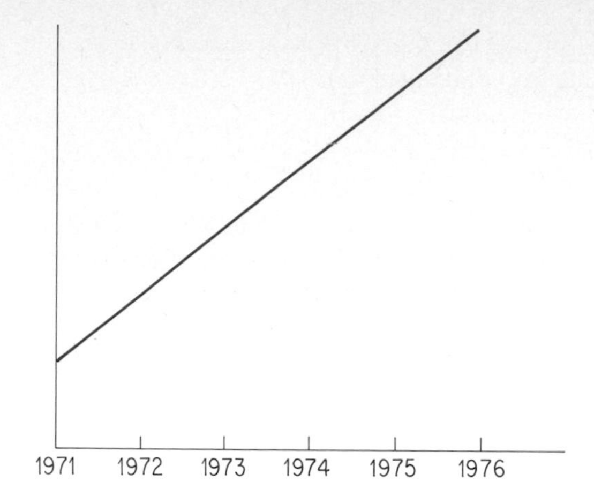

The visual impression given by the graph is that of tremendous growth over a five-year period of time. However without any scale on the vertical axis it is not possible to draw any conclusion other than that there has been a change in the circulation. If the vertical scale is oriented upward and there has been an increase, this increase may be quite small, since we have no assurance that the vertical axis begins with 0. For example, the actual facts may be as in the next graph, where there has been an increase, but only a very modest one, of 400 readers (an increase of 0.4%).

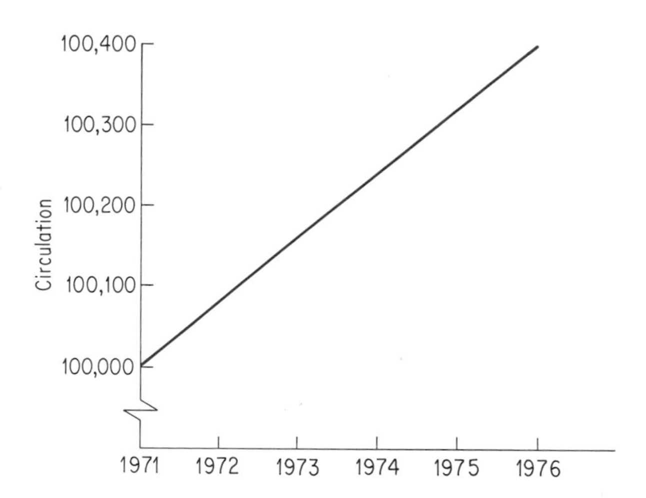

Note that in the preceding graph the scale on the vertical axis did not begin at 0, as indicated by the broken line near the base. The next graph shows how the same data would appear if the vertical scale were set to begin at 0. This time the graph shows the correct picture, namely a very gradual and almost imperceptible increase over the five-year period.

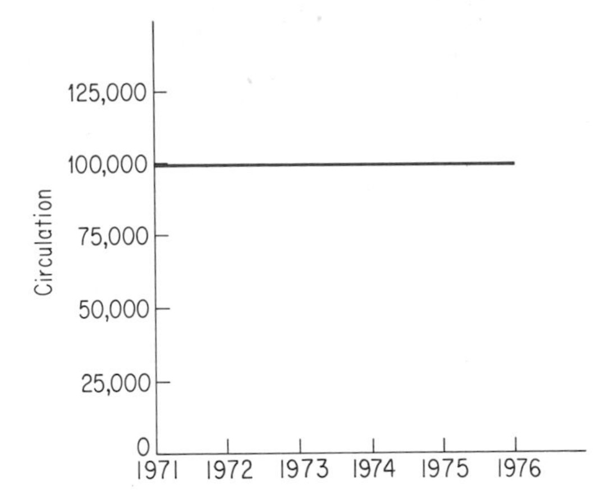

In the following graph, let us assume that the data presented are correct. Note, however, how the picture graph distorts the data by showing the box representing KORNIES to be not only three times as high as the OATIES picture, but also three times as wide. Thus the area of the KORNIES box is nine times as large as the OATIES box, instead of three times as large as indicated by the data. With the numerical facts absent, as is often the case, the picture might present a completely misleading impression.

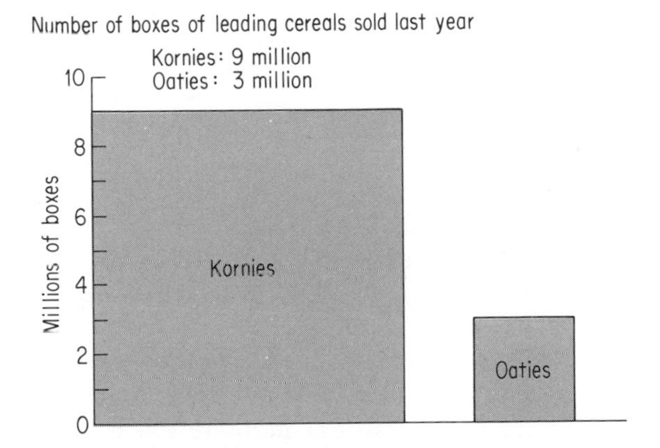

The appearance of a contrast between the sales of KORNIES and OATIES is even more striking when three-dimensional boxes are pictured, as in the graph at the top of the next page. Then the volumes of the boxes are visualized and the volume of the KORNIES box is twenty-seven times that of the OATIES box.

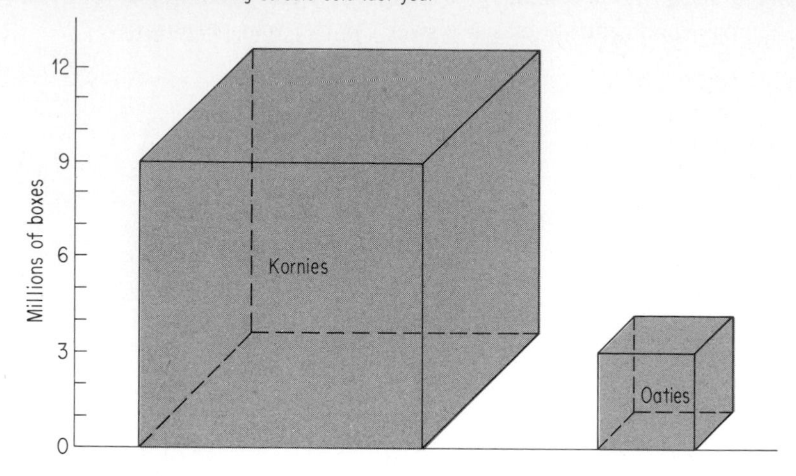

The collecting of actual examples of uses of statistics to mislead readers can be a very interesting activity. In particular, an analysis of advertising claims can be rewarding in the insight it brings to an understanding of basic statistical concepts.

The intelligent consumer needs to ask basic questions about statistical statements that one hears or reads. Unfortunately too many people tend to accept as true any statement that contains numerical data, probably because such statements sound impressive and authentic. Consider, for example, this statement:

> Two out of three adults no longer smoke.

Before accepting and repeating such a statement, one should ask such questions as the following:

> Does the statement sound reasonable?
> What is the source of these data?
> Can the facts be verified?
> Do the facts appear to conform to your own observations?

The reader will undoubtedly think of other questions to raise. The important thing is that one raise these questions and does not accept all statistical facts as true in general. Furthermore this is not to imply that all such facts are biased or untrue. Rather, as we have seen and shall see later in this chapter, even true statements can be presented in many ways so as to provide various impressions and interpretations.

One should examine the logical structure of each statement; for example, is the statement or its converse suggested to the reader? One should also examine each statement to search for ambiguous words or meaningless comparisons. Thus, consider this assertion:

If you take vitamin C daily, your health will be better.

In addition to the questions one might raise about the scientific basis for this statement, note that the word "better" is ambiguous. We need to ask the question: "Better than what?" That is, we need to know the basis for comparison. As an extreme case, we certainly can say that your health will be better than the health of one who is desperately ill! The exercises that follow will provide the reader with an opportunity to test powers of reasoning and questioning.

EXERCISES

Discuss each of the following examples and tell what possible misuse or misinterpretation of statistics each one involves.

1. Most automobile accidents occur near home. Therefore one is safer taking long trips than short ones.
2. Over 95% of the doctors interviewed endorse SMOOTHIES as a safe cigarette to smoke. Therefore it is safe to smoke SMOOTHIES.
3. More college students are now studying mathematics than ever before. Therefore mathematics must be a very popular subject.
4. Professor X gave out more A's last semester than Professor Y. Therefore, one should try to enroll in X's class next semester rather than in Y's class.
5. At a certain college 100% of the students bought *Contemporary Mathematics* last year. Therefore this must be a very popular book at that school.
6. Most accidents occur in the home. Therefore it is safer to be out of the house as much as possible.
7. Over 75% of the people surveyed favor the Democratic candidate. Therefore he is almost certain to win the election.
8. A psychologist reported that the American girl kisses an average of 79 men before getting married.
9. A magazine reported the presence of 200,000 stray cats in New York City.
10. A recent report cited the statistic of the presence of 9,000,000 rats in New York City.
11. Arizona has the highest death rate in the nation for asthma. Therefore if you have asthma, you should not go to Arizona.

12. During World War II more people died at home than on the battlefield. Therefore it was safer to be on the battlefield than to be at home.

13. In a pre-election poll, 60% of the people interviewed were registered Democrats. Thus the Democratic candidate will surely win the election.

14. Over 90% of the passengers who fly to a certain city do so with Airline X. It follows that most people prefer Airline X to other airlines.

15. A newspaper exposé reported that 50% of the children in the local school system were below average in arithmetic skills.

Assume that each of the following statements are made. For each one, list two or three questions that you would wish to raise before accepting the statement as true. Identify words, if any, that are ambiguous or misleading.

16. Most people can swim.

17. Short men are more aggressive than tall men.

18. If you walk three miles a day you will live longer.

19. Brand A aspirin is twice as effective as Brand B.

20. Teen-agers with long hair are happier individuals.

21. People who swim have fewer heart attacks.

22. If you sleep 8 hours the night before an exam, you will do better.

23. Most students have success with this textbook.

24. Teachers tend to be less conscious about social ills.

25. About 3 out of 4 college students marry within one year of graduation.

Use the following graph for Exercises 26 through 27.

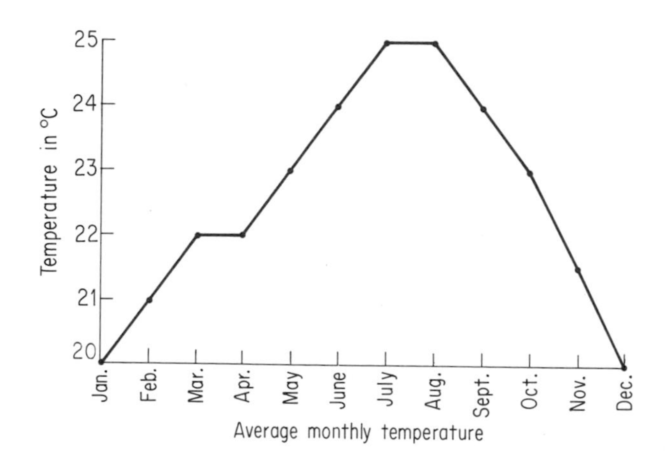

26. Disregarding the scale, what visual impression do you get from this graph about the fluctuation of temperatures during the year?

27. In what ways, if any, is the graph misleading?

Use the following bar graph to answer Exercises 28 through 30.

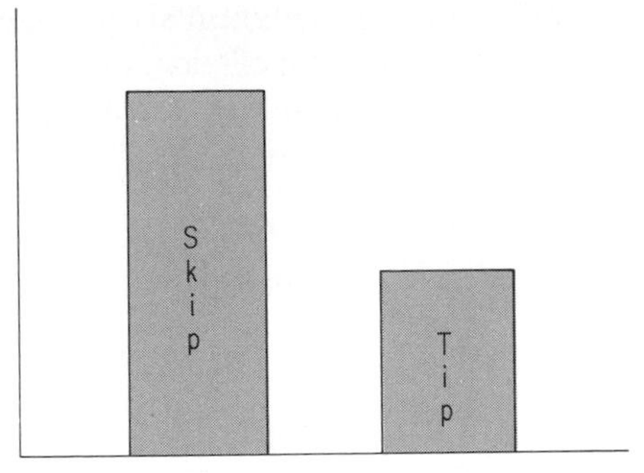

Circulation of two leading magazines

28. Can one deduce from the graph that the circulation of SKIP is greater than the circulation of TIP?

29. Can one deduce from the graph that the circulation of SKIP is approximately twice as great as the circulation of TIP?

30. What, if anything, could possibly be misleading or deceiving about the graph?

PEDAGOGICAL EXPLORATIONS

1. Read *How to Lie With Statistics* by Darrell Huff, W. W. Norton & Company, Inc., 1954 or *Flaws and Fallacies in Statistical Thinking* by Stephen K. Campbell, Prentice-Hall, Inc., 1974.
2. Begin a collection from newspapers, magazines, and other media of examples of misuses of statistics. In particular find examples that use:
   (a) Scales that have been cut off to mislead the reader.
   (b) Areas or volumes used to exaggerate comparisons.
   (c) Parts of the figure enlarged to provide undue emphasis.
3. Discuss the following classic quotes:
   (a) Facts are facts.
   (b) Figures don't lie but liars figure.
4. Assume that ten per cent of your regular class time for teaching is disrupted by extracurricular activities. Make a graph that presents the data correctly. Then make a graph that might be used to mislead the PTA into thinking that the situation was much worse than it actually was.

13-2 collecting and presenting data

Although most consumers need to be able to interpret rather than to present data, a discussion of methods of presentation should serve to help develop skills of interpretation. As an illustration of such methods, let us assume that four coins have been tossed 32 times and that each time the number of heads showing has been recorded. This **raw data** is listed in the order of the occurrence of the events.

Number of Heads

1	2	3	2	4	2	2	1
0	2	4	2	2	1	3	3
2	3	1	1	2	2	3	0
4	1	3	2	4	2	1	2

The information is somewhat more meaningful when the numbers of heads are tallied and summarized in tabular form as a **frequency distribution.**

Number of heads	*Tally*	*Frequency*
0	II	2
1	~~IIII~~ II	7
2	~~IIII~~ ~~IIII~~ III	13
3	~~IIII~~ I	6
4	IIII	4

This information can now be treated graphically in a number of different ways. One common form of presenting data is by means of a **histogram,** a bar graph without spaces between the bars.

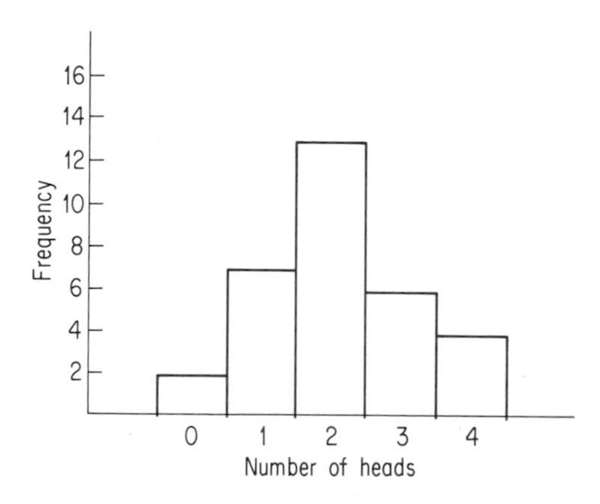

Frequently a histogram is used to construct a **frequency polygon** or **line graph.** Thus the histogram can be approximated by a line graph by connecting the midpoints of the tops of successive bars. The graph is then extended to the base line as in the next figure.

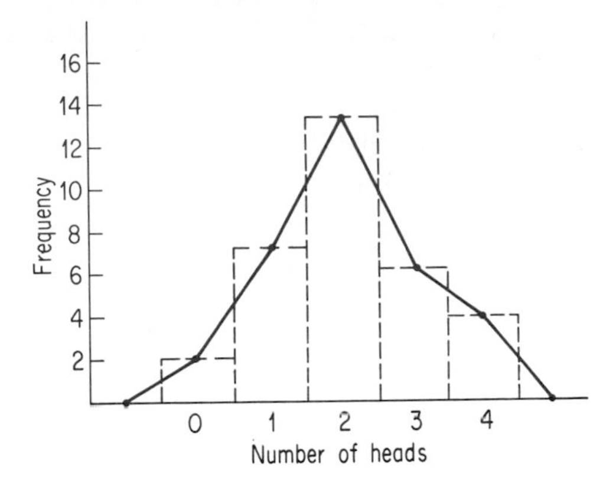

Line graphs appear to associate values with all points on a continuous interval rather than just with points that have integers as coordinates. Thus such line graphs are more appropriate for **continuous data** (all intermediate values have meaning) than for **discrete data** (only isolated values have meaning). For example, consider a graph of the temperatures in a particular city. Although it was 40° at 5 P.M. and 35° at 6 P.M., the temperature must have passed through *all* possible values between 40° and 35° in one hour. This, then, is an example of continuous data. Temperatures may be plotted and connected to produce a line graph as in the figure. Since gradual changes are expected, we "smooth out" the graph as a curve without "sharp turns."

Hour	6 A.M.	7	8	9	10	11	12	1	2	3	4	5	6
Temp.	32	35	35	38	40	42	45	46	45	42	41	40	35

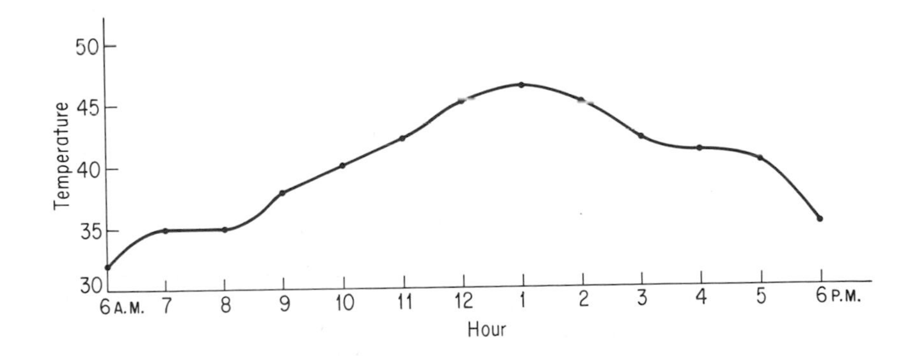

There are many different types of graphs used to present statistical data. One very popular type is the **circle graph**, which is especially effective when one wishes to show how an entire quantity is divided into parts. Local and federal government documents frequently use this type of graph to show the distribution of tax money, and various budget distributions. As an example, consider a family that plans to distribute its income according to the following guide:

Food:	40%
Household:	25%
Recreation:	5%
Savings:	10%
Miscellaneous:	20%
Total:	100%

To draw a circle graph for these data, we first recognize that there are 360° in a circle. Then we find each of the given per cents of 360, and with the aid of a protractor construct central angles of the appropriate sizes as in the figure.

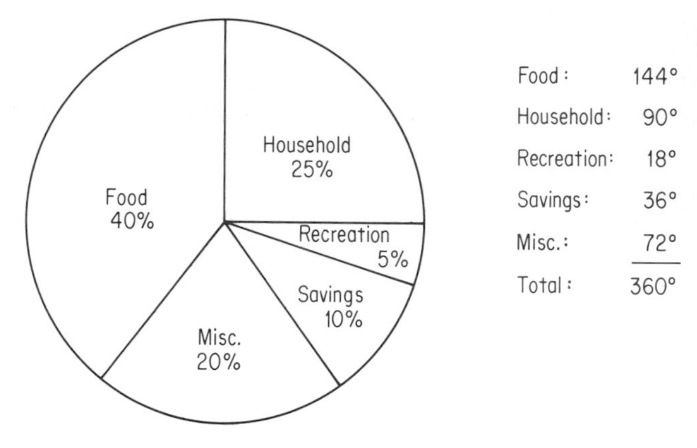

The same data can also be presented in the form of a **divided bar graph.** Here we select an arbitrary unit of length to represent 100%, and divide this in accordance with the given per cents. In practice, even though the rectangular region (bar) has an arbitrary length, it is highly desirable to select that length so that it can be easily subdivided for the desired parts. In the case of the data of our example all parts are integral multiples of 5%; that is, of one-twentieth of the 100% total. Thus it is very convenient to select the length of the bar so that twentieths of that length are easily identified. For example, if the total length were 10 cm, then the lengths of the parts would be 4 cm, 2.5 cm, 0.5 cm, 1 cm, and 2 cm.

Food 40%	Household 25%	Rec. 5%	Savings 10%	Misc. 20%

There is an almost endless supply of graphs that one could use to illustrate this method of presenting data. Indeed, the reader can observe that one is constantly subjected to graphical presentations in the daily newspapers and in other periodicals as well. Other examples of graphs are given in the exercises that follow.

EXERCISES

1. Toss four coins simultaneously and record the number of heads obtained. Repeat this for a total of 32 tosses of four coins. Present your data in the form of a frequency distribution.
2. Present the data for Exercise 1 in the form of a bar graph.
3. Present the data for Exercise 1 in the form of a line graph.
4. Repeat Exercise 1 for 32 tosses of a set of five coins.
5. Present the data for Exercise 4 in the form of a bar graph.
6. Present the data for Exercise 4 in the form of a line graph.
7. Here is the theoretical distribution of heads when four coins are tossed for a total of 64 times. Present these data in the form of a bar graph.

Number of Heads	0	1	2	3	4
Frequency	4	16	24	16	4

8. Here is the theoretical distribution of heads when five coins are tossed for a total of 64 times. Present these data in the form of a bar graph.

Number of Heads	0	1	2	3	4	5
Frequency	2	10	20	20	10	2

9. Construct a circle graph to show this distribution of time spent by one student.

Sleep:	8 hours
School:	6 hours
Homework:	4 hours
Eating:	2 hours
Recreation:	4 hours
Total:	24 hours

10. Construct a divided bar graph for the data of Exercise 9.

Use the following graph to answer Exercises 11 through 17.

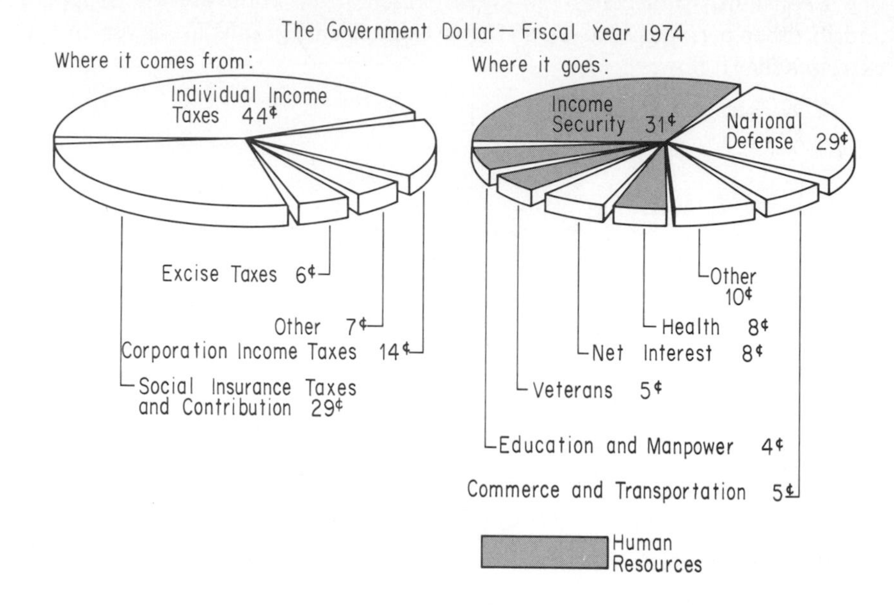

11. What per cent of the government dollar comes from individual income taxes?

12. What per cent of the government dollar goes to national defense?

13. Can you tell from the graphs (**a**) what per cent of the government expenditures were for health matters? (**b**) how many dollars were spent on health matters? Explain your answers.

14. What is the size of the central angle in the graph that shows the amount spent for commerce and transportation?

15. To the nearest degree, what is the size of the central angle that shows the amount obtained from individual income taxes?

16. Of every million dollars collected, how many dollars were collected from excise taxes?

17. Of every million dollars spent, how many dollars were spent for health?

Insurance companies compile data on deaths and use these data to make predictions of life expectancy and to determine insurance rates. Use the two graphs on the next page to answer Exercises 18 through 22.

18. At approximately what age does the largest number of deaths occur?

19. At approximately what ages are there 150,000 deaths?

20. Explain why both graphs show an initial decrease before both start to rise again.

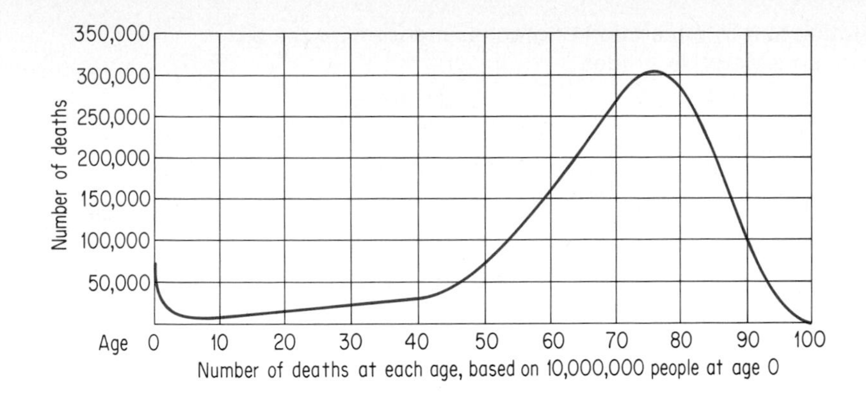

Number of deaths at each age, based on 10,000,000 people at age 0

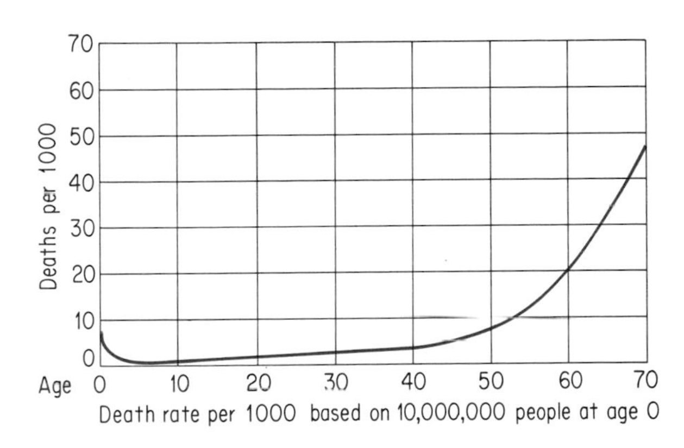

Death rate per 1000 based on 10,000,000 people at age 0

21. Both graphs have the same shape at first. The latter portions, however, are quite different. Explain why this is so.

22. The graph of the death rate per 1000 stops at age 70. What do you think happens to this curve after age 70?

PEDAGOGICAL EXPLORATIONS

1. In one kindergarten room the names of the months of the year were listed across the bottom of a long sheet of paper. Each child put an "X" in a column over the name of the month in which the child's birthday occurred. What kind of graph was being constructed? What are some of the questions that you would raise with the children after the graph was completed?
2. Describe several other primary school activities that result in graphs.

In Explorations 3, 4, and 5 study one or more recently published series of elementary school mathematics textbooks, describe the treatments given to the construction and interpretation of graphs, and identify the different types of graphs considered in grades:

3. K through 3.
4. 4 and 5.
5. 6 through 8.
6. Start a collection of graphs from newspapers, magazines, and other sources. Be sure to include sufficient descriptive material to enable you to use the graphs in an elementary school class. Also see how many different types of graphs you can find.

13-3 measures of central tendency

Most of us have neither the ability nor the desire to digest large quantities of statistical data. Rather, we prefer to see a graphical presentation of such data, or some figure cited as representative of the entire collection. Thus we are often faced by such statements as:

"On the average, 9 out of 10 doctors recommend H_2O."
"The average family earns $14,580 per year and has 1.8 children."
"The average college teacher gives out 21% A's each year."

The word "average" is used loosely and may have a number of different meanings. However, in each case it is used in an effort to provide a capsule summary of a collection of data by means of a single number.

Assume that the following set of data represents the number of dates that each of the ten girls in Alpha Beta Gamma sorority had last semester:

6, 7, 8, 9, 10, 12, 15, 15, 20, 28

Then each of the following statements is correct for the ten girls of this particular sorority:

1. *The average sorority member had 13 dates last semester.*

In this case the average is computed by finding the sum of the given numbers and dividing the sum by 10, the number of girls. This is the most commonly used type of average, is referred to as the **arithmetic mean,** or simply the **mean,** of a set of data, and is generally denoted by the Greek letter μ (read as "mu").

6
7
8
9
10
12
15
15
20
28
130

$$\mu = \frac{130}{10} = 13$$

Arithmetic mean: 13

2. *The average sorority member had 11 dates last semester.*

In this case the word *average* is used to denote the number that divides the data so that half of the scores are above this number and half are below.

$$\left.\begin{matrix}6\\7\\8\\9\\10\end{matrix}\right\}\ \text{Five scores below 11}$$

— Median: 11

$$\left.\begin{matrix}12\\15\\15\\20\\28\end{matrix}\right\}\ \text{Five scores above 11}$$

The number that divides a set of scores in this way is called the **median** of the set of data. In this case it is determined as the number midway between 10 and 12. Note that the median is not necessarily one of the scores and that the data must be considered in order of size before the median can be found.

3. *The average sorority member had 15 dates last semester.*

In this case the average is the number that appears most frequently in the set of data. That is, it is correct to say that more of these girls had 15 dates than any other number of dates. Such an average is referred to as the **mode** of a set of data. Obviously, in this case, the mode is the best average to use if the sorority wishes to impress others with the social success of its members. Actually, however, the median is more representative of the number of dates of the typical member of the sorority.

Each of the three types of averages that we have discussed is known as a **measure of central tendency.** That is, each average is an attempt to describe a set of data by means of a single representative number. Note, however, that not every distribution has a mode, whereas some may have more than one mode. The set of scores {8, 12, 15, 17, 20} has no mode. The set {8, 10, 10, 12, 15, 15, 17} is **bimodal** and has both 10 and 15 as modes.

Example 1: Find (a) the mean; (b) the median; and (c) the mode for this set of test scores:

72, 80, 80, 82, 88, 90, 96

Solution: (a) The sum of the seven scores is 588.

$$\mu = \tfrac{588}{7} = 84$$

(b) The median is 82. This is the middle score; there are three scores below 82 and three above it.

(c) The mode is 80, since this score appears more frequently than any other one.

Example 2: Repeat Example 1 for these scores:

30, 80, 80, 82, 88, 90, 96

Solution: (a) $\mu = \frac{546}{7} = 78.$
(b) The median is still 82.
(c) The mode is still 80.

A comparison of Examples 1 and 2 indicates that the test scores are the same except for the first one. Note that the arithmetic mean is the only one of the three averages that is affected by the one low score. In general, the mean is affected by extreme scores, whereas the median and mode are not. Thus the arithmetic mean should be used as a representative of a set of data when extreme (high or low) scores should be reflected in the average. Otherwise, the arithmetic mean should not be used to describe a set of data that contains extreme scores unless deception is one's major objective.

Consider this example of the earnings of the employees of a small business run by a foreman and three other employees. The foreman earns \$19,000 per year. The others earn \$4800, \$4500, and \$4500, respectively. To impress the union, the owner claims that he pays his employees an average salary of \$8200. He selects the arithmetic mean as the representative salary.

$$19{,}000 + 4800 + 4500 + 4500 = 32{,}800$$
$$32{,}800 \div 4 = 8200$$

Actually, the median, \$4650, would present a fairer picture of the average, or typical, salary.

As a more extreme example, consider a group of 49 people with a mean income of \$5000. Let us see what happens to the average income of the group when an additional person with an income of \$100,000 joins the group.

$$\begin{aligned} 49 \times 5000 &= 245{,}000 \quad &\text{(total income of the 49 people)} \\ 1 \times 100{,}000 &= 100{,}000 \\ \text{Sum} &= 345{,}000 \quad &\text{(total income of the 50 people)} \end{aligned}$$

$$\mu = \frac{345{,}000}{50} = 6900$$

The average (arithmetic mean) salary is now \$6900. The addition of this one extreme score has raised the mean \$1900, whereas the median and the mode of the incomes have not been affected.

In reading or hearing advertisements, the consumer must always raise the question of the type of average being used. Furthermore, one should question the source and plausibility of the data. For example, one might wonder how a psychologist can say that the "average" American girl kisses 79 men before getting married! Another important consideration is whether the entire population is represented in the data or only a sample. If only a sample is involved, the size of the sample and the manner in which it was selected need to be considered.

EXERCISES

Find (a) the mean; (b) the median; and (c) the mode for each set of data.

1. 60, 61, 65, 65, 70, 73, 73, 79, 84

2. 73, 79, 80, 82, 84, 84, 92

3. 10, 15, 18, 19, 21, 24, 26, 27

4. 10, 11, 14, 14, 14, 17, 18

5. 85, 61, 68, 73, 91, 68, 93

In Exercises 6 through 8 tell which one, if any, of the three measures of central tendency seems most appropriate to represent the data that is described.

6. The average salary in a shop staffed by the owner and five employees.

7. The average salary of the workers in a factory that employs 100 people.

8. The average number of cups of coffee ordered by individual diners in a restaurant.

9. What relationship docs the mode have to the use of this word in everyday language?

10. The mean score on a set of 15 tests is 75. What is the sum of the 15 test scores?

11. The mean score on a set of 30 tests is 84. What is the sum of the test scores?

12. The mean score on a set of 40 tests is 86. What is the sum of the test scores?

13. The mean score on nine of a set of ten tests is 70. The tenth score is 50. What is the sum of the test scores?

14. The mean score on 25 of a set of 27 tests is 80. The other two scores are 30 and 35. What is the sum of the scores?

15. Two sections of a course took the same test. In one section the mean score on the 25 tests was 80. In the other section the mean score on the 20 tests was 75. **(a)** What is the sum of the 45 test scores? **(b)** To the nearest integer what is the mean of the 45 test scores?

16. The mean score on five tests is 85. The mean on three other tests is 78. What is the mean score on all eight tests?

17. A student has a mean score of 85 on seven tests taken to date. What score must the student achieve on the eighth test in order to have a mean score of 90 on all eight tests? Comment on your answer.

***18.** An interesting property of the arithmetic mean is that the sum of the deviations (considered as signed numbers) of each score from the mean is 0. Show that this is true for the set of scores: 8, 10, 13, 17, 22.

PEDAGOGICAL EXPLORATIONS

The term **percentile rank** is frequently found in educational literature, especially as it applies to scores on standardized tests. For example, when we say that someone's test score has a percentile rank of 80 we mean that 80% of the scores on this test fall at or below this particular grade. Of course this also means that 20% of all the scores lie above the given grade. A score with a percentile rank of 80 is also said to be at the eightieth percentile.

1. What is the percentile rank of the median of a distribution?
2. A student scores a grade of 127 on an aptitude test. What can you say about his aptitude as a result of this test score? Suppose you are then given the additional information that the score of 127 is at the ninety-fifth percentile in a distribution of scores for all students who have taken this particular test. What can you conclude then about his aptitude?
3. Why does percentile rank appear to be more significant than rank (first, second, third, . . .) in a class?

Describe several situations in which the most informative measure of central tendency appears to be:

4. The arithmetic mean.
5. The median.
6. The mode.

13-4 measures of dispersion

A measure of central tendency describes a set of data through the use of a single number. However, as in the examples considered in §13-3, a single number without other information can be misleading. Some information can be obtained by comparing the arithmetic mean and the median, since the mean is affected by extreme scores and the median is relatively stable. In this section we consider other ways of obtaining information about sets of data without considering all of the elements of the data individually.

A **measure of dispersion** is a number that provides some information on the variability of a set of data. The simplest such measure to use for a set of numbers is the **range,** the difference between the largest and the smallest number in the set.

Consider the test scores of these two students

$$\begin{array}{lll} \text{Betty:} & \{68, 69, 70, 71, 72\}, & \mu = 70 \\ \text{Jane:} & \{40, 42, 70, 98, 100\}, & \mu = 70 \end{array}$$

The range for Betty's test scores is 72 — 68, or 4; the range for Jane's test scores is 100 — 40, or 60. The averages and ranges provide a more meaningful picture of the two students' test results than the averages alone:

$$\begin{array}{lll} \text{Betty:} & \mu = 70, & \text{range, 4} \\ \text{Jane:} & \mu = 70, & \text{range, 60} \end{array}$$

Although the range is an easy measure of dispersion to use, it has the disadvantage of relying on only two extreme scores—the lowest and the highest scores in a distribution. For example, consider Bob's test scores:

$$\{40, 90, 95, 95, 100\}$$

Bob is quite consistent in his work, and there may well be some good explanation for his one low grade. However, in summarizing his test scores one would report a range of 60, since $100 - 40 = 60$.

There is another measure of dispersion that is widely used in describing statistical data. This measure is known as the **standard deviation,** and is denoted by the Greek letter σ (sigma). The standard deviation is relatively difficult to compute but can be very informative. Fortunately, the average consumer of statistics needs to be able to understand and interpret, rather than to compute standard deviations. This need is particularly acute for teachers since they need to interpret the performance of their students on standardized tests.

To understand the significance of standard deviations as measures of dispersion, we first turn our attention to a discussion of **normal distributions.** The line graphs of these distributions are the familiar bell-shaped curves that are used to describe distributions for so many physical phenomena. For example, the distribution of intelligence quotient (IQ) scores in the entire population of the United States can be pictured by a **normal curve.** The area under the curve represents the entire population.

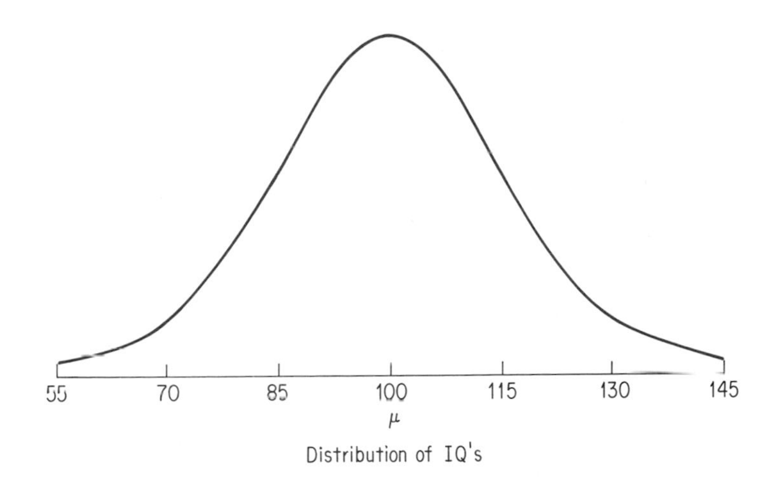

Distribution of IQ's

According to psychologists, a score over 130 is considered to be very superior. At the other end of the scale, a score below 70 generally indicates some degree of academic retardation.

A normal-appearing curve should be expected only for data that involve a large number of elements and are based upon a general population. For example, the IQ scores of the honor society members in a school should not be expected to fit the distribution for the school as a whole.

In a normal distribution the mean, the median, and the mode all have the same value. This common value is associated with the axis of symmetry of the normal curve.

If three standard deviations are added to and subtracted from the mean of a normal distribution, practically all (99.7%) of the data will fall on the interval from $\mu - 3\sigma$ to $\mu + 3\sigma$. If an interval of two standard deviations from the mean is considered, approximately 95% of all data is included. An interval of one standard deviation about the mean includes approximately 68% of all data in a normal distribution. We may summarize these statements as follows:

$$\mu \pm 1\sigma \approx 68\%$$
$$\mu \pm 2\sigma \approx 95\%$$
$$\mu \pm 3\sigma \approx 100\%$$

and as in the following figure:

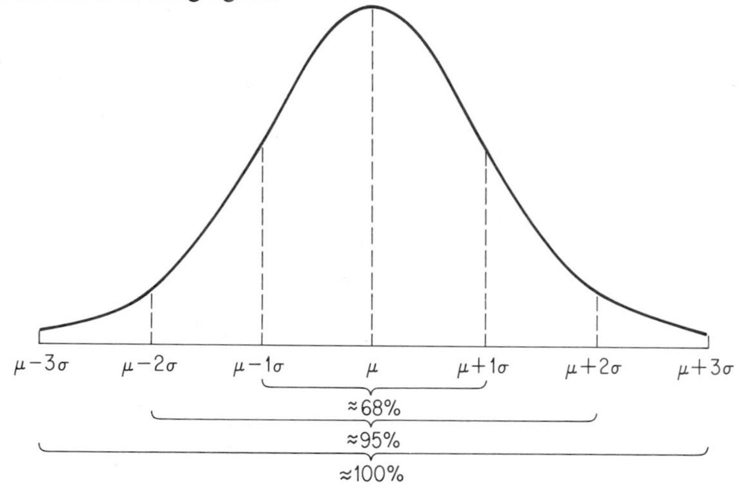

Let us now return to the graph of IQ scores. Suppose we are told that for this distribution $\sigma = 15$. We may then show these standard deviations on the base line of the graph. According to our prior discussion, we may now say that approximately 68% of the population have IQ scores between 85

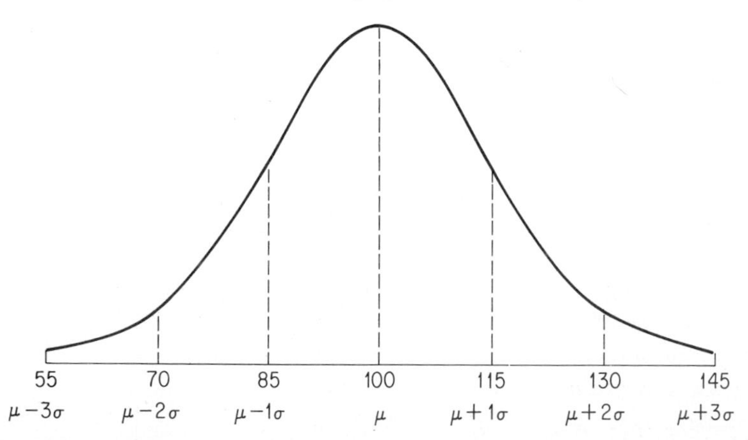

and 115, that is, on the interval $100 \pm 1\sigma$. Approximately 95% of the population have IQ scores between 70 and 130, that is, on the interval $100 \pm 2\sigma$. Finally, almost everyone has an IQ score between 55 and 145, that is, on the interval $100 \pm 3\sigma$. "Almost everyone" really means 99.7%; that is, we might expect 0.3% (3 in 1000) of the population to have scores below 55 or above 145.

Example 1: What per cent of the population have IQ scores less than 115?

Solution: We know that 50% of the population have IQ scores below 100, the mean. Furthermore, 68% have scores on the interval $\mu \pm 1\sigma$. Then by the symmetry of the normal curve, 34% have scores between 100 and 115, and, as in the figure, 84% have scores less than 115.

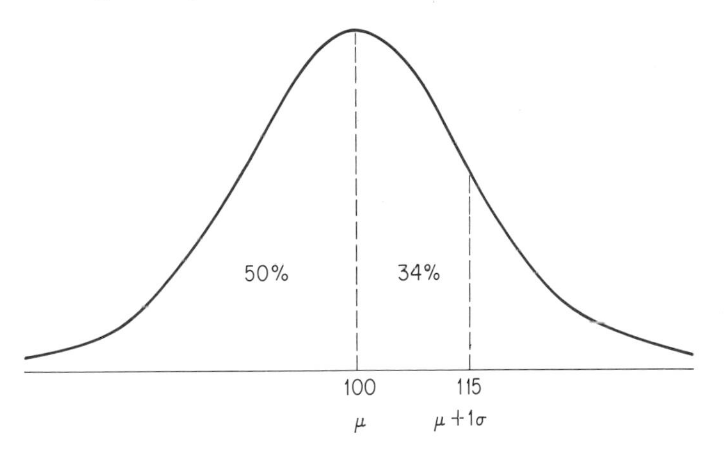

Note that when considering questions such as those in Example 1 we do not concern ourselves with the number of people who make any particular score such as 100 or 115. The scale −3, −2, −1, 0, 1, 2, 3 in standard deviations enables us to use the properties of a normal distribution, such as the 68% within one standard deviation of the mean, for a wide variety of situations. For example, we found in Example 1 that 84% of the population have IQ scores less than 115, that is, less than one standard deviation above the mean. This same result can be used in many other situations. Scholastic Aptitude Test (SAT) scores have a mean of 500 and a standard deviation of 100. Thus 84% of the SAT scores on a particular test are below 600. College Entrance Examination Board (CEEB) scores also have a mean of 500, a standard deviation of 100, and 84% of the scores below 600. If 100,000 students took the test, it is assumed that 84,000 of them had scores below 600 and 16,000 had scores above 600, even though it is known that some students had scores of 600. If a student had a score of 600, that student's test score is at the 84th percentile and has a percentile rank of 84 (see the Pedagogical Explorations of the previous section).

Example 2: What per cent of the population have IQ scores greater than 130?

Solution: The interval $\mu \pm 2\sigma$ includes the scores of 95% of the population, so 47.5% of the population have scores on the interval from 100 to 130. Thus 97.5% of the population have scores below 130, and 2.5% of the population have IQ scores above 130.

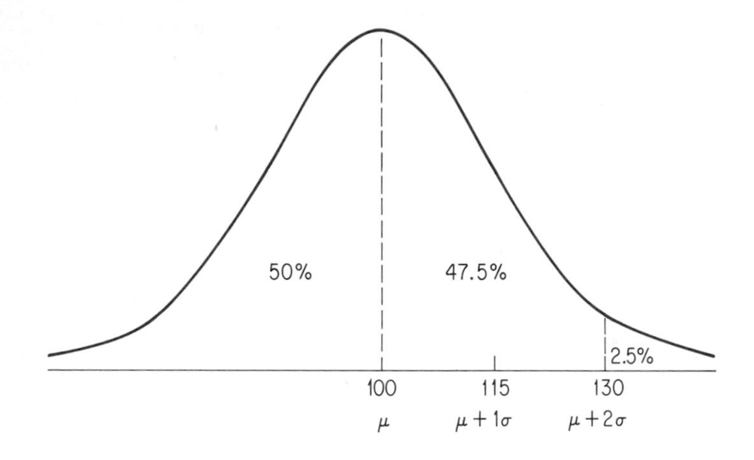

The result obtained in Example 2 as applied to SAT and CEEB scores shows that 97.5% of the scores are below 700 and 2.5% of the scores are over 700.

Note that we have discussed only normal distributions. However, the standard deviation can be computed for other distributions as well. In general, many sets of data tend to be approximately normal, and the standard deviation is a very useful measure of dispersion.

As stated earlier, the standard deviation is relatively cumbersome to compute, especially for data grouped in a table. However, it may be instructive to note the computation of σ for a small set of scores. For such data, we generally use the formula

$$\sigma = \sqrt{\frac{\Sigma d^2}{n}}$$

where Σd^2 represents the sum of the squares of the deviations of each score from the mean, and n is the number of scores.

Consider the set of scores

$$\{8, 10, 12, 16, 19\}$$

We first compute the mean as 13. In the column headed d we find the differences when the mean μ is subtracted from each score. The sum of these deviations is 0 as noted in §13-3, Exercise 18. Thus to obtain a meaningful average of these deviations we may use either absolute values or the squares of the deviations. The standard deviation is based upon the squares of the deviations, as in the column headed d^2.

Scores	d	d^2
8	-5	25
10	-3	9
12	-1	1
16	3	9
19	6	36
$5\overline{)65}$		80
$\mu = 13$		

$$\sigma = \sqrt{\tfrac{80}{5}} = \sqrt{16}$$
$$\sigma = 4$$

Recall that σ is a measure of dispersion. In Example 3 we have a set of scores with the same mean as the above set, but with a much smaller range. Note that the standard deviation is correspondingly smaller.

Example 3: Compute σ: $\{10, 11, 13, 14, 17\}$.

Solution:

Scores	d	d^2
10	-3	9
11	-2	4
13	0	0
14	1	1
17	4	16
$5\overline{)65}$		30
$\mu = 13$		

$$\sigma = \sqrt{\tfrac{30}{5}} = \sqrt{6}$$
$$\sigma \approx 2.5$$

These two sets of five scores for which standard deviations have been computed are much too small to expect a close match with a normal distribution. In each of these two cases all scores are within one and a half standard deviations of the mean.

EXERCISES

1. Find the range for the set of scores:

$$\{55, 67, 80, 85, 90, 92, 98\}$$

2. Give a set of seven scores with the same mean as in Exercise 1 but with a smaller range.
3. Repeat Exercise 2 for a larger range.

4. Give a set of seven scores with the same range as in Exercise 1 but with a smaller mean.

5. Repeat Exercise 4 for a larger mean.

In Exercises 6 through 8 tell which of the distributions can be expected to be approximately normal in shape.

6. The scores of all graduating high school seniors on a particular college board examination.

7. The weights of all college freshmen boys.

8. The number of heads obtained if 100 coins are tossed by each college graduate in the country.

9. For a normal distribution of 10,000 test scores the mean is found to be 500 and the standard deviation is 100.
   (a) What per cent of the scores will be above 700?
   (b) What per cent of the scores will be below 400?
   (c) About how many scores will be above 600?
   (d) About how many scores will be below 300?
   (e) About how many scores will be between 400 and 700?

10. If 100 coins are tossed repeatedly, the distribution of the number of heads is a normal one with a mean of 50 and a standard deviation of 5. What per cent of the number of heads will be: (a) greater than 60? (b) less than 45? (c) between 40 and 60?

11. Compute the standard deviation for the set of scores

$$\{7, 9, 10, 11, 13\}$$

12. Compute the standard deviation for the set of scores

$$\{11, 12, 13, 15, 20, 20, 21\}$$

PEDAGOGICAL EXPLORATIONS

In reading educational literature one frequently encounters the term **coefficient of correlation**, usually denoted by the letter *r*. The coefficient of correlation is given as a decimal from −1.00 to +1.00 and provides an indication of how two variables are related.

A perfect positive correlation of 1.00 indicates that two sets of data are related so that as one increases, so does the other, and does so uniformly. For example, here is a set of ages and weights for a group of six individuals. Note that for each increase of 5 years of age there is a corresponding increase of 10 pounds of weight.

Age	*Weight*	
20	130	
25	140	
30	150	The coefficient of
35	160	correlation is 1.00.
40	170	
45	180	

Now consider the following table for six other individuals. For each increase of 5 years of age, there is a corresponding decrease of 10 pounds of weight.

Age	*Weight*	
20	180	
25	170	
30	160	The coefficient of
35	150	correlation is −1.00.
40	140	
45	130	

A coefficient of correlation of 0 indicates no uniform change of either variable with respect to the other. In general practice such extreme cases seldom occur. Furthermore, one has to read the literature accompanying any particular test or research study to determine whether any particular coefficient of correlation can be considered as significant.

Finally, it is important to note that correlation does not imply causation. For example, there might be a positive correlation between the size of shoe that an elementary school student wears and the student's handwriting ability. This would not mean that big feet improve one's handwriting.

1. There has been a high positive correlation between expenditures for alcohol and for higher education in recent years. Does this mean that drinking alcohol provides one with the thirst for knowledge? Does it mean that education leads one to drink? How can you explain this high correlation?

Explain why you would expect to find a high or a low correlation between:

2. IQ scores and scores on college entrance examinations of all graduating high school seniors.

3. Scores made by elementary school students in reading and in arithmetic.

4. Age and physical abilities of mentally retarded individuals.

5. Manual dexterity and age of normal elementary school children.

6. Grade in arithmetic and number of hours spent watching television by elementary school children.

7. Weight, relative to normal for their heights, of elementary school children and their mothers.
8. Academic grades and extent of participation in extracurricular activities of college students.
9. Effectiveness in teaching and years of college training of teachers.

13-5 pedagogical uses of statistics

Many states require that all students take certain standardized tests. The test scores help the teachers identify the aptitudes and interests of the students. Then teachers have some objective bases for the informal day-to-day guidance and individual counselling that is an important part of teaching. For example, which students should be highly complimented and which should be challenged to seek a deeper understanding of a subject when a certain behavioral objective has been met? Which students are working close to their peak abilities and which are "coasting" unchallenged and unmotivated while performing well above the class average?

Statistical results must be interpreted very carefully. In the case of the fourth toss of a coin that has shown heads three times in a row, either the probability of tails is still one-half on the fourth toss or the coin has a bias. Similarly, in the case of test scores neither a sequence of successes nor a sequence of failures provides more than a mild indication of the capabilities of the individual. Physical limitations (exhaustion, sickness, allergies), emotional limitations (excitement or distress over past or forthcoming events), and attitudes (expectations, desires, compatibility with other students or the teacher) can drastically affect student performance. Indeed, several studies have shown that students tend to perform at the level that they feel their teacher expects of them. Standardized test scores provide one indication of the reasonableness of certain expectations for individual students.

In appraising the performance of students, numerous factors must be considered in addition to test scores. Is the student maturing more slowly than most of the others? If so, is there a poor skill (such as poor reading), a poor attitude (student or parent), a deep frustration caused by an unrealistic goal (student or parent), or simply a special need for encouragement and recognition? Note that the basic problems are often unrelated to the amount of effort exerted by the student. Thus goading the student to "try harder" may not be the answer.

Statistical data have many constructive pedagogical uses. The effectiveness of these uses depends upon the comprehension of the data by the teacher and the educational philosophy of the school system. However, one fact deserves special recognition: *Standardized test scores are not exact rankings* of the students. Each score indicates that the student's actual position is probably on an interval about that score. The size of the interval will vary from one test to another and for any standardized test should be given in the instructions to the person who is to interpret the test. For example, a difference of test scores such as 598 and 601 in different subject areas is insufficient evidence for any major decisions as to the student's preferences or potential success. (Such a difference was once used by a student to decide upon a major when entering college. This was an outright abuse of the statistics.) A score of 598 might mean that the chances are 2 to 1 that the student's true score lies within the interval from 586 to 610; similarly, for 601 the chances might be 2 to 1 that the true score lies within the interval from 588 to 614. Note that these intervals, even for only 2 to 1 probability of correctness, have ranges of 24 and 26 points respectively. Also the intervals overlap for about two-thirds of their extent. Thus test scores provide one indication but are not exact measures of the student's skills or abilities being tested. A variety of scores and indicators are often needed. Even IQ scores may vary ten, fifteen, or more points for the same person on tests taken as little as two weeks apart. Drastic differences are unusual but do occur.

EXERCISES

In Exercises 1 through 4 discuss each question providing reasons for your explicit answers.

1. Jane has an IQ score of 130 and SAT scores of 610 and 590. Does she appear to be producing up to her potential?
2. Jack has an IQ score of 95 and SAT scores of 450 and 500. Does he appear to be producing up to his potential?
3. A student once remarked to the teasing of one of the "bright boys": "I am one of the people that make it possible for you to be in the upper half of the class." How would you explain the situation to an irate parent who considered the fact that half of your students were below average in arithmetic skills to be sufficient evidence for firing you and finding a "good teacher."
4. In a certain year the median SAT verbal score for students at Salem High School was 495. Was this cause for serious concern?
5. Any distribution of data for which there are only *two* possible events is a **binomial distribution.** The distribution of heads in tossing coins is a binomial distribution since each coin is assumed to have exactly one of two possible outcomes, heads or tails. A "fair" coin can be expected to land heads about half the time. Accordingly, we say that the *probability* of heads is $\frac{1}{2}$ on each toss of the coin; that is, the ratio of the number of heads to the total number of tosses is expected to be about $\frac{1}{2}$. If the experiment of tossing a coin n times is repeated over and over, the expected mean of the distribution of heads on these trials of the experiment is $\frac{1}{2}n$. The result of tossing one coin n times is the same as that of tossing n coins once. For example, if one coin is tossed 100 times or 100

coins are tossed once, we have probability $p = \frac{1}{2}$ and $n = 100$. Therefore, the mean or average number of heads expected is $(\frac{1}{2})(100) = 50$. The standard deviation of a binomial distribution is easily found by the formula

$$\sigma = \sqrt{p \times (1 - p) \times n}$$

where p is the probability of success and n is the number of trials.

Find the standard deviation for the number of heads in the experiment of making 100 tosses of an unbiased coin.

6. The expected arithmetic mean of the success when an experiment with probability of success p is performed n times is pn. Suppose that each member of your class tosses an unbiased coin 100 times and counts the number of heads. **(a)** What is the expected mean of the numbers of heads obtained?

On what interval would you expect **(b)** about two-thirds of the numbers of heads to occur? **(c)** 95%? **(d)** practically all?

7. Suppose that you assigned each member of your class the experiment of tossing an unbiased coin 80 times. Then suppose that they reported the numbers of heads obtained as follows:

Alice 45	Bob 40	Charles 30
Doris 35	Eve 25	Fred 50
Gwen 42	Harry 37	Ike 69

Explain why you would suspect that one or more of the reports reflected a failure to do the experiment or the use of a biased coin.

8. Explain why it appears reasonable to think of:

(a) $\mu - 3\sigma$ and $\mu + 3\sigma$ as limits in which you have almost 100% confidence.

(b) $\mu - 2\sigma$ and $\mu + 2\sigma$ as confidence limits in which you have about 95% confidence.

(c) $\mu - \sigma$ and $\mu + \sigma$ as confidence limits in which you have about 68% confidence.

PEDAGOGICAL EXPLORATIONS

1. Think of the "homeroom duties" of many elementary school teachers, identify at least one circumstance in which data are usually collected, and describe how it is usually represented.

Give at least two examples of types of data that could be collected from a primary grade class for the purpose of making:

2. A bar graph.

3. A divided bar graph.

4. A circle graph.

5. A line graph.

Explain why you would or would not expect a normal distribution of quiz (test) scores:

6. For a class of ten students and on a topic that was covered while two students were absent.

7. For a class of ten students and on an arithmetic skill that has been the basis for considerable classroom practice.

8. For all 500 students entering a junior high school and on counting skills up to one thousand.

9. For all 500 students entering a junior high school and on a variety of arithmetic skills to identify areas which should be given special attention.

10. For all 500 students entering a junior high school and on reading skills.

Select an elementary school grade level and prepare a quiz of ten questions on which you would expect the scores to have:

11. A normal distribution.

12. A distribution with most of the scores clustered at the high end of the scale.

13. A *bimodal distribution*, that is, a distribution with two modes.

Select an elementary school grade level and prepare a worksheet, or set of worksheets, for helping the students understand:

14. Measures of central tendency.

15. Measures of dispersion.

CHAPTER TEST

In Exercises 1 and 2 tell what possible misuse or misinterpretation of statistics each of the given statements involves.

1. Mathville's population increased by 100% last year. This is a greater per cent increase than for any other city in the state. Therefore the population of Mathville must be the largest of any city in that state.

2. Statistics show that college students drink more.

3. Describe two common devices for making graphs misleading to the reader.

4. Make a frequency distribution for the grades of a class with grades: C, A, B, A, B, A, C, B, A, B, B, F, B, B, B, B, B, B, B, B, B, A, C, B, B, C, B, B, A, A, B, D, B, B, A, B, B

Represent in the form specified in Exercises 5, 6, and 7 the following distribution of grades given last semester by Professor X.

Grade	A	B	C	D	F
Per cent	20	25	40	10	5

5. Histogram.

6. Circle graph.

7. Divided bar graph.

The given circle graph is based upon a 1972 United Nations estimate of the population of China. Use this data for Exercises 8 through 10.

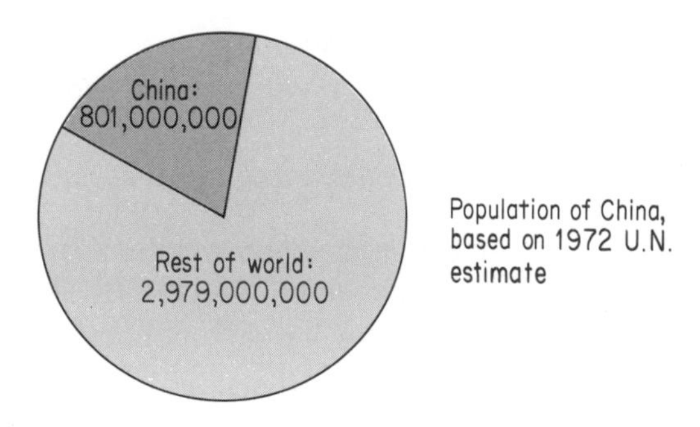

8. The population of China is approximately what per cent (nearest whole number) of the population of the entire world? (Note that the population of the entire world includes the population of China as well.)

9. What is the size of the central angle (nearest degree) that represents the population of China?

10. Assume that you wish to draw a bar graph to show the same facts as does the circle graph. If the population of China is represented by a bar that is one centimeter high, approximately how long a bar (nearest centimeter) would you need to show the population of the rest of the world?

Consider the set of scores: 55, 62, 70, 74, 74, 79. Then find its:

11. Arithmetic mean.

12. Median.

13. Mode.

Find the range for each of the following sets of scores:

14. 75, 82, 64, 98, 79

15. 100, 45, 79, 82, 96, 68

A collection of 1000 scores forms a normal distribution with a mean of 50 and a standard deviation of 10. What per cent of the scores are:

16. Above 70?

17. Below 60?

18. Below 40?

19. Between 40 and 70?

20. Above 80?

An unbiased coin is tossed 64 times.

21. What is the expected number of heads?

22. What is the standard deviation for results of this experiment when it is repeated many times?

23. Within what limits can we say, with almost 100% confidence, that we shall find the total number of heads?

24. If the experiment is repeated many times, what per cent of the times should be expected to produce fewer than 36 heads?

25. As in Exercise 24 what per cent of the times should be expected to produce more than 28 heads?

READINGS AND PROJECTS

1. Write for a copy of the booklet *Sets, Probability, and Statistics* by Paul Clifford, Mildred Keiffer, and Max Sobel. Copies of this booklet may be obtained free of charge from The Institute of Life Insurance, 277 Park Avenue, New York, New York, 10017. Report on the potential usefulness of the booklet for classroom instruction.
2. Read and summarize "A Critical Assessment of Published Tests for Elementary School Mathematics" by L. Ray Carry on pages 14 through 18 of the January 1974 issue of *The Arithmetic Teacher*.
3. Read and summarize "What Tests Don't Tell" by Donald M. Peck and Stanley M. Jencks on pages 54 through 56 of the January 1974 issue of *The Arithmetic Teacher*.
4. Read and summarize "Testing in the Mathematics Classroom" by Marc Swadener and D. Franklin Wright on pages 11 through 17 of the January 1975 issue of *The Mathematics Teacher*.
5. Select an elementary school grade level and prepare a lesson plan for organizing data from measurements made by the students.
6. Select an elementary school grade level. Then prepare a lesson plan in which you involve the students in the gathering of data and summarizing the results.

7. Read "Self-Paced Mathematics Instruction: How Effective Has It Been?" by Harold L. Schoen on pages 90 through 96 of the February 1976 issue of *The Arithmetic Teacher.* Prepare a critique of this article, including a discussion of the statistical presentation used by the author.
8. Read the position stated in the NACOME report concerning the role of statistics in the curriculum. (See Readings and Projects 13 on page 156.) In particular, read pages 44 through 48 and page 145. Prepare a summary of the recommendations offered in this report.

Read and report to your class on the following selections from *Statistics: A Guide to the Unknown.* This book was edited by Judith M. Tanur and others for a joint committee of the American Statistical Association and the National Council of Teachers of Mathematics. The book is published by Holden-Day, Inc., 1972.

9. "Setting Dosage Levels" by W. J. Dixon on pages 34 through 39.
10. "Statistics, Scientific Method, and Smoking" by B. W. Brown, Jr. on pages 40 through 51.
11. "The Importance of Being Human" by W. W. Howells on pages 92 through 100.
12. "Parking Tickets and Missing Women: Statistics and the Law" by Hans Zeisel and Harry Kalven, Jr. on pages 102 through 111.
13. "Deciding Authorship" by Frederick Mosteller and David L. Wallace on pages 164 through 175.

Flaws and Fallacies in Statistical Thinking by Stephen K. Campbell is published by Prentice-Hall, Inc., 1974. Read and report to your class on at least one from each of the following sets of chapters.

14. Dangers of Statistical Ignorance, Some Basic Measurement and Definition Problems, Meaningless Statistics.
15. Far-Fetched Estimates, Cheating Charts, Accommodating Averages.
16. Ignoring Dispersion, Puffing Up a Point with Percents, Improper Comparisons.
17. Jumping to Conclusions, Faulty Thinking About Probability, Faulty Induction.

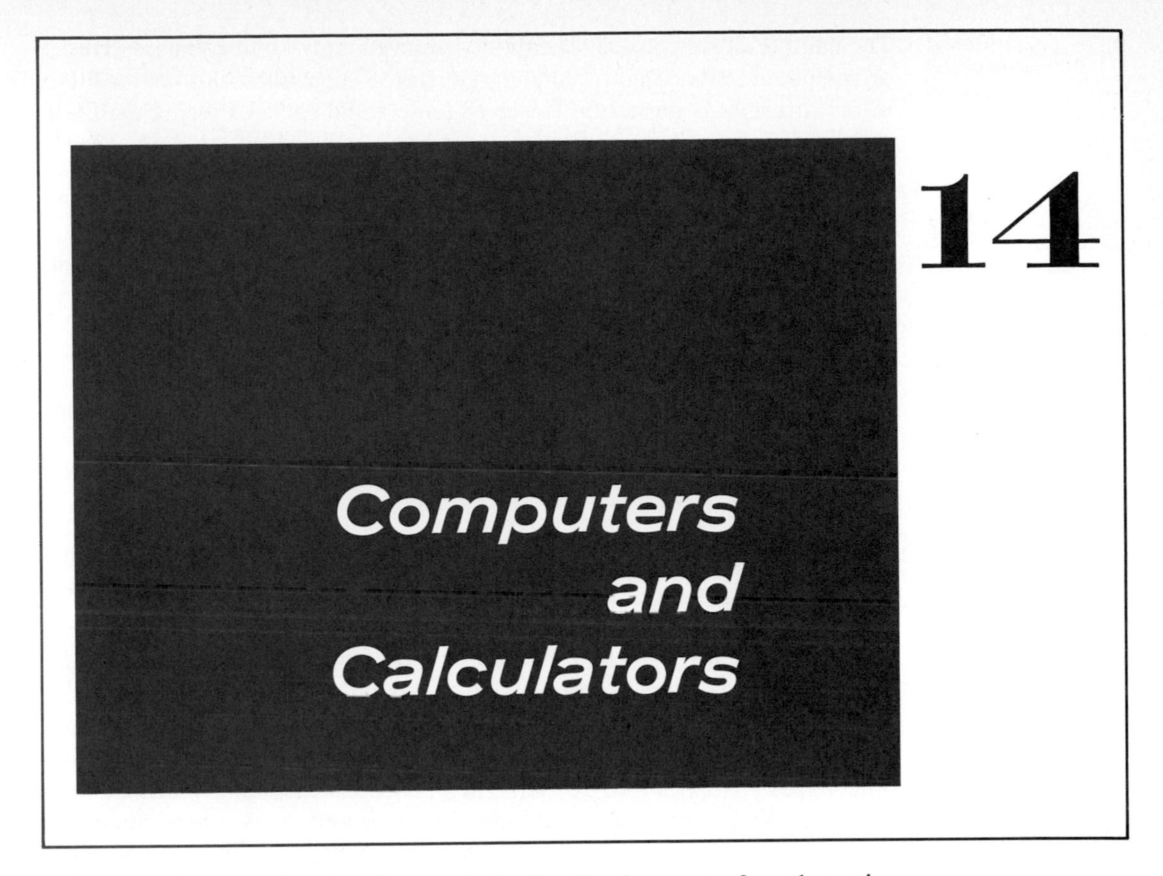

Computation has had a significant role in the development of mathematics, from the early use of tally marks in the sand to the use of modern electronic computers. Computer technology is changing so rapidly that obsolescence is a serious problem. One relatively small electronic calculator about the size of a typewriter cost over four thousand dollars in 1968. Six years later the same manufacturer sold a hand-held model (about 14.5 cm by 8 cm by 3 cm) that had the same capabilities as the 1968 model for less than four hundred dollars. Six months later the price had again dropped, this time by a little over 20%.

Some hand-held calculators are now purchased by many students and schools. Experiments are in progress to explore the usefulness of computers and calculators in developing facility in the use of numbers in elementary school grades. Prolonged and intense arguments over their merits in the classroom may be expected.

Computers are having a rapidly increasing influence on our personal lives. Among the functions of computers that affect many of us are:

Production of salary checks.
Processing of checking account transactions.
Processing of store charge accounts.
Processing of income tax data.
Preparation of class schedules and grade reports.

The number of potential applications is limited only by our imaginative use of computers. Since electronic computers are becoming such a major factor in our future, it is important for us to understand some of their capabilities and limitations.

14-1 computers and calculators

Primitive peoples probably did their computations using their fingers and toes. Several other early computational procedures were considered in §3-1 and §3-2.

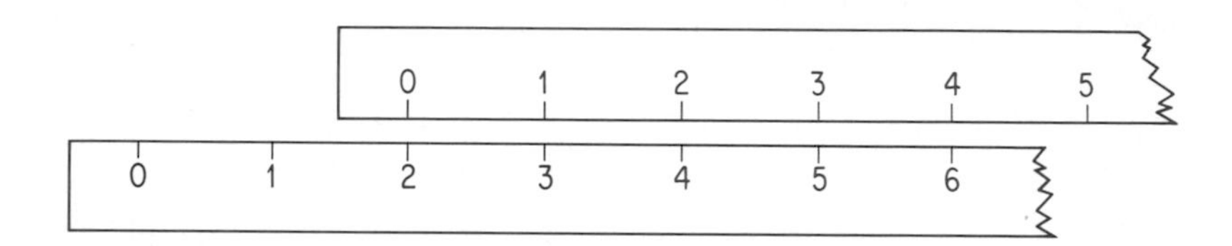

Two ordinary rulers may be used to add (or subtract) numbers. For example, the sum $2 + 3$ may be obtained by first placing the rulers so that their scales match, then moving one ruler so that the origin of its scale matches 2 on the fixed ruler. The number on the fixed ruler that matches 3 on the ruler that was moved is $2 + 3$. Such simple "slide rules" have been extended to use other types of scales and to perform operations including multiplication, division, taking square roots, taking cube roots, and many others. Slide rules are a simple form of **analog computer** because they use physical quantities (distance) to represent numbers. Very extensive advances have been made in the development of analog computers and they are used for many purposes. However, the most striking advances in recent years have been with **digital computers,** that is, computers based upon representations for the digits in a system of numeration for numbers.

Probably the first digital computer was designed by Blaise Pascal in 1642 and used cogged wheels to perform additions. The use of punched cards in 1725 to control automatic looms for weaving fabric provided the basis for Hermann Hollerith's 1880 invention of punched card tabulating machines. An earlier "Analytical Engine" was designed in 1833 by Charles Babbage with an accuracy of 20 decimal places, provision for storing numbers, and provision for storing instructions. Unfortunately, the technology of the time was not sufficiently developed to make it possible to build an effective model of Babbage's machine.

The first electronic computer was the Electronic Numerical Integrator and Computer (ENIAC) completed in 1945 at the University of Pennsylvania for the U. S. Army. This computer occupied a space 30 feet by 50 feet and contained 18,000 vacuum tubes. Computers using vacuum tubes are called *first-generation computers.* Computers using transistors instead of tubes began to appear in 1958–59 and are called *second-generation computers.* Computers using integrated circuits appeared in 1962 and are called *third-*

generation computers. The transistors and the integrated circuits each had the advantage of requiring much less space and much less maintenance. The trend toward miniaturization has continued so that in 1974 there existed microprocessor integrated circuits with over 4500 transistors in an area 4.2 by 4.8 millimeters (about one-thirtieth of a square inch). This dramatic decrease in the size of its components has made our present hand-held calculators possible.

Electronic computers are now available in a wide variety of sizes and designed for a wide variety of purposes. What can they do? Basically, they can only add or subtract numbers expressed in terms of the digits 0 and 1. The effectiveness of electronic computers is based upon the fantastic speed with which they can perform these operations—literally millions of operations per second. The large computers have extensive storage capacity for numbers and instructions. These computers can be programmed to accept instructions in special compiler languages such as FORTRAN, BASIC, PLATO, and others. We shall consider BASIC in §14-3.

Computers are now able to accept a complete sequence of instructions at one time and store large amounts of data for use in completing the assigned problem. Often computers can check the compatibility of the given instructions with the established algorithms under which the computer functions and report any inconsistencies back to the person who prepared the instructions. There are also programmable calculators that accept previously prepared instructions on cards about the size of a credit card. However, all computers and programmable calculators do exactly what they are instructed to do—no more and no less. Thus computers are very efficient and useful machines but they are strictly tools of the people who prepare the instructions for them, that is, the *programmers*. Very slight errors or mistyped instructions that any child would notice are scrupulously followed by the computer, unless instructions to do otherwise have been included in the program. The basic limitation of a computer is that it does what it is told to do.

Computers can be given a long sequence of instructions at one time using a stack of cards, a magnetic tape, or other devices. *Calculators* are given instructions one at a time, usually by pressing various buttons, or keys. Different types of calculators also differ noticeably from each other. Thus it is essential that one follow the special procedures for each type of calculator. It is also essential that students and teachers be very cautious in their acceptance of advertisements such as:

> Operates the way you see the formula written.

Such statements are not correct in the sense in which we ordinarily do arithmetic. A concern for such misleading statements is desirable since calculators are already being used in many elementary schools and even more secondary schools and colleges. Thus the remainder of this section is concerned with *hand-held electronic* calculators. Such calculators are also often called *minicalculators, minicomputers, electronic pocket calculators*, and, for some advanced models, *programmable pocket calculators*.

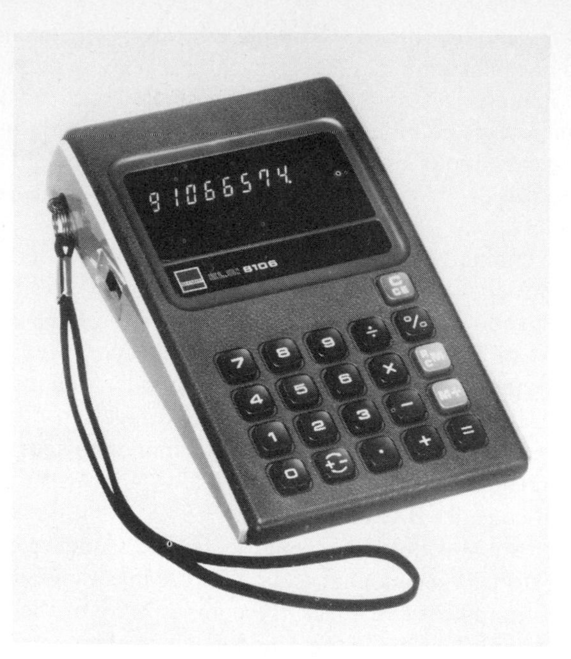

Courtesy of Sharp Electronics Corporation.

In order to illustrate the need for caution in the interpretation of advertisements and also the need for following a set pattern of steps for any given electronic calculator, we consider the evaluation of the expression

$$17 + 25 \times 2$$

On one calculator for which the cited advertisement is made, the following results are obtained:

Instructions	*Displays*	
On	0.	
17	17.	
+	17.	(This is the first addend.)
25	25.	
×	42.	(This is the first factor.)
2	2.	
=	84.	
Off		

Another calculator, and a very popular model, would have produced the same results except that decimal points do not appear in the display when a whole number is first given in the instructions; for example, on lines 2, 4, and 6 of the previous display. Both calculators have evaluated the expression

$$(17 + 25) \times 2$$

Neither calculator used the customary procedure of doing multiplication and division before addition and subtraction. At each step the last number on the

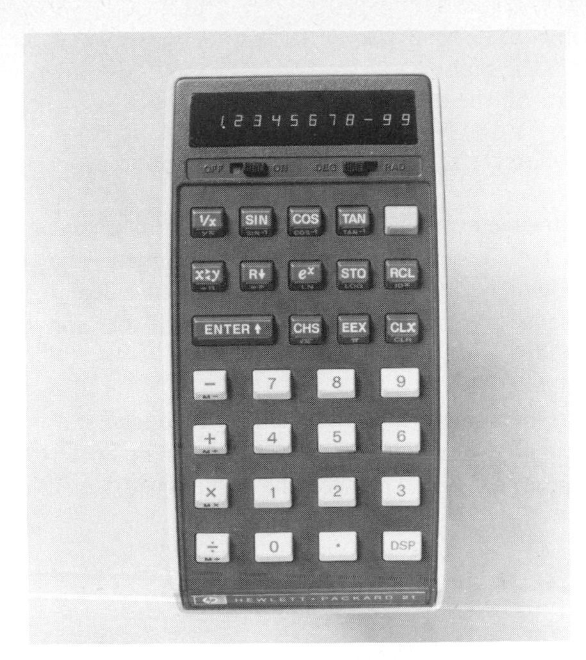

Courtesy of Hewlett-Packard Company.

display is the one used in the next step. In order to have a name for this use of the last number displayed, we call this procedure the **last x procedure.** Most, if not all, electronic pocket calculators follow the last x procedure.

In order to use an electronic calculator to evaluate the expression $17 + 25 \times 2$, we must arrange the sequence of instructions so that the last x procedure can be used. That is, we evaluate the equivalent expression $25 \times 2 + 17$. The instructions and displays on two different models of calculators are shown to illustrate the existence of different procedures.

Instructions	*Displays*	*Instructions*	*Displays*
On	0.	On	0.00
25	25.	25	25.
×	25.	Enter	25.
2	2.	2	2.
+	50.	×	50.00
17	17.	17	17.
=	67.	+	67.00
Off		Off	

We may think of the steps used with the second calculator as follows:

Turn on the calculator.
25, enter 25 into the register of the calculator for use.
2, multiply whatever is in the register by 2 and leave that product in the register.

17, add 17 to whatever is in the register and leave that sum in the register.
Turn off the calculator.

This use of "whatever is in the register" is the basic property of the last x procedure.

The illustrative examples in the instructor's manual for any particular calculator can be expected to emphasize the problems that are easy to do with that calculator. Although the first-mentioned calculator can be useful in many problems, its manual has no examples of the type

$$25 - 3 \times 2$$

Subtraction can be done, but $(3 \times 2)(-1) + 25$ does not appear feasible on that particular calculator. The second calculator has a CHS (change sign) key for the operation $-x$. Then for the problem just mentioned we have:

Instructions	*Displays*
On	0.00
3	3.
Enter	3.00
2	2.
×	6.00
CHS	−6.00
25	25.
+	19.00
Off	

Example: Rewrite each expression so that the new expression can be evaluated using the last x procedure:

(a) $29 + 3 \times 5 + 6$
(b) $20 - (4 \times 3)$
(c) $17 + 11^2$
(d) $19 + 17^3$

Solution: (a) $(3 \times 5) + 29 + 6$
(b) $(4 \times 3) \times (-1) + 20$
(c) $(11 \times 11) + 17$
or $11^2 + 17$ for calculators with x^2 as an operation.
(d) $(17 \times 17) \times 17 + 19$
or $17^3 + 19$ for calculators with y^x as an operation.

Some hand-held electronic calculators have numbered storage registers. Then expressions such as

$$19 \times 75 + 57 \times 23$$

can be evaluated. A typical procedure is shown at the top of the next page.

Instructions	*Displays*
On	0.00
19	19.
Enter	19.00
75	75
×	1425.00
Store, 1	1425.00
57	57
Enter	57.00
23	23
×	1311.00
Recall, 1	1425.00
+	2736.00
Off	—

Flow Chart

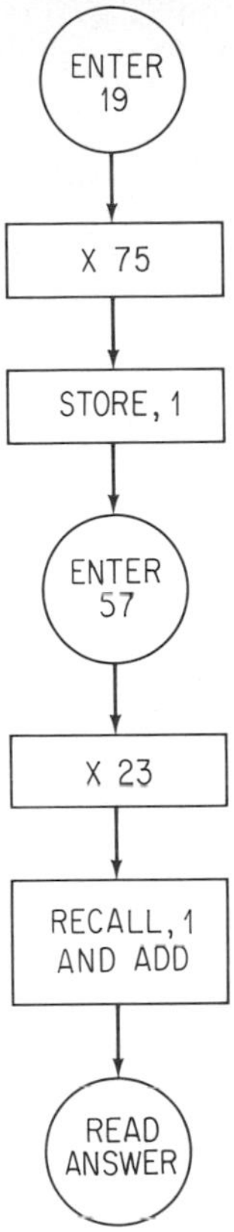

ENTER 19
↓
X 75
↓
STORE, 1
↓
ENTER 57
↓
X 23
↓
RECALL, 1 AND ADD
↓
READ ANSWER

The instruction booklet for each type of hand-held calculator must be carefully studied to determine the precise sequence of instructions that the particular calculator is programmed to follow.

The variety of operations that a particular calculator will perform depends upon its complexity (sophistication) and thus its cost. Among the common operations are $+$, $-$, $\times$, $\div$, $-x$, x^2, $\sqrt{x}$, $x!$, 10^x, y^x, and several special operations needed in secondary school and advanced mathematics. In each case a particular sequencc of steps, called an *algorithm*, must be followed. This is true for both small calculators and large computers. Thus a major implication of computers for elementary school mathematics is the importance of understanding algorithms, or procedures, for doing computations. As noted earlier, the primary limitation of computers is that they are "so dumb" that they do exactly what they are told. Thus, consideration of ways of instructing a calculator or a computer to evaluate different types of expressions can be very useful in demonstrating the value of learning correct procedures.

EXERCISES

Assume that you have a calculator with $+$, $-$, $\times$, $\div$, $-x$, x^2, $\sqrt{x}$, $x!$, $1/x$, *and only the last x storage register. Rewrite each expression so that the new expression can be evaluated using the last x procedure.*

1. $7 + 5 \times 21$

2. $12 + 18 \div 3$

3. $14 - 4 \div 2$

4. $36 + \sqrt{64}$

5. $\sqrt[4]{81}$

6. $5 - 2\sqrt{25}$

7. $7 + \sqrt{121} - 15$

8. $63 - 2\sqrt{49} - 27$

List each step (instruction and display) for evaluating your answer for the specified exercise.

9. Exercise 1.

10. Exercise 2.

11. Exercise 3.

12. Exercise 4.

13. Exercise 5.

14. Exercise 6.

15. Exercise 7.

16. Exercise 8.

Make a flow chart with a sequence of steps for evaluating each given expression. Assume that the calculator has the operations used for Exercises 1 through 16 and also has storage registers at 1, 2, 3, and 4.

17. $79 + \frac{1000}{5!}$

18. $101 - \frac{10{,}080}{7!}$

19. $73 \times 17 - 67 \times 15$

20. $\sqrt{21} - \sqrt{19}$

21. $\sqrt{31 \times 17} - \sqrt{65}$

22. $\sqrt{85 + 17} + \sqrt{101}$

PEDAGOGICAL EXPLORATIONS

1. Prepare a lesson plan on the use of rulers to form a "slide rule" for addition and subtraction. Include a discussion of the use of a single setting of the rulers for all of the sums 2 + 1, 2 + 2, 2 + 3, . . . , 2 + 10.
2. Select a grade level in which you are particularly interested and make up a set of ten arithmetic problems (possibly similar to those in Exercises 1 through 8) that can be used as follows:

   One student serves as the "computer" possibly with a second student serving as a checker.
   Members of the class take turns telling the "computer" what steps to take.
   The "computer" displays the result for each step on the chalk board.

3. Many hand-held calculators have displayed digits that appear as English alphabet letters when viewed upside down. For example, the numeral 7735 may read upside down as SELL. Assume that this matching of numerals and letters is: 0 – 0, 1 – I, 3 – E, 4 – h, 5 – S, 7 – L, and 8 – B. Then make a worksheet of ten arithmetic problems that students can do on an electronic calculator and read their upside down answers as English words. The decimal point may be used to separate two words so that two-word answers can be obtained.

4. Suggest imaginative ways of using minicalculators in elementary schools and discuss the following policy statement of the Board of Directors of the National Council of Teachers of Mathematics.

   With the decrease in cost of the minicalculator, its accessibility to students at all levels is increasing rapidly. Mathematics teachers should recognize the potential contribution of this calculator as a valuable instructional aid. In the classroom, the minicalculator should be used in imaginative ways to reinforce learning and to motivate the learner as he becomes proficient in mathematics.

   For further details read "Minicalculators in Schools" on pages 92 through 94 of the January 1976 issue of *The Mathematics Teacher.*

5. Survey recent publications and report on the points of view of the authors of several articles regarding the use of calculators in elementary school classes.

6. Compare the advantages and disadvantages of using calculators in elementary school classes.

7. One popular calculator's instruction booklet suggests that sums of fractions be calculated using the following statement of equivalence of forms of fractions:

$$\frac{A}{B} + \frac{C}{D} = \frac{\dfrac{A \times D}{B} + C}{D}$$

   Show that this is equivalent to reducing the sum of fractions to a last x procedure. Verify that the method works for some common fractions.

"I COULDN'T DO MY ARITHMETIC LAST NIGHT TEACHER . . . THE BATTERIES IN MY HAND CALCULATOR WERE DEAD!"

Reproduced by permission from the Oklahoma Council of Teachers of Mathematics *Newsletter*, Fall 1974 issue.

14-2 binary notation

Electronic digital computers depend upon a numeration system using only two digits, so that each digit may be represented by either the presence or absence of some electrical signal. The presence or absence of such electric currents is controlled by holes in a punched card or magnetic fields on a magnetic tape. The place-value system of numeration in which only two digits, usually denoted by 0 and 1, are used is called **binary notation.** The arithmetic procedures are similar to those for other place-value numeration systems as considered in the last two sections of Chapter 3. Our interest in this base two notational system arises from its use in modern electronic computers. Such computers are used in many ways besides doing arithmetic. For example, pictures of Mars have been taken from a spacecraft traveling nearby, transmitted back to Earth in binary notation, reassembled with the aid of computers, and displayed on television as well as in other ways.

There is some evidence that the basic concepts of binary notation were known to the ancient Chinese about 2000 B.C. However, it is only in relatively recent years that binary notation has been widely applied in computer mathematics, card sorting operations, and other electronic devices. Some of the place values in the binary system are as follows:

2^7	2^6	2^5	2^4	2^3	2^2	2^1	2^0
128	64	32	16	8	4	2	1

Thus, in binary notation, $11011101_2 = 221$ as shown in the following example.

Example 1: Write 11011101_2 in base ten notation.

Solution:

$$\begin{aligned} 11011101_2 &= (1 \times 2^7) + (1 \times 2^6) + (0 \times 2^5) + (1 \times 2^4) \\ &\quad + (1 \times 2^3) + (1 \times 2^2) + (0 \times 2^1) + (1 \times 2^0) \\ &= 128 + 64 + 16 + 8 + 4 + 1 = 221 \end{aligned}$$

Here are the first 16 counting numbers written in binary notation:

Base Ten	*Base Two*	*Base Ten*	*Base Two*
1	1	9	1001_2
2	10_2	10	1010_2
3	11_2	11	1011_2
4	100_2	12	1100_2
5	101_2	13	1101_2
6	110_2	14	1110_2
7	111_2	15	1111_2
8	1000_2	16	10000_2

Tables for addition and multiplication in binary notation are easy to complete.

+	0	1
0	0	1
1	1	10_2

x	0	1
0	0	0
1	0	1

Example 2: Multiply: $101_2 \times 1101_2$

Solution:

$$\begin{array}{r} 1101_2 \\ \times 101_2 \\ \hline 1101 \\ 11010 \\ \hline 1000001_2 \end{array}$$

Check:

$$\begin{array}{rcl} 1101_2 & = & 13 \\ \times 101_2 & = & \times 5 \\ \hline & & 65 \end{array}$$

$$1000001_2 = (1 \times 2^6) + (1 \times 2^0)$$
$$= 64 + 1 = 65$$

The seven-digit numeral 1000001 is the name for the letter A when that letter is sent over teletype to a computer. The American Standard Code for Information Interchange (ASCII) was adopted in 1967 and is used to convert letters, numerals, and other symbols into binary notation.

Binary numerals can be shown by means of electric lights flashing on and off. If a light is on, the digit 1 is represented; whereas if the light is off, the digit 0 is shown. Similarly the digits 0 and 1 can be used in a card sorting operation. The reader can gain an appreciation of the process used by construction of a small set of punched cards. First prepare a set of sixteen index cards with four holes punched in each and a corner notched as in the figure.

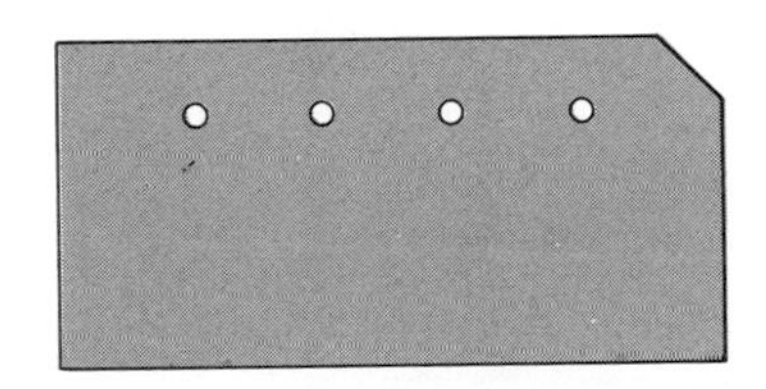

Next represent the numbers 0 through 15 on these cards in binary notation. Cut out the space above each hole to represent 1; leave the hole untouched to represent 0. Several cards are shown in the figure.

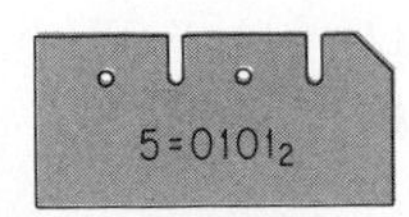

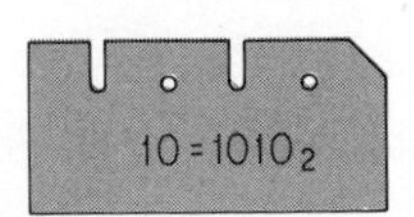

After all the cards have been completed in this manner, shuffle them thoroughly and align them, making certain that they remain "face up." (The notched corners will help indicate when the cards are right side up.) Then, going from right to left, perform the following operation: Stick a pencil or other similar object through the first hole and lift up. Some of the cards will come up, namely, those in which the holes have not been cut through to the edge of the card (that is, those cards representing numbers whose units digit in binary notation is 0).

Place the cards that lift up in front of the other cards and repeat the same operation for the remaining holes in order from right to left. When you have finished, the cards should be in numerical order, 0 through 15.

Note that only four operations are needed to arrange the sixteen cards. As the number of cards is doubled, only one additional operation will be needed each time to place them in order. That is, 32 cards may be placed in numerical order with five of the described card sorting operations; 64 cards may be arranged with six operations; 128 cards with seven operations; and so forth. Thus a large number of cards may be arranged in order with a relatively small number of operations. For example, over one billion cards may be placed in numerical order with only thirty sortings.

Electronic computers are designed to have specified capacities both as to the number of digits that can be represented for a given number and the number of digits that are processed at one time in performing an operation. In the previous example of card sorting, only numbers that required at most four digits in binary notation were represented and the digits were processed one at a time. In "computer language" each binary digit is a *binary bit* and the *word length* for a particular computer is the number of binary bits that are processed simultaneously Calculators process one bit at a time and thus have a word length of one bit. Some early computers had a word length of three bits and essentially worked in octal notation (see Exercises 21 through 33). As computer technology has improved, the word lengths of computers have increased to 8, 16, and even 32 and 64 bits.

EXERCISES

Write each number in binary notation.

1. 38 **2.** 35 **3.** 29

4. 128 **5.** 156 **6.** 425

Change each number to decimal notation.

7. 1110_2 **8.** 10100_2 **9.** 111011_2

10. 101010_2 **11.** 101110_2 **12.** 11011011_2

Perform the indicated operation in binary notation and check in base ten.

13. $\begin{array}{r} 1111_2 \\ +1011_2 \\ \hline \end{array}$ **14.** $\begin{array}{r} 10111_2 \\ +10101_2 \\ \hline \end{array}$ **15.** $\begin{array}{r} 11001_2 \\ -10110_2 \\ \hline \end{array}$

16. $\begin{array}{r} 100101_2 \\ -10111_2 \\ \hline \end{array}$ **17.** $\begin{array}{r} 1101_2 \\ \times 11_2 \\ \hline \end{array}$ **18.** $\begin{array}{r} 10111_2 \\ \times 101_2 \\ \hline \end{array}$

19. In the ASCII code the numbers for *B*, *D*, and *G* are 66, 68, and 71, respectively. Find the binary representation of **(a)** *B*; **(b)** *D*; **(c)** *G*.

20. In the ASCII code capital letters *A*, *B*, *C*, . . . in alphabetical order are assigned numbers 65, 66, 67, ... What is the word transmitted by the code 1010010, 1010101, 1001110?

21. Write the number 214 in base eight and then in base two notation. Can you discover a relationship between these two bases?

Base eight notation is often called **octal notation.** *Write in octal notation.*

22. 101111001010_2 **23.** 11101011001_2

24. 1001101110010_2 **25.** 10010101011011_2

26. 111110101101011001_2 **27.** 1101011001000100110_2

Write in binary notation.

28. 335_8 **29.** 5023_8 **30.** 4357_8

31. 4624_8 **32.** 12345_8 **33.** 76543_8

PEDAGOGICAL EXPLORATIONS

Many recreational items are based on the binary system of notation. Consider, for example, the boxes shown, within which the numbers 1 to 15 are placed according to the following scheme.

D	C	B	A
8	4	2	1
9	5	3	3
10	6	6	5
11	7	7	7
12	12	10	9
13	13	11	11
14	14	14	13
15	15	15	15

In box A place all numbers that have a 1 in the units place when written in binary notation. In box B place those with a 1 in the second position from the right in binary notation. In C and D are those numbers with a 1 in the third and fourth positions, respectively.

Next, ask someone to think of a number and tell you in which box or boxes it appears. You then tell that person his number by finding the sum of the numbers that are first in each of the boxes that he mentions. Thus, if his number is 11, he lists boxes A, B, and D. You then find the sum $1 + 2 + 8$ as the number under discussion.

1. Explain why the method given for finding a number after knowing the boxes in which it appears works as it does.
2. Extend the set of boxes to include all the numbers through 31. (A fifth column, E, will be necessary.) Then explain for the set of five boxes how to find a number if one knows in which boxes it appears.
3. You can use binary notations to identify any one of eight numbers by means of three questions that can be answered as "yes" or "no." Thus consider the numbers from 0 through 7 written in binary notation. The place values are identified by columns A, B, and C.

	C	B	A
0	0	0	0
1	0	0	1
2	0	1	0
3	0	1	1
4	1	0	0
5	1	0	1
6	1	1	0
7	1	1	1

The three questions to be asked are: "Does the number have a 1 in position A?" "Does the number have a 1 in position B?" "Does the number have a 1 in position C?" Suppose the answers are "Yes, No, Yes." Then the number is identified as 101_2; that is, 5. Extend this process to show how you can guess a number that someone is thinking of from 0 through 15 by means of four "yes—no" questions.

14-3 BASIC

As electronic computers developed, special languages were also developed for giving instructions to computers. Actually, a part of the computer is used to translate the information provided into the form that is needed for the computer to perform its operations. For example, one early stage was to program the computer to accept numbers in decimal notation, convert them to binary notation, perform the indicated operation(s), and convert the results back to decimal notation. In retrospect this example appears to describe an essentially trivial start. Gradually computers were programmed to accept instructions in one of several "computer languages" that are closely related to our ordinary mathematical language. The existence of several languages is due to their development for different purposes (scientific research, business applications, school instructional programs) and by different groups of people. Each language has certain advantages and disadvantages relative to each of the others. However, at our level the languages are alike in that they all are based upon very carefully defined algorithms.

The problem-oriented nature of the languages is indicated by their names. Here are a few:

FORTRAN	FORmula TRANslation
COBOL	COmmon Business Oriented Language
ALGOL	ALGorithm Oriented Language
BASIC	Beginner's All-purpose Symbolic Instruction Code

We have selected BASIC for our brief introduction to a computer language. A few of the symbols used in this language are:

+	addition
−	subtraction
*	multiplication
/	division
<	less than
>	greater than
=	equal to (or, is replaced by)
< =	less than or equal to
> =	greater than or equal to
< >	not equal to
↑	exponent (indicates that the next number is an exponent)
SQR( )	square root (indicates that the square root is to be taken of the number stated in the parentheses)

Parentheses are used as in ordinary mathematical notation.

Example 1: Write in BASIC:

(a) $5(3 - 17)$ (b) $3^2 + 5$

(c) $5 \div (2^3 - 17)$ (d) $\sqrt{29} \div 3$

Solution:

(a) **5 * (3 − 17)** (b) **3 ↑ 2 + 5**

(c) **5/(2 ↑ 3 − 17)** (d) **SQR(29)/3**

Example 2: Compute:

(a) **(2 * 3) ↑ 2** (b) **(12/4) ↑ 3**

Solution:

(a) $(2 \times 3)^2 = 6^2 = 36$ (b) $(12/4)^3 = 3^3 = 27$

A set of instructions in BASIC consists of a set of lines called *statements*. Each statement starts with a *line number* and a word that indicates the type of statement. If a statement is too long to fit on a single line, it is split into two statements and a second line number is assigned to the part on the second line. A similar procedure is followed for additional lines. The line number serves both as a serial number for ordering the operations of the computer and as a label for the statement. Usually, consecutive numbers are avoided so that additional data (or instructions) may be inserted. New lines, with appropriate numbers, may be added at the end of the program. Lines that contain errors may be replaced simply by adding new lines with the same number as the lines to be replaced. The computer orders and uses the statements according to their serial numbers. The instruction **LIST** may be used to obtain for the operator a list of the lines of the program in their proper order. Then, if the program is satisfactory, the instruction **RUN** may be used to implement the program.

In the previous examples of notations used in BASIC, note the absence of raised exponents. A special symbol was introduced to indicate that the next number would be an exponent, and thus it was unnecessary to raise the number. Subscripts are also avoided so that everything is presented on a single line; for example, x_1 is denoted by **X1**.

The ability of a computer to perform repetitive tasks any specified number of times or until a specified objective has been accomplished provides one basis for its usefulness. For example, a table of the numbers from 1 through 100 listed on separate lines and with each number followed by its cube can be obtained using the following four instructions, called a **program.**

```
10  FØR X = 1 TØ 100
20  PRINT X, X * X * X
30  NEXT X
40  END
```

To modify the previous program to provide the fourth powers of the numbers, we may simply add, as a replacement, the instruction:

```
20  PRINT X, X * X * X * X
```

The only additional information needed would be the satisfying of the machine that you were a recognized user and then, after presenting the program to the machine, typing **RUN** to indicate that the instructions are to be processed. Note that in communicating with a computer there must be a distinction between the letter O and the numeral 0. This distinction is made by consistently using **Ø** for exactly one of these symbols. In this text **Ø** is used for the letter O.

Programs with hundreds of statements arise in many detailed problems. However, our purpose is simply to illustrate the power of the computer and its significance for elementary school mathematics. For the previous program the computer will run until there are no more data to process; that is, until the number 100 has been processed. In any case the output of the computer will be as specifically requested in the program.

Example 3: State in ordinary notation the printed output for this program.

```
10  FØR X = 1 TØ 5
20  PRINT X, 2 + X
30  NEXT X
40  END
```

Solution:

1	3
2	4
3	5
4	6
5	7

Among the other instructions used in BASIC are:

READ The computer will read the first, or next, specified number of numbers from the data.

DATA Lists of numbers to be used.

LET Used to assign a letter as a name for an expression.

GØ TØ Used to send the computer back to the indicated line of the program.

Many computer programs are very simple. Here is a program for printing the cubes of the first five whole numbers. Note that the program can be modified for any numbers that one wishes to list in the **DATA** statement. We obtain a copy of the proposed program by typing **LIST**. After the computer has listed the program and if the program is acceptable to the computer, we obtain the printed output of the results of the program by typing **RUN**. The printed output might appear as follows:

```
LIST
10  READ X
20  DATA 1, 2, 3, 4, 5
30  LET Y = X ↑ 3
40  PRINT Y
50  GØ TØ 10
60  END
RUN
1
8
27
64
125
        10 ØUT ØF DATA
```

The pattern formed by the differences of the squares of the first eight successive whole numbers from the squares of their predecessors is shown in the output of the next program.

```
1  READ D
2  DATA 1, 2, 3, 4, 5, 6, 7, 8,
3  LET P = D ↑ 2
4  LET Q = (D- 1) ↑ 2
5  LET R = P - Q
6  PRINT R
7  GØ TØ 1
8  END
```

Printed Output
1
3
5
7
9
11
13
15

To test a conjecture regarding the pattern formed by the output of the previous program for some additional numbers, such as the first ten integers greater than one thousand, only the **DATA** statement needs to be changed and the new set of numbers inserted. Computers are frequently used to test conjectures of arithmetic patterns. A practical example of the use of computers for teachers is considered in Example 4.

Example 4: Suppose that you have three grades for each of your students and wish to average these grades. Prepare a BASIC program that may be used to do this.

Solution:

```
10  READ G1, G2, G3
20  DATA  (List the grades for each student being sure to list
          together three grades for each student; use extra
          lines 21, 22, 23, . . . , as needed)
60  LET A = (G1 + G2 + G3)/3
70  Print G1, G2, G3, A   (The computer will print the three
80  GØ TØ 10               grades and the average of each
90  END                    student with a separate line for each
                           student.)
```

EXERCISES

Write in BASIC and evaluate:

1. $2 \times 18 \div 3$

2. 3^3

3. $6(25 + 15)$

4. $1000 \div 2^5$

5. $6 - (\frac{8}{3})^2$

6. $75 - 2^6$

Evaluate:

7. **6 * 7/3**

8. **3 ↑ 4**

9. **12 ↑ 2 - 100**

10. **4 * 5 ↑ 2**

11. SQR(3600) − 3 ↑ 5

12. 1 ↑ 17 − SQR(0.36)

State in ordinary notation the printed output for each program:

13.
```
10  READ X
20  PRINT X, SQR(X)
30  DATA 1, 0.01, 0.04, 0.0009, 0.000016
40  GØ TØ 10
50  END
```

14.
```
10  FØR X = 1 TØ 10
20  PRINT X, X ↑ 2 - 1
30  NEXT X
40  END
```

15.
```
10  FØR X = 1 TØ 6
20  PRINT X, 2 - 6 * X
30  NEXT X
40  END
```

16.
```
10  FØR X = 1 TØ 6
20  LET Y = X + 1
30  PRINT X, Y, X * Y
40  NEXT X
50  END
```

Write a BASIC program to print out:

17. A table of the fifth powers of the counting numbers from 1 through 20.

18. For positive integral values of x from 1 through 20, a table of the coordinates of the points (x, y) on the curve $y = x^2 - 8x + 3$.

19. Repeat Exercise 18 for the curve $y = 2x + \sqrt{x}$.

20. Repeat Exercise 18 for the curve $y = x^2 - \sqrt{7x}$.

**21.* The coordinates of the endpoints and the midpoint of the line segment with endpoints that have coordinates:
(a) 7, 11 **(b)** 75, 123
(c) −567, 891 **(d)** 4357, −5437

**22.* The coordinates of the endpoints and the midpoint of the line segment with endpoints that have coordinates:
(a) (2, 5), (8, 17) **(b)** (75, 17), (−21, 93)
(c) (537, 693), (−23, 4537) **(d)** (756, 898), (−75, −94)

**23.* The coordinates of the endpoints and the midpoint of the line segment with endpoints that have coordinates:
(a) (2, 4, 6), (−2, 6, 4)
(b) (25, 37, 61), (125, 367, −75)
(c) (275, 17, −975), (4532, 6475, 9801)
(d) (7536, 4444, 3323), (5763, −7985, −9873)

PEDAGOGICAL EXPLORATIONS

Many people think of flow charts as a prerequisite to computer programs. Such is not strictly the case although flow charts are often helpful in preparing computer programs.

1. Write down the output obtained by following this flow chart.

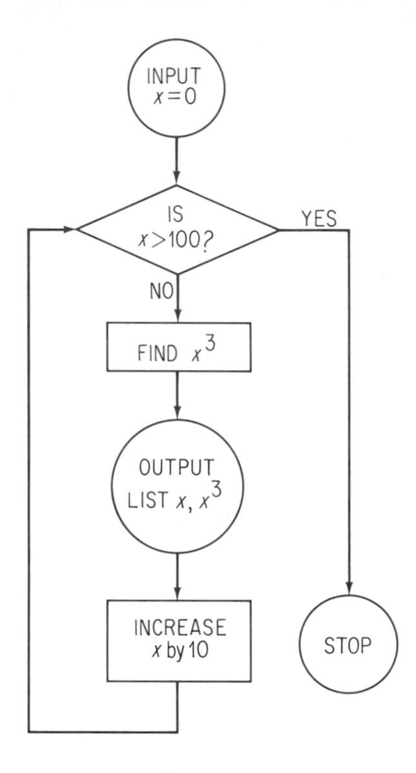

Instructions such as **FØR B = 1 TØ 20** in BASIC may be modified to indicate the jumps (increments) in the value of B that are to be taken. For example,

FØR B = 1 TØ 21 STEP 5

means that B takes on the values 1, 6, 11, 16, 21.

2. Write a BASIC program to produce the list obtained in Exploration 1.
3. Study a recently published series of texts for grades K through 8 and report on the treatment of calculators and computers.

The next program is designed to find the intersection of two sets

$$B = \{1, 3, 5, 7, \ldots, 21\}$$
$$C = \{2, 5, 8, 11, \ldots, 23\}$$

The procedure is to compare each element of C with the first element of B, then compare each element of C with the next element of B, and so forth. In each case if the comparison leads to equality, the element is listed as a member of the intersection; otherwise the next element of C is tried.

```
10  FØR B = 1 TØ 21 STEP 2
20  FØR C = 2 TØ 23 STEP 3
30  IF B = C THEN 50     (Another BASIC instruction; it
                          sends the computer to line 50 if
                          B = C)
40  GØ TØ 60
50  PRINT B
60  NEXT C
70  NEXT B
80  END
```

Note that an **IF __ THEN** statement enables you to jump another line in the program and procede from there; it plays the role of the decision box in a flow chart. The listing of **NEXT C** with a lower line number than **NEXT B** means that each of the entire set of elements in C will be considered before the next B is considered.

Write a BASIC program to find:

4. The intersection of the sets $S = \{1, 3, 5, 7, \ldots, 49\}$
$T = \{1, 6, 11, 16, \ldots, 51\}$

5. The intersection of the sets $S = \{1, 4, 9, 16, 25, 36, 49, 64, 81, 100\}$
$T = \{3, 6, 9, 12, \ldots, 99\}$

6. Discuss the merits of teaching a computer language, such as BASIC, in elementary school classes.

14-4 precision and accuracy

Electronic digital computers make use of at most a predetermined number of digits to represent any number. Thus all real numbers are approximated by rational numbers when they are used by digital computers. For example, one hand-held electronic calculator displays numbers with ten digits. For very large or very small numbers *scientific notation* (§5-5 Pedagogical Explorations) is used.

In scientific notation any positive real number is expressed as the product of a number N where $1 \leq N < 10$ and a power of 10. For example,

$$53{,}467 = 5.3467 \times 10^4$$
$$0.015 = 1.5 \times 10^{-2}$$

One major advantage of scientific notation is the ability to distinguish between digits that are *significant* and those (zeros) that are used solely to place the decimal point. For example, the population of the state of Vermont is about 450,000. Without further information we do not know whether this is a statement of the population to the nearest ten people, to the nearest thousand people, or even to the nearest ten thousand people. In scientific notation zeros are used only when they are significant. Then the population of the state of Vermont would be expressed:

To the nearest person as	$4.500\,00 \times 10^5$
To the nearest ten people as	$4.500\,0 \times 10^5$
To the nearest thousand people as	4.50×10^5
To the nearest ten thousand people as	4.5×10^5

In this manner the intended meaning may be stated explicitly.

Most approximations arise from measurements. Any measurement is to the nearest unit for some arbitrarily specified unit. In general, there is a smallest unit that is used in any approximation and this smallest unit indicates the **precision** of the approximate number. In our discussions each unit of precision will be represented by a place value in decimal notation. In the metric system these decimal place values correspond exactly to the subdivisions and multiples of basic units. However, other units are possible; for example, a measurement might be to the nearest eighth of an inch.

Suppose that the unit of precision for a particular problem is given as 0.001 of some basic unit. Then each measure is expressed to the nearest thousandth; for example,

$$\frac{1}{4} = 0.250$$
$$\frac{3}{8} = 0.375$$
$$\sqrt{200} \approx 14.14213562 \approx 14.142$$
$$\pi \approx 3.141592654 \approx 3.142$$
$$0.0007 \approx 0.001$$
$$0.0002 \approx 0.000$$

In the previous examples numbers were rounded off to the nearest thousandth whenever necessary. Special consideration is required for situations in which the part to be rounded off is a 5 or a 5 followed by zeros. When a particular problem does not indicate a need for rounding in a particular direction, we round off such numbers *to the nearest even digit.* Then, for example,

$$\frac{3}{80} = 0.0375000\ldots \approx 0.038$$
$$\frac{7}{80} = 0.0625000\ldots \approx 0.062$$

This approach is recommended in order to avoid accumulating excessive surpluses (or shortages) by always rounding up (or down).

There is a common misconception that mathematics is an exact science. Actually mathematics is a very comprehensive subject and includes extensive use of approximations and estimations. Computations with approximate numbers are very common and need special emphasis because of their many practical, usually labor-saving, applications.

An approximate number may be considered as a counting number of units. The size of the unit is the *precision* of the approximate number. The number of decimal places used to express the number of units is a measure of the **accuracy** of the approximate number. Both concepts are essential since one is used when the numbers are added or subtracted and the other is used when the numbers are multiplied or divided.

Approximate Number	*Precision*	*Accuracy*
4.5	0.1	2-digit
4.500	0.001	4-digit
3.1416	0.0001	5-digit
2.3×10^6	100,000	2-digit
2.30×10^6	10,000	3-digit

In computations with approximate numbers it is desirable to work with at least one extra digit until the final answer is rounded off. The *rule* for addition and subtraction of approximate numbers is that the *precision* cannot be increased. For approximate numbers

$$\begin{array}{r} 25 \\ +10.37 \\ \hline 35.37 \end{array} \approx 35 \qquad \begin{array}{r} 25 \\ -\ 0.15 \\ \hline 24.85 \end{array} \approx 25$$

These examples may be interpreted in terms of dollars and cents. If you had \$25 to the nearest dollar, you would have between \$24.50 and \$25.50. Suppose that somebody repaid a loan of \$10.37. Then you would have between \$34.87 and \$35.87. However, without any knowledge of how much change you had in the first place, our best answer for the sum is \$35. Possibly, the actual answer is \$36 to the nearest dollar. Thus the precision of working to the nearest dollar cannot be increased; it may be decreased.

The subtraction of approximate numbers is handled in a similar manner. If you have about \$25 and spend 15 cents for a candy bar, then you still have about \$25.

The rule for multiplication, division, raising to powers, and extracting roots is that the *accuracy* cannot be increased. For example,

$1.6 \times 0.3 = 0.48 \approx 0.5$ (one-digit accuracy)

$\frac{5280}{2.5} = 2112 \approx 2.1 \times 10^2$ (two-digit accuracy)

$\sqrt{3.0000} \approx 1.7321$ (five-digit accuracy)

$(2.50)^6 = 2.441406252 \times 10^2 \approx 2.44 \times 10^2$ (three-digit accuracy)

A special situation arises when an **exact number** (a counting number of identifiable units) is used with approximate numbers. For example, consider the two problems

$$\begin{array}{r} 1.23 \\ +1.23 \\ \hline \end{array} \qquad \begin{array}{r} 1.23 \\ \times\ \ 2 \\ \hline \end{array}$$

Since these are two interpretations of the same problem, the same answer should be obtained. Accordingly, any exact number is assumed to have infinite precision; that is, precision and accuracy of exact numbers are not considered in determining the precision or accuracy of computed results.

A knowledge of the correct procedures for working with approximate numbers enables us to identify the measures that need to be made with great care and the ones that can be estimated or even neglected altogether. For example, if several measures are to be added and one of the measurements can only be obtained to the nearest unit, then it is usually futile to make other measurements more precise than the nearest tenth of a unit.

EXERCISES

1. The population of China has been estimated as 801,000,000 people. Express this number in scientific notation under the assumption that it is correct to the nearest:
(**a**) Million people. (**b**) Ten million people.
(**c**) Hundred thousand people. (**d**) Ten thousand people.

Round off each number to express it to (a) four-digit accuracy; (b) three-digit accuracy.

2. 3.1245
3. 75.345
4. $\sqrt{7} \approx 2.64575$
5. $\sqrt{1100} \approx 33.1662$
6. 4.5800×10^3
7. 2.5000×10^4

Assume that the given numbers are approximate and find:

8. $\begin{array}{r} 52.873 \\ +\ 3.56 \\ \hline \end{array}$
9. $\begin{array}{r} 265.0 \\ +\ \ 1.256 \\ \hline \end{array}$
10. $\begin{array}{r} 287.3 \\ -\ \ 0.08 \\ \hline \end{array}$
11. $\begin{array}{r} 356 \\ -\ 0.08 \\ \hline \end{array}$
12. 3.5×0.2
13. 4.12×0.003
14. 25×1.20
15. $656/4$
16. $\sqrt{0.64}$
17. $\sqrt[3]{0.008}$
18. $\sqrt[3]{8.00}$
19. $\sqrt{4 \times 10^6}$

For people with access to hand-held electronic calculators find:

***20.** $\sqrt{19.0}$
***21.** $\sqrt{753}$
***22.** $\sqrt{0.150}$
***23.** $\sqrt[4]{5286}$
***24.** $\sqrt{3.700 \times 10^8}$
***25.** $\sqrt{3.5286 \times 10^5}$

PEDAGOGICAL EXPLORATIONS

1. Explain why a measurement of 80 mm to the nearest millimeter should be represented by 8.0 cm instead of 8 cm.
2. Represent a measurement of 500 liters to the nearest liter in cubic centimeters and explain your answer.
3. A hectare is widely used in Europe as we use an acre to denote areas of parcels of land. Recall that one hectometer is one hundred meters. A **hectare** is a square hectometer. Express an area of 450,000 square meters (to the nearest one thousand square meters) in hectares.
4. Explain, as if to an elementary school class (possibly fifth grade) why addition and subtraction of approximate numbers cannot be expected to increase the precision of the measures.
5. Repeat Exploration 4 for multiplication and the accuracy of the measures.
6. Repeat Exploration 5 for division.
7. Repeat Exploration 5 for the extraction of square roots.
8. Prepare, for use in an elementary school class, a set of twenty practice exercises on rounding off. Include several, such as Exercise 3 of this section, in which the rounding must be done all at once and not step by step. Explain the circumstances in which this is essential; for example, consider

$$15.346 \approx 15.35$$

   where $15.346 \approx 15.3$ but $15.35 \approx 15.4$.

9. Identify each statement as true or false.
   (a) For approximate numbers to the nearest unit, any measure that is less than 0.5 serves as an additive identity.
   (b) In scientific notation if an approximate number is expressed as a multiple of 10^6, then its square root is expressed as a multiple of 10^3.
10. Give three examples or counterexamples to support your answer for each part of Exploration 9.
11. Generalize the statement in Exploration 9(b) for 10^{2n} and repeat Exploration 10 for your new statement.
12. Repeat Exploration 11 for a suitable power of 10 and cube roots.
13. Repeat Exploration 12 for fifth roots.
14. Repeat Exploration 12 for the square of the number and explain why two possible cases need to be considered.
15. Repeat Exploration 12 for the cube of the number and explain why three possible cases need to be considered.

CHAPTER TEST

In Exercises 1 through 4 assume that you have a calculator with +, −, ×, ÷, −x, x^2, and $\sqrt{\ }$. Rewrite each expression so that the new expression can be evaluated using the last x procedure.

1. $35 - 17 \times 2$
2. $5 + 3\sqrt{121}$
3. $12 + 2 \times 13^2 - 25$
4. $23 - 125 \div 50$

5. Write 79 in binary notation.
6. Change 10101011_2 to decimal notation.
7. Add in binary notation:

$$\begin{array}{r} 1010101_2 \\ +\ 111110_2 \\ \hline \end{array}$$

8. Subtract in binary notation:

$$\begin{array}{r} 101010101_2 \\ -\ 11111010_2 \\ \hline \end{array}$$

9. Write in octal notation: 111011010110011110_2.
10. Write in binary notation: 576354_8.
11. Write in BASIC and evaluate: $6 + 2\sqrt{9} - 11^2$.
12. Evaluate: **SQR(49) − 15 * 8/10**
13. State the printed output for the BASIC program:

```
10  FØR X = 1 TØ 6
20  PRINT X, X ↑ 2 - 5
30  NEXT X
40  END
```

14. Write a BASIC program to print each of the even numbers from 2 through 50 with its cube.
15. Express 275,000 in scientific notation to show that the number is correct to the nearest (**a**) thousand; (**b**) ten.
16. Round off $\frac{7}{8}$ to (**a**) two-digit accuracy; (**b**) one-digit accuracy.

In Exercises 17 through 20 assume that the given numbers are approximate and find:

17. $\begin{array}{r} 25.603 \\ -\ 5.60 \\ \hline \end{array}$

18. $\begin{array}{r} 6.7 \\ +0.02 \\ \hline \end{array}$

19. $\sqrt{25.0}$

20. $\sqrt{4.0 \times 10^8}$

READINGS AND PROJECTS

1. Assume that you have enough hand-held electronic calculators to lend one to each member of a class of fourth grade students. Prepare a 20-minute unit to acquaint them with the use of the calculators.
2. Read "Hand-held Calculators: Help or Hindrance" by Frank S. Hawthorne on pages 671 and 672 of the December 1973 issue of *The Arithmetic Teacher*. Discuss some of the issues considered in this article.
3. Complete a set of 32 punched cards to represent the numbers 0 through 31 as in §14-2. Here five holes are needed on each card and five "lifting" operations are necessary to place the set in numerical order.
4. Prepare a bulletin board display for use in an elementary school classroom on one of the following:
   (a) Computers.
   (b) Hand-held electronic calculators.
   (c) Precision and accuracy.
   (d) Applications of mathematics.
5. Read "Introducing the Binary System in Grades Four to Six" by Jan Unenge on pages 182 and 183 of the March 1973 issue of *The Arithmetic Teacher*. Prepare a lesson plan based upon ideas related to those in this article.
6. Read and demonstrate to your class "The Binary Adder: A Flow Chart for the Addition of Binary Numbers" by Alfred Ellison on pages 131 through 134 of the February 1973 issue of *The Mathematics Teacher*.
7. Assume that your elementary school class is about to visit a computer center where there will be terminals available for using BASIC. Prepare a 15-minute orientation for presentation to the class before the visit.
8. Read "Electronic Calculators in the Classroom" by Lowell Stultz on pages 135 through 138 in the February 1975 issue of *The Arithmetic Teacher*. Specify a grade level in which you are interested and make a list of ways in which electronic calculators may be used effectively in a class at that level.
9. Read *Introduction to an Algorithmic Language* (*BASIC*) published by the National Council of Teachers of Mathematics, 1968. Complete at least three of the exercise sets given there.
10. Discuss ways of distinguishing between precision and accuracy in elementary school classes and the importance of doing so.
11. Read "Instant Insanity: that Ubiquitous Baffler" by Dewey C. Duncan on pages 131 through 135 of the February 1972 issue of *The Mathematics Teacher*. Explain to a friend the use of geometric arrays to solve this popular puzzle.
12. See Chapter 10 (Mathematics Projects, Exhibits and Fairs, Games, Puzzles, and Contests) of the thirty-fourth yearbook of the National Council of Teachers of Mathematics for a list of 67 possible topics for

mathematics projects. Select one of these and prepare a model, display, or bulletin board exhibit for your topic.

13. Read "Electronic Calculators—Friend or Foe of Instruction" by Eleanor Machlowitz on pages 104 through 106 of the February 1976 issue of *The Mathematics Teacher.* Describe briefly how you might defend your use of electronic calculators in your classroom if you were asked by the PTA to do so.

14. Read "ENIAC: The First Computer" by Aaron Strauss on pages 66 through 72 of the January 1976 issue of *The Mathematics Teacher.* Prepare a short summary of the items from the article that would be of interest to elementary school students.

15. Read "The Small Electronic Calculator" by Eugene W. McWhorter on pages 88 through 98 of the March 1976 issue of *Scientific American.*

16. Read "Minicalculators in Schools" on pages 72 through 74 of the January 1976 issue of *The Arithmetic Teacher.* Describe at least five additional ways that hand-held calculators may be used in the classroom.

17. Read "The Hand-Held Calculator" by George Immerzeel on pages 230 and 231 of the April 1976 issue of *The Arithmetic Teacher.* Suggest at least five additional ways to use a calculator in a specified elementary mathematics class.

18. Discuss the position taken by the NACOME report concerning the use of calculators in the curriculum. (See Readings and Projects 13 on page 156.) In particular, read pages 40 through 43 of that report.

19. Read one of the following chapters from the thirty-seventh yearbook of the National Council of Teachers of Mathematics:
Chapter 2 (The Curriculum)
Chapter 3 (Research on Mathematics Learning)
Chapter 12 (Directions of Curricular Change)
Prepare a brief summary of the chapter that you read.

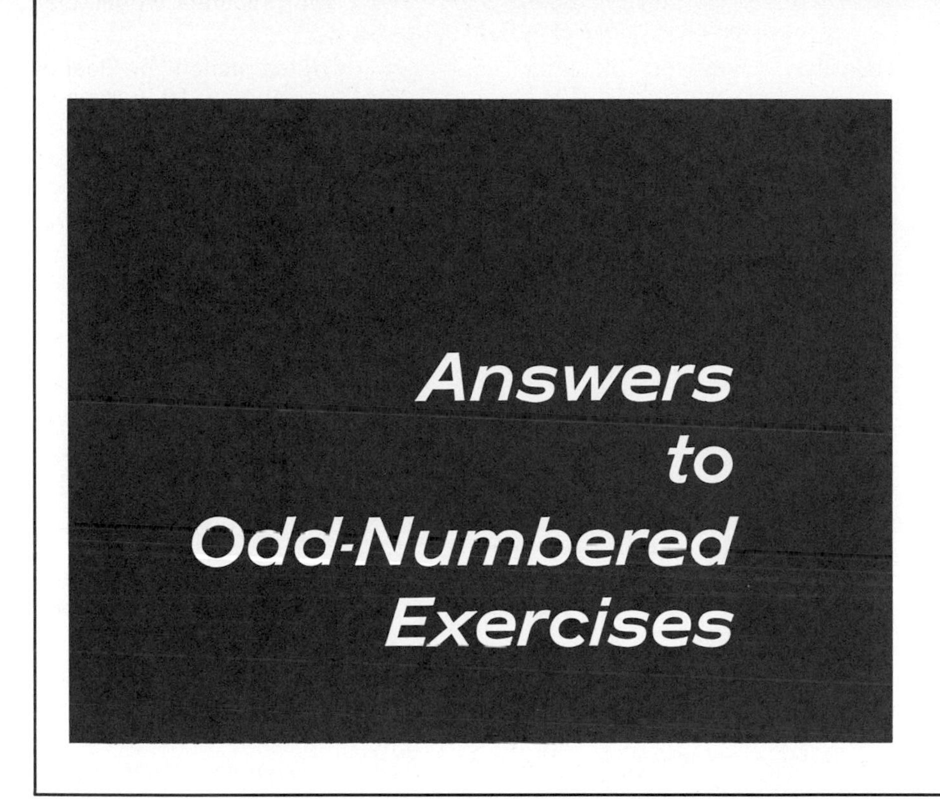

Introduction: EXPLORATIONS WITH MATHEMATICS

1 EXPLORATIONS WITH NUMBER PATTERNS *Page 6*

1. The text shows the diagrams for 3×9, 7×9, and 4×9 respectively. Here are the others:

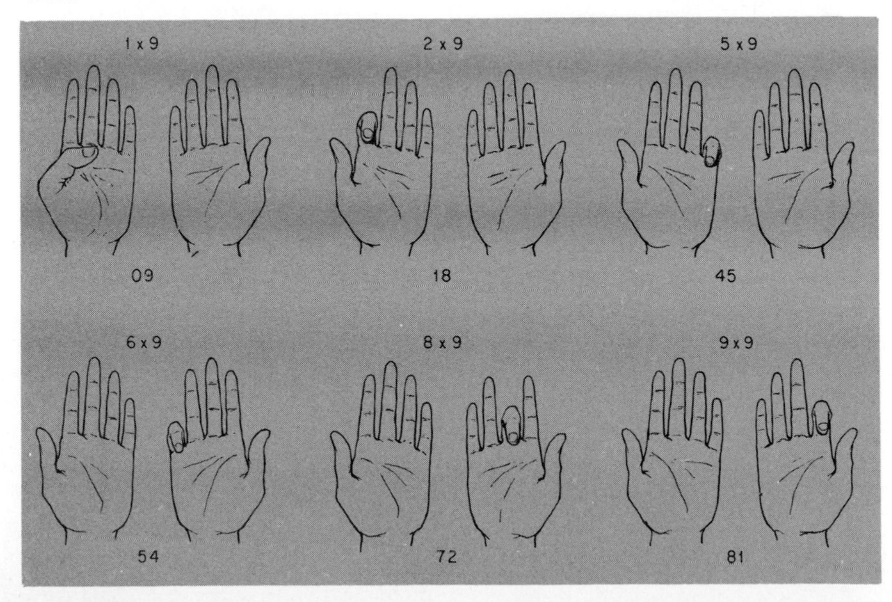

3. $6^2 = 1 + 2 + 3 + 4 + 5 + 6 + 5 + 4 + 3 + 2 + 1$
$7^2 = 1 + 2 + 3 + 4 + 5 + 6 + 7 + 6 + 5 + 4 + 3 + 2 + 1$
$8^2 = 1 + 2 + 3 + 4 + 5 + 6 + 7 + 8 + 7 + 6 + 5 + 4 + 3 + 2 + 1$
$9^2 = 1 + 2 + 3 + 4 + 5 + 6 + 7 + 8 + 9 + 8 + 7 + 6 + 5 + 4 + 3 + 2 + 1$

7. **(a)** $9 \times 47 = 423$

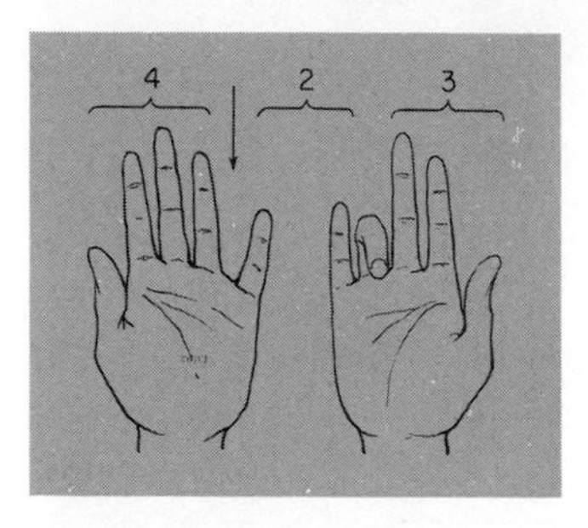

(b) $9 \times 39 = 351$

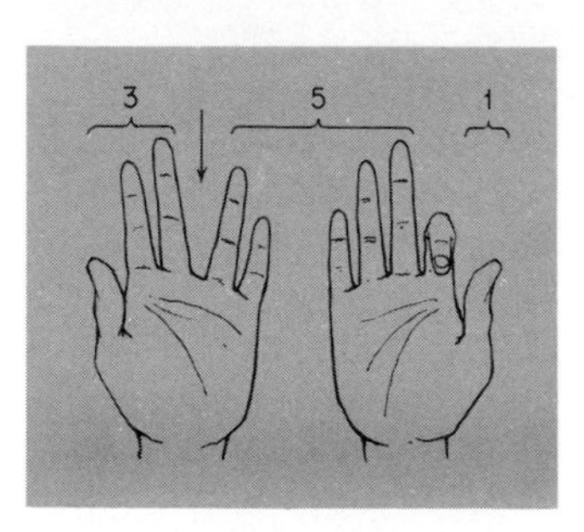

(c) $9 \times 18 = 162$

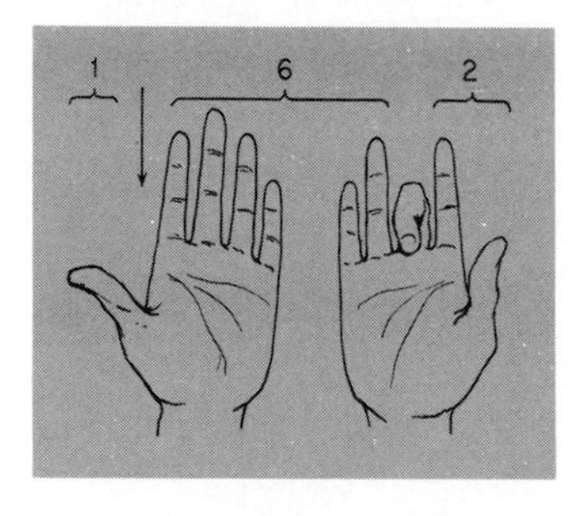

(d) $9 \times 27 = 243$

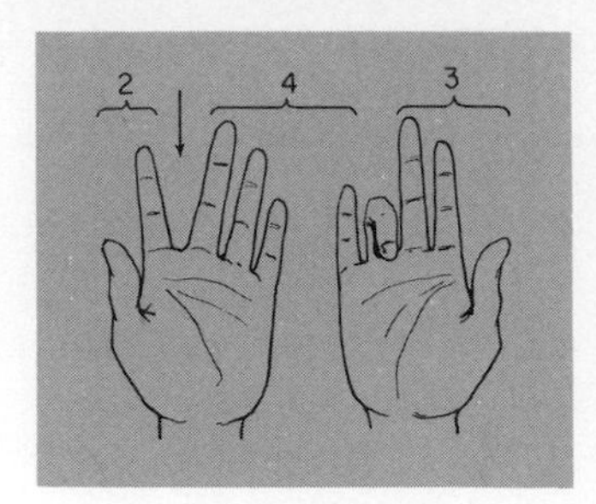

9. **(a)** 40×81, that is, 3240;
(b) 100×201, that is, 20,100;
(c) $\frac{25}{2} \times 50$, that is, 625;
(d) 50×200, that is, 10,000;
(e) 100×402, that is, 40,200.

11. **(a)** $\frac{2}{3}$; $\frac{3}{4}$.
(b) The last fraction in each row is of the form $1/[n(n + 1)]$ and the sum is $n/(n + 1)$. Thus, since the last fraction in the original series is $1/(9 \times 10)$, the sum is $\frac{9}{10}$.
(c) $\frac{4}{5}$.
(d) $\frac{99}{100}$.

13. **(a)** $\frac{3}{4}$;
(b) $\frac{7}{8}$;
(c) $\frac{15}{16}$. In general: $(2^n - 1)/2^n$.

2 EXPLORATIONS WITH GEOMETRIC PATTERNS *Page 11*

1. There will be 4 holes produced by 4 folds, and 2^{n-2} holes with n folds.

3. $V + R = A + 2$: the sum of the number of vertices and the number of regions is 2 more than the number of arcs.

5.

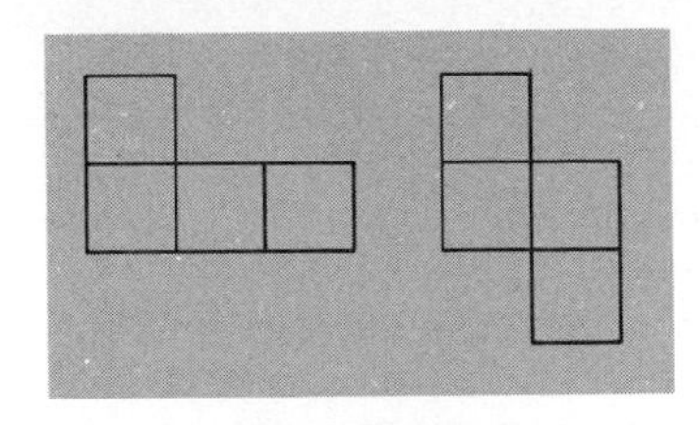

7. Dotted lines show segments removed.

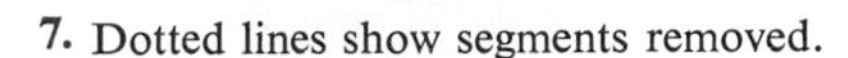

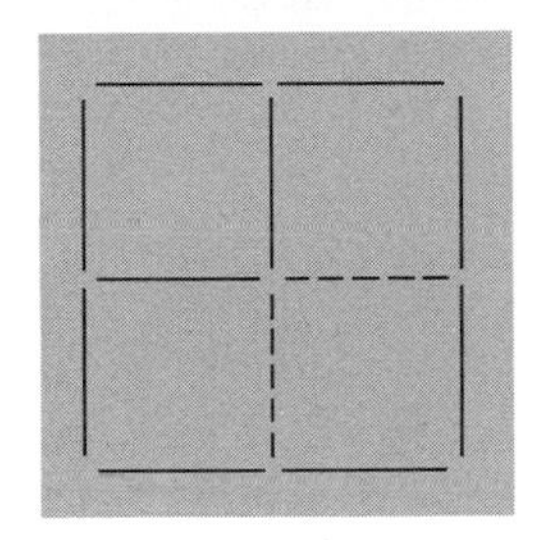

9.

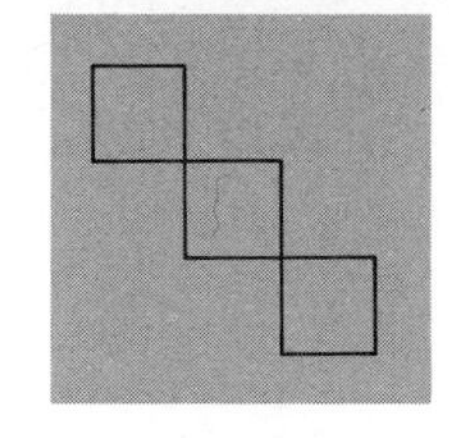

11. Approximately **(a)** 7; **(b)** 50,000,000 miles.

3 EXPLORATIONS WITH MATHEMATICAL RECREATIONS *Page 18*

1. (a) 12.
(b) Only one, if it's long enough.
(b) Only halfway; then you start walking out.
(d) *One* of them is not a nickel, but the other one is.
(e) There is no dirt in a hole.
(f) Brother-sister.

3. There are eleven trips needed. First one cannibal and one missionary go over; the missionary returns. Then two cannibals go over and one returns. Then two missionaries go over; one missionary and one cannibal return. Two missionaries go over next and one cannibal returns. Then two cannibals go over and one of them returns. Finally, the last two cannibals go over.

5. After 27 days the cat still has 3 feet to go. It does this the next day and is at the top after 28 days.

7.

9. Both *A*'s and *B*'s will say that they are *B*'s. Therefore, when the second man said that the first man said he was a *B*, the second man was telling the truth. Thus the first two men told the truth and the third man lied.

11. If the penny is in the left hand and the dime in the right hand, the computation will give $3 + 60 = 63$, an odd number. If the coins are reversed, we have $30 + 6 = 36$, an even number.

13. There really is no missing dollar. The computation may be done in one of two ways: $(30 - 3) - 2 = 25$, or $25 + 2 + 3 = 30$. In the problem the arithmetic was done in a manner that is not legitimate; that is, $(30 - 3) + 2$.

15. Eight moves are needed. The coins are identified in the following diagram as well as the squares which may be used. The moves are as follows, where the first numeral indicates the position of the coin and the second one tells you where to move it: D_1: 4–3; P_2: 2–4; P_1: 1–2; D_1: 3–1; D_2: 5–3; P_2: 4–5; P_1: 2–4; D_2: 3–2.

1	2	3	4	5
P_1	P_2		D_1	D_2

17. *E*; *N*. These are the first letters of the names (one, two, three, etc.) of the counting numbers.

19. Use H for the half-dollar, Q for the quarter, and N for the nickel. Assume that each coin is on a larger coin. The seven moves may be made in the order listed:

N: A to C; Q: A to B; N: C to B;
H: A to C; N: B to A; Q: B to C;
N: A to C.

For the four coins use P for the penny. Then the moves are:

P: A to B; N: A to C; P: B to C;
Q: A to B; P: C to A; N: C to B;
P: A to B; H: A to C; P: B to C;
N: B to A; P: C to A; Q: B to C;
P: A to B; N: A to C; P: B to C.

For a discussion of the number of moves required for 64 discs, see page 171 of *Mathematics and the Imagination* by Edward Kasner and James Newman, Simon and Schuster, 1940. They estimate that it would take more than 58 billion centuries to complete the task.

21. The sum of the values on two opposite faces of any ordinary die is always 7. Therefore for three dice this sum is 21. Thus the difference between 21 and the value on the top face of the top die is the sum of the values on the other specified faces.

Chapter 1: THE LANGUAGE OF MATHEMATICS

1-1 SETS *Page 26*

1. **(a)** Equivalent; **(b)** equal.

3. **(a)** Not equivalent; **(b)** not equal.

5. **(a)** Equivalent; **(b)** not equal.

7. $\in$

9. $\notin$

11. Well-defined.

13. {January, February, March, April, May, June, July, August, September, October, November, December}.

15. **(a)** $\{r, p\}$; **(b)** $\{r\}$; **(c)** $\{\ \}$.

17. False; for example, see sets in Exercises 2, 4, and 5.

19. Among others: the counting numbers less than 6.

21. Among others: the whole numbers greater than 100.

1-2 OPERATIONS WITH SETS *Page 31*

1. **(a)** $\{2, 4, 6, 8\}$ **(b)** $\{2, 3, 5, 6, 7, 8\}$

3. **(a)** $\{2, 4\}$ **(b)** $\{7\}$

5. **(a)** $\{2, 4, 6, \ldots\}$ **(b)** $\varnothing$

7. **(a)** $\{1, 3, 4, 5, 7\}$ **(b)** $\{1, 3\}$

9. **(a)** $\{2, 4, 6, 7, 8\}$ **(b)** $\{4, 6, 8\}$

11. **(a)** $\{1, 2, 3, \ldots\}$ **(b)** $\varnothing$

13. **(a)** $\{3, 4, 5\}$ **(b)** $\{2, 4\}$ **(c)** $\{2, 3, 4, 5\}$ **(d)** $\{4\}$

15. **(a)** $\{2, 4, 6, \ldots\}$ **(b)** $\{1, 3, 5, \ldots\}$ **(c)** $\{1, 2, 3, \ldots\}$ **(d)** $\varnothing$

17. **(a)** $\{2, 3\}$ **(b)** $\{1, 2\}$ **(c)** $\{1, 2, 3\}$ **(d)** $\{2\}$

19. $\{1, 6, 8\}$

21. $\{1, 2, 3, 4, 5, 6, 7, 8, 9\}$

23. $\{10\}$

25. **(a)** $\{b, c\}$, $\{a, c\}$, $\{a, b\}$, $\{a\}$, $\{b\}$, $\{c\}$, $\varnothing$.
*(b) $\{b, c, d, e\}$, $\{a, c, d, e\}$, $\{a, b, d, e\}$, $\{a, b, c, e\}$, $\{a, b, c, d\}$, $\{a, b, c\}$, $\{a, b, d\}$, $\{a, b, e\}$, $\{a, c, d\}$, $\{a, c, e\}$, $\{a, d, e\}$, $\{b, c, d\}$, $\{b, c, e\}$, $\{b, d, e\}$, $\{c, d, e\}$, $\{a, b\}$, $\{a, c\}$, $\{a, d\}$, $\{a, e\}$, $\{b, c\}$, $\{b, d\}$, $\{b, e\}$, $\{c, d\}$, $\{c, e\}$, $\{d, e\}$, $\{a\}$, $\{b\}$, $\{c\}$, $\{d\}$, $\{e\}$, $\varnothing$.
*(c) None exist.

1-3 RELATIONS AMONG SETS *Page 36*

1.

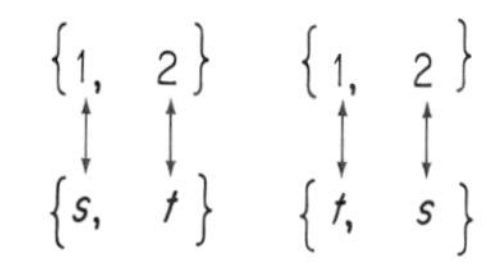

3. Among others:

{C, A, T}
{1, 2, 3}

5. Among others:

{0, 1, 4, P, A, T, R, I, C, K}
↕ ↕ ↕ ↕ ↕ ↕ ↕ ↕ ↕ ↕
{1, 2, 3, 4, 5, 6, 7, 8, 9, 10}

7. 8

9. 0

11. 51

13. (a) $\{(a, 2), (b, 1), (c, 3)\}$
(b)

3	$(a, 3)$	$(b, 3)$	**$(c, 3)$**
2	**$(a, 2)$**	$(b, 2)$	$(c, 2)$
1	$(a, 1)$	**$(b, 1)$**	$(c, 1)$
	a	b	c

15. (a) $\{(1, x), (2, x), (2, y), (3, z)\}$
(b)

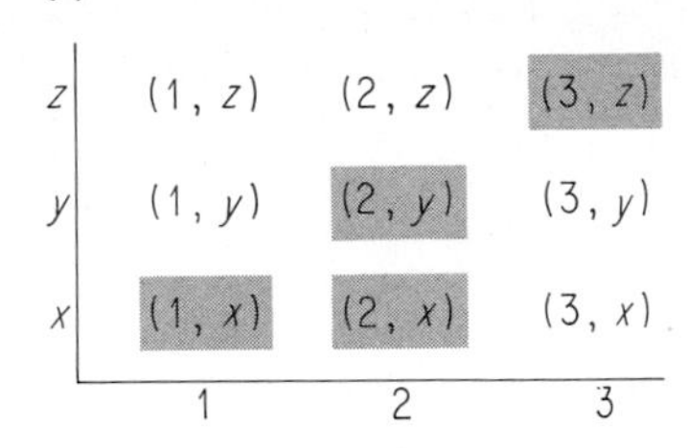

17. 12, 13.

19. (a) Not always true; for example, if $A = \{1\}$ and $B = \{1, 2\}$. In general, the statement is false if $A \subseteq B$.
(b) Always true.

21. (a) $A \cap B = \varnothing$; (b) $A \cap B \neq \varnothing$.

23. (a) $B \subseteq A$; (b) $B \nsubseteq A$.

25. (a) Always true; (b) never false.

27. (a) $A - B$; (b) $A \neq B$.

1-4 SETS OF NUMBERS *Page 42*

1. Ordinal.

3. Cardinal.

5. Identification (also ordinal).

7. Ordinal.

9. Identification (also ordinal).

11. $\{3, 6, 9, 12, \ldots\}$

13. $\{\ldots, -15, -10, -5, 0, 5, 10, 15, \ldots\}$

15. $\{\ldots, -21, -14, -7\}$

17. (a) $\{15, 30, 45, \ldots\}$;
(b) the set of positive integral multiples of 15.

19. (a) $\{6, 12, 18, \ldots\}$;
(b) the set of positive integral multiples of 6.

***21.** $\{1, 2, 3, 6\}$

***23.** $\{1, 17\}$

***25.** $\{1, 3, 9, 27\}$

1-5 SETS OF POINTS *Page 50*

1.

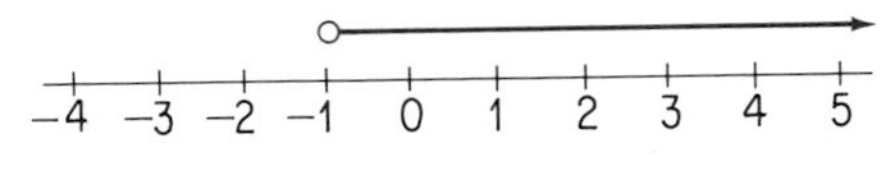

3.

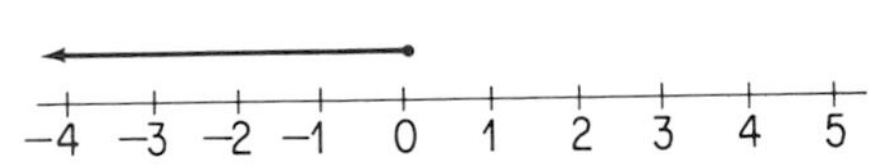

5.

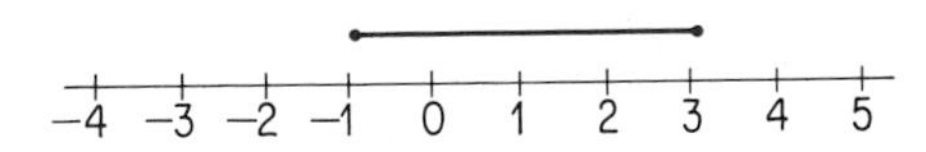

7. Among others:

9. Among others:

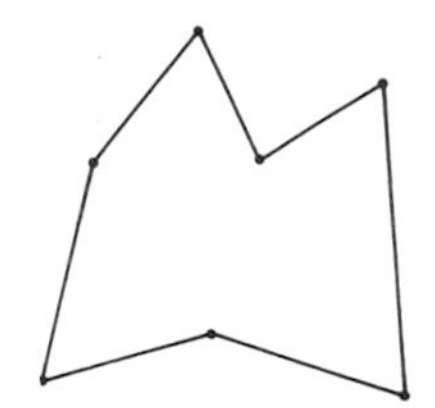

11. $\overline{AD}$

13. $\overline{BC}$

15. $\overline{BC}$

17. $\overline{BC}$

19. $\overrightarrow{BC}$ (or $\overrightarrow{BD}$)

21. B

23. $\overline{BC}$

25. $\overrightarrow{CE}$

27. $\overleftrightarrow{AD} \cup \overrightarrow{CE}$
29. $\varnothing$
31. $\{B, D\}$
33. $\overline{BF}$
35. Interior $\triangle CBD$
37. $\overleftrightarrow{CD}$

1-6 EULER DIAGRAMS AND VENN DIAGRAMS *Page 58*

1. **(a)** 3; **(b)** 7; **(c)** 5; **(d)** 12.

3.

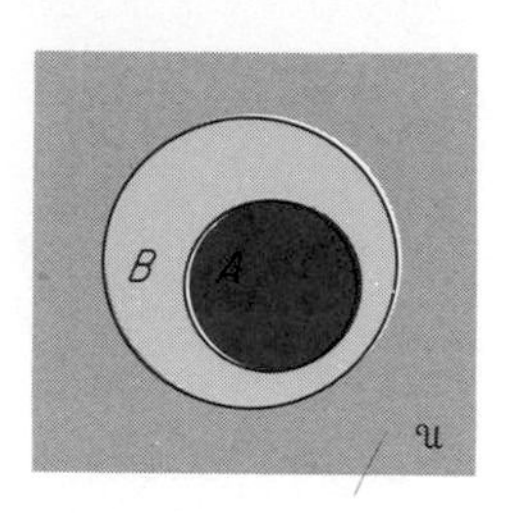

5.

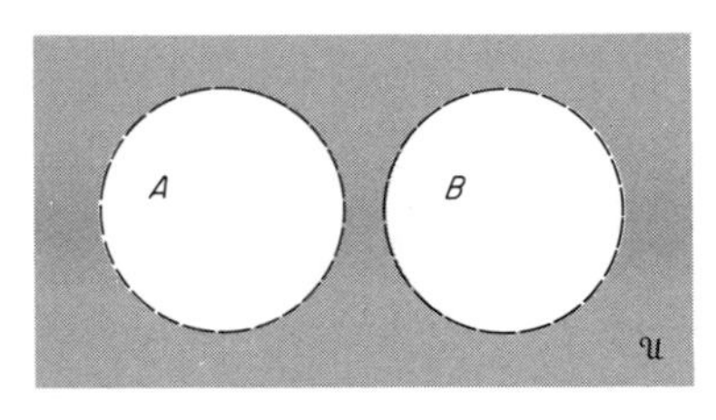

7.

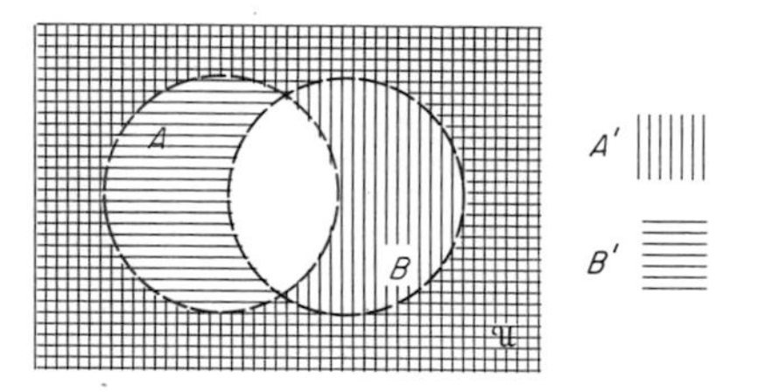

9. A' is shaded with horizontal lines; B is shaded with vertical lines. The union of these two sets is the subset of 𝔘 that is shaded with horizontal lines, vertical lines, or lines in both directions.

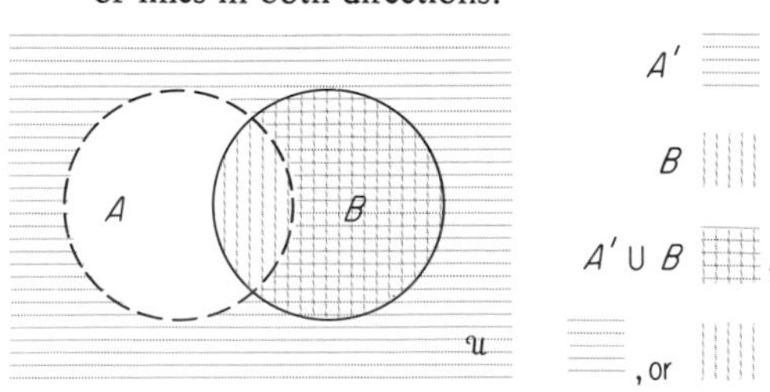

11. A is shaded with vertical lines; B' is shaded with horizontal lines. The intersection of these two sets is the subset of 𝔘 that is shaded with lines in both directions.

13. In the following pair of diagrams, the final result is the same, showing the equivalence of the statements given.

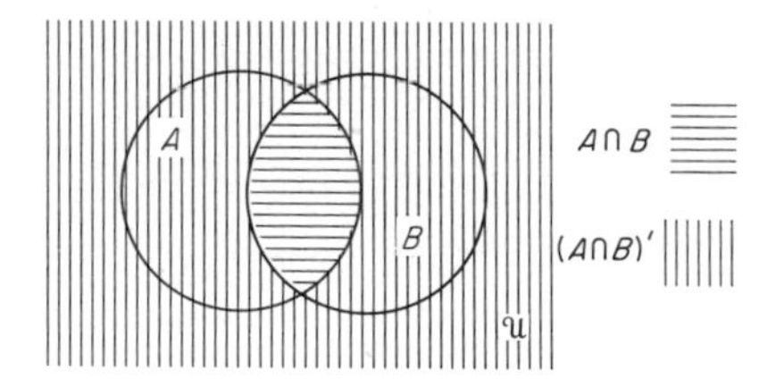

The set $(A \cap B)$ is shaded with horizontal lines. Its complement, $(A \cap B)'$, is the remaining portion of 𝔘 shaded with vertical lines.

The set A' is shaded with vertical lines; the set B' is shaded with horizontal lines. Their union, $A' \cup B'$, is the portion of 𝔘 shaded with vertical lines, horizontal lines, or lines in both directions.

15. **(a)** 2; **(b)** 4; **(c)** 6; **(d)** 15; **(e)** 28; **(f)** 29.

17. **(a)** 33; **(b)** 10; **(c)** 10; **(d)** 39; **(e)** 46; **(f)** 8.

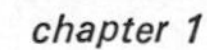

19. **(a)**

$A \cap B \cap C$

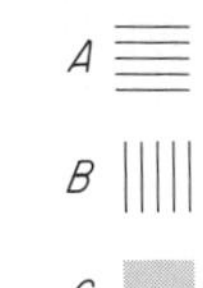

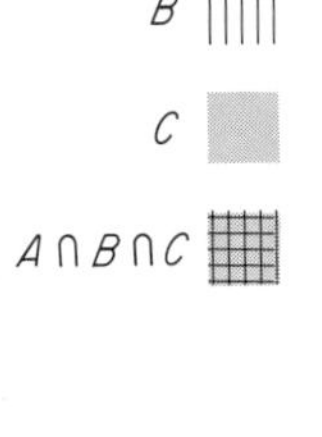

(b)

$A \cap B \cap C'$

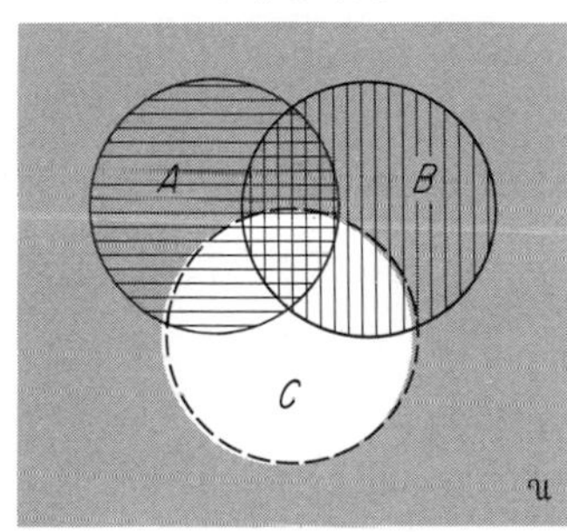

(c)

$A \cap B' \cap C$

(d)

$A \cap B' \cap C'$

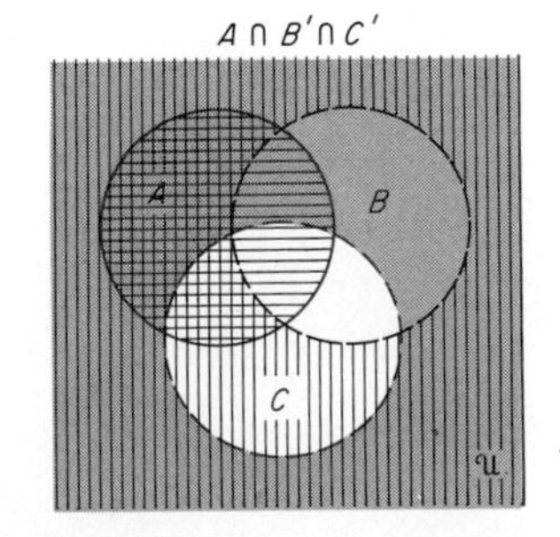

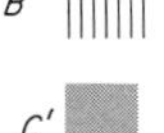

21. $(A \cup B) \cap C'$

23. $A \cap B' \cap C$

25. As shown in the following diagram:
(a) 12; **(b)** 9; **(c)** 3.

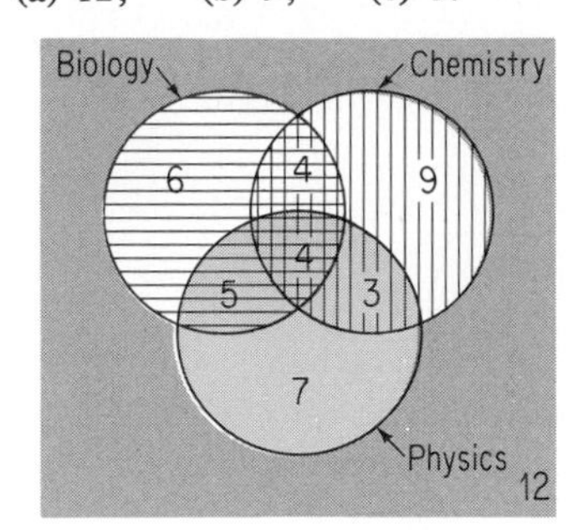

27. As shown in the following diagram:
(a) 4; **(b)** 0.

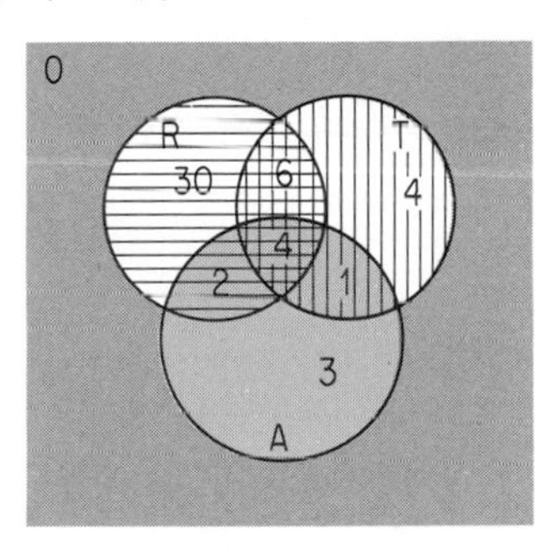

29. As shown in the following diagram the data would require -2 students in $A' \cap B' \cap C$ which is impossible.

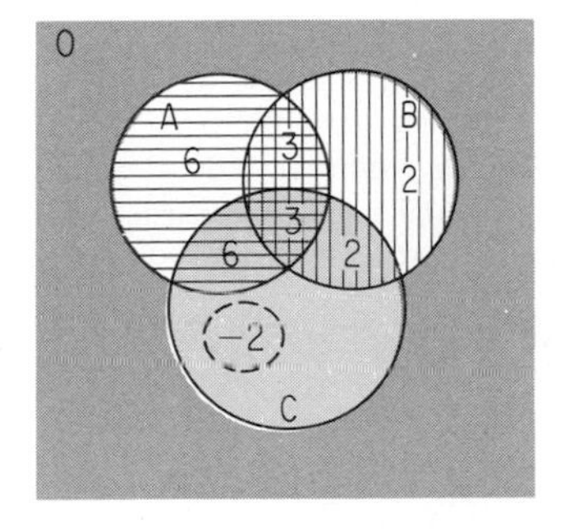

1-7 FLOW CHARTS *Page 65*

1. (a) 2075; (b) 5275.
3. (a) 56; (b) 65.
5. 23
7. 29
9. 37
11. 12, 19, 26, 33, 40, 47, 54, 61.

1-8 SETS OF STATEMENTS *Page 71*

1. 4
3. 14
5. 5
7. (a) (ii), (iii);
(b) (i), (iv), (v), (vi);
(c) (iv).
9. (a) (iii), (iv), (vi);
(b) (i), (ii), (v);
(c) none.
11. Among others:
(a) $(\sim p) \wedge (\sim q)$;
(b) $(\sim p) \wedge q$;
(c) $(\sim p) \wedge q$;
(d) $\sim[p \wedge (\sim q)]$;
(e) $p \vee (\sim q)$.
13. Among others:
(a) $p \wedge (\sim q)$;
(b) $p \vee (\sim q)$;
(c) $(\sim p) \wedge q$;
(d) $\sim[(\sim p) \wedge q]$;
(e) $\sim[(\sim p) \wedge q]$.
15. (a) I like this book and I like mathematics.
(b) I do not like mathematics.
(c) I do not like this book.
(d) I do not like this book and I do not like mathematics.
17. 15(a), 16(b) and 16(d).
***19.** (a) True; (b) true;
(c) true; (d) true.

Chapter 2: LOGIC

2-1 LANGUAGE *Page 81*

1. Among others: Alice: Do Exercises 5 or (7 and 10 or 12).
Bill: Do Exercises (5 or 7) and (10 or 12).
Charles: Do Exercises 5 or (7 and 10) or 12.
Don: Do Exercises (5 or 7) and (10 or 12).
Ed: Do Exercises (5 or 7 and 10) or 12.

3. None.

5. Among others: Not all numbers are whole numbers; there exists at least one number that is not a whole number.

7. Among others: Some fractions do not represent rational numbers; not all fractions represent rational numbers.

9. Among others: All real numbers are rational numbers; if a number is a real number, then it is a rational number.

11. Among others: No rational numbers are integers; all rational numbers are not integers.

2-2 TRUTH VALUES OF STATEMENTS *Page 85*

1.

p	q	$(\sim p) \wedge q$
T	T	F
T	F	F
F	T	T
F	F	F

3.

p	q	$(\sim p) \vee (\sim q)$
T	T	F
T	F	T
F	T	T
F	F	T

5.

p	q	$\sim(p \wedge q)$
T	T	F
T	F	T
F	T	T
F	F	T

7.

p	q	$\sim$	$[p$	$\vee$	$(\sim q)]$
T	T	F	T	T	F
T	F	F	T	T	T
F	T	T	F	F	F
F	F	F	F	T	T
		(d)	(a)	(c)	(b)

9.

p	q	$\sim$	$[(\sim p)$	$\wedge$	$(\sim q)]$
T	T	T	F	F	F
T	F	T	F	F	T
F	T	T	T	F	F
F	F	F	T	T	T
		(d)	(a)	(c)	(b)

11.

p	q	$p \veebar q$
T	T	F
T	F	T
F	T	T
F	F	F

13. $\sim(p \wedge q)$

2-3 CONDITIONAL STATEMENTS *Page 91*

1. If you pass the examination, then you pass the course.

3. If you do not pass the examination, then you do not pass the course.

5. (1) If John drives a red car, then John lives in a white house.
   (2) If John lives in a white house, then John drives a red car.
   (3) If John does not drive a red car, then John does not live in a white house.
   (4) If John does not live in a white house, then John does not drive a red car.

7. **(a)** T; **(b)** T; **(c)** T.

9. *Converse:* If we buy a new car, then we can afford it.
   Inverse: If we cannot afford it, then we do not buy a new car.
   Contrapositive: If we do not buy a new car, then we cannot afford it.

11. *Converse:* If the triangles are congruent, then two sides and the included angle of one are congruent to two sides and the included angle of the other.
   Inverse: If two sides and the included angle of one triangle are not congruent to two sides and the included angle of another triangle, then the triangles are not congruent.
   Contrapositive: If two triangles are not congruent, then two sides and the included angle of one are not congruent to two sides and the included angle of the other.

13. *Converse:* If $x = 1$, then $x(x - 1) = 0$.
   Inverse: If $x(x - 1) \neq 0$, then $x \neq 1$.
   Contrapositive: If $x \neq 1$, then $x(x - 1) \neq 0$.

15. (11) True; (12) true; (13) false.

17. (11) True; (12) false; (13) true.

19. $\{7\}$

*21. The set of all real numbers.

*23. The set of all real numbers different from 2.

2-4 CONDITIONAL AND BICONDITIONAL STATEMENTS *Page 96*

1. If the creature is a duck, then it is a bird.

3. If two angles are complements of the same angle, then the two angles are congruent.

5. If two lines are parallel to the same line, then the two lines are coplanar.

7. If a geometric figure is a circle, then the figure is round.

9. If a person is a teacher, then that person is boring.

11. If you like this book, then you like mathematics.

13. If you like mathematics, then you like this book.

15. If you like mathematics, then you will like this book.

17. $q \longrightarrow p$

19. $q \longrightarrow p$

21. $p \longrightarrow (\sim q)$

23. $p \longleftrightarrow q$

25. $p \longrightarrow q$

27. $p \longleftrightarrow q$

29. If $9 + 3 < 10$, then $11 - 3 > 8$; true.

31. If $5 + 3 = 8$, then $7 \times 4 = 25$; false.

33. If $7 \times 6 = 42$, then $8 \times 5 \neq 40$; false.

35. The assertion is equivalent to the statement: "If you give me $10,000, then I will marry your daughter." If he received the money, then he should have married the girl. Thus he should be sued for breach of promise.

2-5 IMPLICATIONS *Page 102*

1. Bill is driving a Ford, Bill is not driving a Ford; not logically true.

3. Logically true since $7 \times 8 = 56$ is always true.

5. Logically true since one or the other of the parts is always true.

7. Ginny is 16 years old, Ginny is not 16 years old and is at most 21 years old, Ginny is over 21 years old; not logically true.

9. My house number is 41, my house number is not 41; logically true.

11. Implication.

13. Not an implication.

15. Not an implication.

17. Implication.

19. Implication.

21. Not an implication.

23. Implication.

25. Equivalence.

27. Equivalence.

29. Equivalence.

2-6 MATHEMATICAL PROOFS *Page 107*

1. For

p: Elliot is a freshman.
q: Elliot takes mathematics.

the argument has the form

$$[(p \longrightarrow q) \wedge p] \longrightarrow q.$$

This statement is an implication (§2-5, Example 5) and the argument is valid.

3. For

p: The Yanks win the game.
q: The Yanks win the pennant.

the argument has the form

$$[(p \longrightarrow q) \wedge (\sim q)] \longrightarrow (\sim p).$$

This statement is an implication (§2-5, Exercise 20) and the argument is valid.

5. For

p: You work hard.
q: You are a success.

the argument has the form

$$[(p \longrightarrow q) \wedge (\sim q)] \longrightarrow (\sim p).$$

This statement is an implication (§2-5, Exercise 20) and the argument is valid.

7. For

p: You are reading this book.
q: You like mathematics.

the argument has the form

$$[(p \longrightarrow q) \wedge (\sim p)] \longrightarrow (\sim q).$$

This statement is not an implication (§2-5, Exercise 22) and the argument is not valid.

9. This argument is of the form $[(p \longrightarrow q) \wedge (q \longrightarrow r)] \longrightarrow (r \longrightarrow p)$ and is not valid.

11. You do not drink milk.

13. If you like to fish, then you are a mathematician.

15. If you like this book, then you will become a mathematician.

2-7 WHAT WAS ACTUALLY SAID? *Page 113*

1. Someone is home.

3. I am passing at least one of my courses.

5. $x = -5$

7. Valid.

9. Not valid.

11. **(a)** Valid; **(b)** valid; **(c)** valid; **(d)** not valid.

3-1 EGYPTIAN NUMERATION *Page 124*

1.

∩∩|||||

3.

𓆼𓆼𓆼𓍢𓍢𓍢𓍢∩|||||||

5.

𓂭𓆼𓆼𓍢𓍢𓍢𓍢|||||||

7. 22

9. 1102

11. 1324

13.

∩∩∩∩∩∩∩∩∩|||
\+ ∩∩∩||
―――――――
𓍢 ∩ |||||

15.

𓍢𓍢∩∩∩||||||||
\+ 𓍢∩∩∩|||||
―――――――
𓍢𓍢𓍢∩∩∩∩∩∩∩|||

17.

𓆼𓆼𓍢𓍢𓍢𓍢𓍢𓍢∩∩∩∩∩|
− 𓍢𓍢𓍢𓍢𓍢 ∩∩||||||||
―――――――

After exchanging:

||||||||||
𓆼𓆼𓍢𓍢𓍢𓍢𓍢𓍢∩∩∩∩|
− 𓍢𓍢𓍢𓍢𓍢 ∩∩||||||||
―――――――
𓆼𓆼 𓍢 ∩∩|||

19.
$$\begin{aligned}
&\textcircled{1} \times 45 = \textcircled{45}\\
&2 \times 45 = 90\\
&4 \times 45 = 180\\
&8 \times 45 = 360\\
&\textcircled{16} \times 45 = \textcircled{720}\\
&17 = 1 + 16\\
&17 \times 45 = (1 + 16) \times 45\\
&\quad = 45 + 720 = 765
\end{aligned}$$

21.
$$\begin{aligned}
&\textcircled{1} \times 41 = \textcircled{41}\\
&2 \times 41 = 82\\
&\textcircled{4} \times 41 = \textcircled{164}\\
&\textcircled{8} \times 41 = \textcircled{328}\\
&\textcircled{16} \times 41 = \textcircled{656}\\
&29 = 1 + 4 + 8 + 16\\
&29 \times 41 = (1 + 4 + 8 + 16) \times 41\\
&\quad = 41 + 164 + 328 + 656\\
&\quad = 1189
\end{aligned}$$

23.
$$\begin{aligned}
&29 \longrightarrow \textcircled{44}\\
&14 \quad 88\\
&7 \longrightarrow \textcircled{176}\\
&3 \longrightarrow \textcircled{352}\\
&1 \longrightarrow \textcircled{704}\\
&29 \times 44 = 44 + 176 + 352 + 704\\
&\quad = 1276
\end{aligned}$$

25.
$$\begin{aligned}
&31 \longrightarrow \textcircled{22}\\
&15 \longrightarrow \textcircled{44}\\
&7 \longrightarrow \textcircled{88}\\
&3 \longrightarrow \textcircled{176}\\
&1 \longrightarrow \textcircled{352}\\
&31 \times 22 = 22 + 44 + 88 + 176 + 352\\
&\quad = 682
\end{aligned}$$

3-2 EARLY METHODS OF COMPUTATION *Page 129*

1. 5 × 785, that is, 3925.

3. 7 × 387, that is, 2709.

5. 9 × 279, that is, 2511.

7.

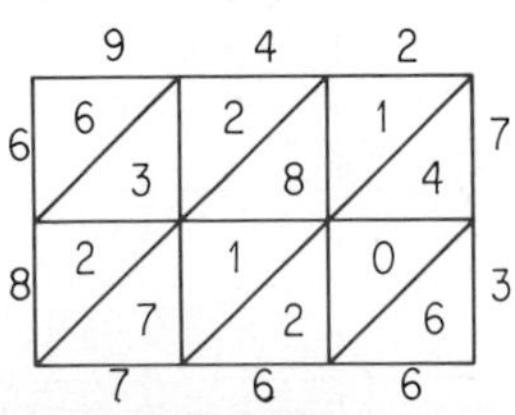

Answer: 68,766

9.

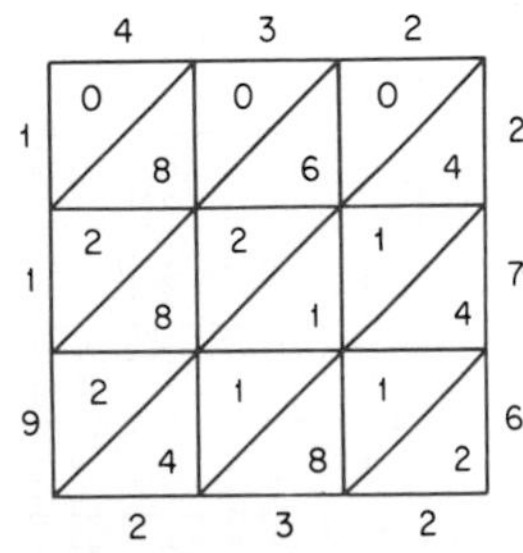

Answer: 119,232

11.

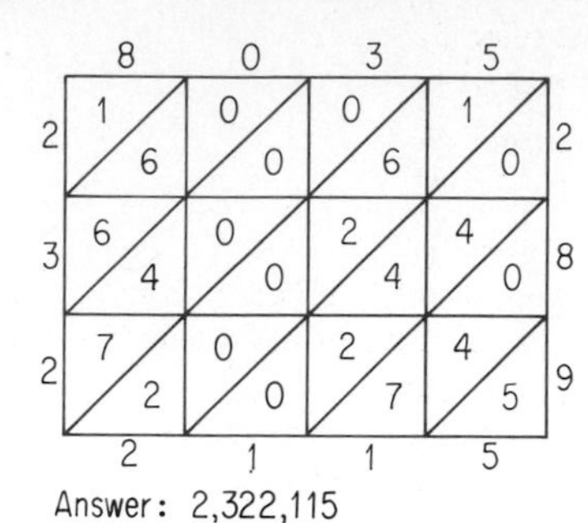

13.

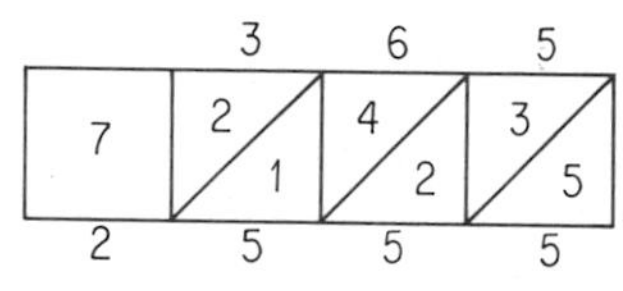

Answer: 2555

15.

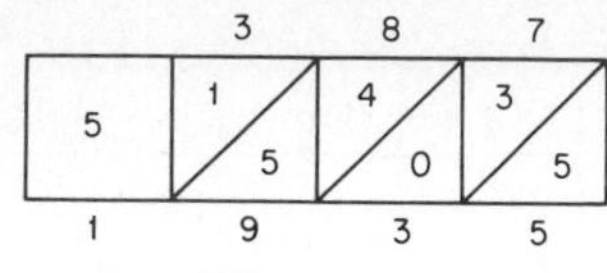

Answer: 1935

17.

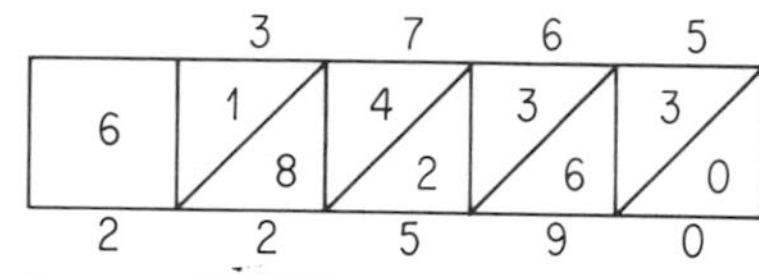

Answer: 22,590

3-3 CLOCK ARITHMETIC *Page 133*

1. 5
3. 7
5. 12
7. 7
9. 4
11. 12
13. 4
15. 4
17. 12
19. 9
21. 7
23. 1, 5, or 9
25. 11
***27.** ∅; an impossible equation, that is, there is no value of t for which this equation is true.
***29.** 𝒰; an identity, that is, this equation is true for all possible replacements of t.

31.

X	1	2	3	4	5	6	7	8	9	10	11	12
1	1	2	3	4	5	6	7	8	9	10	11	12
2	2	4	6	8	10	12	2	4	6	8	10	12
3	3	6	9	12	3	6	9	12	3	6	9	12
4	4	8	12	4	8	12	4	8	12	4	8	12
5	5	10	3	8	1	6	11	4	9	2	7	12
6	6	12	6	12	6	12	6	12	6	12	6	12
7	7	2	9	4	11	6	1	8	3	10	5	12
8	8	4	12	8	4	12	8	4	12	8	4	12
9	9	6	3	12	9	6	3	12	9	6	3	12
10	10	8	6	4	2	12	10	8	6	4	2	12
11	11	10	9	8	7	6	5	4	3	2	1	12
12	12	12	12	12	12	12	12	12	12	12	12	12

3-4 MODULAR ARITHMETIC *Page 136*

1.

×	0	1	2	3	4
0	0	0	0	0	0
1	0	1	2	3	4
2	0	2	4	1	3
3	0	3	1	4	2
4	0	4	3	2	1

3. Among others:
$(3 \times 2) \times 4 = 3 \times (2 \times 4) = 4$
$(4 \times 3) \times 4 = 4 \times (3 \times 4) = 3$

5. The inverse of 1 is 1, of 2 is 3, of 3 is 2, and of 4 is 4. Note that 0 does not have an inverse with respect to multiplication.

7. 3 (mod 5)

9. 4 (mod 5)

11. 4 (mod 5)

13. 2 (mod 5)

15. 3 (mod 5)

17. 4 (mod 5)

19. 2 (mod 5)

21. 3 (mod 5)

23. An impossible equation; that is, there is no value of x for which this equation is true.

25. $2 \times 6 \equiv 0 \pmod{12}$
$3 \times 4 \equiv 0 \pmod{12}$
$3 \times 8 \equiv 0 \pmod{12}$
$4 \times 6 \equiv 0 \pmod{12}$
$4 \times 9 \equiv 0 \pmod{12}$
$6 \times 6 \equiv 0 \pmod{12}$
$6 \times 8 \equiv 0 \pmod{12}$
$6 \times 10 \equiv 0 \pmod{12}$
$8 \times 9 \equiv 0 \pmod{12}$

Thus the zero divisors in arithmetic modulo 12 are:
2 and 6, 3 and 4, 3 and 8, 4 and 6, 4 and 9, 6 and 6, 6 and 8, 6 and 10, and 8 and 9.

3-5 DECIMAL NOTATION AND DIVISIBILITY *Page 140*

1. $(2 \times 10^2) + (5 \times 10^1) + (7 \times 10^0)$

3. $(3 \times 10^3) + (5 \times 10^2) + (0 \times 10^1) + (4 \times 10^0)$

5. $(2 \times 10^5) + (3 \times 10^4) + (5 \times 10^3) + (1 \times 10^2) + (0 \times 10^1) + (0 \times 10^0)$

7. $(5 \times 10^5) + (0 \times 10^4) + (0 \times 10^3) + (2 \times 10^2) + (0 \times 10^1) + (0 \times 10^0)$

9. 8165

11. 609,502

13. 3,000,002

15. 8,000,000,000

	(a)	(b)	(c)	(d)	(e)	(f)	(g)	(h)	(i)
17.	Yes	yes	yes	yes	yes	yes	no	no	no
19.	Yes	yes	yes	no	yes	yes	yes	no	no
21.	Yes	no	yes	yes	no	no	no	no	no
23.	Yes	yes	yes	yes	yes	yes	yes	yes	yes
25.	Yes	no	yes	yes	no	no	no	no	no

27. A whole number is divisible by 6 if it is divisible by 2 and by 3.

29. A whole number is divisible by 9 if the sum of its digits is divisible by 9.

31. A whole number is divisible by 100 if its tens digit and its units digit are both 0.

3-6 OTHER SYSTEMS OF NUMERATION *Page 146*

1. 22_{five}

3. 30_{four}

5.

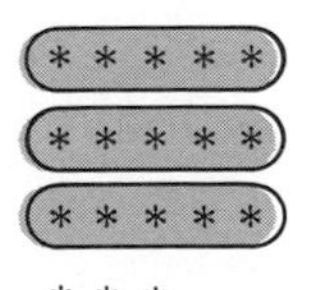

7.

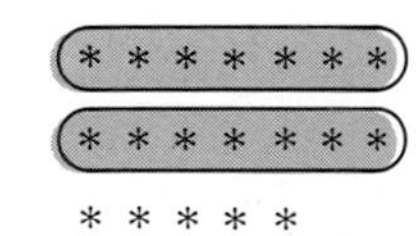

9.

11. 19
13. 35
15. 23
17. 108
19. 124
21. 111
23. 517
25. 596
27. 1212_5
29. 12412_5
31. 10000_5
33. 2322_5
35. 4412_5
***37.** 223
***39.** 27
***41.** 862
***43.** 694

3-7 COMPUTATION IN BASE FIVE NOTATION *Page 152*

1. 131_5
3. 1132_5
5. 12320_5
7. 212_5
9. 144_5
11. 113_5
13. 212_5
15. 2131_5
17. 10332_5
19. 3113_5
21. 24333_5
23. 22_5
25. 224_5 R 2
27. 4_5
29. 32_5
31.

X	0	1	2	3	4
0	0	0	0	0	0
1	0	1	2	3	4
2	0	2	4	11_5	13_5
3	0	3	11_5	14_5	22_5
4	0	4	13_5	22_5	31_5

Chapter 4: WHOLE NUMBERS: PROPERTIES AND OPERATIONS

4-1 COUNTING NUMBERS *Page 161*

1. {(1, 1), (1, 2), (1, 3), (1, 4), (2, 1), (2, 2), (2,3), (2,4)}.

3. Use any counterexample such as $3 - 2 \neq 2 - 3$.

5. No, $8 - (3 - 2) = 8 - 1 = 7$; $(8 - 3) - 2 = 3$; no.

7. Use any counterexample such as
$$2 + (3 \times 5) \neq (2 + 3) \times (2 + 5);$$
$$17 \neq 5 \times 7.$$

9. Commutative, +.

11. Associative, ×.

13. Distributive.

15. Commutative, +.

17. Distributive.

19. $92 + (50 + 8)$
$= 92 + (8 + 50)$ Commutative, +.
$= (92 + 8) + 50$ Associative, +.

21. $37 \times (1 + 100)$
$= 37 \times (100 + 1)$ Commutative, +.
$= (37 \times 100) + (37 \times 1)$ Distributive.
$= 37 \times 100 + 37$ Identity, ×.
$= 3700 + 37$ Number fact.

23. $(73 + 19) + (7 + 1)$
$= [(73 + 19) + 7] + 1$ Associative, +.
$= [73 + (19 + 7)] + 1$ Associative, +.
$= [(73 + (7 + 19)] + 1$ Commutative, +.
$= [(73 + 7) + 19] + 1$ Associative, +.
$= (73 + 7) + (19 + 1)$ Associative, +.

25. No.

27. $7 \times (80 - 1) = (7 \times 80) - (7 \times 1) = 560 - 7 = 553$

29. $8 \times (90 + 2) = (8 \times 90) + (8 \times 2) = 720 + 16 = 736$

31. 5

33. 5

35. 3

37. 6

39. Any counting number.

4-2 PRIME NUMBERS *Page 168*

1. The set of numbers divisible by 15 is a proper subset of the set of numbers divisible by 3.

3. The intersection of the set of numbers divisible by 3 and the set of numbers divisible by 5 is the set of numbers divisible by 15.

5. The set of numbers divisible by 12 is a proper subset of the set of numbers divisible by 3.

7. {21, 22, 24, 25, 26, 27, 28, 30, 32, 33, 34, 35, 36, 38, 39, 40, 42, 44, 45, 46, 48, 49}

9. No, for example, 9 is not a prime number. No, 2 is a prime number.

11. There are other possible answers in many cases. $4 = 2 + 2$, and

$6 = 3 + 3$; $8 = 3 + 5$;
$10 - 3 + 7$; $12 = 5 + 7$;
$14 = 7 + 7$; $16 = 3 + 13$;
$18 = 5 + 13$; $20 = 7 + 13$;
$22 = 5 + 17$; $24 = 7 + 17$;
$26 = 3 + 23$; $28 = 5 + 23$;
$30 = 7 + 23$; $32 = 3 + 29$;
$34 = 5 + 29$; $36 = 7 + 29$;
$38 = 7 + 31$; $40 = 3 + 37$.

13. Any three consecutive odd numbers includes 3 or a multiple of 3. Each multiple of 3 that is greater than 3 is composite, and 1 is by definition not a prime. Therefore 3, 5, 7 is a set of three consecutive odd numbers that are all prime numbers and 1 is by definition not a prime. There-composite number that is a multiple of 3, this is the only prime triplet.

***15.** **(a)** 13; **(b)** 19; **(c)** 31.

17. 1×15; 3×5.

19. 1×24; 2×12; 3×8; 4×6.

21. 1×31

23. 2^6

25. 3×71

27. $5^2 \times 97$

29. $2 \times 3 \times 103$

4-3 APPLICATIONS OF PRIME FACTORIZATIONS *Page 173*

1. {1, 2, 4, 5, 10, 20}

3. {1, 5, 25}

5. {1, 2, 4, 7, 14, 28}

7. {1, 2, 3, 4, 5, 6, 10, 12, 15, 20, 30, 60}

9. $68 = 2^2 \times 17$; $96 = 2^5 \times 3$; the G.C.F. is 2^2, that is, 4.

11. $96 = 2^5 \times 3$; $1425 = 3 \times 5^2 \times 19$; the G.C.F. is 3.

13. $215 = 5 \times 43$; $1425 = 3 \times 5^2 \times 19$; the G.C.F. is 5.

15. $12 = 2^2 \times 3$; $15 = 3 \times 5$;
$20 = 2^2 \times 5$; the G.C.F. is 1.

17. $12 = 2^2 \times 3$; $18 = 2 \times 3^2$;
$36 = 2^2 \times 3^2$; the G.C.F. is 2×3, that is, 6.

19. {7, 14, 21, 28, 35}

21. {15, 30, 45, 60, 75}

23. See Exercise 9; the L.C.M. is $2^5 \times 3 \times 17$, that is, 1632.

25. See Exercise 11; the L.C.M. is $2^5 \times 3 \times 5^2 \times 19$, that is, 45,600.

27. See Exercise 13; the L.C.M. is $3 \times 5^2 \times 19 \times 43$, that is, 61,275.

29. See Exercise 15; the L.C.M. is $2^2 \times 3 \times 5$, that is, 60.

31. See Exercise 17; the L.C.M. is $2^2 \times 3^2$, that is, 36.

33. $\frac{123}{215}$

35. $\frac{29}{24}$

37. $\frac{5}{6}$

39. $\frac{5}{12}$

41. $\frac{1,259}{26,445}$

***43.** 1

4-4 WHOLE NUMBERS *Page 176*

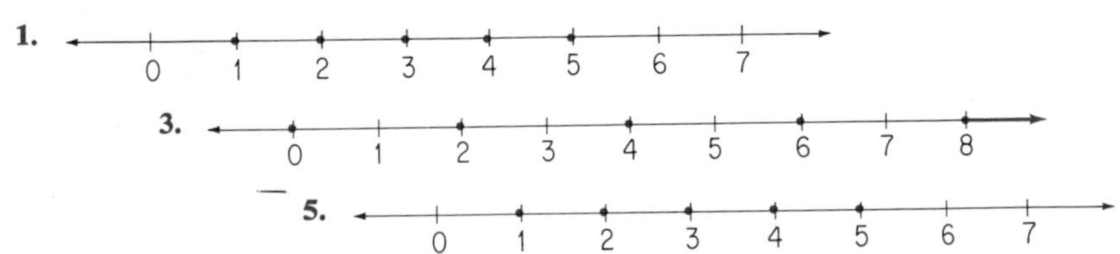

7.

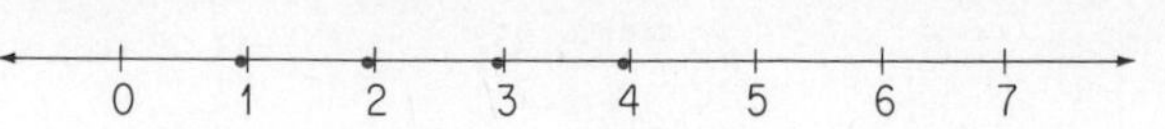

9. The graph is the empty set; there are no points in the graph.

11.

$$\begin{array}{ccccc} \{0, & 1, & 2, & 3, & 4\} \\ \updownarrow & \updownarrow & \updownarrow & \updownarrow & \updownarrow \\ \{1, & 2, & 3, & 4, & 5\} \end{array}$$

13. If $a \times b = 1$, then both $a = 1$ and $b = 1$. If $a \times b = 2$, then either $a = 1$ and $b = 2$ or $a = 2$ and $b = 1$. If $a \times b = 3$, then either $a = 1$ and $b = 3$ or $a = 3$ and $b = 1$. If $a \times b = 4$, then $a = 1$ and $b = 4$, or $a = 2$ and $b = 2$, or $a = 4$ and $b = 1$.

15. None.

17.

$$\begin{array}{lcccccc} W = \{0, & 1, & 2, & 3, & 4, \ldots, & n, & \ldots\} \\ \quad\ \updownarrow & \updownarrow & \updownarrow & \updownarrow & \updownarrow & \updownarrow & \\ C = \{1, & 2, & 3, & 4, & 5, \ldots, & n+1, & \ldots\} \end{array}$$

4-5 ADDITION AND SUBTRACTION *Page 182*

1.
$$\begin{array}{rl} 45 & (4 \times 10) + (5 \times 1) \\ +\ 38 & (3 \times 10) + (8 \times 1) \\ \hline & (7 \times 10) + (13 \times 1) = (8 \times 10) + (3 \times 1) = 83 \end{array}$$

3.
$$\begin{array}{rl} 375 & (3 \times 10^2) + (7 \times 10) + (5 \times 1) \\ +\ 287 & (2 \times 10^2) + (8 \times 10) + (7 \times 1) \\ \hline & (5 \times 10^2) + (15 \times 10) + (12 \times 1) \\ & = (6 \times 10^2) + (6 \times 10) + (2 \times 1) = 662 \end{array}$$

5.
$$\begin{array}{rl} 1309 & (1 \times 10^3) + (3 \times 10^2) + (0 \times 10) + (9 \times 1) \\ +\ 2578 & (2 \times 10^3) + (5 \times 10^2) + (7 \times 10) + (8 \times 1) \\ \hline & (3 \times 10^3) + (8 \times 10^2) + (7 \times 10) + (17 \times 1) \\ & = (3 \times 10^3) + (8 \times 10^2) + (8 \times 10) + (7 \times 1) = 3887 \end{array}$$

7.
$$\begin{array}{rl} 95 & (9 \times 10) + (5 \times 1) \\ -\ 32 & (3 \times 10) + (2 \times 1) \\ \hline & (6 \times 10) + (3 \times 1) = 63 \end{array}$$

9.
$$\begin{array}{rll} 304 & (3 \times 10^2) + (0 \times 10) + (4 \times 1) & = (2 \times 10^2) + (9 \times 10) + (14 \times 1) \\ -\ 128 & (1 \times 10^2) + (2 \times 10) + (8 \times 1) & = (1 \times 10^2) + (2 \times 10) + (8 \times 1) \\ \hline & & (1 \times 10^2) + (7 \times 10) + (6 \times 1) = 176 \end{array}$$

11.
$$\begin{array}{rl} 5023 & (5 \times 10^3) + (0 \times 10^2) + (2 \times 10) + (3 \times 1) \\ -\ 2709 & (2 \times 10^3) + (7 \times 10^2) + (0 \times 10) + (9 \times 1) \\ \hline & (4 \times 10^3) + (10 \times 10^2) + (1 \times 10) + (13 \times 1) \\ & (2 \times 10^3) + (7 \times 10^2) + (0 \times 10) + (9 \times 1) \\ \hline & (2 \times 10^3) + (3 \times 10^2) + (1 \times 10) + (4 \times 1) = 2314 \end{array}$$

4-6 MULTIPLICATION AND DIVISION *Page 189*

1. $20 + 20 + 20 + 20 + 20 = 100$

3. $7 + 7 + 7 + 7 = 28$

5. $40 - 5 = 35$; $35 - 5 = 30$;
$30 - 5 = 25$; $25 - 5 = 20$;
$20 - 5 = 15$; $15 - 5 = 10$;
$10 - 5 = 5$; $5 - 5 = 0$;
therefore, $40 \div 5 = 8$.

7. $120 - 20 = 100$; $100 - 20 = 80$;
$80 - 20 = 60$; $60 - 20 = 40$;
$40 - 20 = 20$; $20 - 20 = 0$;
therefore, $120 \div 20 = 6$.

9.
$$\begin{aligned} 8 \times 15 &= 8 \times (10 + 5) \\ &= (8 \times 10) + (8 \times 5) \\ &= 80 + 40 = 120 \end{aligned}$$

11.
$$\begin{aligned} 9 \times 36 &= 9 \times (30 + 6) \\ &= (9 \times 30) + (9 \times 6) \\ &= 270 + 54 = 324 \end{aligned}$$

13.
$$\begin{aligned} 12 \times 15 &= (10 + 2) \times (10 + 5) \\ &= (10 \times 10) + (10 \times 5) \\ &\quad + (2 \times 10) + (2 \times 5) \\ &= 100 + 50 + 20 + 10 \\ &= 180 \end{aligned}$$

15. $35 \times 45 = (30 + 5) \times (40 + 5)$
$= (30 \times 40) + (30 \times 5) + (5 \times 40) + (5 \times 5)$
$= 1200 + 150 + 200 + 25$
$= 1575$

17.

```
      1 ⎫
     40 ⎬ 241 (quotient)
    200 ⎭
32)7712
   6400
   1312
   1280
     32
     32
```

19.

```
       2 ⎫
      30 ⎬ 432 (quotient)
     400 ⎭
54)23328
   21600
    1728
    1620
     108
     108
```

21.

```
     10 ⎫
    200 ⎭ 210 (quotient)
27)5683
   5400
    283
    270
     13  (remainder)
```

23. The division is correct; $(48 \times 23) + 3 = 1107$.

25. The division is incorrect; $3163 \div 37 = 85$, remainder 18.

4-7 EQUIVALENCE AND ORDER RELATIONS *Page 193*

1. (a), (b), (c), (e), (f), (i).
3. (a), (d), (e), (g), (h), (i).
5. **(a)** Yes; **(b)** yes; **(c)** yes; **(d)** yes.
7. **(a)** Yes; **(b)** no; **(c)** yes; **(d)** no.
9. **(a)** No; **(b)** no; **(c)** yes; **(d)** no.
11. **(a)** No; **(b)** yes; **(c)** no; **(d)** no.
13. **(a)** $>$; **(b)** $<$; **(c)** $=$; **(d)** $>$; **(e)** $<$.

Chapter 5: INTEGERS: PROPERTIES AND OPERATIONS

5-1 THE INTEGERS AS A COMMUTATIVE GROUP *Page 203*

1.

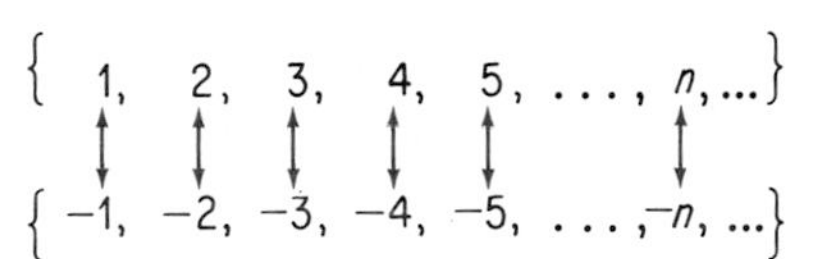

3.

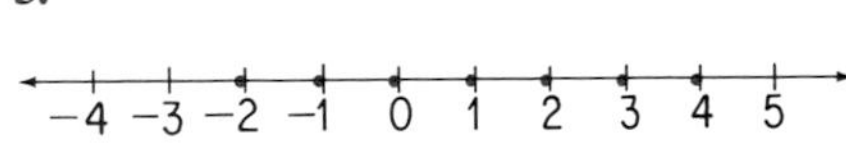

5.

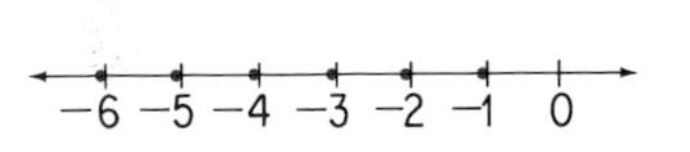

7. True.
9. False.
11. True.
13. False; -0 is not a negative integer.
15. False; for example, $4 \div 3$ is not an integer.
17. False; the set is not closed since the sum of two odd integers is an even integer; also there is no identity element.
19. No, zero is an integer but is neither positive nor negative and thus is not in the union of the set of positive integers and the set of negative integers.
21. Among others:

$\{1, 2, 3, 4, 5, 6, 7, \ldots, 2n, 2n+1, \ldots\}$
$\{0, 1, -1, 2, -2, 3, -3, \ldots, n, -n, \ldots\}$

23. Let $2k$ and $2m$ represent two even integers. Then

$$2k \times 2m = 2 \times 2 \times k \times m = 2(2km),$$

where $2km$ is an integer. Thus $2(2km)$ is an even integer.

25. Any integer is either even or odd. If an integer is even, then its square is even (as in Example 5). If an integer is odd, then its square is odd (as in Exercise 24). If the square of an integer is odd, the integer cannot be even and thus must be odd. If the square of an integer is even, the integer cannot be odd and thus must be even.

5-2 ADDITION AND SUBTRACTION *Page 207*

1. $(+5) + (-7) = -2$

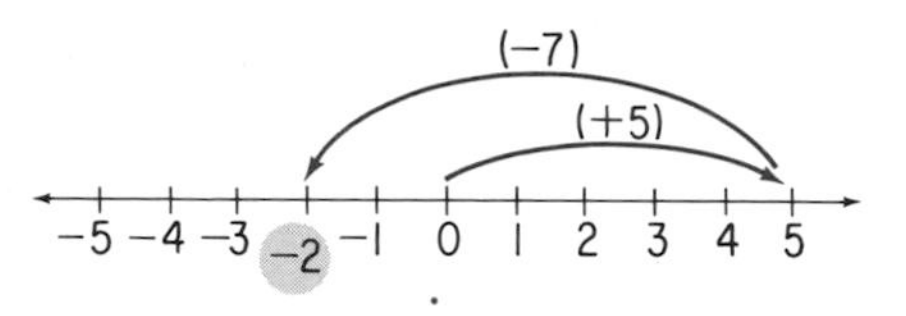

3. $(+3) + (+4) = +7$

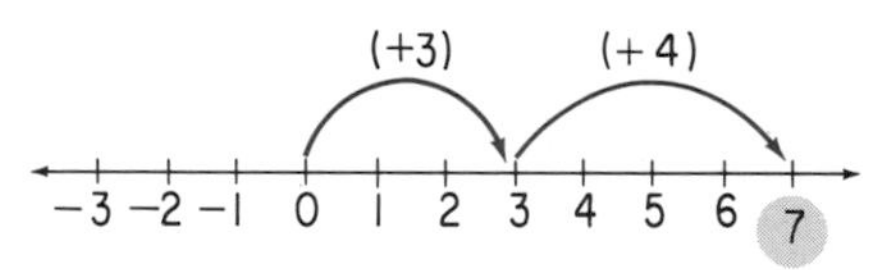

5. -4
7. $+4$
9. -27
11. -15
13. -9
15. 0
17. -3
19. $+12$
21. $+3$
23. $+7$
25. -27
27. -10
29. $+16$
31. -1
33. -31
35. -11
37. -11
39. $+7$
41. Among others: $(+5) - (+3) \neq (+3) - (+5)$; that is, $2 \neq -2$.

5-3 MULTIPLICATION AND DIVISION *Page 213*

1. -45
3. -45
5. -96
7. $+625$
9. -170
11. -105
13. 0
15. $+24$
17. -8
19. -8
21. $+2$
23. $+12$
25. -1
27. -5
29. $+8$
31. -2
33. -20
35. $+40$
37. -4
39. Division of integers is not associative; for example,

$$24 \div (6 \div 2) = 24 \div 3 = 8,$$
$$(24 \div 6) \div 2 = 4 \div 2 = 2.$$

5-4 ABSOLUTE VALUE *Page 218*

1. $+9$
3. -8
5. $+13$
7. $+13$
9. $+3$
11. -13
13. -13
15. $+12$
17. $+2$
19. $+54$
21. -54

23. The greater of +5 and −5 is +5; $+5 = |5|$. The greater of −2 and +2 is +2; $+2 = |-2|$. The greater of 0 and 0 is 0; $0 = |0|$. (Many other examples may be given.)

25. 6, −12.

27. 12, −6.

5-5 INTEGERS AS EXPONENTS *Page 223*

1. a^8
3. 3^5, that is, 243.
5. 10^3, that is, 1000.
7. $6x^9$
9. $10a^7b^7$
11. x^5
13. 10^4, that is, 10,000.
15. 5^3, that is, 125.
17. $5a^2b^3$
19. $4b$
21. n^9
23. a^{12}
25. $x^{10} + y^{10}$
27. $\frac{1}{64}$
29. $-\frac{1}{125}$
31. $\frac{1}{m^2n^2}$
33. $\frac{1}{x^2y^3}$

Chapter 6: RATIONAL NUMBERS: PROPERTIES AND OPERATIONS

6-1 THE SYSTEM OF RATIONAL NUMBERS *Page 230*

1. <
3. >
5. <
7. (a) 2; (b) 1; (c) 2; (d) 0.

	Counting numbers	Whole numbers	Integers	Positive rationals	Rational numbers
9.	✓	✓	✓	✓	✓
11.	x	✓	✓	x	✓
13.	✓	✓	✓	✓	✓
15.	✓	✓	✓	✓	✓
17.	✓	✓	✓	✓	✓
19.	✓	✓	✓	✓	✓
21.	✓	✓	✓	✓	✓

6-2 RATIONAL NUMBERS AND THE NUMBER LINE *Page 234*

1. True.
3. True.
5. False.

7., 9.

Set	8	$\frac{2}{3}$
Counting numbers	✓	x
Whole numbers	✓	x
Integers	✓	x
Rational numbers	✓	✓

11. Every rational number except 0 has a multiplicative inverse; the set of rational numbers is dense.

13. Among others: $\frac{1}{101}, \frac{1}{102}, \frac{1}{103}, \frac{1}{200}, \frac{1}{400}$.

15. No, the density property of the set of rational numbers states that between any two elements of the set there is always another element of the set.

	Sentence	Counting numbers	Whole numbers	Integers	Rational numbers
17.	$n + 5 = 5$	none	0	0	0
19.	$5n = 3$	none	none	none	$\frac{3}{5}$

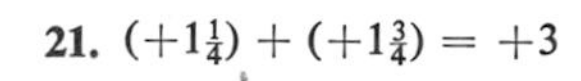

21. $(+1\frac{1}{4}) + (+1\frac{3}{4}) = +3$

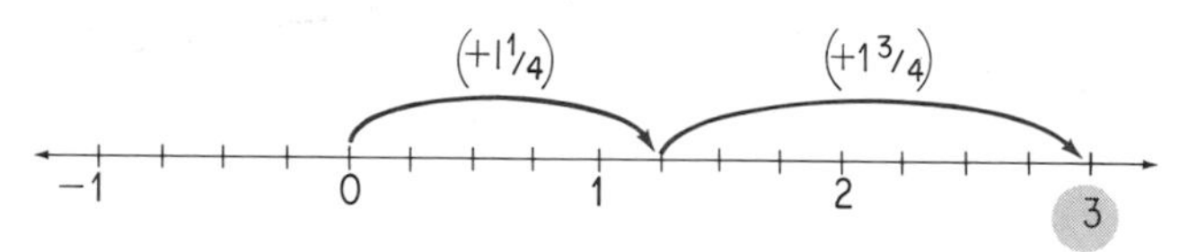

23. $(+1\frac{1}{3}) + (-3\frac{2}{3}) = -2\frac{1}{3}$

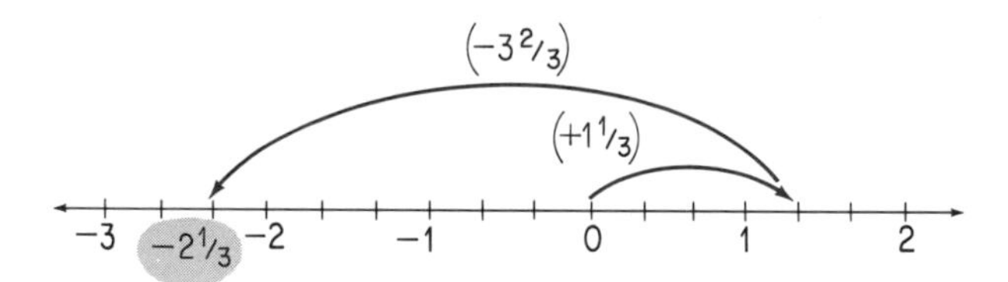

6-3 FRACTIONS: MULTIPLICATION AND DIVISION *Page 242*

1. $\frac{3}{10}$
3. $\frac{3}{4}$
5. $-\frac{1}{5}$
7. $\frac{10}{15}, \frac{9}{15}$
9. $-\frac{8}{12}, -\frac{9}{12}$
11. $\frac{3}{5}$
13. 1
15. $\frac{40}{21}$
17. $-\frac{6}{25}$
19. $-\frac{3}{10}$
21. $-\frac{4}{3}$
23. $-\frac{5}{2}$
25. $\frac{2}{5}$
27. $\frac{5}{12}$
29. $\frac{7}{6}$
31. Among others:
$\frac{2}{3} \div (\frac{1}{4} \div \frac{1}{2}) \neq (\frac{2}{3} \div \frac{1}{4}) \div \frac{1}{2}$; $\frac{4}{3} \neq \frac{16}{3}$.
***33.** Among others:
$\frac{2}{3} \div (\frac{1}{4} \times \frac{1}{2}) \neq (\frac{2}{3} \div \frac{1}{4}) \times \frac{1}{2}$; $\frac{16}{3} \neq \frac{4}{3}$.
In general: $\frac{a}{b} \div \left(\frac{c}{d} \times \frac{e}{f}\right) = \frac{adf}{bce}$ and $\left(\frac{a}{b} \div \frac{c}{d}\right) \times \frac{e}{f} = \frac{ade}{bcf}$
35. $\frac{6}{5}$
37. $\frac{24}{35}$

6-4 FRACTIONS: ADDITION AND SUBTRACTION *Page 249*

1. $\frac{11}{12}$
3. $\frac{14}{9}$, that is, $1\frac{5}{9}$.
5. $\frac{1}{6}$
7. $\frac{7}{8}$
9. $-\frac{3}{2}$, that is, $-1\frac{1}{2}$.
11. $-\frac{2}{5}$
13. $\frac{7}{2}$
15. $\frac{23}{8}$
17. $-\frac{19}{10}$
19. $3\frac{1}{4}$
21. $-2\frac{1}{7}$
23. $7\frac{7}{8}$

25. $6\frac{3}{4}$
27. $3\frac{19}{24}$
29. $\frac{55}{24}$, that is, $2\frac{7}{24}$.
31. $-\frac{3}{8}$
33. $8\frac{5}{12}$
35. $\frac{9}{14}$
37. $\frac{5}{2}$
*39. $\dfrac{adf + bcf + bde}{bdf}$
*41. $\dfrac{acf + ade}{bdf}$
*43. $\dfrac{adf}{bcf + bde}$

6-5 OPERATIONS WITH DECIMALS *Page 255*

1. $(9 \times 10^0) + (3 \times 10^{-1}) + (4 \times 10^{-2})$
3. $(2 \times 10^2) + (3 \times 10^1) + (5 \times 10^0) + (7 \times 10^{-1}) + (8 \times 10^{-2})$
5. $(0 \times 10^0) + (0 \times 10^{-1}) + (0 \times 10^{-2}) + (9 \times 10^{-3})$
7. $(0 \times 10^0) + (0 \times 10^{-1}) + (0 \times 10^{-2}) + (0 \times 10^{-3}) + (2 \times 10^{-4})$
9. $(0 \times 10^0) + (4 \times 10^{-1}) + (0 \times 10^{-2}) + (4 \times 10^{-3})$
11. 0.685
13. 90.053
15. 6901.9
17. 1.4
19. 0.13
21. 0.5
23. 0.05
25. 0.15
27. 0.065
29. 3
31. 0.4
33. 9.9015
35. 0.93805
37. 43.7
39. 0.85
41. 16.18
43. 7.83564
45. 41.5875
47. 10.93
49. (a)
51. (c)
53. (a)
55. (a)

6-6 RATIONAL NUMBERS AND REPEATING DECIMALS *Page 260*

1. 0.6
3. 0.875
5. 1.8
7. $0.\overline{714285}$
9. $0.\overline{692307}$
11. $\frac{8}{11}$
13. $\frac{47}{111}$
15. $\frac{59}{111}$
17. $\frac{55}{9}$
19. $\frac{311}{99}$
21. $\frac{107}{333}$
23. **(a)** 0.4545454545; **(b)** 0.3572572572; **(c)** 0.8000000000; **(d)** 0.4289898989
25. \$2.08, \$2.37, \$2.59, \$2.65, \$2.89
27. $1.77\overline{7}$, 1.780, $1.7\overline{87}$, $1.78\overline{8}$, $1.88\overline{8}$
*29. Among others: 0.2343
*31. Among others: 0.231
*33. **(a)** 15; **(b)** 120.

6-7 RATIONAL NUMBERS AS PER CENTS *Page 265*

1. 0.57
3. 0.03
5. 0.0095
7. 2.50
9. 1.00
11. $\frac{1}{2}$
13. $\frac{19}{20}$
15. $\frac{99}{100}$
17. $\frac{3}{2}$
19. $\frac{2}{25}$
21. 35%
23. 45%
25. 90%
27. 145%
29. 0.1%
31. 9%

33. 92%

35. 150%

37. 76%

39. 170%

41. $\frac{40}{100} = \frac{n}{60}$; 24.

43. $\frac{35}{100} = \frac{n}{80}$; 28.

45. $\frac{20}{160} = \frac{n}{100}$; $12\frac{1}{2}\%$.

47. $\frac{120}{160} = \frac{n}{100}$; 75%.

49. $\frac{25}{n} = \frac{20}{100}$; 125.

51. $\frac{80}{n} = \frac{125}{100}$; 64.

53. $2.00

55. $8.00

57. $14\frac{2}{7}\%$

***59.** $23\frac{1}{2}\%$

Chapter 7: REAL NUMBERS: PROPERTIES AND OPERATIONS

7-1 THE REAL NUMBER LINE *Page 276*

1.

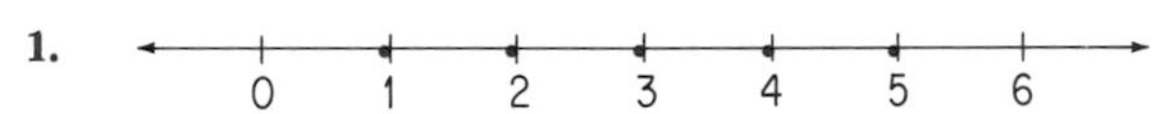

3.

5.

7.

9.

11.

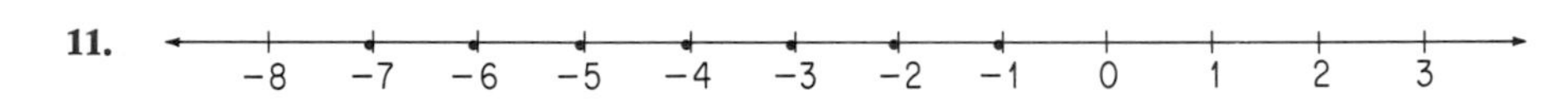

13.

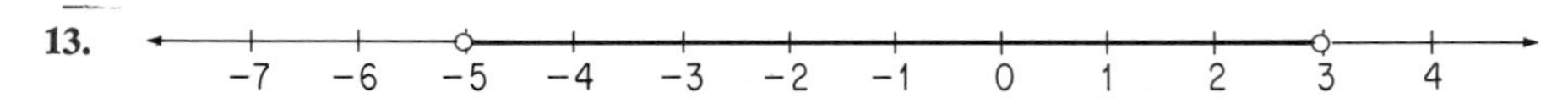

15.

17.

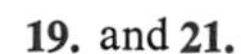

19. and **21.**

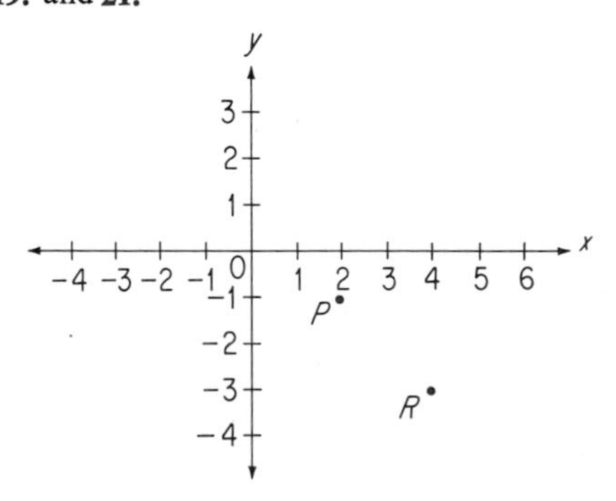

23. (−4, 3)

25. (2, 5)

27. (−3, −4)

7-2 LENGTHS OF LINE SEGMENTS *Page 282*

1. True.

3. False.

5. True.

7. True.

9. **11.**

Set	$-\frac{2}{3}$	0
Counting numbers	x	x
Whole numbers	x	✓
Integers	x	✓
Rational numbers	✓	✓
Real numbers	✓	✓

13. If $\sqrt{2} + 2$ were a rational number a/b, then $\sqrt{2}$ would be a rational number $(a/b) - 2$ (impossible). Therefore $\sqrt{2} + 2$ cannot be a rational number and must be an irrational number.

15. If $\sqrt{2} - (1/2)$ were a rational number a/b, then $\sqrt{2}$ would be a rational number $(a/b) + (1/2)$ (impossible). Therefore $\sqrt{2} - (1/2)$ cannot be a rational number and must be an irrational number.

17. If $5 - 2\sqrt{2}$ were a rational number a/b, then $\sqrt{2}$ would be a rational number $(1/2)[5 - (a/b)]$ (impossible). Therefore $5 - 2\sqrt{2}$ cannot be a rational number and must be an irrational number.

19.

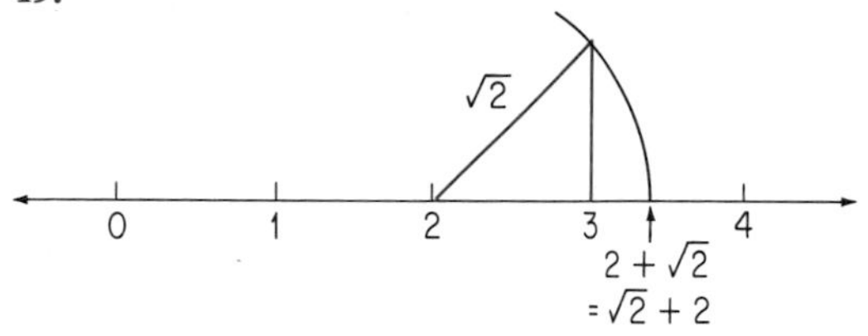

21.

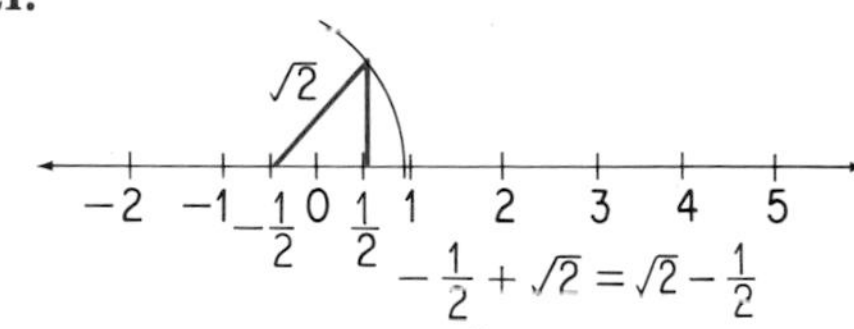

23.

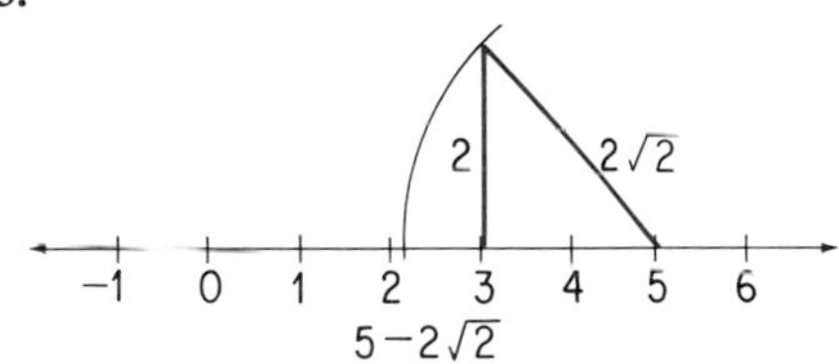

7-3 DECIMAL REPRESENTATIONS OF REAL NUMBERS *Page 286*

1. Rational number.

3. Irrational number.

5. Rational number.

7. A terminating decimal, 0.375.

9. A nonterminating, nonrepeating decimal.

11. A terminating decimal, 10.

13. 0.131331333133331

15. 9

17. $2.89 < 3 < 3.24$

19. $4.84 < 5 < 5.29$

21. 2.6

23. 3.3

25. 0.45, $0.\overline{45}$, 0.45455, 0.454554555 . . . , $0.4\bar{5}$.

27. 0.06, $0.0\bar{6}$, 0.067, 0.067677677 . . . , $0.06\bar{7}$.

29. (c) and (d).

31. Among others: 0.48010010001 . . . , 0.48565565556. . . .

7-4 PROPERTIES OF THE REAL NUMBERS *Page 290*

	Sentence	*Counting numbers*	*Whole numbers*	*Integers*	*Rational numbers*	*Real numbers*
1.	$n + 3 = 1$	none	none	−2	−2	−2
3.	$n \times n = 2$	none	none	none	none	$\sqrt{2}, -\sqrt{2}$

5. (a) An irrational number; (b) an irrational number.

7. (a) A rational number; (b) an irrational number.

9. Among others:
(a) $\sqrt{2} \times (3\sqrt{2})$, $(2 - \sqrt{2})(2 + \sqrt{2})$, $\sqrt{3} \times \sqrt{3}$;
(b) $\sqrt{2} \times \sqrt{3}$, $(3 + \sqrt{2}) \times \sqrt{5}$, $\sqrt{3} \times \sqrt{2}$.

11. Among others:
(a) $\sqrt{2} - \sqrt{2}$, $(1 + \sqrt{2}) - (3 + \sqrt{2})$, $\sqrt{3} - \sqrt{3}$;
(b) $3\sqrt{2} - \sqrt{2}$, $(1 + \sqrt{2}) - (3 - \sqrt{2})$, $\sqrt{3} - \sqrt{2}$.

13. False; among others: 3/2.

15. True; among others: $-\sqrt{3}$.

17. True; among others: $(\sqrt{5}) + (-\sqrt{5}) = 0$.

19. True; among others: $(1/2)(\sqrt{2} + \sqrt{3})$ is between $\sqrt{2}$ and $\sqrt{3}$.

	Counting numbers	Whole numbers	Integers	Rational numbers	Real numbers
21.	✓	✓	✓	✓	✓
23.	✓	✓	✓	✓	✓
25.	✓	✓	✓	✓	✓
27.	x	✓	✓	✓	✓
29.	x	x	✓	✓	✓
31.	✓	✓	✓	✓	✓
33.	x	x	x	✓	✓

7-5 POWERS AND ROOTS *Page 298*

1. $11^{1/2}$
3. $5^{2/3}$
5. $2 + 7^{1/3}$
7. $\sqrt{19}$
9. $\sqrt[3]{7^2}$
11. $6 - \sqrt[3]{3^2}$
13. $(\sqrt{5})/3$
15. $(\sqrt[3]{15})/3$
17. -2
19. $-(\sqrt[3]{245})/7$
21. 4
23. 13
25. 3.27
27. 7.07
29. 4.83
31. 3.35
33. 8.29
***35.** 2.61

Chapter 8: THE METRIC SYSTEM AND MEASUREMENT

8-1 A SHORT HISTORY OF MEASURMENT *Page 307*

1. False.
3. False.
5. True.
7. False.
9. True.
11. 500
13. 40
15. 8.350
17. 3.58
19. 3.785

21. 85 000

23. 58

8-2 THE METRIC SYSTEM *Page 313*

1. Unlikely.
3. Likely.
5. Unlikely.
7. Likely.
9. Likely.
11. (b)
13. (c)
15. (b)
17. 8000 m
19. 8.0 cm
21. 15 km
23. 7000 g
25. 2.500 liters
27. (a) 35 mm; (b) 3.5 cm.
29. (a) 80 mm; (b) 8.0 cm.
31. 32 °F
33. 77 °F
35. 212 °F
37. 12.5
39. 240
41. (a)1 000 000; (b) 1 000 000; (c) 1000; (d) 5

8-3 LINEAR MEASURES *Page 319*

1. 4
3. 3
5. 6
7. $m(\overline{P_2P_5}) + m(\overline{P_5P_7}) = m(\overline{P_2P_7})$
9. $m(\overline{P_1P_7}) = m(\overline{P_1P_4}) + m(\overline{P_2P_7}) - m(\overline{P_1P_4} \cap \overline{P_2P_7})$
11. 14 mm
13. 73 m
15. $3\frac{1}{2}$
17. (14) $\frac{1}{2}$ unit; (15) $\frac{1}{4}$ unit; (16) $\frac{1}{8}$ unit.
19. Approximate.
21. Exact.

8-4 ANGULAR MEASURES *Page 324*

1. Among others:

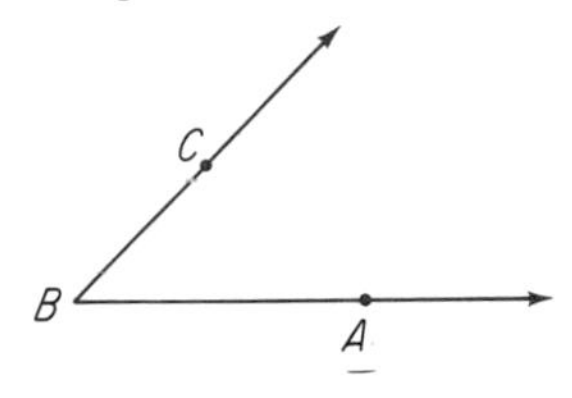

3.

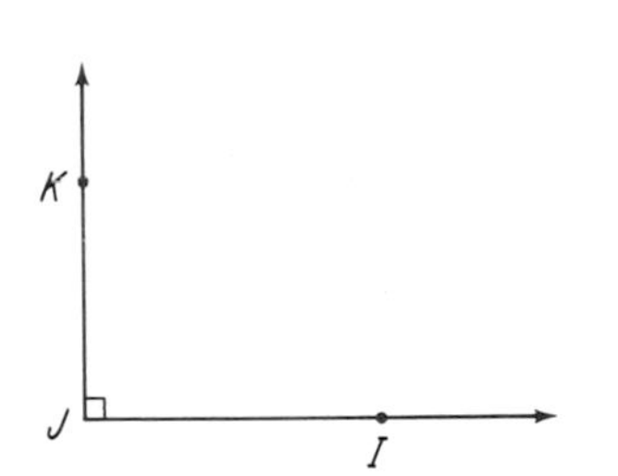

5.

T S R

7. Impossible; any two straight angles each have degree-measure 180 and thus are congruent.

9.

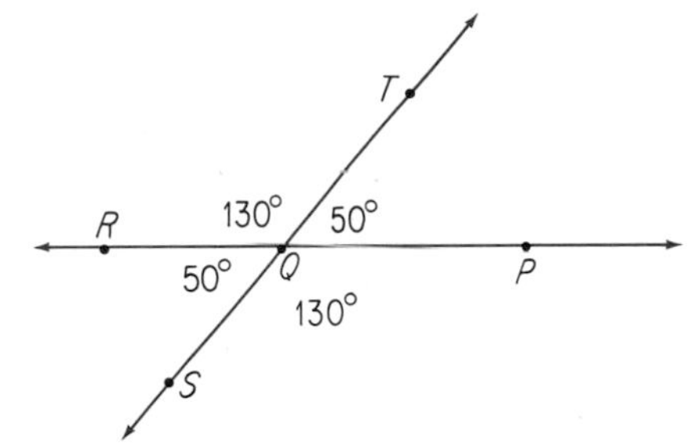

11. $\pi/4$
13. π
15. 2π
17. $5\pi/4$
19. $-7\pi/4$
21. 90
23. 270
25. 180
27. 120
29. −225
31. $180 \div 3.14 \approx 57.3$

8-5 AREA MEASURES *Page 329*

1. 80 cm^2

3. 45 cm^2

5. 68 cm^2

7. 78 cm^2

9. The area measure of the square is multiplied by 4.

11. The area measure of the rectangle is multiplied by **(a)** 4; **(b)** 9; **(c)** $\frac{1}{4}$; **(d)** k^2.

13. The area measure of the polygon is multiplied by **(a)** 4; **(b)** k^2.

15. 314 cm^2

17. 78.5 cm^2

19. Formula for area of a triangle.

21. $r = \sqrt{A/\pi}$

8-6 VOLUME MEASURES *Page 333*

1. **(a)** 150 cm^2; **(b)** 125 cm^3

3. **(a)** 294 cm^2; **(b)** 343 cm^3

5. **(a)** 122 cm^2; **(b)** 84 cm^3

7. $S = 6e^2$

9. 100 cm^3

11. The volume measure of the cube is multiplied by 8.

13. The volume measure is multiplied by **(a)** 8; **(b)** 27; **(c)** $\frac{1}{8}$; **(d)** k^3.

***15.** $B_1 = 0, \; M = \pi r^2, \; B_2 = 0, \; h = 2r$;
$V = \frac{2r}{6}(0 + 4\pi r^2 + 0) = \frac{4}{3}\pi r^3$.

***17.** $B_1 = \pi r^2, M = \frac{1}{4}\pi r^2, B_2 = 0$;
$V = \frac{h}{6}[\pi r^2 + 4(\frac{1}{4}r\pi^2) + 0] = \frac{1}{3}\pi r^2 h$.

Chapter 9: SENTENCES IN ONE VARIABLE

9-1 SENTENCES AND STATEMENTS *Page 345*

1. Identity.

3. Empty set.

5. Identity.

7. Empty set.

9. {5}

11. {0, 1, 2, 3}

13. {0, 1, 2, 3, 4, 5, 6}

15. {. . . , −2, −1, 0, 1, 2, 3}

17. {. . . , −2, −1, 0, 1, 2}

19. All real numbers less than 2.

21. All real numbers less than 6.

23. All real numbers except 5.

25. Point.

27. Ray.

29. Line segment.

31. Line.

33. Half-line.

35. Ray.

37.

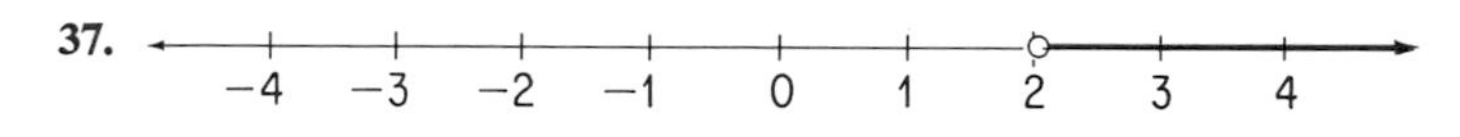

39.

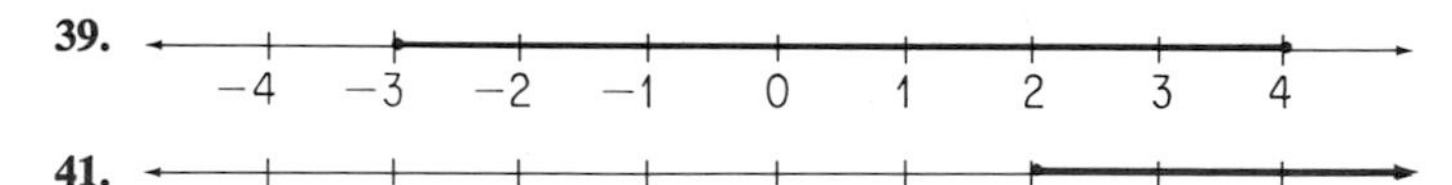

41.

43.

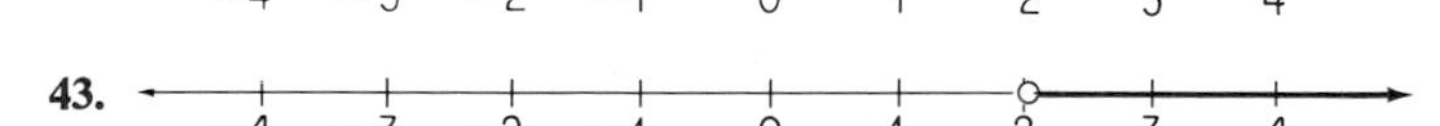

45.

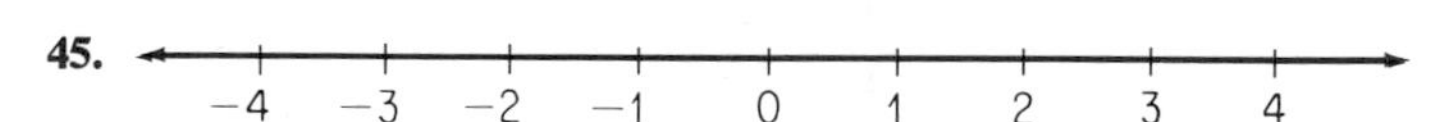

47.

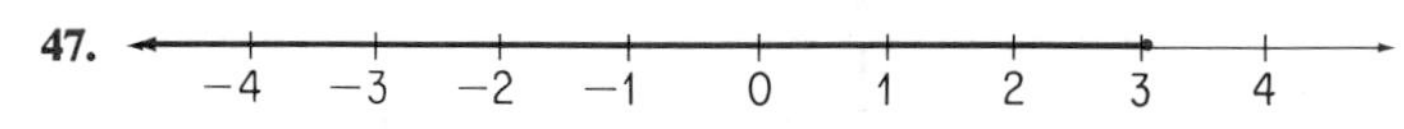

49.

***51.** −4 −3 −2 −1 0 1 2 3 4

***53.** −8 −6 −4 −2 0 2 4 6 8

***55.** −4 −3 −2 −1 0 1 2 3 4

9-2 COMPOUND SENTENCES *Page 349*

1. {1, 2, 3, 4, 5}

3. {2, 3, 4}

5.

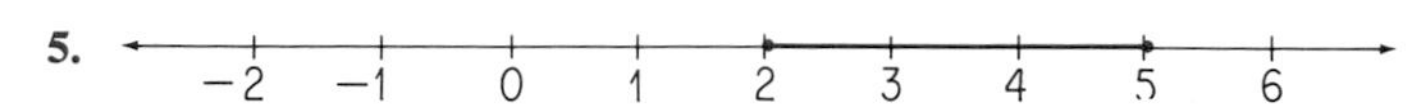

7.

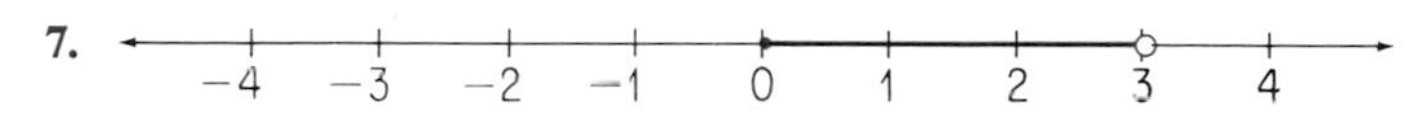

9. −4 −3 −2 −1 0 1 2 3 4

11. ∅

13. ∅

15. −4 −3 −2 −1 0 1 2 3 4

17. −2 −1 0 1 2 3 4 5 6

19. −2 −1 0 1 2 3 4 5 6

21. −6 −5 −4 −3 −2 −1 0 1 2

23. −6 −5 −4 −3 −2 −1 0 1 2

***25.** −6 −5 −4 −3 −2 −1 0 1 2

***27.** −6 −5 −4 −3 −2 −1 0 1 2

9-3 SENTENCES OF THE FIRST DEGREE *Page 354*

1. **(a)** Addition, $=$;
(b) associative, $+$;
(c) addition;
(d) zero, $+$.

3. **(a)** Addition, $=$;
(b) associative, $+$;
(c) addition;
(d) zero, $+$;
(e) multiplication, $=$;
(f) associative, $\times$;
(g) multiplication;
(h) one, $\times$.

5. $x = 4$

7. $x = -3$

9. $x = 22$

11. $x < 3$

13. $x > 3$

15. $x > -4$

17. $x > -8$

19. $x < -16$

21. $x = -3$

23. $x > 3$

9-4 SENTENCES INVOLVING ABSOLUTE VALUE *Page 359*

1. {−5, 5}
3. {−2, −1, 0, 1, 2}
5. {−3, 5}
7. {−4, 0}
9. 9
11. 8
13. 36
15. Line segment.
17. Union of two rays.

19.
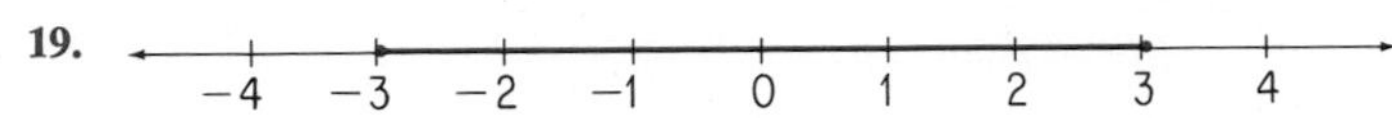

21.
−4 −3 −2 −1 0 1 2 3 4

23.

27.
−6 −5 −4 −3 −2 −1 0 1 2

29.
−4 −3 −2 −1 0 1 2 3 4

***31.**
−4 −3 −2 −1 0 1 2 3 4

***33.**
−4 −3 −2 −1 0 1 2 3 4

***35.**
−4 −3 −2 −1 0 1 2 3 4

Chapter 10: SENTENCES IN TWO VARIABLES

10-1 SOLUTION SETS *Page 365*

1. {(1, 3), (2, 2), (3, 1)}
3. {(1, 1), (1, 2), (2, 1)}
5. ∅
7. {(1, 5), (2, 4), (3, 3), (4, 2), (5, 1)}
9. {(1, 1), (1, 2), (1, 3), (2, 1), (2, 2), (3, 1)}
11. {(−3, 3), (−2, 2), (−1, 1), (0, 0), (1, 1), (2, 2), (3, 3)}
13. {(−3, −3), (−2, −2), (−1, −1), (0, 0), (1, −1), (2, −2), (3, −3)}
15. {(−1, 2), (0, 1), (1, 2) }

10-2 ORDERED PAIRS *Page 368*

1. {(1, 1)}

3.

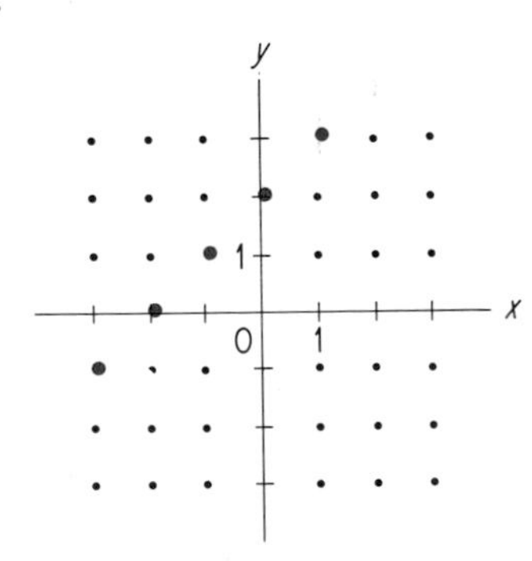

5.

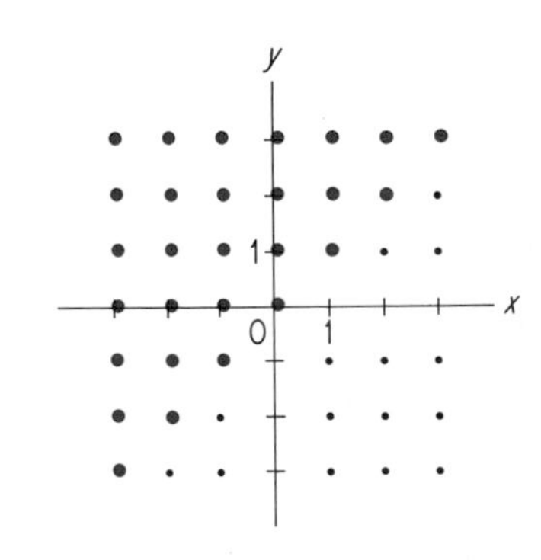

7.

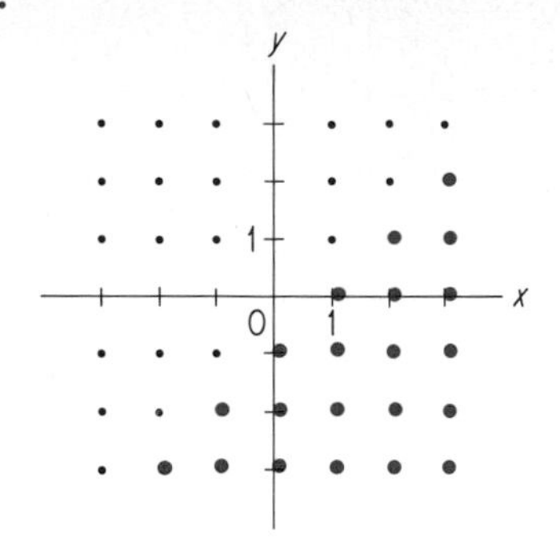

11.

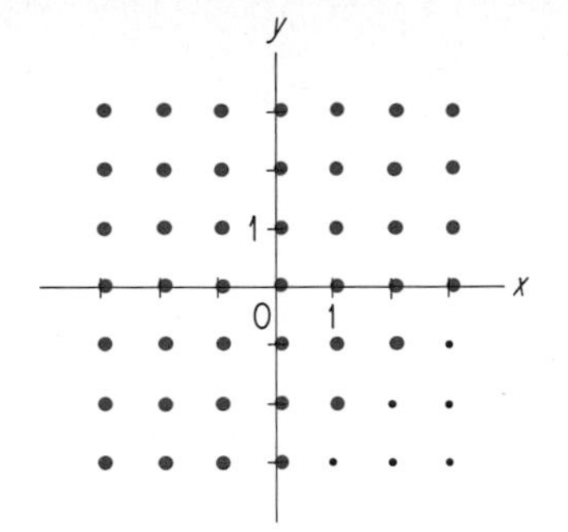

9.

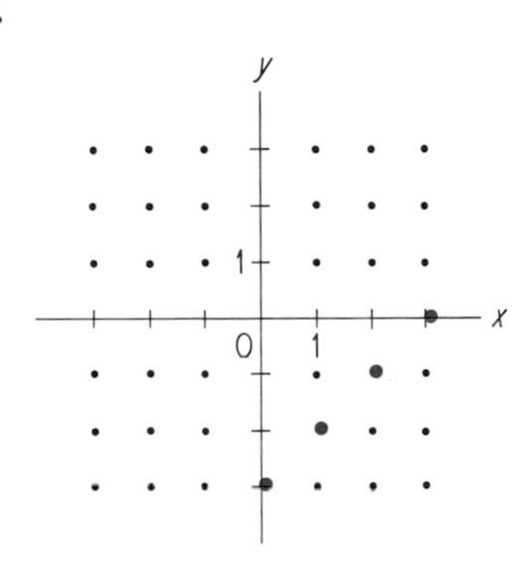

13. {(1, 3), (2, 2), (3, 1)}
15. {(1, 3), (1, 4), (2, 4)}
*17. {(1, 1, 1), (1, 1, 2), (1, 2, 1), (2, 1, 1), (1, 2, 2), (2, 1, 2), (2, 2, 1), (2, 2, 2)}

10-3 RELATIONS AND FUNCTIONS *Page 371*

1. (a) -2; (b) 0; (c) 7.
3. (a) 3; (b) -13; (c) -32.
5.

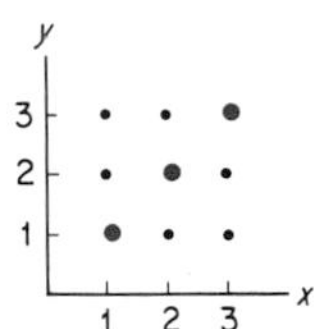

A function.
Domain: {1, 2, 3}
Range: {1, 2, 3}

7.

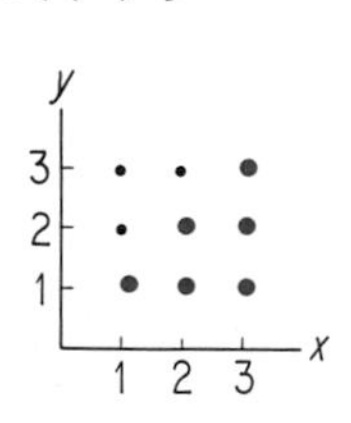

Not a function.

9.

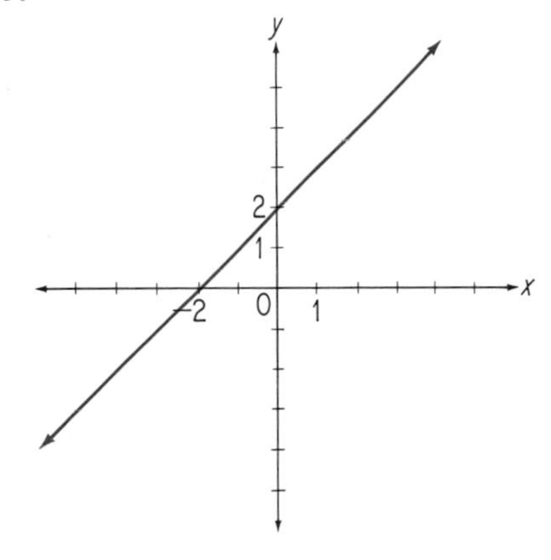

A function. Both the domain and the range are the set of real numbers.

11.

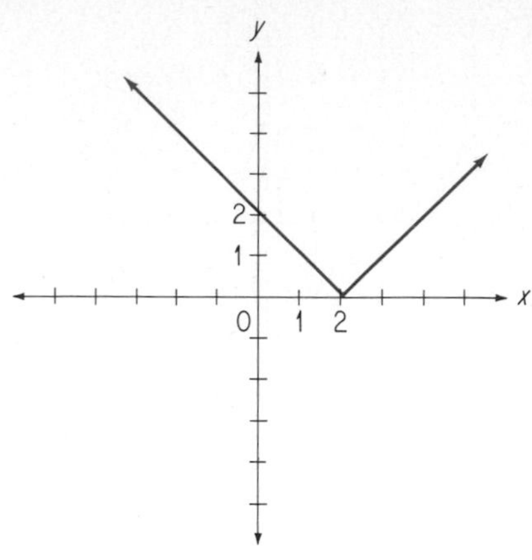

A function. The domain is the set of real numbers; the range is the set of non-negative real numbers.

13.

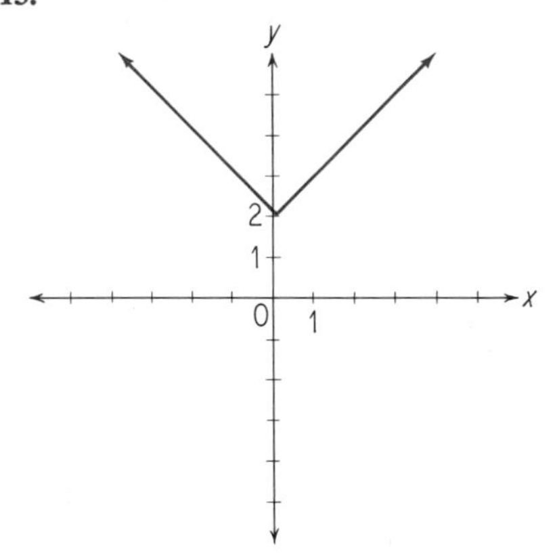

A function. The domain is the set of real numbers; the range is the set of real numbers greater than or equal to 2.

10-4 GRAPHS ON A PLANE *Page 378*

1. **(a)** 8; **(b)** 8.

3. **(a)** 6; **(b)** -4.

5. **(a)** $-2\frac{1}{2}$; **(b)** 5.

7. **(a)** -4; **(b)** 8.

9. **(a)** 2; **(b)** $2\frac{1}{2}$.

11. $y = -2x + 9$

13. $y \geq -2x + 8$

15. $y \geq 2x - 8$

15.

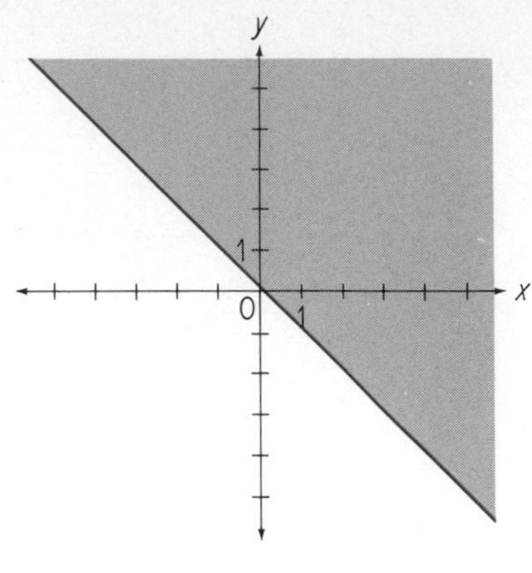

Not a function.

17.

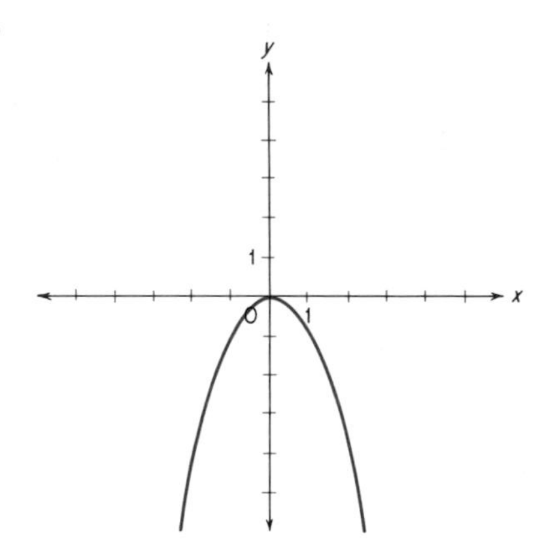

A function. The domain is the set of real numbers; the range is the set of real numbers less than or equal to 0.

17.

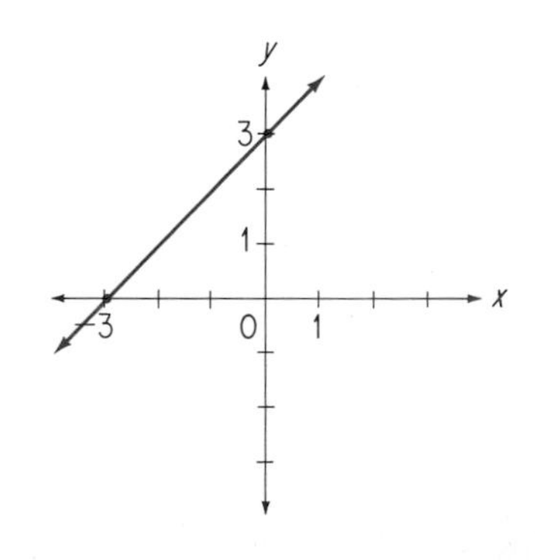

19.

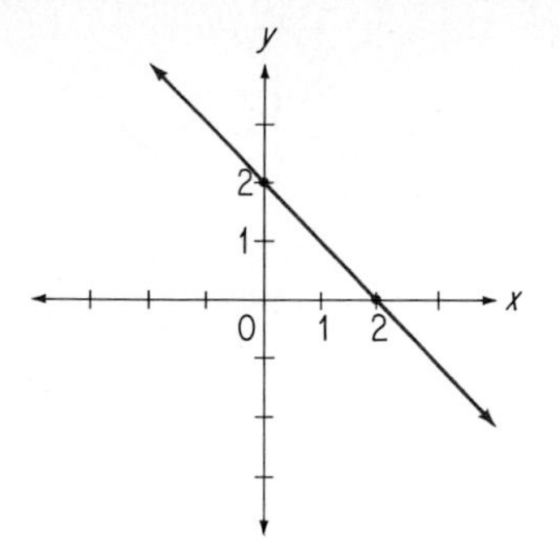

21.

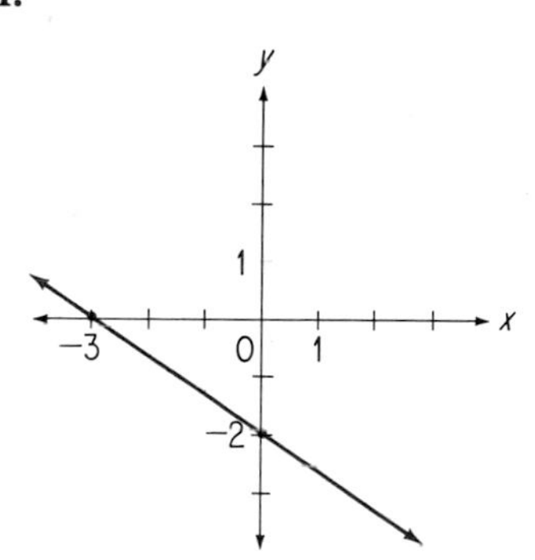

23.

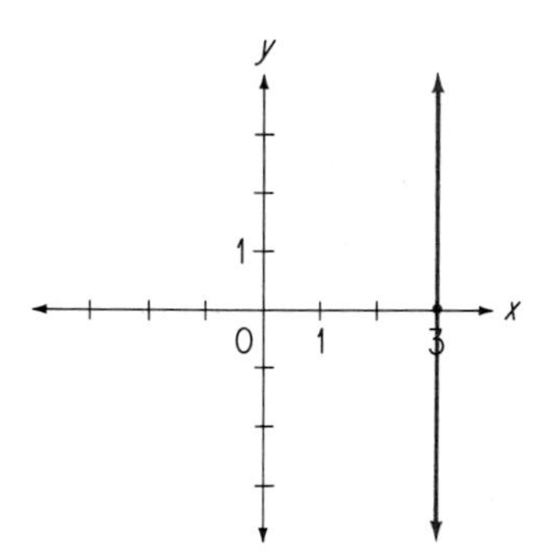

25.

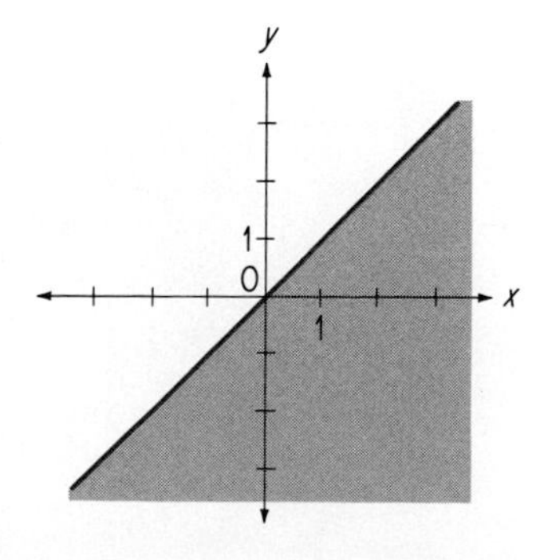

27.

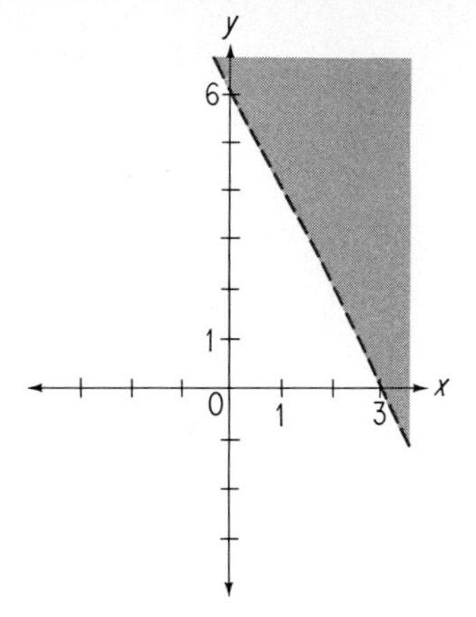

29.

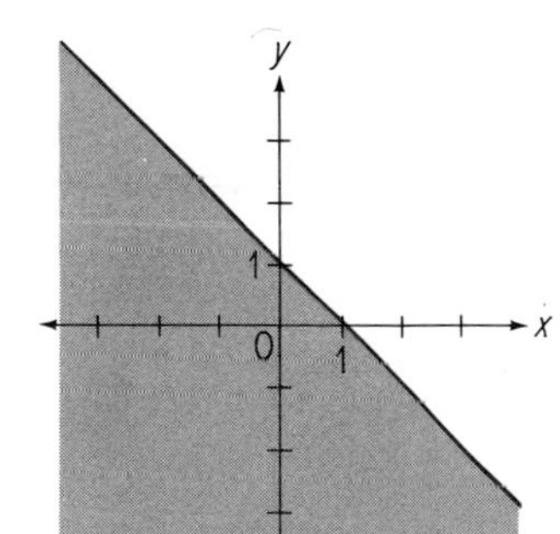

10-5 GRAPHS INVOLVING ABSOLUTE VALUE *Page 382*

1.

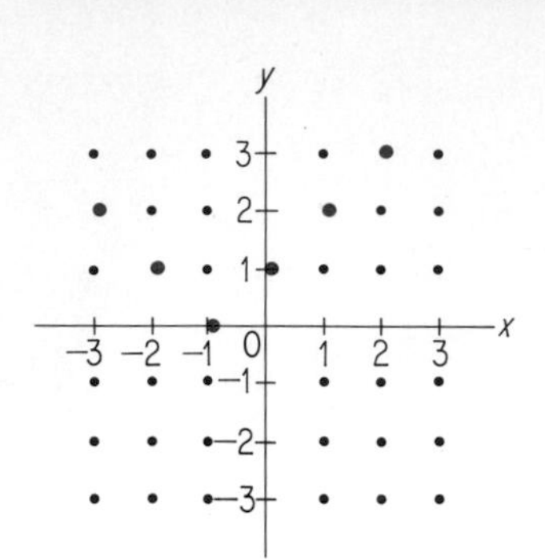

3.

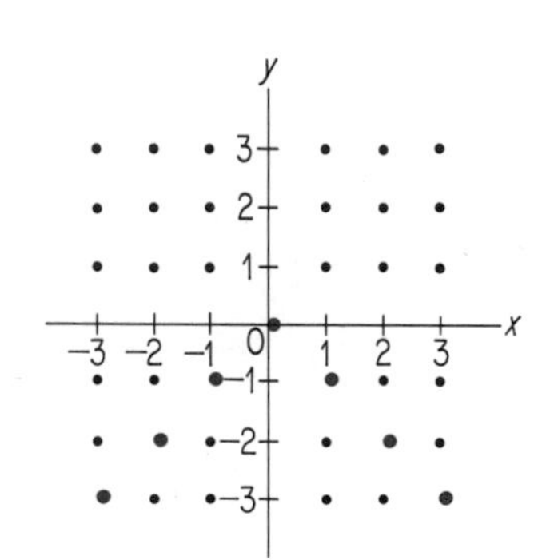

5.

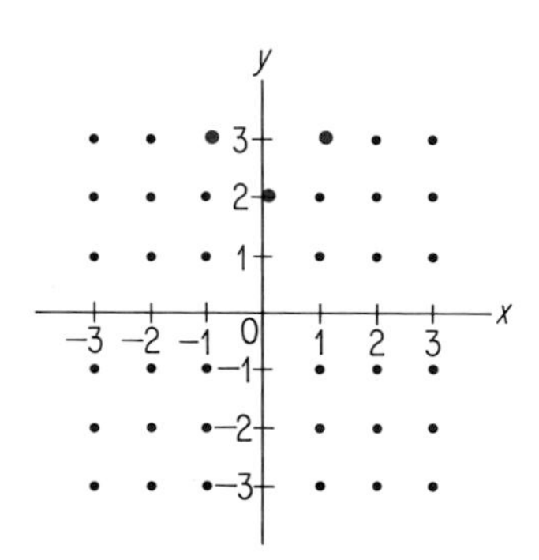

7.

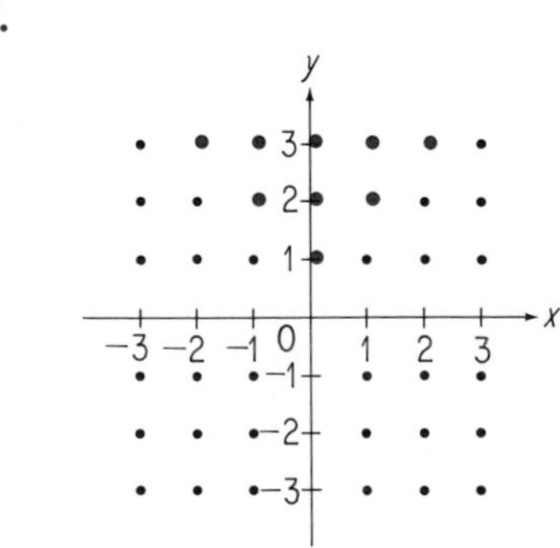

9.

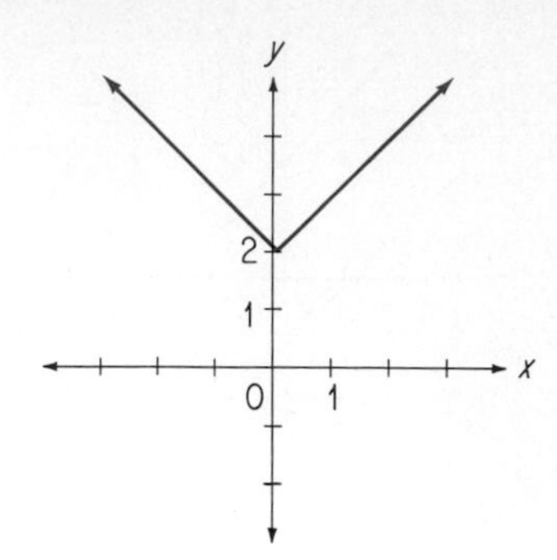

11.

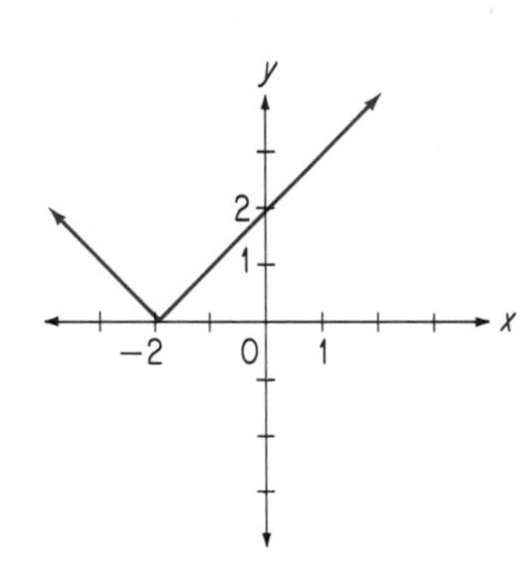

13.

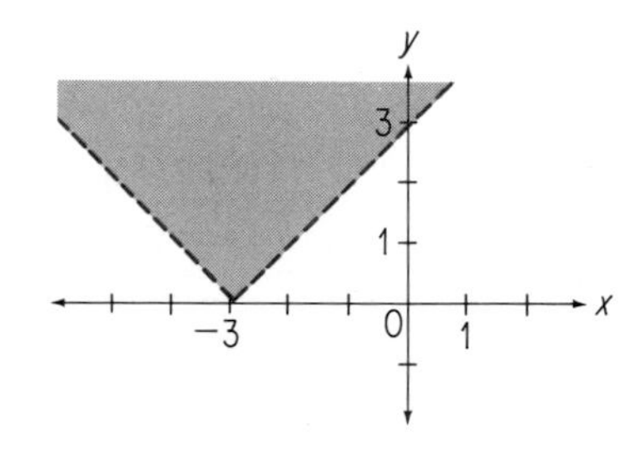

15.

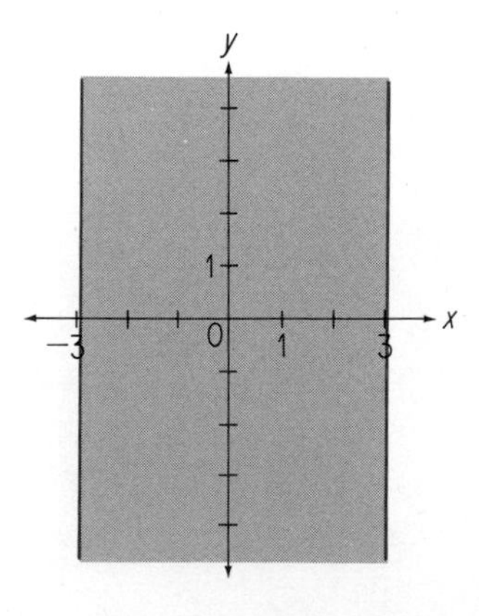

***17.**

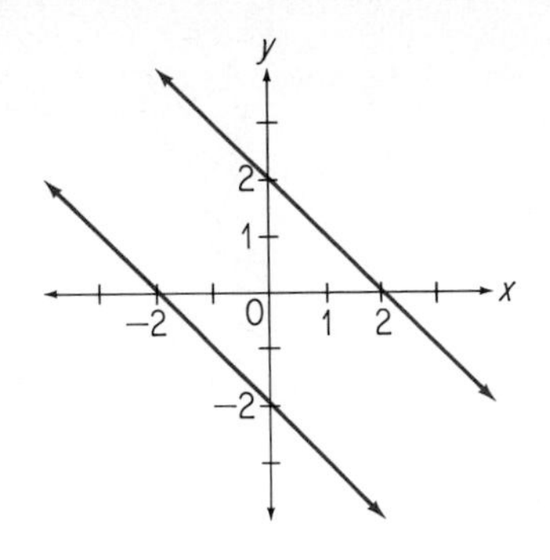

10-6 LINEAR SYSTEMS *Page 388*

1.

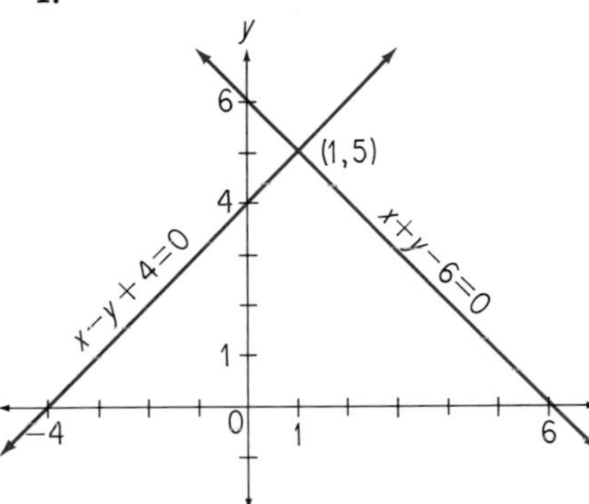

3.

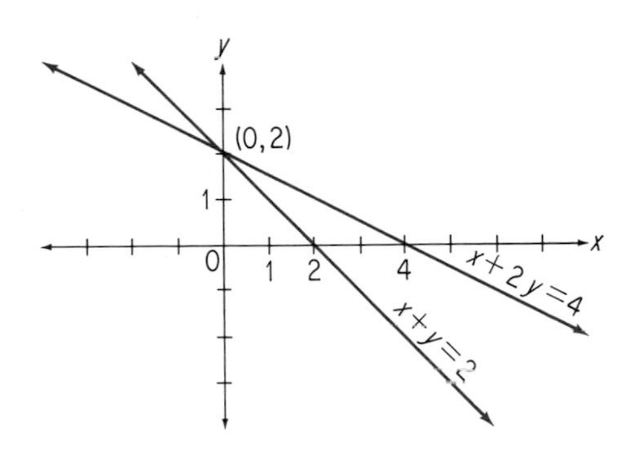

5.

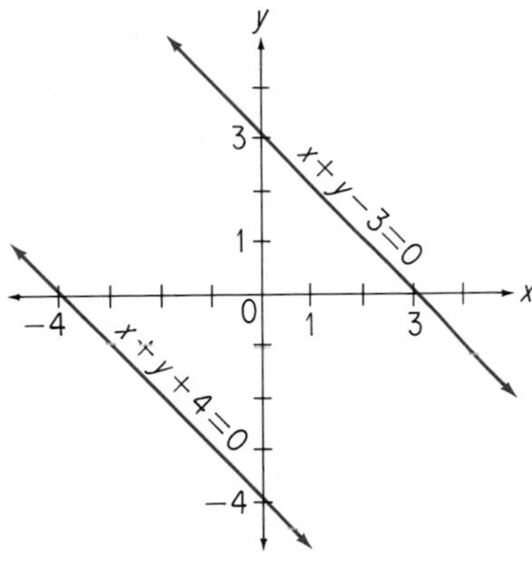

The graph consists of the union of the two lines.

7.

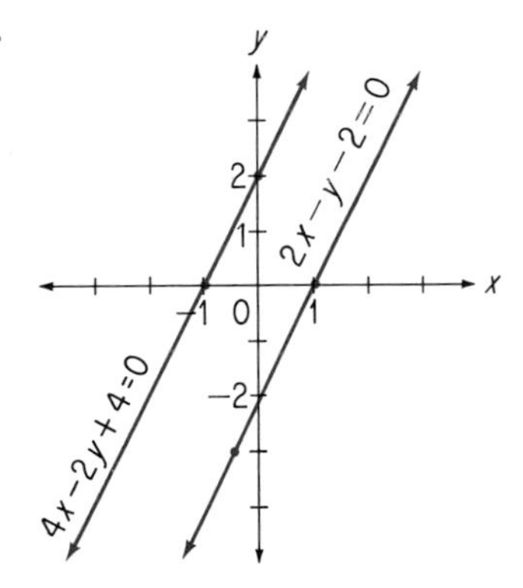

The lines are parallel; the solution set is the empty set.

9.

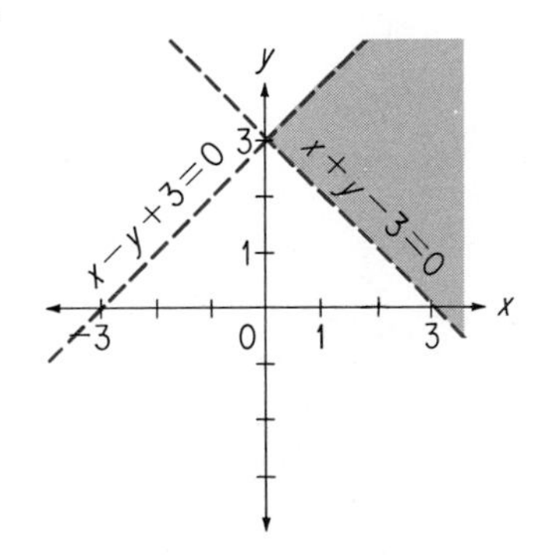

11.

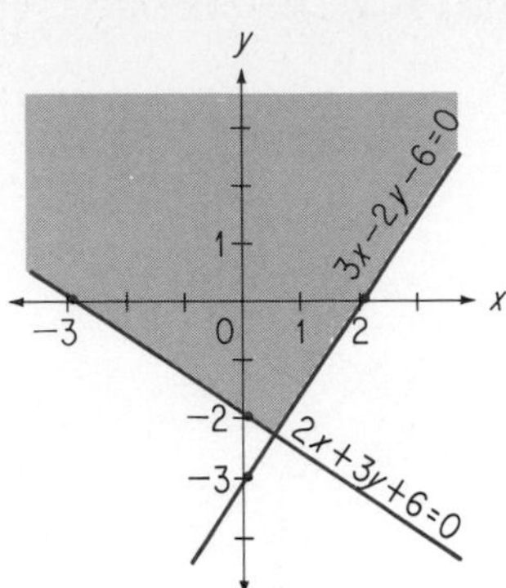
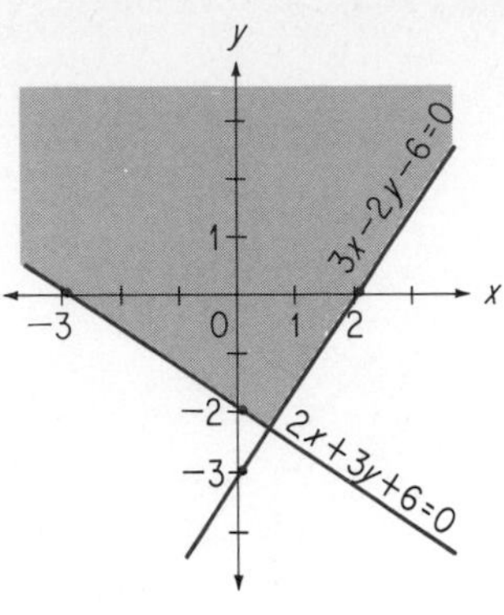

13.

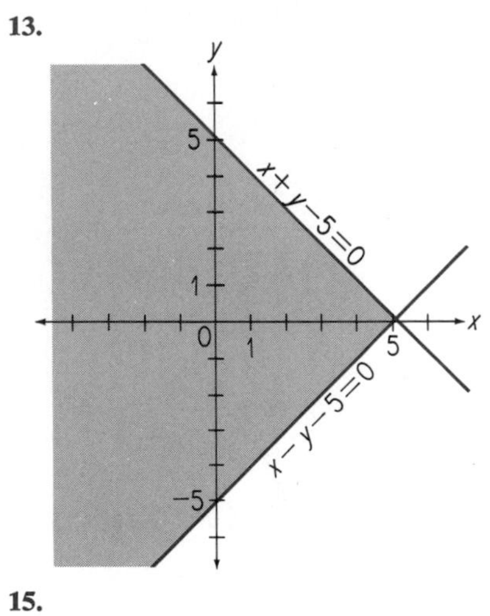

15.

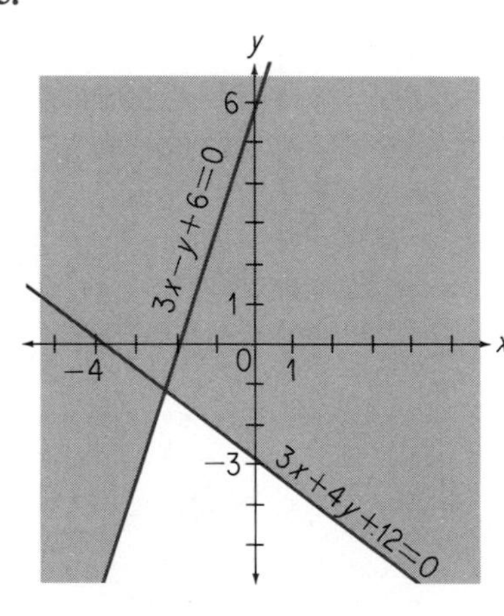

***17.**

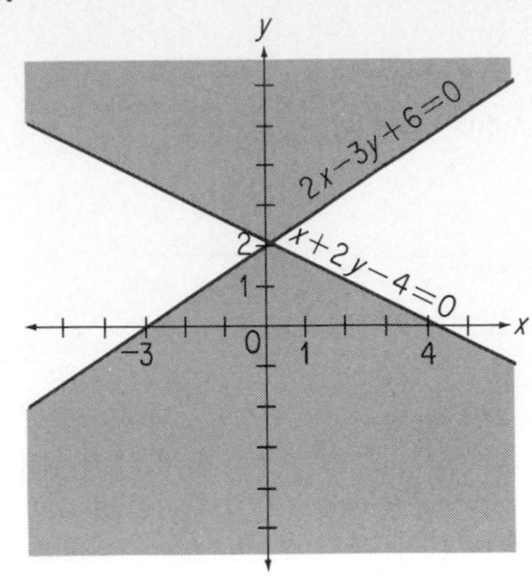

***19.**

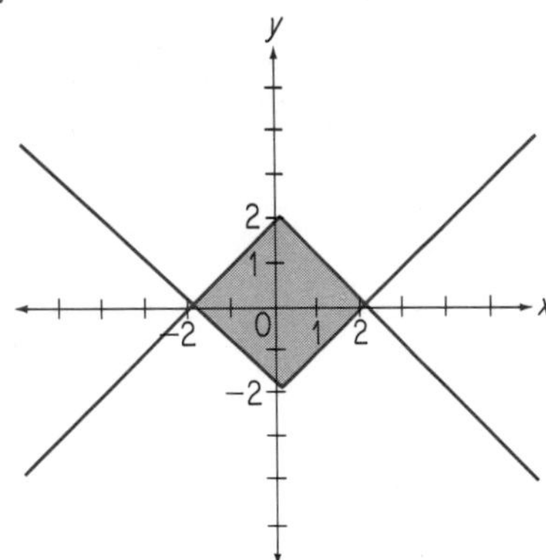

***21.**

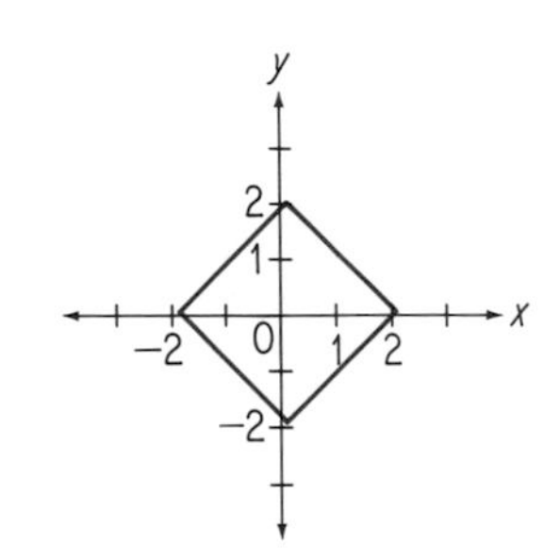

***23.**

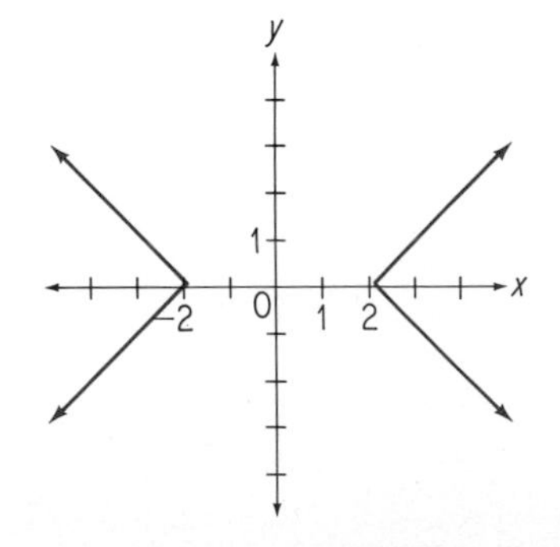

11-1 GEOMETRIC FIGURES *Page 394*

Among others:

1.

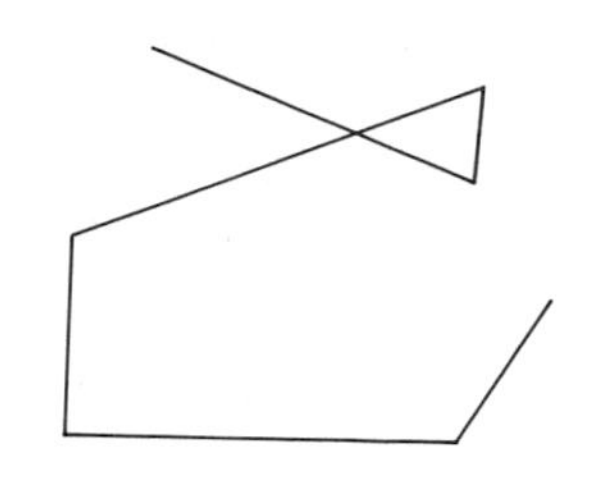

3.

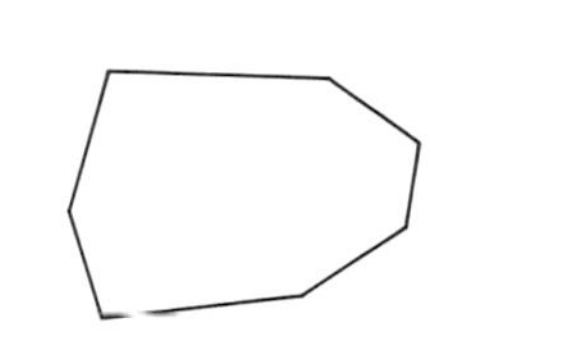

5.

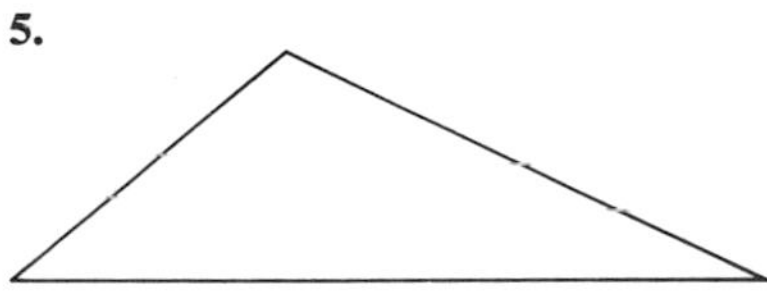

7.

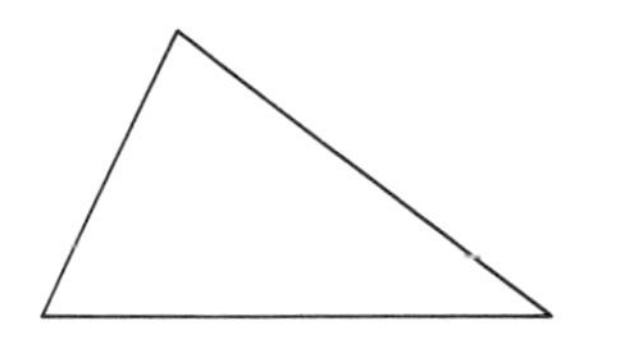

9.

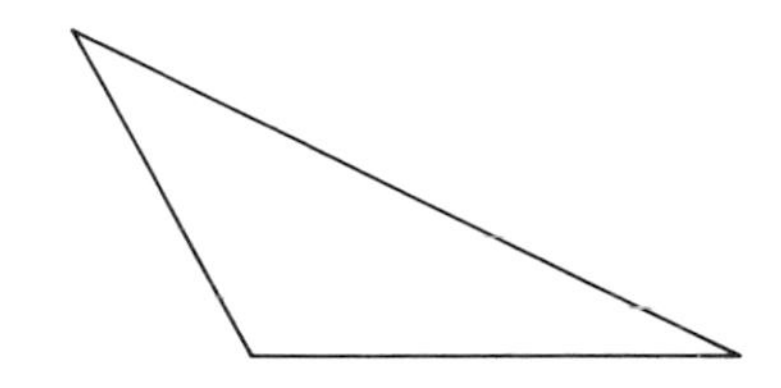

11.

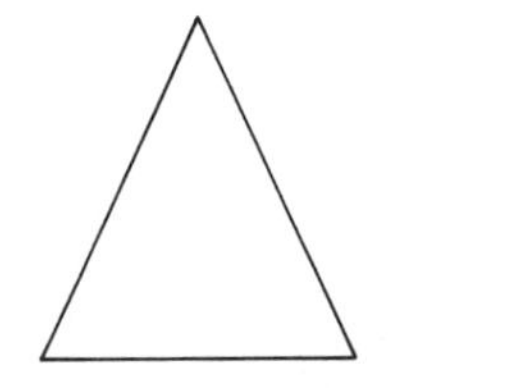

13.

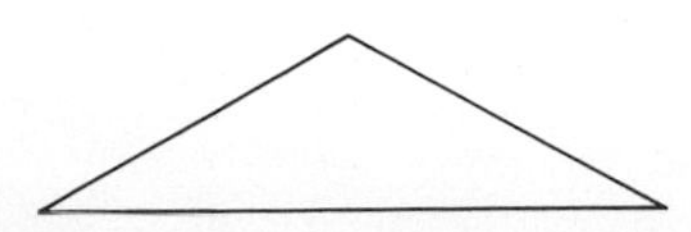

15.

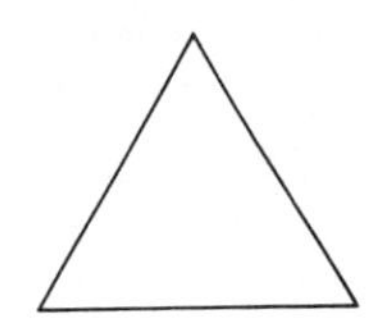

17.

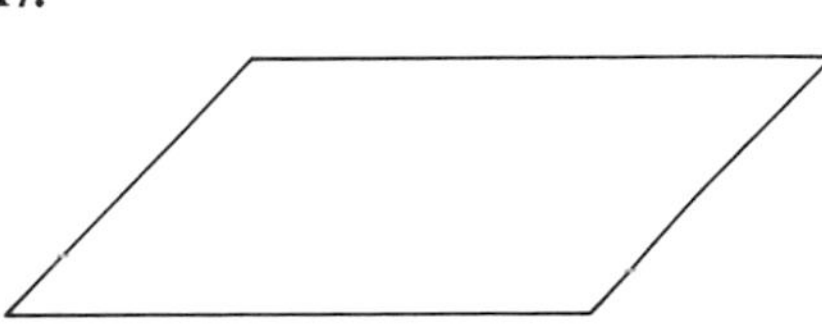

19.

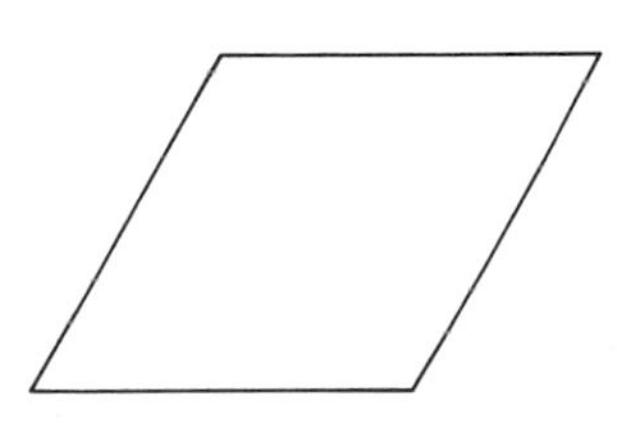

21. Impossible; a rhombus is defined as a special type of parallelogram.

23.

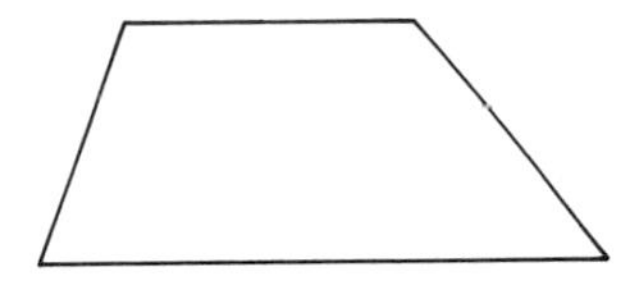

***25.**

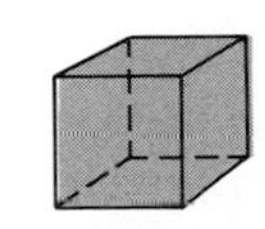

***27.**

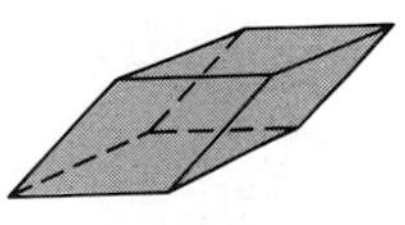

***29.**

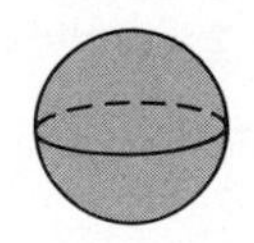

11-2 CONSTRUCTIONS OF GEOMETRIC FIGURES *Page 402*

1. Among others:

(a) $\angle Q = \angle RQT$, $\angle QRS = \angle SRQ$;

(b) $\angle Q \cong \angle R$, $\angle Q \cong \angle T$;

(c) $\overline{QR} = \overline{RQ}$, $\overline{QT} = \overline{TQ}$;

(d) $\overline{QR} \cong \overline{TS}$, $\overline{QT} \cong \overline{RS}$;

(e) $\square QRST = \square RSTQ$,
$\square QRST = \square STQR$.

3. Among others:

(a) $\triangle RQP$;

(b) $\triangle TPS$;

(c) $\triangle PQR \cong \triangle PST$.

5. (a)

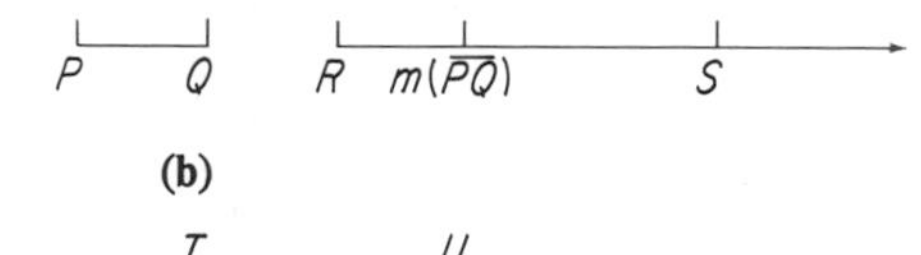

(b)

$m(\overline{TU}) = 2\, m(\overline{PQ})$

7.

9.

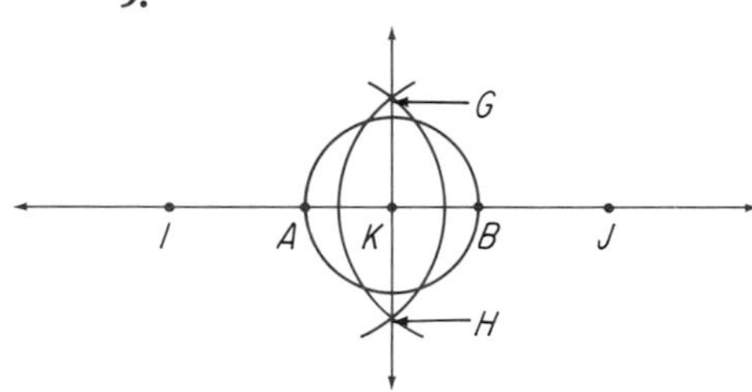

11.

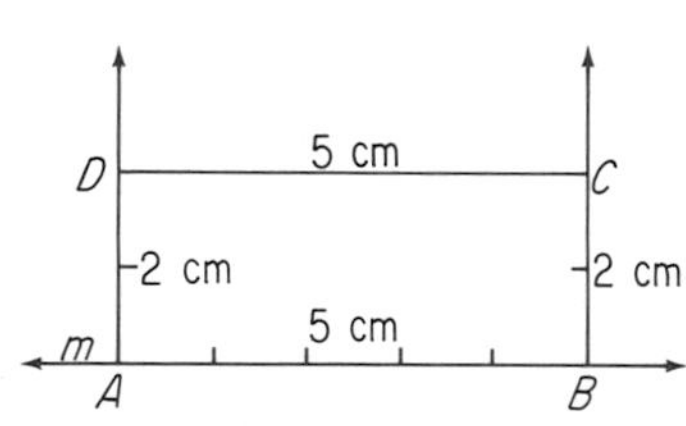

On any line m mark off $\overline{AB}$ of length 5 cm. As in Exercise 9 construct lines perpendicular to m at A and B. On these two lines and on the same side of m, mark off points D and C as in the figure such that $m(\overline{AD}) = m(\overline{BC}) = 2$ in centimeters. Draw $\overline{CD}$. Then $ABCD$ is the desired rectangle.

No; any two such rectangles will look alike (be congruent).

13. Among others:

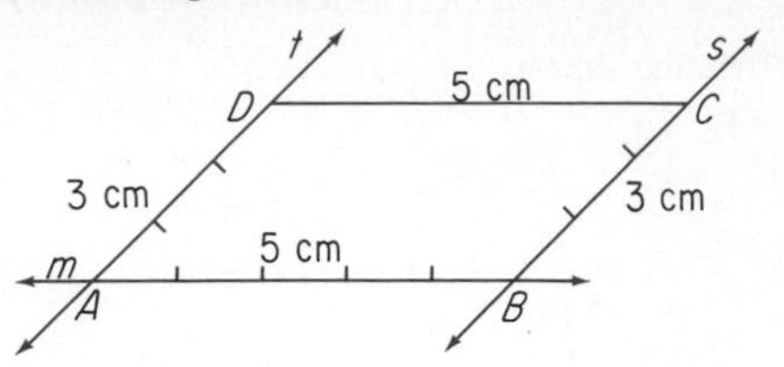

On any line m mark off $\overline{AB}$ of length 5 cm. Draw any line $t \neq m$ through A and mark off $\overline{AD}$ of length 3 cm on t. At B copy $\angle BAD$ with one side on $\overrightarrow{AB}$ and so that the line s on the other side of the angle does not intersect (is parallel to) $\overleftrightarrow{AB}$. Mark off $\overline{BC}$ of length 3 cm so that B and C are on the same side of m. Draw $\overline{CD}$. Then $ABCD$ is a parallelogram that satisfies the given conditions.

Yes; $\angle BAD$ may have any degree measure between 0 and 180.

15.

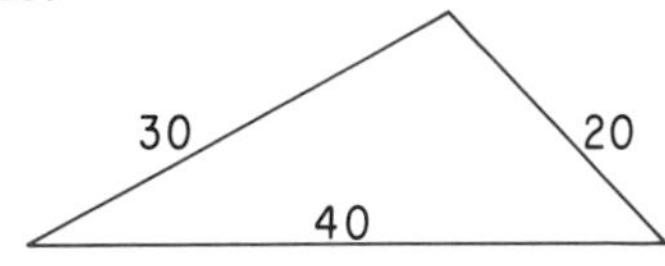

17. Impossible, since $2 + 4 = 6$.

19. Impossible, since $2 + 3 < 6$.

21. Think of $\frac{15}{60}$, $\frac{20}{60}$, and $\frac{12}{60}$.

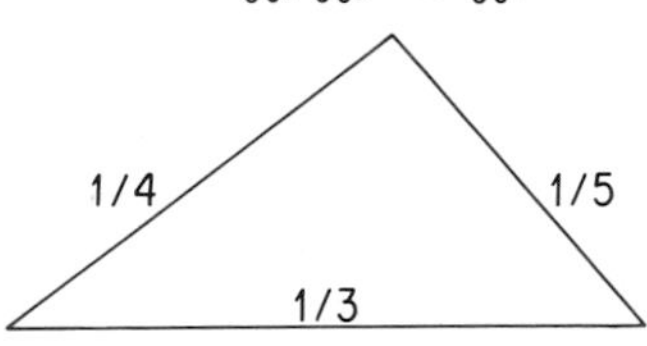

23.

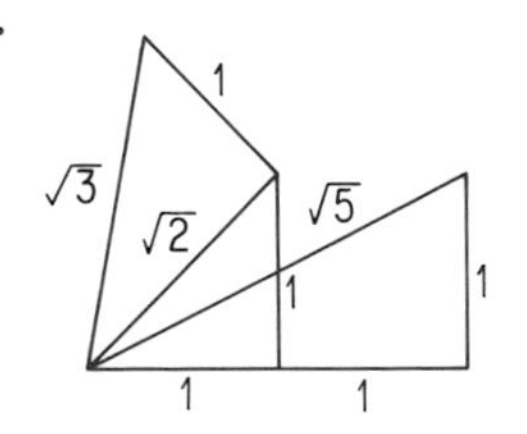

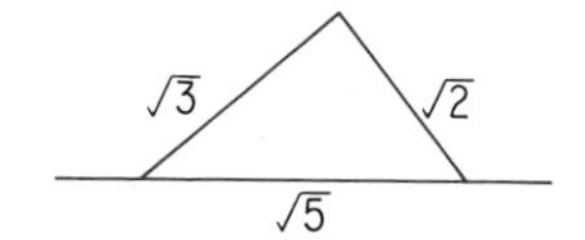

11-3 RELATIONS AMONG GEOMETRIC FIGURES *Page 409*

1.

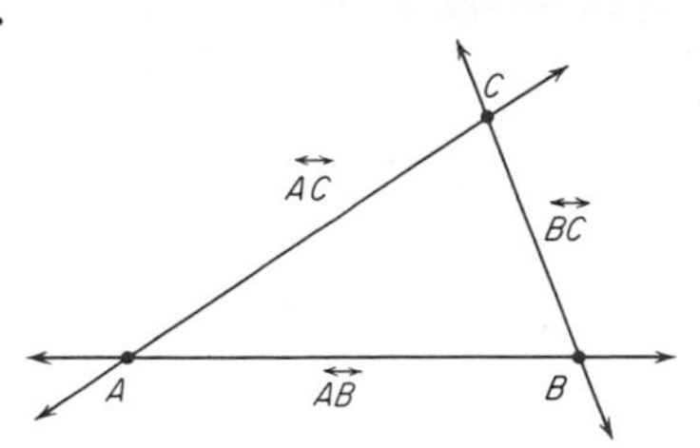

3. *ABC, ABD, ACD, BCD.*

5. (a) 1;
(b) 3;
(c) 6;
(d) 10;
(e) 15;
*(f) 45;
*(g) $n(n - 1)/2$.

7. True.

9. False.

11. False.

13. False.

15. False.

17. True.

19. Make a crease in the paper. Select a point on this crease to correspond to itself and fold the rest of the crease onto other parts of the crease; repeat for a a second point on the original crease and for another point on one of the other creases obtained.

21. The resulting figure consists of lines that would be tangent to a circle having the same center and one-half the radius of the given circle.

11-4 NETWORKS *Page 414*

1. (a) 4;
(b) 0;
(c) traversable, *A, B, C, D.*

3. (a) 2;
(b) 2;
(c) traversable, *K, M.*

5. (a) 4;
(b) 0;
(c) traversable, *U, V, W, X.*

7. (a) 0;
(b) 4;
(c) not traversable.

9. The inspector can use a map for the highways involved as a network, determine the number of odd vertices, and know that each section can be traversed exactly once in a single trip if there are at most two odd vertices.

11.

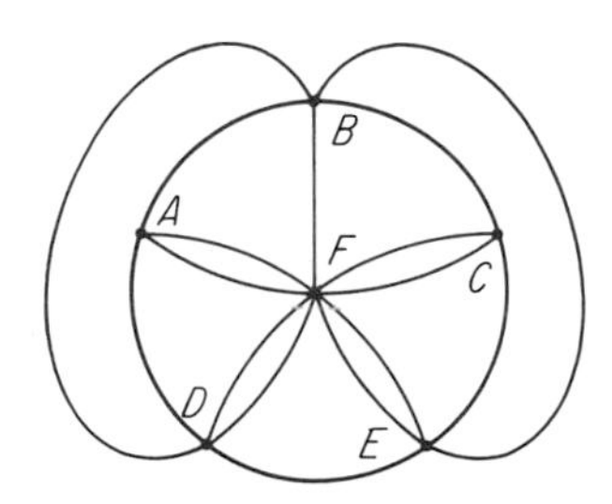

Think of the six regions as labeled and note the line segments that are needed as represented by arcs in the network. Since the network has four odd vertices (*B, D, E,* and *F*), the network is not traversable and the suggested broken line cannot be drawn.

***13.** Add one more bridge joining any two of the points *A, B, C, D.*

11-5 FIGURES ON A COORDINATE PLANE *Page 423*

1. (a) 7, (3.5, 5); (b) 5, (5.5, 1).

3. (a) $|b|$; (b) $|k|$.

5. 5

7. 25

9. $3\sqrt{2}$

11. (a) (4, −2); (b) (2, 4); (c) (−3, −6).

13. (−5, 8)

15. C: (a, a)

***17.** S: $(a - b, c)$

19.

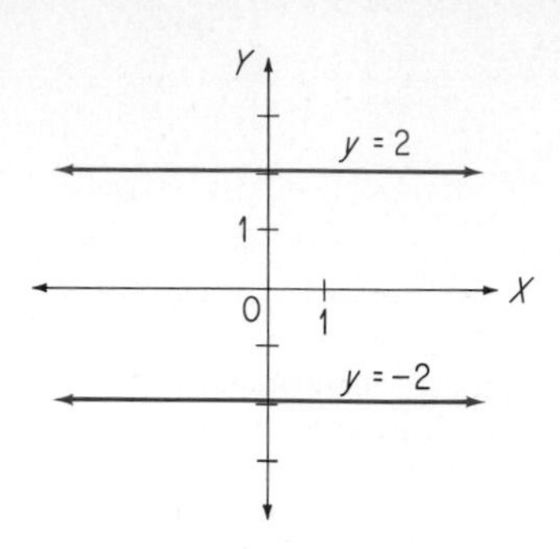

21.

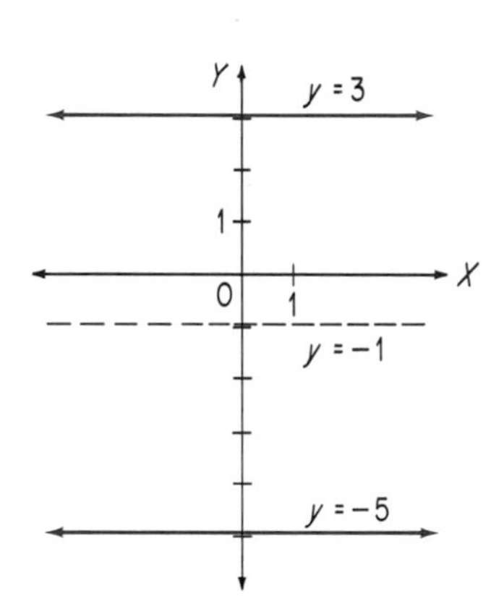

23. The points of the circular region with center (0, 0) and radius 5.

25. The interior points of the circle with center (−3, 1) and radius 2.

27. The exterior points of the circle with center (−3, −1) and radius 3.

29. $(x + 3)^2 + (y + 1)^2 \leq 25$

31. $(x + 4)^2 + (y - 3)^2 > 49$

11-6 FIGURES ON A COORDINATE SPACE *Page 429*

1.

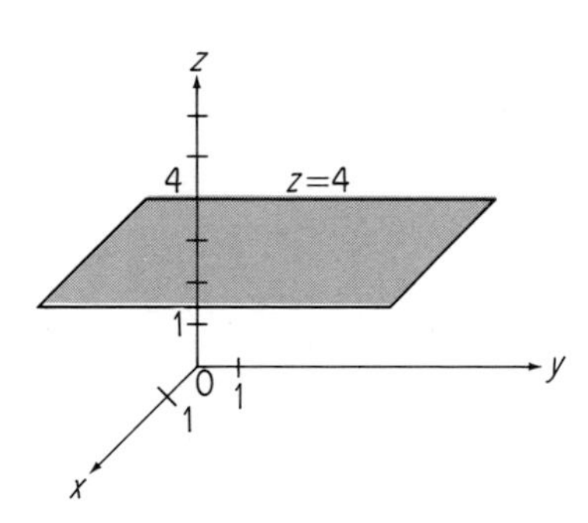

3.

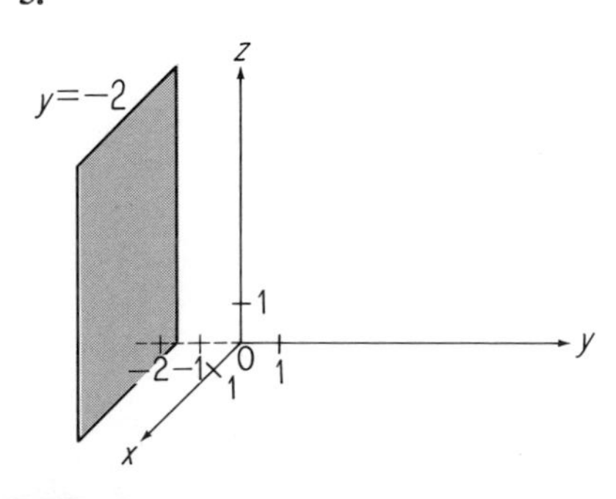

5.

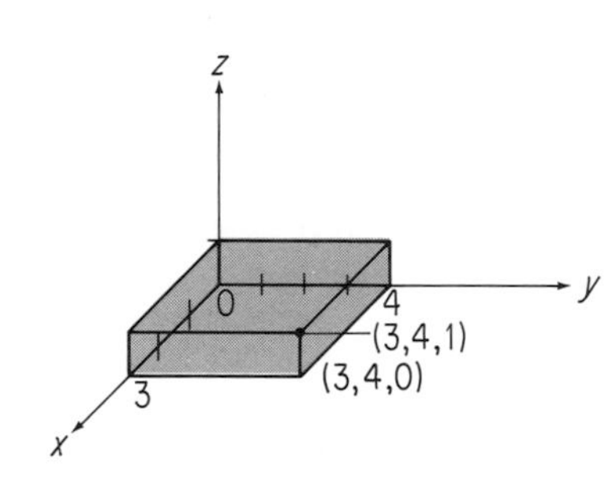

7.

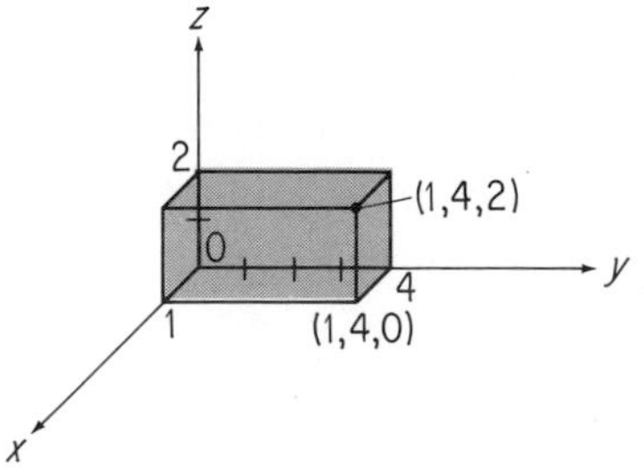

9. The yz-plane.

11. The xy-plane.

13. The plane that is parallel to the xz-plane and intersects the y-axis at b.

15. The z-axis.

17. The y-axis.

19. A: (4, 0, 0), B: (4, 2, 0), C: (4, 2, 3), D: (4, 0, 3), E: (0, 2, 0), F: (0, 0, 0), G: (0, 2, 3), H: (0, 0, 3).

21. A: (4, −2, 0), B: (4, 0, 0), C: (4, 0, 3), D: (4, −2, 3), E: (0, 0, 0), F: (0, −2, 0), G: (0, 0, 3), H: (0, −2, 3).

23. A: (7, 0, 0), B: (7, 5, 0), C: (7, 5, 2), D: (7, 0, 2), E: (0, 5, 0), F: (0, 0, 0), G: (0, 5, 2), H: (0, 0, 2).

25. A: (7, −5, 0), B: (7, 0, 0), C: (7, 0, 2), D: (7, −5, 2), E: (0, 0, 0), F: (0, −5, 0), G: (0, 0, 2), H: (0, −5, 2).

27. $\sqrt{(1-1)^2+(4-0)^2+(5-2)^2}$; that is, 5.

29. $\sqrt{(3-2)^2+[1-(-1)]^2+(2-0)^2}$; that is, 3.

31. $(x-1)^2+(y+2)^2+(z-5)^2=9$

33. $(x+1)^2+(y+2)^2+(z-3)^2=25$

35. A sphere with center (0, 0, 0) and radius 2.

37. A sphere with center (2, −1, −2) and radius 5.

39. $z=3$

41. $(x-1)^2+(y+2)^2+(z-3)^2=9$

43. $(x-2)^2+(y-3)^2+(z-4)^2 \geq 16$

45. $z \geq 3$

*47. $|x| < 4$

*49. $9 \leq (x-1)^2+(y-6)^2+(z-4)^2 \leq 25$

Chapter 12: AN INTRODUCTION TO PROBABILITY

12-1 COUNTING PROBLEMS *Page 439*

1. (a) 24; (b) 64.

3. (a) 120; (b) 216.

5. 15

7. 24

9. (a) 25; (b) 20.

11. 2730

13. 27,216

15. (a) 12; (b) 3.

*17. (a) 450; (b) 180; (c) 648; (d) 200.

*19. $26 \times 25 \times 9 \times 9 \times 8 \times 7$; that is, 2,948,400.

12-2 DEFINITION OF PROBABILITY *Page 443*

1. 1/2

3. 2/3

5. 5/6

7. 0

9. 1/13

11. 1/4

13. 10

15. 3/10

17. 1/12

19. 7/8

21. (a) 3/5; (b) 3/10.

12-3 SAMPLE SPACES *Page 447*

1. 1/10

3. 2/5

5. 3/5

7. 1/2

9. 

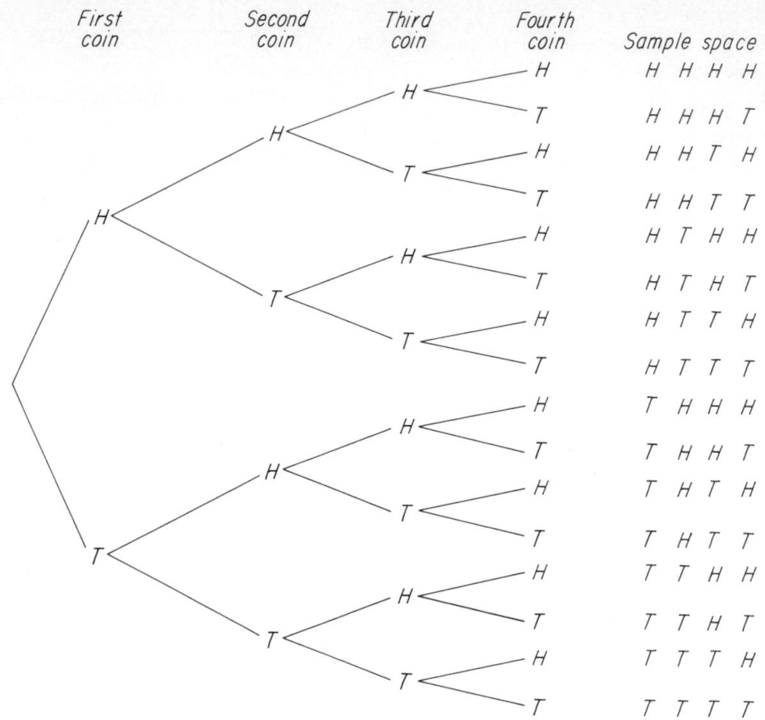

The sample space may also be given as: {*HHHH*, *HHHT*, *HHTH*, *HHTT*, *HTHH*, *HTHT*, *HTTH*, *HTTT*, *THHH*, *THHT*, *THTH*, *THTT*, *TTHH*, *TTHT*, *TTTH*, *TTTT*}.

11. **(a)** 5/16; **(b)** 5/16.

13. See figure on next page.

15. **(a)** 1/36; **(b)** 35/36.

17. **(a)** 1/18; **(b)** 17/18.

19. **(a)** 1/12; **(b)** 1/6.

21.
R_1R_2, R_2R_1, W_1R_1, W_2R_1,
R_1W_1, R_2W_1, W_1R_2, W_2R_2,
R_1W_2, R_2W_2, W_1W_2, W_2W_1; $\frac{1}{6}$.

23.
R_1R_2, R_2R_1, R_3R_1, W_1R_1, W_2R_1,
R_1R_3, R_2R_3, R_3R_2, W_1R_2, W_2R_2,
R_1W_1, R_2W_1, R_3W_1, W_1R_3, W_2R_3,
R_1W_2, R_2W_2, R_3W_2, W_1W_2, W_2W_1; $\frac{3}{10}$.

12-4 COMPUTATION OF PROBABILITIES *Page 453*

1. 2/3

3. 2/3

5. 1/2

7. 1

9. 1/2

11. 1/52

13. 19/52

15. 0

17. **(a)** 13/204; **(b)** 1/663.

19. **(a)** 1/16; **(b)** 1/2704.

21. 1/64

23. Consider the sample space:
*H*1, *H*2, *H*3, *H*4, *H*5, *H*6,
*T*1, *T*2, *T*3, *T*4, *T*5, *T*6.
(a) 1/12; **(b)** 1/4; **(c)** 7/12; **(d)** 3/4.

25. $\frac{13}{52} \times \frac{12}{51} \times \frac{11}{50} \times \frac{10}{49} \times \frac{9}{48}$, that is, $\frac{33}{66{,}640}$.

27. **(a)** 27/2197; **(b)** 72/2197;
(c) 343/2197; ***(d)** 307/2197.

29. **(a)** 1/8; **(b)** 1/8; **(c)** 1/8; **(d)** 3/8.

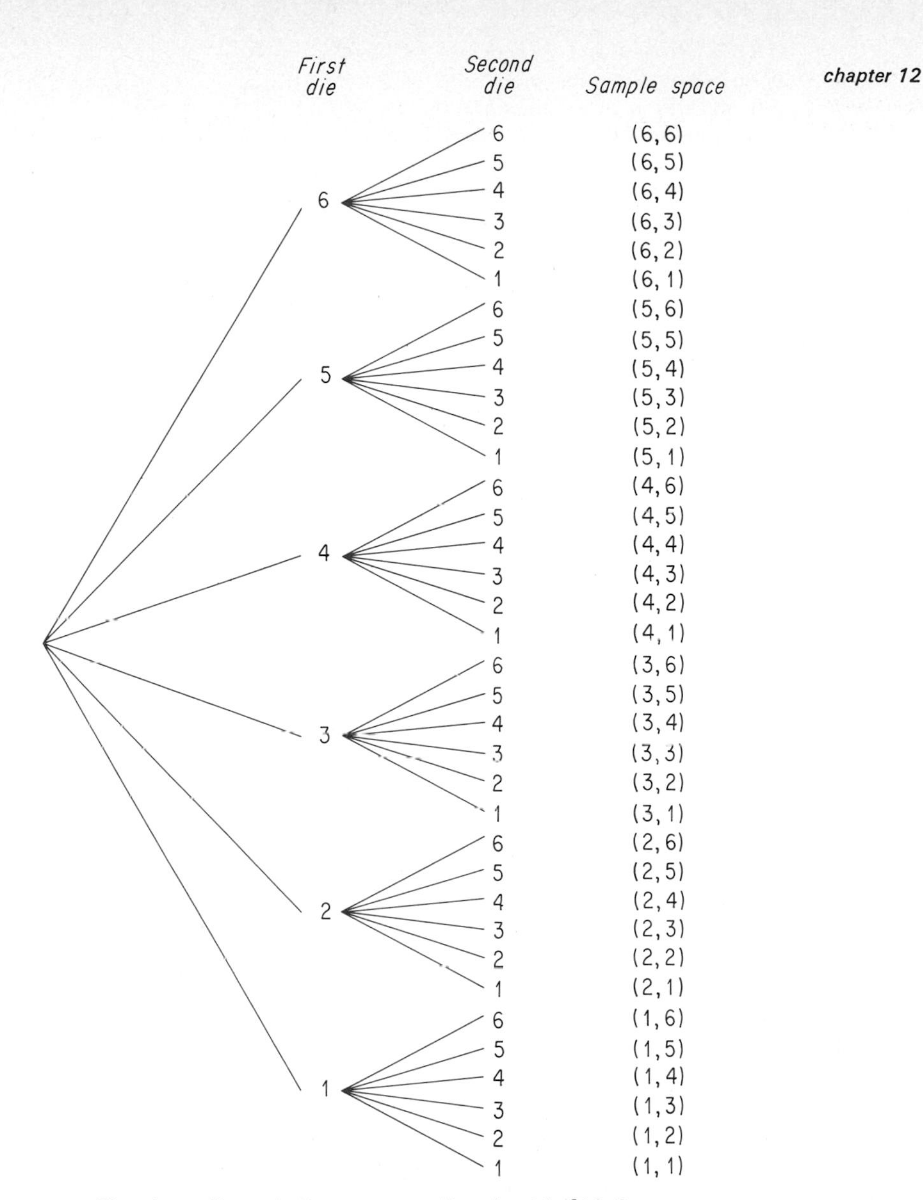

The above figure is the answer to Exercise 13, §12-3.

12-5 ODDS AND MATHEMATICAL EXPECTATION *Page 457*

1. 1 to 3.

3. 1 to 3.

5. 3 to 1.

7. 11 to 2.

9. 2 to 7.

11. \$4

13. \$1.67

15. \$2.00

***17.** The probability that both of the bills drawn will be tens is $\frac{2}{5} \times \frac{1}{4} = \frac{1}{10}$. The probability that both will be fives is $\frac{3}{5} \times \frac{2}{4} = \frac{3}{10}$. The probability that one will be a five and one a ten is found as $(\frac{3}{5} \times \frac{2}{4}) + (\frac{2}{5} \times \frac{3}{4}) = \frac{3}{5}$. The mathematical expectation is then found to be

$$(\$20)(\tfrac{1}{10}) + (\$10)(\tfrac{3}{10}) + (\$15)(\tfrac{3}{5}) = \$14.$$

***19.** $30\frac{2}{3}$ cents since the probability of 2 cents is 2/30; the probabilities of 6, 11, 26, and 51 cents are 4/30; and the probabilities of 15, 30, 55, 35, 60, and 75 cents are each 2/30.

***21.** **(a)** p to $1 - p$; **(b)** $1 - p$ to p.

***23.** $\frac{1}{n}(5000) + \frac{1}{n-1}(1000) + \left[1 - \left(\frac{1}{n} + \frac{1}{n-1}\right)\right](-1)$, that is, $\frac{5000}{n} + \frac{1000}{n-1} + \frac{2n-1}{n(n-1)} - 1$ dollars.

12-6 PERMUTATIONS *Page 462*

1. **(a)** bat, ball; **(b)** (bat, ball), (ball, bat).
Note: Parentheses are often used for ordered sets (arrangements, permutations).

3. **(a)** A, B, C, D;
(b) $AB, AC, AD, BA, BC, BD, CA, CB, CD, DA, DB, DC$;
(c) $ABC, ACB, ABD, ADB, ACD, ADC, BAC, BCA, BAD, BDA, BCD, BDC, CAB, CBA, CAD, CDA, CBD, CDB, DAB, DBA, DAC, DCA, DBC, DCB$.
(d) Note: Simply add the remaining element at the end of each permutation in part **(c)**.
$ABCD, ACBD, ABDC, ADBC, ACDB, ADCB, BACD, BCAD, BADC, BDAC, BCDA, BDCA, CABD, CBAD, CADB, CDAB, CBDA, CDBA, DABC, DBAC, DACB, DCAB, DBCA, DCBA$.

5. 720

7. 7920

9. 210

11. 10!, that is, 3,628,800.

13. 1320

15. 604,800

17. $n!$

19. $n!/2$

21. $n!/n!$, that is, 1.

23. 21

25. 5

27. 6

29. **(a)** 120; **(b)** 48; **(c)** 2/5; **(d)** 3/5.

12-7 COMBINATIONS *Page 466*

1. 35

3. 120

5. 165

7. *ab* and *ba*, *bc* and *cb*, *ac* and *ca*; ${}_3C_2 = 3$.

9.

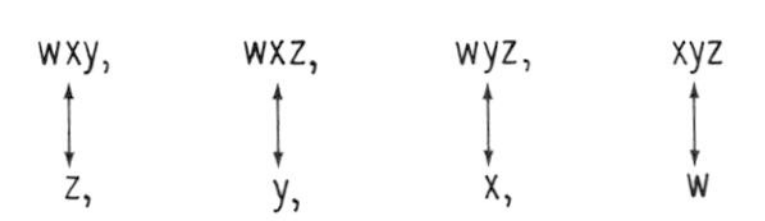

11. ${}_nC_0 = 1$; the only possible combination of n things 0 at a time is obtained when none are selected.

13. ${}_5C_0 + {}_5C_1 + {}_5C_2 + {}_5C_3 + {}_5C_4 + {}_5C_5$
$= 1 + 5 + 10 + 10 + 5 + 1$
$= 32 = 2^5$

15. 32

17. 1848

19. ${}_7C_3$, that is, 35.

21. ${}_5C_3 \times {}_{10}C_7$, that is, 1200.

23. **(a)** ${}_{15}C_4$, that is, 1365;
(b) ${}_{10}C_3 \times {}_5C_1$, that is 600.

25. $\frac{{}_4C_4 \times {}_{48}C_1}{{}_{52}C_5}$, that is, $\frac{1}{54,145}$.

27. ${}_{10}C_4$, that is, 210.

29. ${}_{10}C_2$, that is, 45.

31. **(a)** ${}_{52}C_4$, that is, 270,725.
(b) ${}_{52}C_7$, that is, 133,784,560.

***33.** $(1/2){}_8C_4$, that is, 35.

Chapter 13: AN INTRODUCTION TO STATISTICS

13-1 USES AND ABUSES OF STATISTICS *Page 477*

1. The fact that most accidents occur near home probably means that most of the miles driven are near the home of the driver. It does not necessarily mean that long trips are safer than short trips.

3. The conclusion that mathematics is

very popular is not justified by the previous statement even though the conclusion may be true. The fact that more students are studying mathematics reflects the fact that there are more students in college and that colleges are requiring more mathematics.

5. The 100% sale probably indicates that the book was the required text in a course that all students had to elect, but the given evidence does not justify the given conclusion.

7. The survey may not have been a representative sample of the voting population. Details of the sampling procedure would be needed to feel confident of the conclusion.

9. One wonders how such a count could possibly have been made. Details of the sampling procedure would be needed to feel confident of the conclusion.

11. Many people who have asthma go to Arizona because of the climate. Probably a larger part of the people in Arizona have asthma than for any other state. Thus the conclusion is not justified.

13. People do not necessarily vote for the candidate of the party in which they are registered.

15. It should be stated what average means in this statement. If average denotes the median, there would always be 50% at or below average.

17. How short? How tall? What do you mean by aggressive? How do you measure aggressiveness?

19. Effective for what conditions? How does brand A compare to other brands? What do you mean by effectiveness? How do you determine effectiveness?

21. Fewer than whom? What age groups are under consideration? How much swimming is considered?

23. What do you mean by success? What is implied by "student"?

25. Are those who marry during their last year of college included? What is the source of this information?

27. The small changes in temperature appear very large due to the fact that the temperature scale does not start at zero.

29. No; a scale is needed before any such conclusions can be made.

13-2 COLLECTING AND PRESENTING DATA *Page 483*

7.

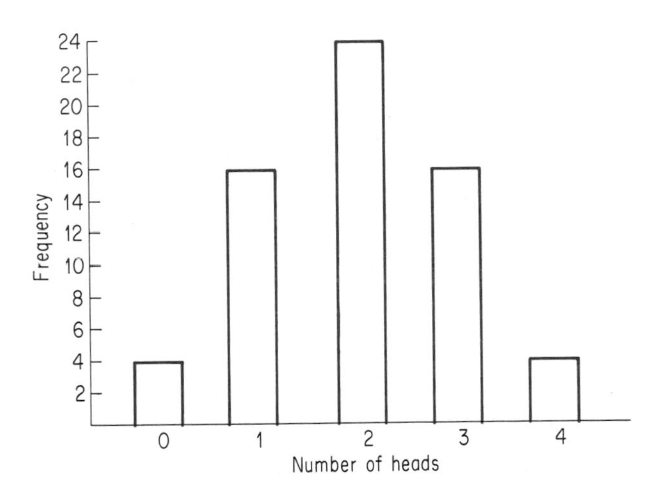

9.

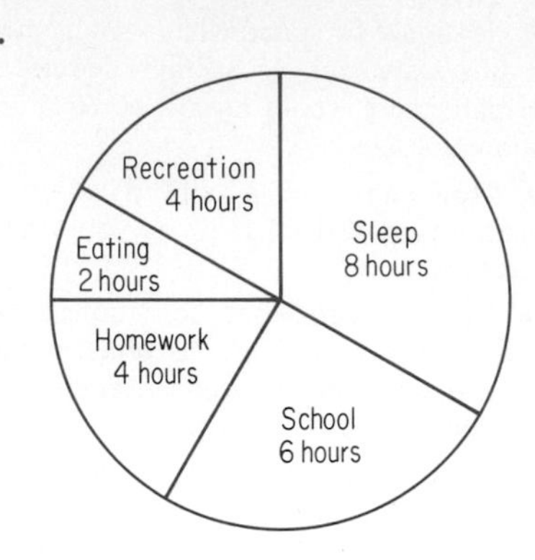

11. 44

13. **(a)** Yes, 8;
(b) no, per cents but not amounts are shown.

15. 158°

17. $80,000

19. 59 and 88.

21. The number of deaths each year decreases after age 75 because there are fewer people still alive then. However, on a percentage basis, the number of deaths per 1,000 increases.

13-3 MEASURES OF CENTRAL TENDENCY *Page 489*

1. **(a)** 70; **(b)** 70; **(c)** 65, 73.

3. **(a)** 20; **(b)** 20;
(c) there is no mode; all numbers occur with the same frequency (all numbers might be considered modes).

5. **(a)** 77; **(b)** 73; **(c)** 68.

7. Mean.

9. Mode is often used as related to style; that which is done or worn most often by most people.

11. 2520

13. 680

15. **(a)** 3500; **(b)** 78.

17. 125; the student cannot obtain a 90 average if the tests are on a 100-point basis.

13-4 MEASURES OF DISPERSION *Page 495*

1. 43

3. Among others:
{50, 67, 85, 85, 90, 92, 98}.

5. Among others:
{55, 70, 85, 85, 90, 92, 98}.

7. Normal.

9. **(a)** 2.5%; **(b)** 16%; **(c)** 1600; **(d)** 250; **(e)** 8150.

11.

Scores:	d	d^2
7	−3	9
9	−1	1
10	0	0
11	1	1
13	3	9
50		20

$\mu = 50/5 = 10$

$\sigma = \sqrt{\frac{20}{5}} = \sqrt{4} = 2$

13-5 PEDAGOGICAL USES OF STATISTICS *Page 499*

1. No.

3. Among other things emphasize that *no single class should be expected to have a normal distribution.* Also clarify the meaning of the term "average."

5. 5

7. $\mu = 40$, $\sigma \approx 4.5$. Thus it appears doubtful that Eve and Ike used unbiased coins. Note that this answer is true whether they failed to do the experiment at all or they used biased coins.

Chapter 14: COMPUTERS AND CALCULATORS

14-1 COMPUTERS AND CALCULATORS *Page 512*

1. Among others: $5 \times 21 + 7$

3. $(4/2)(-1) + 14$

5. $\sqrt{\sqrt{81}}$

7. $\sqrt{121} + 7 - 15$

9.

Instructions	Displays
On	0.00
5	5.
Enter	5.00
21	21.
×	105.00
7	7.
+	112.00
Off	

11.

Instructions	Displays
On	0.00
4	4.
Enter	4.00
2	2.
÷	2.00
CHS	−2.00
14	14.
+	12.00
Off	

13.

Instructions	Displays
On	0.00
81	81.
$\sqrt{x}$	9.00
$\sqrt{x}$	3.00
Off	

15.

Instructions	Displays
On	0.00
121	121.
$\sqrt{x}$	11.00
7	7.
+	18.00
15	15.
—	3.00
Off	

17.

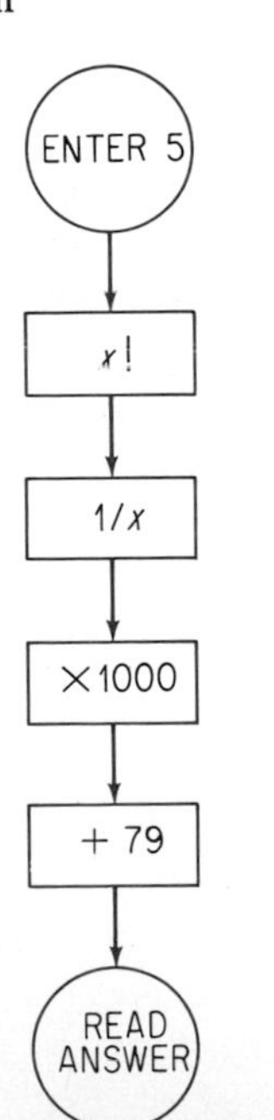

19.

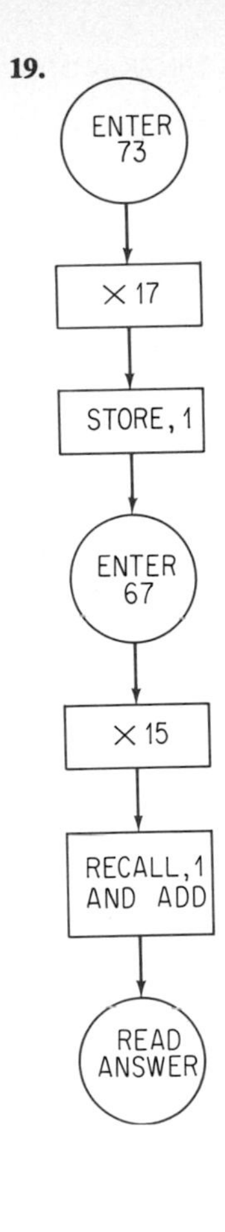

21.

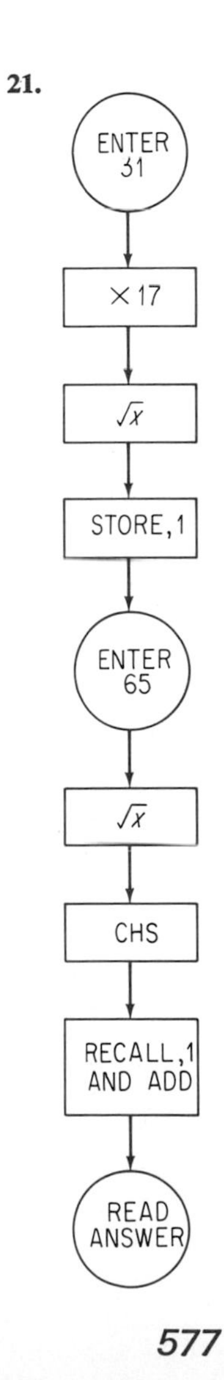

14-2 BINARY NOTATION *Page 516*

1. 100110_2
3. 11101_2
5. 10011100_2
7. 14
9. 59
11. 46
13. 11010_2
15. 11_2
17. 100111_2
19. (a) 1000010_2;
    (b) 1000100_2;
    (c) 1000111_2.
21. $214 = 326_8 = 11010110_2$. When placed in groups of three, starting from the units digit, the binary representation can be translated into the octal system, and conversely. Thus

$$\underbrace{011}_{3}\ \underbrace{010}_{2}\ \underbrace{110_2}_{6} = 326_8.$$

23. 3531_8
25. 22533_8
27. 1531046_8
29. 101000010011_2
31. 100110010100_2
33. 111110101100011_2

14-3 BASIC *Page 522*

1. **2 * 18/3**; 12.
3. **6 * (25 + 15)**; 240.
5. **6 − (8/3) ↑ 2**; $-10/9$, that is, $-1.\bar{1}$.
7. 14
9. 44
11. $60 - 243$, that is, -183.
13.

1	1
0.01	0.1
0.04	0.2
0.0009	0.03
0.000016	0.004

15.

1	−4
2	−10
3	−16
4	−22
5	−28
6	−34

17. Among others:

```
10  FØR X = 1 TØ 20
20  LET Y = X ↑ 5
30  PRINT X, Y
40  NEXT X
50  END
```

19. Among others:

```
10  FØR X = 1 TØ 20
20  LET Y = 2 * X + SQR(X)
30  PRINT X, Y
40  NEXT X
50  END
```

*21. Among others:

```
10  READ X1, X2
20  DATA 7, 11, 75, 123
21  DATA -567, 891, 4357, -5437
30  LET M = (X1 + X2)/2
40  PRINT X1, X2, M
50  GØ TØ 10
60  END
```

*23. Among others:

[Note **DATA** statements are broken up because in BASIC each line must be numbered. Some of the following **DATA** statements can be combined for use on most computers. Statement 64 is inserted simply for spacing.]

```
10  READ X1, Y1, Z1, X2, Y2, Z2
20  DATA 2, 4, 6, -2, 6, 4
21  DATA 25, 37, 61, 125, 367, -75
22  DATA 275, 17, -975
23  DATA 4532, 6475, 9801
24  DATA 7536, 4444, 3323
25  DATA 5763, -7985, -9873
30  LET X = (X1 + X2)/2
40  LET Y = (Y1 + Y2)/2
50  LET Z = (Z1 + Z2)/2
60  PRINT X1, Y1, Z1
61  PRINT X2, Y2, Z2
62  PRINT X, Y, Z
64  PRINT
70  GØ TØ 10
80  END
```

14-4 PRECISION AND ACCURACY *Page 528*

1. **(a)** 8.01×10^8; **(b)** 8.0×10^8; **(c)** 8.010×10^8; **(d)** 8.0100×10^8.

3. **(a)** 75.34; **(b)** 75.3.

5. **(a)** 33.17; **(b)** 33.2.

7. **(a)** 2.500×10^4; **(b)** 2.50×10^4.

9. 266.3

11. 356

13. 0.01

15. 2×10^2

17. 0.2

19. 2×10^3

***21.** 27.4

***23.** 8.527

***25.** $\sqrt{35.286 \times 10^4}$, that is, 594.02

A

B

C

D

E

F

G

H

I

J

K

L

M

N

O

P

Q

R

S

T

U

V

W

X

Y

Z